In the 2012–13 academic year, the Mathematical Sciences Research Institute, Berkeley, hosted programs in Commutative Algebra (Fall 2012 and Spring 2013) and Noncommutative Algebraic Geometry and Representation Theory (Spring 2013). There have been many significant developments in these fields in recent years; what is more, the boundary between them has become increasingly blurred. This was apparent during the MSRI program, where there were a number of joint seminars on subjects of common interest: birational geometry, D-modules, invariant theory, matrix factorizations, noncommutative resolutions, singularity categories, support varieties, and tilting theory, to name a few. These volumes reflect the lively interaction between the subjects witnessed at MSRI.

The Introductory Workshops and Connections for Women Workshops for the two programs included lecture series by experts in the field. The volumes include a number of survey articles based on these lectures, along with expository articles and research papers by participants of the programs.

Mathematical Sciences Research Institute
Publications

67

Commutative Algebra and Noncommutative Algebraic Geometry

Volume I: Expository Articles

Mathematical Sciences Research Institute Publications

1 Freed/Uhlenbeck: *Instantons and Four-Manifolds*, second edition
2 Chern (ed.): *Seminar on Nonlinear Partial Differential Equations*
3 Lepowsky/Mandelstam/Singer (eds.): *Vertex Operators in Mathematics and Physics*
4 Kac (ed.): *Infinite Dimensional Groups with Applications*
5 Blackadar: *K-Theory for Operator Algebras*, second edition
6 Moore (ed.): *Group Representations, Ergodic Theory, Operator Algebras, and Mathematical Physics*
7 Chorin/Majda (eds.): *Wave Motion: Theory, Modelling, and Computation*
8 Gersten (ed.): *Essays in Group Theory*
9 Moore/Schochet: *Global Analysis on Foliated Spaces*, second edition
10–11 Drasin/Earle/Gehring/Kra/Marden (eds.): *Holomorphic Functions and Moduli*
12–13 Ni/Peletier/Serrin (eds.): *Nonlinear Diffusion Equations and Their Equilibrium States*
14 Goodman/de la Harpe/Jones: *Coxeter Graphs and Towers of Algebras*
15 Hochster/Huneke/Sally (eds.): *Commutative Algebra*
16 Ihara/Ribet/Serre (eds.): *Galois Groups over* $\mathbb{Q}$
17 Concus/Finn/Hoffman (eds.): *Geometric Analysis and Computer Graphics*
18 Bryant/Chern/Gardner/Goldschmidt/Griffiths: *Exterior Differential Systems*
19 Alperin (ed.): *Arboreal Group Theory*
20 Dazord/Weinstein (eds.): *Symplectic Geometry, Groupoids, and Integrable Systems*
21 Moschovakis (ed.): *Logic from Computer Science*
22 Ratiu (ed.): *The Geometry of Hamiltonian Systems*
23 Baumslag/Miller (eds.): *Algorithms and Classification in Combinatorial Group Theory*
24 Montgomery/Small (eds.): *Noncommutative Rings*
25 Akbulut/King: *Topology of Real Algebraic Sets*
26 Judah/Just/Woodin (eds.): *Set Theory of the Continuum*
27 Carlsson/Cohen/Hsiang/Jones (eds.): *Algebraic Topology and Its Applications*
28 Clemens/Kollár (eds.): *Current Topics in Complex Algebraic Geometry*
29 Nowakowski (ed.): *Games of No Chance*
30 Grove/Petersen (eds.): *Comparison Geometry*
31 Levy (ed.): *Flavors of Geometry*
32 Cecil/Chern (eds.): *Tight and Taut Submanifolds*
33 Axler/McCarthy/Sarason (eds.): *Holomorphic Spaces*
34 Ball/Milman (eds.): *Convex Geometric Analysis*
35 Levy (ed.): *The Eightfold Way*
36 Gavosto/Krantz/McCallum (eds.): *Contemporary Issues in Mathematics Education*
37 Schneider/Siu (eds.): *Several Complex Variables*
38 Billera/Björner/Green/Simion/Stanley (eds.): *New Perspectives in Geometric Combinatorics*
39 Haskell/Pillay/Steinhorn (eds.): *Model Theory, Algebra, and Geometry*
40 Bleher/Its (eds.): *Random Matrix Models and Their Applications*
41 Schneps (ed.): *Galois Groups and Fundamental Groups*
42 Nowakowski (ed.): *More Games of No Chance*
43 Montgomery/Schneider (eds.): *New Directions in Hopf Algebras*
44 Buhler/Stevenhagen (eds.): *Algorithmic Number Theory: Lattices, Number Fields, Curves and Cryptography*
45 Jensen/Ledet/Yui: *Generic Polynomials: Constructive Aspects of the Inverse Galois Problem*
46 Rockmore/Healy (eds.): *Modern Signal Processing*
47 Uhlmann (ed.): *Inside Out: Inverse Problems and Applications*
48 Gross/Kotiuga: *Electromagnetic Theory and Computation: A Topological Approach*
49 Darmon/Zhang (eds.): *Heegner Points and Rankin L-Series*
50 Bao/Bryant/Chern/Shen (eds.): *A Sampler of Riemann–Finsler Geometry*
51 Avramov/Green/Huneke/Smith/Sturmfels (eds.): *Trends in Commutative Algebra*
52 Goodman/Pach/Welzl (eds.): *Combinatorial and Computational Geometry*
53 Schoenfeld (ed.): *Assessing Mathematical Proficiency*
54 Hasselblatt (ed.): *Dynamics, Ergodic Theory, and Geometry*
55 Pinsky/Birnir (eds.): *Probability, Geometry and Integrable Systems*
56 Albert/Nowakowski (eds.): *Games of No Chance 3*
57 Kirsten/Williams (eds.): *A Window into Zeta and Modular Physics*
58 Friedman/Hunsicker/Libgober/Maxim (eds.): *Topology of Stratified Spaces*
59 Caporaso/M[C]Kernan/Mustaţă/Popa (eds.): *Current Developments in Algebraic Geometry*
60 Uhlmann (ed.): *Inverse Problems and Applications: Inside Out II*
61 Breuillard/Oh (eds.): *Thin Groups and Superstrong Approximation*
62 Eguchi/Eliashberg/Maeda (eds.): *Symplectic, Poisson, and Noncommutative Geometry*
63 Nowakowski (ed.): *Games of No Chance 4*
65 Deift/Forrester (eds.): *Random Matrix Theory, Interacting Particle Systems, and Integrable Systems*
67–68 Eisenbud/Iyengar/Singh/Stafford/Van den Bergh (eds.): *Commutative Algebra and Noncommutative Algebraic Geometry*

Commutative Algebra and Noncommutative Algebraic Geometry

Volume I: Expository Articles

Edited by

David Eisenbud
University of California, Berkeley and Mathematical Sciences Research Institute

Srikanth B. Iyengar
University of Utah

Anurag K. Singh
University of Utah

J. Toby Stafford
University of Manchester

Michel Van den Bergh
Fonds Wetenschappelijk Onderzoek - Vlaanderen and Universiteit Hasselt

CAMBRIDGE UNIVERSITY PRESS

David Eisenbud
de@msri.org

Srikanth B. Iyengar
iyengar@math.utah.edu

Anurag K. Singh
singh@math.utah.edu

J. Toby Stafford
toby.stafford@manchester.ac.uk

Michel Van den Bergh
michel.vandenbergh@uhasselt.be

Silvio Levy (*Series Editor*)
Mathematical Sciences Research Institute
levy@msri.org

The Mathematical Sciences Research Institute wishes to acknowledge support by the National Science Foundation for the publication of this series.

CAMBRIDGE
UNIVERSITY PRESS

32 Avenue of the Americas, New York, NY 10013-2473, USA

Cambridge University Press is part of the University of Cambridge.

It furthers the University's mission by disseminating knowledge in the pursuit of education, learning and research at the highest international levels of excellence.

www.cambridge.org
Information on this title: www.cambridge.org/9781107065628

First published 2015

A catalog record for this publication is available from the British Library.

ISBN 978-1-107-06562-8 Hardback

Commutative Algebra and Noncommutative Algebraic Geometry, I
MSRI Publications
Volume **67**, 2015

Contents

Preface

In the 2012–13 academic year, the Mathematical Sciences Research Institute, in Berkeley, hosted programs in Commutative Algebra (Fall 2012 and Spring 2013) and Noncommutative Algebraic Geometry and Representation Theory (Spring 2013). The programs had 174 participants visiting for periods ranging between one and nine months, and many others for shorter periods and for week-long workshops.

There have been many significant developments in these fields in recent years; what is more, the once rather strict boundary between them has become increasingly blurred. This was apparent during the MSRI program, where there were a number of joint seminars on subjects of common interest: birational geometry, $\mathscr{D}$-modules, invariant theory, matrix factorizations, non-commutative resolution of singularities, singularity categories, support varieties, and tilting theory, to name a few. This volume is intended to reflect, and stimulate, the lively interaction between the two subjects that we witnessed at MSRI.

The Introductory Workshops and Connections for Women Workshops for the two programs included lecture series by experts in the field; the volume includes a number of survey articles based on these lectures. There are also expository articles and research papers by some of the other participants of the programs.

In addition to the editors of this volume, the organizers of the programs and the Introductory and Connections for Women workshops were Mike Artin, Georgia Benkart, Victor Ginzburg, Bernard Keller, Ellen Kirkman, Ezra Miller, Claudia Polini, Idun Reiten, Sue Sierra, Karen E. Smith, Catharina Stroppel, Alexander Vainshtein, Lauren Williams, and Efim Zelmanov. We take this opportunity to express our thanks to the participants, our co-organizers, the MSRI staff, and the National Science Foundation, which supported the programs under grant DMS 0932078000, and the National Security Agency, which supported the workshops through grants H98230-12-1-0236/0256/0296/0298.

David Eisenbud
Srikanth B. Iyengar
Anurag K. Singh
J. Toby Stafford
Michel Van den Bergh

Commutative Algebra and Noncommutative Algebraic Geometry, I
MSRI Publications
Volume **67**, 2015

Growth functions

JASON P. BELL

We give a survey of the use of growth functions in algebra. In particular, we define Gelfand–Kirillov dimension and give an overview of some of the main results about this dimension, including Bergman's gap theorem, the solution of the Artin–Stafford conjecture by Smoktunowicz, and the characterization of groups of polynomially bounded growth by Gromov. In addition, we give a summary of the main ideas employed in the proof of Gromov's theorem and discuss the work of Lenagan and Smoktunowicz, which gives a counterexample to Kurosh's conjecture with polynomially bounded growth.

1. Introduction

The notion of growth is a fundamental object of study in the theory of groups and algebras, due to its utility in answering many basic questions in these fields. The concept of growth was introduced by Gelfand and Kirillov [1966] for algebras and by Milnor [1968] for groups, who showed that there is a strong relation between the growth of the fundamental group of a Riemannian manifold and its curvature. After the seminal works of Gelfand and Kirillov and of Milnor, the study of growth continued and many important advances were made. In particular, Borho and Kraft [1976] further developed the theory of growth in algebras, giving a systematic study of the theory of Gelfand–Kirillov dimension. In addition to this, Milnor [1968] and Wolf [1968] gave a complete characterization of solvable groups with polynomially bounded growth (see Section 2 for relevant definitions).

The reason for the importance of Gelfand–Kirillov dimension, Gelfand–Kirillov transcendence degree, and corresponding notions in the theory of groups is that it serves as a natural noncommutative analogue of Krull dimension (resp. transcendence degree) and thus provides a suitable notion of dimension for noncommutative algebras. Indeed, the first application of Gelfand–Kirillov dimension was to show that the quotient division algebras of the m-th and n-th Weyl algebras are isomorphic if and only if $m = n$, by showing that their transcendence degrees differed when $n \neq m$. Since this initial application,

The author was supported by NSERC grant 31-611456.

the theory of growth has expanded considerably and this notion now plays a fundamental role in both geometric group theory and noncommutative projective geometry, where it serves as a natural notion of dimension.

The objective of these notes is to give a survey of the foundational results of Gelfand–Kirillov dimension as well as related growth functions in the theory of groups and rings which have been used to answer difficult questions. In Section 2, we give an overview of the basic terminology that we will be using throughout. In Section 3, we define Gelfand–Kirillov dimension and give a survey of the most important results in the theory of growth, and in Section 4 we give some of the important results in the theory of combinatorics on words and their application to growth of algebras; in particular, we prove Bergman's gap theorem, which asserts that no algebras whose growth is subquadratic but faster than linear can exist.

In Section 5, we show that Gelfand–Kirillov dimension is a noncommutative analogue of Krull dimension for finitely generated algebras and discuss algebras of low Gelfand–Kirillov dimension. In particular, we discuss the Small–Stafford–Warfield theorem [Small et al. 1985] saying that finitely generated algebras of Gelfand–Kirillov dimension satisfy a polynomial identity and we give a brief discussion of what is known about algebras of Gelfand–Kirillov dimension two. In Section 6, we discuss the ingredients in the proof of Gromov's theorem, which beautifully characterizes the finitely generated groups whose group algebras have finite Gelfand–Kirillov dimension; namely, Gromov's theorem asserts that such groups must have a finite-index subgroup that is nilpotent. In Section 7 and Section 8, we discuss two relatively recent advances in the study of growth; in Section 7, we discuss constructions, mostly developed by Smoktunowicz, which show how to construct pathological examples of algebras of finite Gelfand–Kirillov dimension; in Section 8, we discuss Zhang's so-called lower transcendence degree and its applications to the study of division algebras. Finally, in Section 9, we give a brief overview of Artin's conjecture on the birational classification of noncommutative surfaces and its relation to growth.

2. Preliminaries

We begin by stating the basic definitions we will be using. We let $\mathscr{C}$ denote the class of maps $f : \mathbb{N} \to \mathbb{N}$ that are monotonically increasing and have the property that there is some positive number C such that $f(n) < C^n$. We say that $f \in \mathscr{C}$ has *polynomially bounded* growth if there is some $d > 0$ such that $f(n) \le n^d$ for all n sufficiently large; we say that f has *exponential growth* if there exists a constant $C > 1$ such that $f(n) > C^n$ for all n sufficiently large. If $f \in \mathscr{C}$ has neither polynomially bounded nor exponential growth then we say it has *intermediate growth*. If $f(n) = \exp(o(n))$ we say that $f(n)$ has *subexponential growth*. Note that according to these definitions, an element of $\mathscr{C}$ can have intermediate growth without having subexponential growth. As an example, let T be the union of sets of the form $\{(2i)!, \ldots, (2i+1)! - 1\}$ as i ranges over the natural numbers. We may define a weakly increasing map $f(n)$ by declaring that $f(0) = 1$ and $f(n+1) = 2f(n)$ if $n \in T$ and $f(n+1) = f(n)$ if $n \notin T$. Then $f((2m)!) \le 2^{(2m-1)!}$ and thus f cannot have exponential growth. On the other hand, $f((2m+1)!) \ge 2^{(2m+1)!-(2m)!}$. Thus when $n = (2m+1)!$ and $m \ge 1$, we have $f(n) \ge (3/2)^n$ and so f does not have subexponential growth according to our definition. (We note that some, perhaps even most, authors take subexponential growth to include growth types such as the one given in this example.)

Given $f \in \mathscr{C}$ of polynomially bounded growth. We define the *degree* of growth to be

$$\deg(f) := \limsup_{n\to\infty} \frac{\log f(n)}{\log n}.$$

We note that if $f(n)$ is asymptotic to Cn^α then this quantity is equal to α, and so this notion coincides with our usual notion of degree in this case.

Given $f, g \in \mathscr{C}$, we say that f is *asymptotically dominated* by g if there natural numbers $k_1, k_2 \ge 1$ such that $f(n) \le k_1 g(k_2 n)$. If f is asymptotically dominated by g and g is asymptotically dominated by f, then we say that the functions are *asymptotically equivalent*. Asymptotically equivalent functions need not be asymptotic to one another in the conventional sense, but polynomial, exponential, intermediate, and subexponential growth are preserved under this notion of asymptotic equivalence. Furthermore, asymptotically equivalent functions of polynomially bounded growth have the same degree of growth. Henceforth, we will only consider functions up to this notion of asymptotic equivalence.

Given a finitely generated group G and a generating set S with the properties that $1 \in S$ and if $s \in S$ then $s^{-1} \in S$, we can construct a *growth function* of G with respect to the generating set S as follows. We let $d_S(n)$ denote the number of distinct elements of G that can be written as a product of n elements of S. Then $d_S(n)$ is an element of $\mathscr{C}$ since $1 \in S$. For example, if $G = \mathbb{Z}^2$ with generators x, y, then if we take $S = \{1, x, x^{-1}, y, y^{-1}\}$ then we have $d_S(n) = \#\{x^i y^j \ : \ |i| + |j| \le n\} = (n+1)^2$. We note that if T is another generating set then since $T^k \supseteq S$ and $S^k \supseteq T$ for some natural number k we have $d_S(n) \le d_T(kn)$ and $d_T(n) \le d_S(kn)$ for all $n \ge 0$. Thus, although two different growth functions need not be equal, they are equal up to asymptotic equivalence. Thus we may speak unambiguously of the growth function of a finitely generated group G.

Similarly, if k is a field and A is a finitely generated k-algebra, we can associate a growth function as follows. Let V be a finite-dimensional subspace of A that generates A as a k-algebra. Then we define

$$d_V(n) := \dim_k\left(\sum_{j=1}^{n} V^j\right).$$

Unless otherwise specified, we assume that our algebras have an identity; in this case we also assume that $1 \in V$ and so we have $d_V(n) = \dim_k(V^n)$. As in the case with groups, if W is another generating subspace then we have that $d_W(n)$ and $d_V(n)$ are asymptotically equivalent and so we again speak unambiguously of the growth function of A. We make the remark that the growth of a group G is equal to the growth of its group algebra $k[G]$ and so it is enough to consider growth of algebras.

3. General results for algebras of polynomially bounded growth

Given a finitely generated k-algebra A of polynomially bounded growth. We recall that we have a degree function associated to its growth. In this setting, the degree function is called the *Gelfand–Kirillov* dimension and is denoted by $\mathrm{GKdim}(A)$. More formally, we have

$$\mathrm{GKdim}(A) := \limsup_{n\to\infty} \frac{\log \dim(V^n)}{\log n},$$

where V is a finite-dimensional vector space containing 1 that generates A as a k-algebra.

A related quantity was first used by Gelfand and Kirillov [1966] to show that $D_n \cong D_m$ if and only if $n = m$ where D_n and D_m are respectively the quotient division algebras of the n-th and m-th Weyl algebras. In addition they conjectured that the quotient division algebra of the enveloping algebra of a finite-dimensional algebraic Lie algebra is isomorphic to the quotient division algebra of a Weyl algebra. (This was ultimately shown to be false [Alev et al. 1996].) We note that in the nonfinitely generated case, one simply defines the GK dimension to be the supremum of the GK dimensions of all finitely generated subalgebras.

We now discuss the foundational results in the theory of growth. Before discussing these results in greater detail, we give a quick summary of these results for the reader's convenience. We let k be a field and we let A be a finitely generated k-algebra. Then we have the following:

(i) if $\mathrm{GKdim}(A) \in [0, 1)$ then A is finite-dimensional;

(ii) (Bergman (see [Krause and Lenagan 2000, Theorem 2.5])) if $\mathrm{GKdim}(A) \in [1, 2)$ then A has GK dimension 1;

(iii) (Small, Stafford and Warfield [Small et al. 1985]) if A has GK dimension one then A satisfies a polynomial identity;

(iv) (Small, Stafford, Warfield [Small et al. 1985] and van den Bergh) if A is a domain of GK dimension one and k is algebraically closed then A is commutative;

(v) (Smoktunowicz [2005; 2006]) if A is a graded domain with GK dimension in $[2, 3)$ then A has GK dimension two (and in fact has quadratic growth).

(vi) (Artin and Stafford [1995]) if A is a graded complex domain of GK dimension 2 that is generated in degree one then A is — up to a finite-dimensional vector space — equal to the twisted homogeneous coordinate ring of a curve;

(vii) (Borho and Kraft [1976]) for each $\alpha \in [2, \infty]$ there is an algebra of GK dimension α;

(viii) if A satisfies a polynomial identity and A is semiprime then its GK dimension is equal to a nonnegative integer [Krause and Lenagan 2000, Chapter 10];

(ix) if A is commutative then the GK dimension is equal to the Krull dimension;

(x) (Gromov [1981]) If $A = k[G]$, where G is a finitely generated group, then A has finite GK dimension if and only if G is nilpotent-by-finite;

(xi) (Bass and Guivarch [Krause and Lenagan 2000, Theorem 11.14]) If $A = k[G]$, where G is a finitely generated nilpotent-by-finite group, then the GK dimension of A is an integer given by

$$\sum_i i \cdot d_i,$$

where d_i is the rank of the of the i-th quotient of the lower central series of G.

We note that (i) is immediate since if A is an algebra then we either have $V^i = V^{i+1}$ for some i or we have $V^n \geq n + 1$ for all $n \geq 0$. We note that for (x), Gromov [1981] proved that a finitely generated group of polynomially bounded growth is virtually nilpotent and Bass and Guivarch had given the formula for the degree of growth earlier.

4. Combinatorics on words

In this section, we discuss the values that can arise as the GK dimension of an algebra. Many of the foundational results for algebras and groups of low growth come from the theory of combinatorics on words. The reason for this is that given a finitely generated k-algebra A one can associate a monomial algebra

B that has the same growth as A and to determine the growth of a monomial algebra depends on estimating the number of words of length n that avoid a given set of forbidden subwords. We now make this more precise.

A finitely generated algebra A can be written in the form $k\{x_1, \ldots, x_d\}/I$ for some ideal I in the free algebra $k\{x_1, \ldots, x_d\}$. We may put a degree lexicographic ordering on the monomials of the free algebra by declaring that $x_1 \prec x_2 \prec \cdots \prec x_d$. Given an element $f \in k\{x_1, \ldots, x_d\}$, we write f as a linear combination of words in $\{x_1, \ldots, x_d\}$. We then define $\mathrm{in}(f)$, the initial monomial of f, to be the degree lexicographic word w that appears with nonzero coefficient in our expression for f. Then we may associate a monomial ideal J to I by taking the ideal generated by all initial words of elements of I. (Note: it is not sufficient to take the initial words of a generating set for I, as anyone who has worked with Gröbner bases will understand.)

The monomial algebra $B := k\{x_1, \ldots, x_d\}/J$ and A then have identical growth functions, but B has the advantage of having a more concrete way of studying its growth. We record this observation now.

Remark 4.1. Given a finitely generated associative algebra A, there is a finitely generated monomial algebra $B = k\{x_1, \ldots, x_d\}/I$ with identical growth; moreover, if V is the image of the vector space spanned by $\{1, x_1, \ldots, x_d\}$ in B then the dimension of V^n is precisely the number of words over the alphabet $\{x_1, \ldots, x_d\}$ of length at most n that are not in I.

We recall two classical results in the theory of combinatorics of words. These deal with *right infinite words*, which are as one might expect just infinite sequences over some alphabet Σ. The first result is generally called König's infinity lemma, which is very easy to prove, but is nevertheless incredibly useful.

Theorem 4.2 (König). *Let Σ be a finite alphabet and let S be an infinite subset of Σ^*. Then there is a right infinite word w over Σ such that every subword of w is a subword of some word in S.*

The second result is Furstenberg's theorem, which is really part of a more general theorem relating to dynamical systems. We recall that a right infinite word w is *uniformly recurrent* if for any finite subword u that appears in w, there is a natural number $N = N(u)$ with the property that in any block of N consecutive letters in w there must be at least one occurrence of u.

Theorem 4.3 (Furstenberg). *Let Σ be a finite alphabet and let w be a right infinite word over Σ. Then there is a right infinite uniformly recurrent word u over Σ such that every subword of u is also a subword of w.*

The first significant use of the ideas from combinatorics of words in the theory of growth is due to Bergman (see [Krause and Lenagan 2000]), who showed that there is a gap theorem for growth. It is clear that there are no algebras with GK dimension strictly between 0 and 1, but Bergman showed that there are in fact no algebras with GK dimension strictly between 1 and 2. We give a short proof of this theorem.

Theorem 4.4 (Bergman). *There are no algebras of GK dimension strictly between* 1 *and* 2.

Proof. Suppose that A is a finitely generated algebra of GK dimension $\alpha \in (1, 2)$. It is no loss of generality to assume that $A = k\{x_1, \ldots, x_d\}/I$ where I is generated by monomials. Let $\mathcal{S}$ denote the set of words over $\{x_1, \ldots, x_d\}$ that are not in I but have the property that all sufficiently long right extensions are in I. By König's infinity lemma, $\mathcal{S}$ must be a finite set. We let J denote the ideal generated by I and the elements of $\mathcal{S}$. Then $B = k\{x_1, \ldots, x_d\}/J$ and A have the same growth since there is only finite set of words over $\{x_1, \ldots, x_d\}$ that are not in I but are in J. By construction, if W is a word that survives mod J then W has arbitrarily long right extensions that survive mod J.

Let $f(n)$ denote the number of words of length n over $\{x_1, \ldots, x_d\}$ that survive mod J. If V is the span of the images of $1, x_1, \ldots, x_d$ in B then $V^n = 1 + f(1) + \cdots + f(n)$.

There are now two quick cases to consider. If $f(n+1) > f(n)$ for all n then we have $f(n) \geq n+1$ for every natural number n and so $\dim(V^n) \geq \binom{n+2}{2}$ which gives that B has GK dimension at least two, a contradiction.

If $s = f(i+1) = f(i)$ for some i, then we let $W_1, \ldots, W_s$ denote the set of distinct words of length i that are not in J. By our construction of J, each W_j is an initial subword of a word of length $i+1$ that is not in J. Moreover, since there are only s words of length i, we see that each W_j has a unique right extension to a word of length $i+1$ that is not in J. But now we use the fact that each word of length $i+1$ can be written in the form $x_k W_j$ for some $k \in \{1, \ldots, d\}$ and $j \in \{1, \ldots, s\}$. Since W_j has a unique right extension to a word of length $i+1$ that is not in J we see each word of length $i+1$ has a unique extension to a word of length $i+2$ that is not in J. In particular, $f(i+2) = s$. Continuing in this manner, we see that $f(n) = s$ for all $n \geq i$, and so B has linear growth, which is again a contradiction. □

On the other hand, Borho and Kraft [1976] showed that no additional gaps exist in general; that is, any real number that is at least two can be realized as the GK dimension of a finite generated algebra. As an example, we show how one can get an algebra of GK dimension 2.5. We note that this example can be easily modified to get any GK dimension between two and three. Taking polynomial rings over these algebras, one can then construct examples of any GK dimension greater than or equal to two.

We let $A = k\{x, y\}/I$, where I is the ideal generated by all words that have at least three copies of x and all words of the form xy^jx with j not a perfect square. Then the set of words of length at most n over $\{x, y\}$ that are not in I is given by

$$\mathcal{S}_n := \{y^i x y^{j^2} x y^k : i + j^2 + k \le n-2\} \cup \{y^i x y^j : i + j \le n-1\} \cup \{y^j : j \le n\}.$$

It is straightforward to check that $\{y^i x y^j : i + j \le n-1\} \cup \{y^j : j \le n\}$ has size $\binom{n+1}{2} + n + 1$. We note that the set of nonnegative integers for which $i + j^2 + k \le n - 2$ has size at most $2n^{5/2}$ since $i, k \le n-1$ and $j \le 2\sqrt{n} - 1$. Thus

$$\#\{y^i x y^{j^2} x y^k : i + j^2 + k \le n-2\} \le 2n^{5/2}.$$

Similarly, since any $i \le (n-2)/4$, $j \le \sqrt{(n-2)/4}$, $k \le (n-2)/4$ satisfies $i + j^2 + k \le n - 2$ we see that

$$\#\{y^i x y^{j^2} x y^k : i + j^2 + k \le n-2\} \ge (n-2)^{5/2}/32.$$

Thus

$$\limsup_{n\to\infty} \log(\#\mathcal{S}_n)/\log n = 2.5.$$

It follows that the GK dimension of A is precisely 2.5.

We note that the examples of [Borho and Kraft 1976] are very far from being Noetherian or even Goldie. Smoktunowicz [2005; 2006] showed that if one considers graded domains of GK dimension less than three, then there is a gap.

Theorem 4.5 (Smoktunowicz). *Let A be a finitely generated graded algebra whose GK dimension is in* $[2, 3)$. *Then A has GK dimension* 2.

A partial result of this type had earlier been obtained by Artin and Stafford [1995], who conjectured that the Smoktunowicz gap theorem should hold. Artin and Stafford proved that finitely generated graded algebras whose GK dimension lies in $(2, 11/5)$ could not exist.

5. Small Gelfand–Kirillov dimension

As pointed out earlier, Gelfand–Kirillov dimension can be viewed as a noncommutative analogue of Krull dimension. Much as in the commutative setting special attention has been paid to algebras of small Krull dimension (and, correspondingly, to the study of curves and surfaces and threefolds), there has also been considerable work devoted to the study of algebras of low GK dimension. We first show that GK dimension can be viewed as a reasonable analogue of Krull dimension.

Proposition 5.1. *Let A be a finitely generated commutative k-algebra. Then* $\mathrm{GKdim}(A) = \mathrm{Kdim}(A)$.

Proof. Let d denote the Krull dimension of A. By Noether normalization, there exists a subalgebra $B \cong k[x_1, \ldots, x_d]$ of A such that A is a finite B-module. It is straightforward to check that A and B have the same GK dimension. Thus it is enough to prove that $k[x_1, \ldots, x_d]$ has GK dimension d.

Let $C = k[x_1, \ldots, x_d]$ and let $V = k + kx_1 + \cdots + kx_d$. Then V^n has a basis given by all monomials in $x_1, \ldots, x_d$ of total degree at most n. Observe that the monomials $x_1^{i_1} \cdots x_d^{i_d}$ with $i_1 + \cdots + i_d \leq n$ are in one-to-one correspondence with subsets of $\{1, 2, \ldots, n+d\}$ of size d via the rule

$$x_1^{i_1} \cdots x_d^{i_d} \mapsto \{i_1 + 1, i_2 + 2, \ldots, i_d + d\}.$$

Thus V^n has dimension $\binom{n+d}{d}$ which is asymptotic to $n^d/d!$ as $n \to \infty$. Thus the GK dimension of the polynomial ring in d variables is precisely d. The result follows. □

We have seen that algebras of GK dimension 0 are finite-dimensional. While the class of finite-dimensional algebras is not well-understood, the Artin–Wedderburn theorem says that in the prime case all such algebras are given by a matrix ring over a division algebra that is finite-dimensional over its center. In particular, a domain of GK dimension zero over an algebraically closed field is equal to the algebraically closed field. Small, Stafford, and Warfield [1985] proved that a finitely generated algebra of GK dimension one satisfies a polynomial identity. A particularly nice consequence of this, apparently first observed by van den Bergh, shows that a domain of GK dimension one over an algebraically closed field is necessarily commutative.

Theorem 5.2. *Let k be an algebraically closed field and let A be a finitely generated k-algebra that is a domain of GK dimension one. Then A is commutative.*

Proof. Let $t \in A \setminus k$. Then t is not algebraic over k and hence $k[t]$ must be a polynomial ring in one variable. A theorem of Borho and Kraft [1976] shows that A has a quotient division ring D and that D is a finite-dimensional left vector space over $k(t)$. It is straightforward to check that $k(t)$ has GK dimension 1 as a k-algebra and thus D has GK dimension one, since it is a finite module over $k(t)$. We let D act on itself, regarded as a finite-dimensional $k(t)$-vector space, by left multiplication. This gives an embedding of D into $\mathrm{End}_{k(t)}(D)$, which is a matrix ring over $k(t)$. It follows that D satisfies a polynomial identity. Thus D is finite-dimensional over its center. But the center Z of D has the same GK dimension as D since $[D : Z] < \infty$. Hence Z is a field of transcendence degree one. By Tsen's theorem we see that $D = Z$. □

As Stafford and van den Bergh point out, intuitively, this result makes perfect sense: a one-dimensional algebra should be essentially generated by one element and since an element commutes with itself, it is quite reasonable that such algebras should be commutative.

For algebras of Gelfand–Kirillov dimension two, the picture becomes significantly more complicated. For GK dimension two, there is a natural subclass of algebras: algebras of *quadratic growth*. These are finitely generated algebras A of GK dimension two with the property that there is some finite-dimensional generating subspace V of A that contains 1 with the property that the growth of the dimension of V^n is bounded above by Cn^2 for some positive constant C for n sufficiently large.

It is known that once one abandons quadratic growth and considers all algebras of GK dimension two, pathologies arise (see, for example, [Bell 2003]). On the other hand, there are no known examples of prime Noetherian algebras of GK dimension two that do not have quadratic growth. In the case of quadratic growth, algebras appear to be very well-behaved. In [Bell 2010] we showed that a complex domain of quadratic growth is either primitive or it satisfies a polynomial identity. This says that the algebra is either very close to being commutative or, in some sense, as far from being commutative as possible.

6. Gromov's theorem

Gromov's theorem states that every finitely generated group of polynomially bounded growth is nilpotent-by-finite. In this section we will give a brief overview of the ideas used in the proof and discuss possible extensions. We first note that the case of solvable groups of subexponential growth had already been considered by Milnor [1968] and Wolf [1968].

Theorem 6.1 [Milnor 1968; Wolf 1968]. *Let G be a finitely generated solvable group of subexponential growth. Then G is nilpotent-by-finite.*

This result has a completely elementary proof. The first main idea is that a combinatorial argument gives that if G is a finitely generated group of subexponential growth then G' is also finitely generated. From this, one obtains that G is polycyclic. This was Milnor's contribution to the theorem. One now uses induction on the solvable length of G and the fact that conjugation by elements of G on a characteristic finitely generated abelian subgroup gives a linear map. By looking at the eigenvalues of this map, one sees that a dichotomy arises: if one has an eigenvalue whose modulus is strictly greater than 1 then one gets exponential growth; if all eigenvalues have modulus one then Kronecker's theorem gives that they are roots of unity and one can deduce nilpotence of a finite-index subgroup from this. This eigenvalue analysis argument was Wolf's contribution to the argument.

The second thing we point out is that the linear case of Gromov's theorem is a consequence of a well-known alternative due to Tits [1972] and the above result of Milnor and Wolf.

Theorem 6.2 (the Tits alternative). *Let K be a field and let G be a finitely generated subgroup of* $\mathrm{GL}_n(K)$. *Then G is either solvable-by-finite or G contains a free subgroup on two generators.*

The proof of the Tits alternative is very difficult and makes use of the so-called "ping-pong" lemma to construct free subgroups. We only consider the case when K is the complex numbers. We recall that a matrix A has a *dominant* eigenvalue α if $|\alpha| > |\beta|$ whenever β is another eigenvalue of A and the kernel of $(A - \alpha I)^d$ is one-dimensional for every $d \geq 1$. One first shows that if the group G has an element A such that both A and A^{-1} have a *dominant* eigenvalue and B is an element of G such that neither B nor its inverse send the corresponding dominant eigenvectors of A and A^{-1} into some proper A-invariant subspace, then there is some n such that A^n and BA^nB^{-1} generate a free group.

From here, one uses different absolute values of the complex numbers and different representations of G to show that if G does not contain such a matrix A then either G has a solvable normal subgroup N such that G/N embeds in a subdirect product of smaller linear groups (one obtains the result by an induction on the size of the linear group in this case) or G has the property that every element of G has all of its eigenvalues equal to roots of unity. In this case, one can use arguments to Burnside and Schur to show that such a group is necessarily solvable-by-finite.

We note that given a finitely generated group G and a finite symmetric generating set S that includes 1, we can create an associated undirected Cayley graph $\Gamma = \Gamma(G, S)$ with vertices given by the elements of G and in which edges x and y are adjacent exactly when $xs = y$ for some $s \in S$. (Our choice of S creates loops in Γ.) We note that Γ has the property that each vertex in Γ has degree $|S|$ (and is adjacent to itself) and so the adjacency matrix has the property that each row has exactly $|S|$ ones and in particular $|S|$ as an eigenvalue. As it turns out, when G is infinite there is a large eigenspace associated to the eigenvalue $|S|$.

We make this precise now. We consider the complex vector space V of maps f from G to $\mathbb{C}$ with the properties that $\sum_{s \in S} f(xs) = |S| \cdot f(x)$ and such that for each $s \in S$ the map $f(x) - f(xs)$ is in $L^\infty(G)$. Such functions are called the *Lipschitz harmonic* functions on G with respect to S. Kleiner [2010] shows that when G is a finitely generated infinite group of polynomially bounded growth, the vector space V is finite-dimensional and has dimension at least two. We record this result now.

Theorem 6.3 [Kleiner 2010]. *Let G be a finitely generated infinite group of polynomially bounded growth. Then the space of Lipshitz harmonic functions on G with respect to S is finite-dimensional of dimension at least two.*

This is a particularly important part of the proof and appears to be the most vulnerable in terms of trying to extend Gromov's theorem to periodic groups whose growth functions are bounded by $\exp(n^{\epsilon})$ with $\epsilon > 0$ very small. The fact that V has dimension at least two is something that is true for all finitely generated infinite groups; it is the finite-dimensionality of V that is really the difficult point.

We now see how this quickly gives Gromov's theorem. Before we begin, we make the remark that it is no loss of generality to replace G by a finite-index subgroup since a finite-index subgroup of a finitely generated group is still finitely generated and has the same growth as the larger group.

Proof of Gromov's theorem using Kleiner's theorem. Let S be a symmetric generating set for G. We let α denote the supremum of all real numbers β with the property that the conclusion to Gromov's theorem holds for all finitely generated groups G with growth degree β. If $\alpha = \infty$ then we are done, and so we may assume that $\alpha < \infty$. Then there exists some $d < \alpha + 1$ and some finitely generated group G of growth degree d that is not nilpotent-by-finite.

We let V denote the space of Lipschitz harmonic functions on G with respect to S. Note that G acts on V via $g \cdot f(x) = f(g^{-1}x)$ and this gives a homomorphism from G into $\mathrm{GL}(V)$. We let N denote the kernel of this homomorphism. Then G/N is a linear group of subexponential growth and so by the Tits' alternative the image is solvable-by-finite (since a free group on two generators has exponential growth).

We note that G/N must be infinite, since otherwise N acts trivially on V and this gives that all functions in V are constant on cosets of N. In particular, each $f \in V$ takes at most $[G : N]$ values. But it is easy to see that this forces f to be constant, since if $x \in G$ is chosen so that $|f(x)|$ is maximal then the equality $\sum_{s \in S} f(xs) = |S| f(x)$, gives that $f(xs) = f(x)$ for all $s \in S$ and since S generates G we see that f is constant on G. This contradicts the fact that V has dimension at least two.

Thus G/N is an infinite solvable-by-finite group. By replacing G by a finite-index subgroup if necessary, we may assume that G/N is solvable and that $(G/N)/(G/N)'$ is infinite. In this case G has a normal subgroup H such that $H \supseteq N$ and $G/H \cong \mathbb{Z}$. Moreover, H is finitely generated by a combinatorial argument due to Milnor and it is easy to check that the growth of H has degree at most $d - 1 < \alpha$. Consequently, H is nilpotent-by-finite. By assumption, G/H is solvable since it is a homomorphic image of G/N. It is straightforward to check that G is solvable-by-finite, since both G/H and H are finitely generated and solvable-by-finite. Thus G has a finite-index subgroup that is solvable in this case. Thus the theorem reduces to the solvable case of Gromov's theorem, which is handled by the Milnor–Wolf theorem. □

If one wishes to use the methods from the proof of Gromov's theorem in other contexts there are a few things that should be pointed out.

(A) First, the proof relies on the construction of a finite-dimensional vector space V of dimension at least two on which no finite-index subgroup can act trivially.

(B) Second, the proof shows that G has a finite-index subgroup that surjects onto $\mathbb{Z}$ and uses an induction on the degree of growth of G to finish the proof.

These two points show us that it is unreasonable to use the method to try to extend Gromov's theorem to include groups of "slow" subexponential growth, since the induction is not available in this case. The second point does lend itself to the study of periodic (torsion) groups, however. In this case the existence of a surjection onto $\mathbb{Z}$ gives an immediate proof that G is not a periodic group and so the first point is the only obstruction in this case.

Another interesting question involves Noetherian group algebras. We note that an immediate consequence of Gromov's theorem is that if G is a finitely generated group of polynomially bounded growth then the group algebra of G is Noetherian. It is conjectured that a group algebra is Noetherian if and only if the group G is polycyclic-by-finite. It seems plausible that one can use the Noetherian property to handle (A); the problem, however, is that the induction step in (B) is not available. Thus we pose the following question, which in theory makes the induction step in (B) doable.

Question 1. Suppose that $k[G]$ is a Noetherian group algebra of finite Krull dimension. Is G polycyclic-by-finite?

7. The Kurosh problem and growth

Due to constructions of Smoktunowicz, there has been renewed interest in the construction of finitely generated algebraic algebras that are not finite-dimensional (see, for example, [Bell and Small 2002; Bell et al. 2012; Lenagan and Smoktunowicz 2007; Lenagan et al. 2012; Smoktunowicz 2000; 2002; 2009]). The first examples of such algebras were constructed by Golod and Sharafervich [Golod 1964; Golod and Shafarevich 1964], who used a simple combinatorial criterion that guaranteed that algebras with certain presentations are infinite-dimensional. Their construction provided a counterexample to Kurosh's conjecture, which asserts that finitely generated algebraic algebras should be finite-dimensional over their base fields. By using their construction in a clever way, Golod and Shafarevich were also able to give a counterexample to the celebrated Burnside problem, which is the group-theoretic analogue of the Kurosh conjecture and asks whether or not finitely generated torsion groups are necessarily finite.

The connection between Burnside-type problems in the theory of groups and Kurosh-type problems in ring theory has led to many interesting conjectures in both fields, which have arisen naturally from results in one field or the other.

As mentioned in the preceding section, Gromov's theorem gives a concrete description of groups of polynomially bounded growth. As finitely generated nilpotent torsion groups are finite, Gromov's theorem thus immediately gives the result that a finitely generated torsion group of polynomially bounded growth is finite.

In light of the result of Gromov and its consequence for the Burnside problem, Small asked whether a finitely generated algebraic algebra of polynomially bounded growth should be finite-dimensional [Lenagan et al. 2012]. Surprisingly, Lenagan and Smoktunowicz [2007] were able to give a counterexample, by constructing a finitely generated nil algebra of Gelfand–Kirillov dimension at most 20. We point out that their construction only works over countable base fields, and it is still an open question as to whether the Kurosh problem for algebras of polynomially bounded growth should hold over uncountable fields.

Recently, Lenagan, Smoktunowicz, and Young [Lenagan et al. 2012] showed that the bound on Gelfand–Kirillov dimension could be lowered from 20 to 3. On the other hand, it is known that the bound cannot be made lower than 2, as finitely generated algebras of Gelfand–Kirillov dimension strictly less than two satisfy a polynomial identity [Small et al. 1985; Krause and Lenagan 2000, Theorem 2.5, p. 18] and the Kurosh conjecture holds for the class of algebras satisfying a polynomial identity [Herstein 1994, Section 6.4].

The fact that these constructions do not work over an uncountable base field is not surprising, as many results have appeared over the years which show there is a real dichotomy that exists regarding Kurosh-type problems when one considers base fields. For example, algebraic algebras over uncountable fields have what is known as the *linearly bounded degree* property (see, for example, [Jacobson 1964, p. 249, Definition 1]). The linearly bounded degree property simply says that given a fixed finite-dimensional subspace of an algebraic algebra, there is a natural number d, depending on the subspace, such that all elements in this subspace have degree at most d. Smoktunowicz [2000] has given an example of a nil algebra over a countable base field with the property that the ring of polynomials over this algebra is not nil and hence this algebra cannot have linearly bounded degree.

This distinction, and the fact that the elements in a finitely generated algebra over a countable base field can be enumerated, has led to a relative dearth of interesting examples of algebraic algebras over uncountable base fields. Indeed, over uncountable fields there has not been much progress since the original construction of Golod and Shafarevich.

In a [Bell and Young 2011], counterexamples to the Kurosh conjecture of subexponential growth over general fields were found.

Theorem 7.1 [Bell and Young 2011]. *Let K be a field and let*

$$\alpha : [0, \infty) \to [0, \infty)$$

be a weakly increasing function tending to ∞. Then there is a finitely generated connected graded infinite-dimensional K-algebra

$$B = \bigoplus_{n \geq 0} B(n)$$

such that the homogeneous maximal ideal $\oplus_{n\geq 1} B(n)$ is nil, and $\dim(B(n)) \leq n^{\alpha(n)}$ *for all sufficiently large n.*

Equivalently, Theorem 7.1 says that if $\beta(n)$ is any monotonically increasing function that grows subexponentially but superpolynomially in n, in the sense that $n^d/\beta(n) \to 0$ as $n \to \infty$ for every $d \geq 0$, then one can find a connected graded K-algebra B whose homogeneous maximal ideal is nil and has the property that the growth function of B is asymptotically dominated by $\beta(n)$.

One should contrast this situation with the situation in group theory, where considerably less is known about the possible growth types of finitely generated torsion groups of superpolynomial growth. There have been many constructions of *branch groups*, which provide examples of groups that have subexponential but superpolynomial growth. The first such construction was done by Grigorchuk [1980], and estimates of Bartoldi [2005] show that the growth of Branch group constructions is at least $\exp(\sqrt{n})$, but significantly less than $\exp(cn)$ for any $c > 0$. We now discuss the method used by Lenagan and Smoktunowicz [2007].

7.1. ***The bottleneck method.*** The bottleneck method is a technique for producing nil algebras that has been largely honed by Smoktunowicz [Lenagan and Smoktunowicz 2007; Lenagan et al. 2012; Smoktunowicz 2000; 2002; 2009] in producing a sequence of counterexamples to Kurosh-type problems. It is the modification of this method given in her paper with Lenagan, however, that has inspired this choice of name. To understand the basic philosophy of the technique, let us suppose that one wishes to construct an algebra of polynomially bounded growth with certain additional properties. Typically, people would do this by imposing a finite number of cleverly chosen relations of small degree that had the effect of curtailing the growth. This method is often used, but it is of apparently no help if one wishes to construct nil algebras of polynomially bounded growth.

Lenagan and Smoktunowicz take a more Malthusian approach to manufacturing growth. In this setting, they take a sparse sequence of natural numbers $n_1 < n_2 < n_3 < \cdots$ with n_1 very large. They then take a free algebra $A = k\{x, y\}$ over a countable field k. Since k is countable, we can enumerate the elements of A. We let $A = \{f_1, f_2, \ldots\}$. We note that A has exponential growth and there are 2^n words over the alphabet $\{x, y\}$ of length n. They then impose a large number of homogeneous relations of degree n_1 with the property that f_1 is nilpotent modulo these relations. In fact, more relations are imposed than necessary and this has the effect of creating a "bottleneck" at degree n_1. While many relations are imposed, the relations are chosen so that if no further relations were imposed, the algebra would sill have exponential growth. In general, at each level n_i, they create another bottleneck by imposing homogeneous relations of degree n_i such that f_i is nilpotent modulo these relations.

There is a lot of care required: if the bottlenecks are too sparse, the algebra will have intermediate growth; if the bottlenecks are too narrow, the algebra will be finite-dimensional. The fact that one is dealing with two-sided ideals, which are much more difficult to control than their one-sided counterparts makes the construction especially difficult.

Despite these obstacles, Lenagan and Smoktunowicz were able to construct a nil algebra of polynomially bounded growth over any countable field. Later, in a paper with Young, they showed that one can construct a nil algebra whose GK dimension is at most three.

Let K be a field and let $A = K\{x, y\}$ denote the free K-algebra on two generators x and y. Then A is an $\mathbb{N}$-graded algebra, and we let $A(n)$ denote the K-subspace spanned by all words over x and y of length n. We give the key proposition of [Lenagan and Smoktunowicz 2007, Theorem 3], which we have expressed in a slightly more general form (see [Bell and Young 2011]).

Proposition 7.2. *Let $f, g : \mathbb{N} \to \mathbb{N}$ be two maps satisfying*

(i) $f(i-1) < f(i) - g(i) - 1$ *for all natural numbers i,*

(ii) *for each natural number i, there is a subspace $W_i \subseteq A(2^{f(i)})$ whose dimension is at most $2^{2^{g(i)}} - 2$,*

and let

$$T = \bigcup_i \{f(i) - g(i) - 1,\, f(i) - g(i),\, \ldots,\, f(i) - 1\}.$$

Then for each natural number n, there exist K-vector subspaces $U(2^n)$ and $V(2^n)$ of $A(2^n)$ satisfying the following properties:

(1) $U(2^n) \oplus V(2^n) = A(2^n)$ *for every natural number* n;

(2) $\dim(V(2^n)) = 2$ *whenever* $n \notin T$;

(3) $\dim(V(2^{n+j})) = 2^{2^j}$ *whenever* $n = f(i) - g(i) - 1$ *and* $0 \le j \le g(i)$;

(4) *for each natural number* n, $V(2^n)$ *has a basis consisting of words over* x *and* y;

(5) *for each natural number* i, $W_i \subseteq U(2^{f(i)})$;

(6) $A(2^n)U(2^n) + U(2^n)A(2^n) \subseteq U(2^{n+1})$ *for every natural number* n;

(7) $V(2^{n+1}) \subseteq V(2^n)V(2^n)$ *for every natural number* n;

(8) *if* $n \notin T$ *then there is some word* $w \in V(2^n)$ *such that* $wA(2^n) \subseteq U(2^{n+1})$.

One should think of the subspaces $U(n)$ and $V(n)$ as follows. Condition (6) says that the sum of the $U(n)$ is in some sense very close to being a two-sided ideal. It is not a two-sided ideal, but Lenagan and Smoktunowicz show that there is a homogeneous two-sided ideal I which is a close approximation to this space. Then one should think of the image of the sum of the $V(n)$ when we mod out by this ideal as being very close to a basis for the factor ring A/I.

The fact that there are infinitely many $n \notin T$, and conditions (2) and (3), say that the growth of A/I should be small if $g(n)$ grows sufficiently slowly compared to $f(n)$. The role of the subspaces W_i is that they correspond to homogeneous relations introduced. Thus if we are not introducing too many relations and we have that the dimension of W_i is bounded by $2^{2^{g(i)}} - 2$, then we can hope to find an infinite-dimensional algebra with slow growth in which the images of all relations coming from the subspaces W_i are zero.

The above proposition allows one, over countable fields, to construct infinite-dimensional nil algebras of finite Gelfand–Kirillov dimension by picking sparse small sets of relations that imply nilpotence of elements in the algebra and augmenting this with sets that give slow growth.

8. Lower transcendence degree

Transcendence degree for fields is an important invariant and has proved incredibly useful in algebraic geometry. In the noncommutative setting, many different transcendence degrees have been proposed [Gelfand and Kirillov 1966; Borho and Kraft 1976; Resco 1980; Schofield 1984; Stafford 1983; Zhang 1998; 1996; Yekutieli and Zhang 2006], many of which possess some of the desirable properties that one would hope for a noncommutative analogue of transcendence degree to possess. None of these invariants, however, has proved as versatile as the ordinary transcendence degree has in the commutative setting, as there has always been the fundamental problem: they are either difficult to compute in practice or are not powerful enough to say anything meaningful about division subalgebras.

The first such invariant was defined by Gelfand and Kirillov [1966], who, as we noted earlier, used their Gelfand–Kirillov transcendence degree to prove that if the quotient division algebras of the n-th and m-th Weyl algebras are isomorphic, then $n = m$. Gelfand–Kirillov transcendence degree is obtained from Gelfand–Kirillov dimension in a natural way.

The *Gelfand–Kirillov transcendence degree* for a division algebra D with center k is defined to be

$$\mathrm{Tdeg}(A) = \sup_V \inf_b \limsup_{n\to\infty} \frac{\log \dim_k (k + bV)^n}{\log n},$$

where V ranges over all finite-dimensional k-vector subspaces of D and b ranges over all nonzero elements of D.

Zhang [1998] introduced a combinatorial invariant, which he called the *lower transcendence degree* of a division algebra D, which he denoted $\mathrm{Ld}(D)$. To give a concrete description of the main principle behind lower transcendence degree, one can consider the problem of assigning a dimension to a geometric object. If one has a d-dimensional hypercube, one can recover d by noting that upon dilating the hypercube by a factor of 2, the volume increases by a factor of 2^d. In some sense, Gelfand–Kirillov dimension works according to this principle. There is, however, another way of extracting d. One can note that if S is the surface area of the d-dimensional hypercube and V is its volume, then S is proportional to $V^{(d-1)/d}$. This is the basic principle upon which lower transcendence degree rests.

Zhang showed that this degree had many of the basic properties that one would expect a transcendence degree to have. In particular, he showed that if k is a field, A is a k-algebra that is an Ore domain of finite GK dimension, and D is the quotient division algebra of A, then

$$\mathrm{GKdim}(A) \geq \mathrm{Ld}(D),$$

and thus the invariant is well-behaved under localization.

We now define lower transcendence degree. Given a field k and a k-algebra A, we say that a k-vector subspace V of A is a *subframe* of A if V is finite-dimensional and contains 1; we say that V is a *frame* if V is a subframe and V generates A as a k-algebra.

If V is a subframe of A, we define $\mathrm{VDI}(V)$ to be the supremum over all nonnegative numbers d such that there exists a positive constant C such that

$$\dim_k(VW) \geq \dim_k(W) + C\dim(W)^{(d-1)/d}$$

for every subframe W of D. (If no nonnegative d exists, we take $\mathrm{VDI}(V)$ to be zero.) VDI stands for "volume difference inequality" and it gives a measure of the growth of an algebra. Note that

$$\dim_k(VW) - \dim_k(W)$$

is really just $\dim_k(VW/W)$ and thus this is in some sense giving the dimension of the boundary of W with respect to V. In terms of the hypercube analogy, this quantity corresponds to the surface area. The quantity $\dim(W)$ corresponds to the volume under this analogy and so the exponent $(d-1)/d$ is telling us what the right notion of "dimension" should be in this case.

We then define the *lower transcendence degree* of A by

$$\mathrm{Ld}(A) = \sup_V \mathrm{VDI}(V),$$

where V ranges over all subframes of A.

The definition, while technical, gives a powerful invariant that Zhang [1998] has used to answer many difficult problems about division algebras. Zhang showed that if A is an Ore domain of finite GK dimension and D is the quotient division algebra of A then $\mathrm{Ld}(D) \le \mathrm{GKdim}(A)$. Moreover, equality holds for many classes of rings. In particular, if A is a commutative domain over a field k, then equality holds and so Lower transcendence degree agrees with ordinary transcendence degree.

Lower transcendence degree has the nice additional property that if D is a division algebra and E is a division subalgebra of D then $\mathrm{Ld}(E) \le \mathrm{Ld}(D)$. Using this fact, Zhang was able to answer a conjecture of Small's by proving the following theorem.

Theorem 8.1 (Zhang). *Let A be a finitely generated k-algebra that is a domain of Gelfand–Kirillov dimension d. If K is a maximal subfield of $Q(A)$, then K has transcendence degree at most d as an extension of k.*

Proof. We have

$$\mathrm{Trdeg}(K) = \mathrm{Ld}(K) \le \mathrm{Ld}(D) \le \mathrm{GKdim}(A) = d.$$

The result follows. □

The author [Bell 2012], extended Zhang's result by showing that if A does not satisfy a polynomial identity then one in fact has $\mathrm{Trdeg}(K) \le d-1$. In particular, this shows that quotient division algebras of domains of GK dimension two are either finite-dimensional over their centers or have the property that all maximal subfields have transcendence degree one.

Despite some of the successes of lower transcendence degree, there are still some basic questions about the invariant that remain unanswered. We give a few of the basic questions.

Question 2. Is there a Bergman-style gap theorem for lower transcendence degree? More specifically, if D is a division ring and $\mathrm{Ld}(D) \in [1, 2)$, does it follow that $\mathrm{Ld}(D) = 1$?

Question 3. Let A be a finitely generated Ore domain and let $D = Q(A)$. If $\mathrm{Ld}(D) = 1$, is it true that D is finite-dimensional over its center?

We note that a positive solution to these questions would immediately give Smoktunowicz's graded gap theorem. The reason for this is that if A is a finitely generated graded domain whose GK dimension lies in $[2, 3)$ then A has a graded quotient division ring $Q_{\mathrm{gr}}(A) = D[t, t^{-1}; \sigma]$. (Here D is obtained by taking the degree zero part of the algebra obtained by inverting the nonzero homogeneous elements of A.) Zhang proves that in such situations one has $\mathrm{Ld}(D) \le \mathrm{Ld}(A) - 1 \le \mathrm{GKdim}(A) - 1$. Thus $\mathrm{Ld}(D) = 1$ and so D is a finite-dimensional over its center, which is a field of transcendence degree one. The arguments of [Artin and Stafford 1995] now show that A has GK dimension exactly two.

9. Division algebras of transcendence degree two: Artin's conjecture

One of the truly difficult problems that pertains to growth is to give a so-called birational classification of finitely generated complex domains of Lower transcendence degree two. This is strongly related to Artin's conjecture, which we now briefly describe.

Artin [1997] gives a proposed birational classification of a certain class of graded domains of GK dimension 3. We note that Artin warns the reader that "everything should be taken with a grain of salt," as far as his proposed classification is concerned, and some people believe there are additional division rings yet to be found that are missing from the list.

To begin, we let A be a graded Noetherian domain of GK dimension 3 that is generated in degree 1 and we let $\mathscr{C}$ denote the category of finitely generated graded right A-modules modulo the subcategory of torsion modules. (This can be thought of as the category of "tails" of finitely generated graded A-modules.) We let $\mathrm{Proj}(A)$ denote the triple $(\mathscr{C}, \mathcal{O}, s)$, where $\mathcal{O}$ is the image of the right module A in $\mathscr{C}$ and s is the autoequivalence of $\mathscr{C}$ defined by the shift operator on graded modules.

We note that A has a *graded quotient division ring*, denoted by $Q_{\mathrm{gr}}(A)$, which is formed by inverting the nonzero homogeneous elements of A. Then there is a division ring D and automorphism σ of D such that

$$Q_{\text{gr}}(A) \cong D[t, t^{-1}; \sigma].$$

We think of D as being the function field of $X = \text{Proj}(A)$.

Artin gives a proposed classification of the type of division rings D that can occur when A is a complex Noetherian domain of GK dimension 3. (There are also a few other technical homological assumptions that he assumes the algebra possesses, but we shall ignore these and instead refer the interested reader to Artin's paper [1997].) If A has these properties and $Q_{\text{gr}}(A) \cong D[t, t^{-1}; \sigma]$, then Artin asserts that up to isomorphism the possible division rings D must satisfy at least one of the conditions on the list:

(1) D is finite-dimensional over its center, which is a finitely generated extension of $\mathbb{C}$ of transcendence degree 2;

(2) D is birationally isomorphic to a quantum plane; that is, D is isomorphic to the quotient division ring of the complex domain generated by x and y with relation $xy = qyx$ for some nonzero $q \in \mathbb{C}$;

(3) D is isomorphic to the Skylanin division ring (see [Artin 1997] for relevant definitions);

(4) D is isomorphic to the quotient division ring of the Weyl algebra;

(5) D is birationally isomorphic to $K[t; \sigma]$ or $K[t; \delta]$, where K is a finitely generated field extension of $\mathbb{C}$ of positive genus and σ and δ are respectively an automorphism and a derivation of K.

In some cases one can homogenize the relations in a domain of GK dimension 2 using a central indeterminate and obtain a graded domain of GK dimension 3 satisfying the conditions that Artin assumes. We also note that many of the division rings on Artin's list are quotient division rings of Noetherian domains of GK dimension 2. In this sense, there is a strong relationship between a birational classification of Noetherian domains of GK dimension 2 and graded Noetherian domains of GK dimension 3.

We summarize some of the work that has been obtained via growth methods on Artin's conjecture.

Theorem 9.1. *Let A be a graded Noetherian complex domain of GK dimension* 3 *that is generated in degree* 1 *and let D be the degree zero part of the homogeneous quotient division ring of A. If D is not finite-dimensional over its center, then*:

(1) *all maximal subfields of D are finitely generated and of transcendence degree one over $\mathbb{C}$*;

(2) *D contains a free algebra on two generators.*

Proof. The fact that the subfields are finitely generated follows from the fact that A is a finitely generated complex Noetherian domain and a theorem of the author [Bell 2007] which shows that all subfields of $Q(A)$ are finitely generated in this case. We note that if one uses the *strong lower transcendence degree*, Ld^*, one can show that $\mathrm{Ld}^*(D) \le \mathrm{Ld}^*(A) - 1 \le 2$, since $Q_{\mathrm{gr}}(A) \cong D[t, t^{-1}; \sigma]$. Since any subfield K of D necessarily has the property that D is infinite-dimensional as a left and right K-vector space, we see that $\mathrm{Ld}(K) \le 1$ by [Bell 2010]. It follows that K has transcendence degree one over $\mathbb{C}$.

The fact that D contains a free algebra on two generators follows from work in progress of the author and Dan Rogalski [Bell and Rogalski $\ge$ 2015] (see also [Bell and Rogalski 2014]), since all subfields of D are finitely generated and the base field is the complex numbers. □

We note that in the case that D is the quotient division algebra of the Weyl algebra, (3) had been proven by Makar-Limanov [1983; 1996].

One of the questions, which could provide an important invariant is the types of subfields that can occur in the division algebras on Artin's list. By Theorem 9.1, all subfields have transcendence degree one over $\mathbb{C}$ and are finitely generated. Mironov [2011] has shown that one can find subfields of all genera inside the quotient division algebra of the Weyl algebra. We ask whether an analogous result holds for the quantum plane.

Question 4. Let D_q denote the quotient division algebra of the quantum plane. If q is not a root of unity, can D_q contain the function field of a smooth curve X of positive genus over $\mathbb{C}$?

References

[Alev et al. 1996] J. Alev, A. Ooms, and M. Van den Bergh, "A class of counterexamples to the Gel′fand–Kirillov conjecture", *Trans. Amer. Math. Soc.* **348**:5 (1996), 1709–1716.

[Artin 1997] M. Artin, "Some problems on three-dimensional graded domains", pp. 1–19 in *Representation theory and algebraic geometry* (Waltham, MA, 1995), edited by A. Martsinkovsky and G. Todorov, London Math. Soc. Lecture Note Ser. **238**, Cambridge Univ. Press, Cambridge, 1997.

[Artin and Stafford 1995] M. Artin and J. T. Stafford, "Noncommutative graded domains with quadratic growth", *Invent. Math.* **122**:2 (1995), 231–276.

[Bartholdi 2005] L. Bartholdi, "Lie algebras and growth in branch groups", *Pacific J. Math.* **218**:2 (2005), 241–282.

[Bell 2003] J. P. Bell, "Examples in finite Gel′fand–Kirillov dimension", *J. Algebra* **263**:1 (2003), 159–175.

[Bell 2007] J. P. Bell, "Noetherian algebras over algebraically closed fields", *J. Algebra* **310**:1 (2007), 148–155.

[Bell 2010] J. P. Bell, "A dichotomy result for prime algebras of Gelfand–Kirillov dimension two", *J. Algebra* **324**:4 (2010), 831–840.

[Bell 2012] J. P. Bell, "Transcendence degree of division algebras", *Israel J. Math.* **190** (2012), 195–211.

[Bell and Rogalski 2014] J. P. Bell and D. Rogalski, "Free subalgebras of division algebras over uncountable fields", *Math. Z.* **277**:1-2 (2014), 591–609.

[Bell and Rogalski $\geq$ 2015] J. P. Bell and D. Rogalski, "Free subalgebras of quotient division rings of Noetherian domains", in preparation.

[Bell and Small 2002] J. P. Bell and L. W. Small, "A question of Kaplansky", *J. Algebra* **258**:1 (2002), 386–388.

[Bell and Young 2011] J. P. Bell and A. A. Young, "On the Kurosh problem for algebras of polynomial growth over a general field", *J. Algebra* **342** (2011), 265–281.

[Bell et al. 2012] J. P. Bell, L. W. Small, and A. Smoktunowicz, "Primitive algebraic algebras of polynomially bounded growth", pp. 41–52 in *New trends in noncommutative algebra*, edited by P. Ara et al., Contemp. Math. **562**, Amer. Math. Soc., Providence, RI, 2012.

[Borho and Kraft 1976] W. Borho and H. Kraft, "Über die Gelfand–Kirillov-Dimension", *Math. Ann.* **220**:1 (1976), 1–24.

[Gelfand and Kirillov 1966] I. M. Gelfand and A. A. Kirillov, "Sur les corps liés aux algèbres enveloppantes des algèbres de Lie", *Inst. Hautes Études Sci. Publ. Math.* 31 (1966), 5–19.

[Golod 1964] E. S. Golod, "On nil-algebras and finitely approximable p-groups", *Izv. Akad. Nauk SSSR Ser. Mat.* **28** (1964), 273–276. In Russian.

[Golod and Shafarevich 1964] E. S. Golod and I. R. Shafarevich, "On the class field tower", *Izv. Akad. Nauk SSSR Ser. Mat.* **28** (1964), 261–272. In Russian.

[Grigorčuk 1980] R. I. Grigorčuk, "On Burnside's problem on periodic groups", *Funktsional. Anal. i Prilozhen.* **14**:1 (1980), 53–54.

[Gromov 1981] M. Gromov, "Groups of polynomial growth and expanding maps", *Inst. Hautes Études Sci. Publ. Math.* **53** (1981), 53–73.

[Herstein 1994] I. N. Herstein, *Noncommutative rings*, Reprint of the 1968 ed., Carus Mathematical Monographs **15**, Mathematical Association of America, Washington, DC, 1994.

[Jacobson 1964] N. Jacobson, *Structure of rings*, Revised ed., American Mathematical Society Colloquium Publications **37**, American Mathematical Society, Providence, R.I., 1964.

[Kleiner 2010] B. Kleiner, "A new proof of Gromov's theorem on groups of polynomial growth", *J. Amer. Math. Soc.* **23**:3 (2010), 815–829.

[Krause and Lenagan 2000] G. R. Krause and T. H. Lenagan, *Growth of algebras and Gelfand–Kirillov dimension*, Revised ed., Graduate Studies in Mathematics **22**, American Mathematical Society, Providence, RI, 2000.

[Lenagan and Smoktunowicz 2007] T. H. Lenagan and A. Smoktunowicz, "An infinite dimensional affine nil algebra with finite Gelfand–Kirillov dimension", *J. Amer. Math. Soc.* **20**:4 (2007), 989–1001.

[Lenagan et al. 2012] T. H. Lenagan, A. Smoktunowicz, and A. A. Young, "Nil algebras with restricted growth", *Proc. Edinb. Math. Soc.* (2) **55**:2 (2012), 461–475.

[Makar-Limanov 1983] L. Makar-Limanov, "The skew field of fractions of the Weyl algebra contains a free noncommutative subalgebra", *Comm. Algebra* **11**:17 (1983), 2003–2006.

[Makar-Limanov 1996] L. Makar-Limanov, "On the hypersurface $x + x^2y + z^2 + t^3 = 0$ in $\mathbf{C}^4$ or a $\mathbf{C}^3$-like threefold which is not $\mathbf{C}^3$", *Israel J. Math.* **96**:part B (1996), 419–429.

[Milnor 1968] J. Milnor, "Growth of finitely generated solvable groups", *J. Differential Geometry* **2** (1968), 447–449.

[Mironov 2011] A. E. Mironov, "Self-adjoint commuting differential operators and commutative subalgebras of the Weyl algebra", preprint, 2011. arXiv 1107.3356

[Resco 1980] R. Resco, "A dimension theorem for division rings", *Israel J. Math.* **35**:3 (1980), 215–221.

[Schofield 1984] A. H. Schofield, "Questions on skew fields", pp. 489–495 in *Methods in ring theory* (Antwerp, 1983), edited by F. Van Oystaeyen, NATO Adv. Sci. Inst. Ser. C Math. Phys. Sci. **129**, Reidel, Dordrecht, 1984.

[Small et al. 1985] L. W. Small, J. T. Stafford, and R. B. Warfield, Jr., "Affine algebras of Gel′fand–Kirillov dimension one are PI", *Math. Proc. Cambridge Philos. Soc.* **97**:3 (1985), 407–414.

[Smoktunowicz 2000] A. Smoktunowicz, "Polynomial rings over nil rings need not be nil", *J. Algebra* **233**:2 (2000), 427–436.

[Smoktunowicz 2002] A. Smoktunowicz, "A simple nil ring exists", *Comm. Algebra* **30**:1 (2002), 27–59.

[Smoktunowicz 2005] A. Smoktunowicz, "The Artin–Stafford gap theorem", *Proc. Amer. Math. Soc.* **133**:7 (2005), 1925–1928.

[Smoktunowicz 2006] A. Smoktunowicz, "There are no graded domains with GK dimension strictly between 2 and 3", *Invent. Math.* **164**:3 (2006), 635–640.

[Smoktunowicz 2009] A. Smoktunowicz, "Makar-Limanov's conjecture on free subalgebras", *Adv. Math.* **222**:6 (2009), 2107–2116.

[Stafford 1983] J. T. Stafford, "Dimensions of division rings", *Israel J. Math.* **45**:1 (1983), 33–40.

[Tits 1972] J. Tits, "Free subgroups in linear groups", *J. Algebra* **20** (1972), 250–270.

[Wolf 1968] J. A. Wolf, "Growth of finitely generated solvable groups and curvature of Riemanniann manifolds", *J. Differential Geometry* **2** (1968), 421–446.

[Yekutieli and Zhang 2006] A. Yekutieli and J. J. Zhang, "Homological transcendence degree", *Proc. London Math. Soc.* (3) **93**:1 (2006), 105–137.

[Zhang 1996] J. J. Zhang, "On Gel′fand–Kirillov transcendence degree", *Trans. Amer. Math. Soc.* **348**:7 (1996), 2867–2899.

[Zhang 1998] J. J. Zhang, "On lower transcendence degree", *Adv. Math.* **139**:2 (1998), 157–193.

jpbell@uwaterloo.ca

Department of Pure Mathematics, University of Waterloo, 200 University Avenue, Waterloo ON N2L 3G1, Canada

Commutative Algebra and Noncommutative Algebraic Geometry, I
MSRI Publications
Volume **67**, 2015

Syzygies, finite length modules, and random curves

CHRISTINE BERKESCH AND FRANK-OLAF SCHREYER

We apply the theory of Gröbner bases to the computation of free resolutions over a polynomial ring, the defining equations of a canonically embedded curve, and the unirationality of the moduli space of curves of a fixed small genus.

Introduction

While a great deal of modern commutative algebra and algebraic geometry has taken a nonconstructive form, the theory of Gröbner bases provides an algorithmic approach. Algorithms currently implemented in computer algebra systems, such as Macaulay2 [Grayson and Stillman] and Singular [Decker et al. 2011], already exhibit the wide range of computational possibilities that arise from Gröbner bases.

In these lectures, we focus on certain applications of Gröbner bases to syzygies and curves. In Section 1, we use Gröbner bases to give an algorithmic proof of Hilbert's syzygy theorem, which bounds the length of a free resolution over a polynomial ring. In Section 2, we prove Petri's theorem about the defining equations for canonical embeddings of curves. We turn in Section 3 to the Hartshorne–Rao module of a curve, showing by example how a module M of finite length can be used to explicitly construct a curve whose Hartshorne–Rao module is M. Section 4 then applies this construction to the study of the unirationality of the moduli space $\mathfrak{M}_g$ of curves of genus g.

1. Hilbert's syzygy theorem

Let $R := \mathbb{K}[x_1, \dots, x_n]$ be a polynomial ring in n variables over a field $\mathbb{K}$. A *free resolution* of a finitely generated R-module M is a complex of free modules

This article consists of extended notes from lectures by Schreyer at the Joint Introductory Workshop: Cluster Algebras and Commutative Algebra at MSRI during the fall of 2012. This was supported by the National Science Foundation under Grant No. 0932078 000.

MSC2010: 13P10, 13D05, 13C40, 14H10, 14-04, 14M20, 14Q05.

$\cdots \to R^{\beta_2} \to R^{\beta_1} \to R^{\beta_0}$ such that the following is exact:

$$\cdots \to R^{\beta_2} \to R^{\beta_1} \to R^{\beta_0} \to M \to 0.$$

Hilbert's syzygy theorem (Theorem 1.1). *Let $R = \mathbb{K}[x_1, \dots, x_n]$ be a polynomial ring in n variables over a field $\mathbb{K}$. Every finitely generated R-module M has a finite free resolution of length at most n.*

In this section, we give an algorithmic Gröbner basis proof of Hilbert's syzygy theorem, whose strategy is used in modern computer algebra systems like Macaulay2 and Singular for syzygy computations. Gröbner bases were introduced by Gordan [1899] to provide a new proof of Hilbert's basis theorem. We believe that Gordan could have given the proof of Hilbert's syzygy theorem presented here.

Definition 1.2. A (*global*) *monomial order* on R is a total order $>$ on the set of monomials in R such that:

(1) if $x^\alpha > x^\beta$, then $x^\gamma x^\alpha > x^\gamma x^\beta$ for all $\gamma \in \mathbb{N}^n$; and

(2) $x_i > 1$ for all i.

Given a global monomial order, the *leading term* of a nonzero polynomial $f = \sum_\alpha f_\alpha x^\alpha \in R$ is defined to be

$$\mathbf{L}(f) := f_\beta x^\beta, \quad \text{where } x^\beta := \max_\alpha \{x^\alpha \mid f_\alpha \neq 0\}.$$

For convenience, set $\mathbf{L}(0) := 0$.

Theorem 1.3 (division with remainder). *Let $>$ be a global monomial order on R, and let $f_1, \dots, f_r \in R$ be nonzero polynomials. For every $g \in R$, there exist uniquely determined $g_1, \dots, g_r \in R$ and a remainder $h \in R$ such that:*

(1) $g = g_1 f_1 + \cdots + g_r f_r + h$.

(2a) *No term of $g_i\, \mathbf{L}(f_i)$ is divisible by any $\mathbf{L}(f_j)$ with $j < i$.*

(2b) *No term of h is a multiple of $\mathbf{L}(f_i)$ for any i.*

Proof. The result is obvious if $f_1, \dots, f_r$ are monomials, or more generally, if each f_i has only a single nonzero term. Thus there is always a unique expression

$$g = \sum_{i=1}^{r} g_i^{(1)}\, \mathbf{L}(f_i) + h^{(1)},$$

if we require that $g_1^{(1)}, \dots, g_r^{(1)}$ and $h^{(1)}$ satisfy (2a) and (2b). By construction, the leading terms of the summands of the expression

$$\sum_{i=1}^{r} g_i^{(1)} f_i + h^{(1)}$$

are distinct, and the leading term in the difference $g^{(1)} = g - (\sum_{i=1}^{r} g_i^{(1)} f_i + h^{(1)})$ cancels. Thus $\mathbf{L}(g^{(1)}) < \mathbf{L}(g)$, and recursion applies. □

The remainder h of the division of g by $f_1, \dots, f_r$ depends on the order of $f_1, \dots, f_r$, since the partition of the monomials in R given by (2a) and (2b) depends on this order. Even worse, it might not be the case that if $g \in \langle f_1, \dots, f_r \rangle$, then $h = 0$. A Gröbner basis is a system of generators for which this desirable property holds.

Definition 1.4. Let $I \subset R$ be an ideal. The *leading ideal* of I (with respect to a given global monomial order) is

$$\mathbf{L}(I) := \langle \mathbf{L}(f) \mid f \in I \rangle.$$

A finite set $f_1, \dots, f_r$ of polynomials is a *Gröbner basis* when

$$\langle \mathbf{L}(\langle f_1, \dots, f_r \rangle) \rangle = \langle \mathbf{L}(f_1), \dots, \mathbf{L}(f_r) \rangle.$$

Gordan's proof of Hilbert's basis theorem now follows from the easier statement that monomial ideals are finitely generated. In combinatorics, this result is called Dickson's lemma [1913].

If $f_1, \dots, f_r$ is a Gröbner basis, then by definition, a polynomial g lies in $\langle f_1, \dots, f_r \rangle$ if and only if the remainder h under division of g by $f_1, \dots, f_r$ is zero. In particular, in this case, the remainder does not depend on the order of $f_1, \dots, f_r$, and the monomials $x^\alpha \notin \langle \mathbf{L}(f_1), \dots, \mathbf{L}(f_r) \rangle$ represent a $\mathbb{K}$-vector space basis of the quotient ring $R/\langle f_1, \dots, f_r \rangle$, a fact known as Macaulay's theorem [1916]. For these reasons, it is desirable to have a Gröbner basis on hand.

The algorithm that computes a Gröbner basis for an ideal is due to Buchberger [1965; 1970]. Usually, Buchberger's criterion is formulated in terms of so-called S-pairs. In the treatment below, we do not use S-pairs; instead, we focus on the partition of the monomials of R induced by $\mathbf{L}(f_1), \dots, \mathbf{L}(f_r)$ via (2a) and (2b) of Theorem 1.9.

Given polynomials $f_1, \dots, f_r$, consider the monomial ideals

$$M_i := \langle \mathbf{L}(f_1), \dots, \mathbf{L}(f_{i-1}) \rangle : \mathbf{L}(f_i) \quad \text{for } i = 1, \dots, r. \tag{1-1}$$

For each minimal generator x^α of an M_i, let $h^{(i,\alpha)}$ denote the remainder of $x^\alpha f_i$ divided by $f_1, \dots, f_r$ (in this order).

Buchberger's criterion (Theorem 1.5) [Buchberger 1970]. *Let $f_1, \dots, f_r \in R$ be a collection of nonzero polynomials. Then $f_1, \dots, f_r$ form a Gröbner basis if and only if all of the remainders $h^{(i,\alpha)}$ are zero.*

We will prove this result after a few more preliminaries.

Example 1.6. Consider the ideal generated by the 3×3 minors of the matrix

$$\begin{pmatrix} x_1 & x_2 & x_3 & x_4 & x_5 \\ y_1 & y_2 & y_3 & y_4 & y_5 \\ z_1 & z_2 & z_3 & z_4 & z_5 \end{pmatrix}.$$

Using the lexicographic order on $\mathbb{K}[x_1, \ldots, z_5]$, the leading terms of the maximal minors of this matrix and the minimal generators of the corresponding monomial ideals M_i are listed in the following table.

$x_1y_2z_3$	$M_1 = 0$
$x_1y_2z_4$	$M_2 = \langle z_3 \rangle$
$x_1y_3z_4$	$M_3 = \langle y_2 \rangle$
$x_2y_3z_4$	$M_4 = \langle x_1 \rangle$
$x_1y_2z_5$	$M_5 = \langle z_3, z_4 \rangle$
$x_1y_3z_5$	$M_6 = \langle y_2, z_4 \rangle$
$x_2y_3z_5$	$M_7 = \langle x_1, z_4 \rangle$
$x_1y_4z_5$	$M_8 = \langle y_2, y_3 \rangle$
$x_2y_4z_5$	$M_9 = \langle x_1, y_3 \rangle$
$x_3y_4z_5$	$M_{10} = \langle x_1, x_2 \rangle$

Note that only 15 of the possible $\binom{10}{2} = 45$ S-pairs are needed to test Buchberger's criterion.

Exercise 1.7. Show that the maximal minors of the matrix in Example 1.6 form a Gröbner basis by using the Laplace expansions of suitable 4×4 matrices.

In order to prove Hilbert's syzygy theorem and Buchberger's criterion, we now extend the notion of a monomial order to vectors of polynomials.

Definition 1.8. A monomial of a free module R^r with basis $e_1, \ldots, e_r$ is an expression $x^\alpha e_i$. A (*global*) *monomial order* on R^r is a total order of the monomials of R^r such that:

(1) if $x^\alpha e_i > x^\beta e_j$, then $x^\gamma x^\alpha e_i > x^\gamma x^\beta e_j$ for all i, j and $\gamma \in \mathbb{N}^n$;

(2) $x^\alpha e_i > e_i$ for all i and $\alpha \neq 0$.

Usually, it is also the case that $x^\alpha e_i > x^\beta e_i$ if and only if $x^\alpha e_j > x^\beta e_j$, i.e., the order on the monomials in the components induce a single monomial order on R.

Thanks to Definition 1.8, we may now speak of the leading term of a vector of polynomials. In this situation, the division theorem still holds.

Theorem 1.9 (division with remainder for vectors of polynomials). *Let $>$ be a global monomial order on R^{r_0}, and let $F_1, \ldots, F_r \in R^{r_0}$ be nonzero polynomial* *vectors. For every $G \in R^{r_0}$, there exist uniquely determined $g_1, \dots, g_r \in R$ and a remainder $H \in R^{r_0}$ such that:*

(1) $G = g_1 F_1 + \cdots + g_r F_r + H$.

(2a) *No term of $g_i\, \mathbf{L}(F_i)$ is a multiple of an $\mathbf{L}(F_j)$ with $j < i$.*

(2b) *No term of H is a multiple of $\mathbf{L}(F_i)$ for any i.* □

Definition 1.10. Generalizing the earlier definition, given a global monomial order on R^r, the *leading term* of a nonzero vector of polynomials $F = (f_1, \dots, f_r)$ is defined to be the monomial

$$\mathbf{L}(F) := f_{\beta_i} x^{\beta_i} e_i, \quad \text{where } x^{\beta_i} = \max_{\alpha_i}\{x^{\alpha_i} \mid f_{\alpha_i} x^{\alpha_i} \text{ is a nonzero term of } f_i\}.$$

A finite set $F_1, \dots, F_s$ of vectors of polynomials in R^r is a *Gröbner basis* when

$$\langle \mathbf{L}(\langle F_1, \dots, F_s \rangle) \rangle = \langle \mathbf{L}(F_1), \dots, \mathbf{L}(F_s) \rangle.$$

Proof of Buchberger's criterion. The forward direction follows by definition. For the converse, assume that all remainders $h^{(i,\alpha)}$ vanish. Then for each minimal generator x^α in an M_i, there is an expression

$$x^\alpha f_i = g_1^{(i,\alpha)} f_1 + \cdots + g_r^{(i,\alpha)} f_r \tag{1-2}$$

such that no term of $g_j^{(i,\alpha)}\mathbf{L}(f_j)$ is divisible by an $\mathbf{L}(f_k)$ for every $k < j$, by condition (2a) of Theorem 1.3. (Of course, for a suitable $j < i$, one of the terms of $g_j^{(i,\alpha)}\mathbf{L}(f_j)$ coincides with $x^\alpha \mathbf{L}(f_i)$. This is the second term in the usual S-pair description of Buchberger's criterion.) Now let $e_1, \dots, e_r \in R^r$ denote the basis of the free module, and let $\varphi : R^r \to R$ be defined by $e_i \mapsto f_i$. Then by (1-2), elements of the form

$$G^{(i,\alpha)} := -g_1^{(i,\alpha)} e_1 + \cdots + (x^\alpha - g_i^{(i,\alpha)}) e_i + \cdots + (-g_r^{(i,\alpha)}) e_r \tag{1-3}$$

are in the kernel of ϕ. In other words, the $G^{(i,\alpha)}$'s are syzygies between $f_1, \dots, f_r$.

We now proceed with a division with remainder in the free module R^r, using the induced monomial order $>_1$ on R^r defined by

$$x^\alpha e_i >_1 x^\beta e_j \iff \begin{array}{l} x^\alpha \mathbf{L}(f_i) > x^\beta \mathbf{L}(f_j) \text{ or} \\ x^\alpha \mathbf{L}(f_i) = x^\beta \mathbf{L}(f_j) \text{ (up to a scalar) with } i > j. \end{array} \tag{1-4}$$

With respect to this order,

$$\mathbf{L}(G^{(i,\alpha)}) = x^\alpha e_i$$

because the term $cx^\beta \mathbf{L}(f_j)$ that cancels against $x^\alpha \mathbf{L}(f_i)$ in (1-2) satisfies $j < i$, and all other terms of any $g_k^{(i,\alpha)}\mathbf{L}(f_k)$ are smaller.

Now consider an arbitrary element

$$g = a_1 f_1 + \cdots + a_r f_r \in \langle f_1, \ldots, f_r \rangle.$$

We must show that $\mathbf{L}(g) \in \langle \mathbf{L}(f_1), \ldots, \mathbf{L}(f_r) \rangle$. Let $g_1 e_1 + \cdots + g_r e_r$ be the remainder of $a_1 e_1 + \cdots + a_r e_r$ divided by the collection of $G^{(i,\alpha)}$ vectors. Then

$$g = a_1 f_1 + \cdots + a_r f_r = g_1 f_1 + \cdots + g_r f_r$$

because the $G^{(i,\alpha)}$ are syzygies, and $g_1, \ldots, g_r$ satisfy (2a) of Theorem 1.3 when $a_1, \ldots, a_n$ are divided by $f_1, \ldots, f_r$, by the definition of the M_i in (1-1). Therefore, the nonzero initial terms

$$\mathbf{L}(g_j f_j) = \mathbf{L}(g_j)\mathbf{L}(f_j)$$

are distinct and no cancellation can occur among them. The proof is now complete because

$$\mathbf{L}(g) := \max_j \{\mathbf{L}(g_j)\mathbf{L}(f_j)\} \in \langle \mathbf{L}(f_1), \ldots, \mathbf{L}(f_r) \rangle. \qquad \square$$

Corollary 1.11 [Schreyer 1980]. *If $F_1, \ldots, F_{r_1} \in R^{r_0}$ are a Gröbner basis, then the $G^{(i,\alpha)}$ of* (1-3) *form a Gröbner basis of* $\ker(\varphi_1 \colon R^{r_1} \to R^{r_0})$ *with respect to the induced monomial order $>_1$ defined in* (1-4). *In particular, $F_1, \ldots, F_{r_1}$ generate the kernel of φ_1.*

Proof. As mentioned in the proof of Buchberger's criterion, the coefficients $g_1, \ldots, g_r$ of a remainder $g_1 e_1 + \cdots + g_r e_r$ resulting from division by the $G^{(i,\alpha)}$ satisfy condition (2a) of Theorem 1.3 when divided by $f_1, \ldots, f_r$. Hence, no cancellation can occur in the sum $g_1 \mathbf{L}(f_1) + \cdots + g_r \mathbf{L}(f_r)$, and $g_1 f_1 + \cdots g_r f_r = 0$ only if $g_1 = \ldots = g_r = 0$. Therefore, the collection of $\mathbf{L}(G^{(i,\alpha)})$ generate the leading term ideal $\mathbf{L}(\ker \varphi_1)$. $\square$

We have reached the goal of this section, an algorithmic proof of Hilbert's syzygy theorem.

Proof of Hilbert's syzygy theorem. Let M be a finitely generated R-module with presentation

$$R^r \xrightarrow{\varphi} R^{r_0} \to M \to 0.$$

Regard φ as a matrix and, thus, its columns as a set of generators for $\operatorname{im} \varphi$. Starting from these generators, compute a minimal Gröbner basis $F_1, \ldots, F_{r_1}$ for $\operatorname{im} \varphi$ with respect to some global monomial $>_0$ order on R^{r_0}. Now consider the induced monomial order $>_1$ on R^{r_1}, and let $G^{(i,\alpha)} \in R^{r_1}$ denote the syzygies obtained by applying Buchberger's criterion to $F_1, \ldots, F_{r_1}$. By Corollary 1.11, the $G^{(i,\alpha)}$ form a Gröbner basis for the kernel of the map $\varphi_1 \colon R^{r_1} \to R^{r_0}$, so we may now repeat this process.

Let ℓ be the maximal k such that the variable x_k occurs in some leading term $\mathbf{L}(F_j)$. Sort $F_1, \ldots, F_{r_1}$ so that whenever $j < i$, the exponent of x_ℓ in $\mathbf{L}(F_j)$ is less than or equal to the exponent of x_ℓ in $\mathbf{L}(F_i)$. In this way, none of the variables $x_\ell, \ldots, x_n$ will occur in a leading term $\mathbf{L}(G^{(i,\alpha)})$. Thus the process will terminate after at most n steps. □

Note that there are a number of choices allowed in the algorithm in the proof of Hilbert's syzygy theorem. In particular, we may order each set of Gröbner basis elements as we see fit.

Example 1.12. Consider the ideal $I = \langle f_1, \ldots, f_5 \rangle \subset R = \mathbb{K}[w, x, y, z]$ generated by the polynomials

$$f_1 = w^2 - xz, \quad f_2 = wx - yz, \quad f_3 = x^2 - wy, \quad f_4 = xy - z^2, \quad f_5 = y^2 - wz.$$

To compute a finite free resolution of $M = R/I$ using the method of the proof of Hilbert's syzygy theorem, we use the degree reverse lexicographic order on R. The algorithm successively produces three syzygy matrices φ_1, φ_2, and φ_3, which we present in a compact way as follows.

$$\begin{array}{r|cccccc}
\mathbf{w^2} - xz & -x & y & 0 & z & 0 & -y^2 + wz \\
\mathbf{wx} - yz & \mathbf{w} & -x & -y & 0 & z & z^2 \\
\mathbf{x^2} - wy & -z & \mathbf{w} & 0 & -y & 0 & 0 \\
\mathbf{xy} - z^2 & 0 & 0 & \mathbf{w} & \mathbf{x} & -y & yz \\
\mathbf{y^2} - wz & 0 & 0 & -z & -w & \mathbf{x} & \mathbf{w^2} \\
\hline
 & 0 & y & -x & \mathbf{w} & -z & 1 \\
 & -y^2 + wz & z^2 & -wy & yz & -w^2 & \mathbf{x}
\end{array}$$

All initial terms are printed in bold. The first column of this table is the transpose of the matrix φ_1. It contains the original generators for I which, as Buchberger's criterion criterion shows, already form a Gröbner basis for I. The syzygy matrix φ_2 resulting from the algorithm is the 5×6 matrix in the middle of our table. Note that, for instance, $M_4 = \langle w, x \rangle$ can be read from the 4th row of φ_2.

By Corollary 1.11, we know that the columns of φ_2 form a Gröbner basis for $\ker(\varphi_1)$ with respect to the induced monomial order on R^5. Buchberger's criterion criterion applied to these Gröbner basis elements yields a 6×2 syzygy matrix φ_3, whose transpose is printed in the two bottom rows of the table above. Note that there are no syzygies on the two columns of φ_3 because the initial terms of these vectors lie with different basis vectors.

To summarize, we obtain a free resolution of the form

$$0 \longrightarrow R^2 \xrightarrow{\varphi_3} R^6 \xrightarrow{\varphi_2} R^5 \xrightarrow{\varphi_1} R \longrightarrow R/I \longrightarrow 0.$$

Observe that, in general, once we have the initial terms of a Gröbner basis for I, we can easily compute the initial terms of the Gröbner bases for all syzygy modules, that is, all bold face entries of our table. This gives us an idea on the amount of computation that will be needed to obtain the full free resolution.

If the polynomial ring is graded, say $R = S = \mathbb{K}[x_0, \ldots, x_n]$ is the homogeneous coordinate ring of $\mathbb{P}^n$, and M is a finitely generated graded S-module, then the resolution computed through the proof of Hilbert's syzygy theorem is homogeneous as well. However, this resolution is typically not minimal. In Example 1.12, the last column of φ_2 is in the span of the previous columns, as can be seen from the first row of φ_3^t.

Example 1.13. Recall that in Example 1.6, we considered the ideal I of 3×3 minors of a generic 3×5 matrix over $S = \mathbb{K}[x_1, \ldots, z_5]$ with the standard grading. The algorithm in the proof of Definition 1.2 produces a resolution of S/I of the form

$$0 \longrightarrow S(-5)^6 \longrightarrow S(-4)^{15} \longrightarrow S(-3)^{10} \longrightarrow S \longrightarrow S/I \longrightarrow 0$$

because I is generated by 10 Gröbner basis elements, there are altogether 15 minimal generators of the M_i ideals, and 6 of the monomial ideal M_i have 2 generators. In this case, the resolution is minimal for degree reasons.

Exercise 1.14. Let I be a Borel-fixed monomial ideal. Prove that in this case, the algorithm in the proof of Hilbert's syzygy theorem produces a minimal free resolution of I. Compute the differentials explicitly and compare your result with the complex of S. Eliahou and Kervaire [1990] (see also [Peeva and Stillman 2008]).

2. Petri's Theorem

One of the first theoretical applications of Gröbner bases is Petri's analysis of the generators of the homogeneous ideal of a canonically embedded curve. Petri was the last student of Max Noether, and he acknowledges help from Emmy Noether in his thesis. As Emmy Noether was a student of Gordan, it is quite possible that Petri became aware of the concept of Gröbner bases through his communication with her, but we do not know if this was the case.

Let C be a smooth projective curve of genus g over $\mathbb{C}$. Let

$$\omega_1, \ldots, \omega_g \in H^0(C, \omega_C)$$

be a basis of the space of holomorphic differential forms on C and consider the canonical map

$$\iota : C \to \mathbb{P}^{g-1} \quad \text{given by} \quad p \mapsto [\omega_1(p) : \cdots : \omega_g(p)].$$

The map ι is an embedding unless C is hyperelliptic. We will assume that C is not hyperelliptic. Let $S := \mathbb{C}[x_1, \dots, x_g]$ be the homogeneous coordinate ring of $\mathbb{P}^{g-1}$, and let $I_C \subset S$ be the homogeneous ideal of C.

Petri's theorem (Theorem 2.1) [Petri 1923]. *The homogeneous ideal of a canonically embedded curve is generated by quadrics unless*

- *C is trigonal (i.e., there is a* 3:1 *holomorphic map $C \to \mathbb{P}^1$) or*
- *C is isomorphic to a smooth plane quintic. In this case, $g = 6$.*

Petri's theorem received much attention through the work of Mark Green [1984], who formulated a conjectural generalization to higher syzygies of canonical curves in terms of the Clifford index. We will not report here on the impressive progress made on this conjecture in the last two decades, but refer instead to [Aprodu and Farkas 2011; Aprodu and Nagel 2010; Aprodu and Voisin 2003; Green and Lazarsfeld 1986; Hirschowitz and Ramanan 1998; Mukai 1992; Schreyer 1986; 1991; 2003; Voisin 1988; 2002; 2005] for further reading.

In the cases of the exceptions in Petri's theorem, also Babbage [1939] observed that the ideal cannot be generated by quadrics alone. If $D := p_1 + \cdots + p_d$ is a divisor of degree d on C, then the linear system $|\omega_C(-D)|$ is cut out by hyperplanes through the span $\bar{D}$ of the points $p_i \in C \subset \mathbb{P}^{g-1}$. Thus Riemann–Roch implies that

$$h^0(C, \mathcal{O}_C(D)) = d + 1 - g + \operatorname{codim} \bar{D} = d - \dim \bar{D}.$$

Hence the three points of a trigonal divisor span only a line, and by Bézout's theorem, we need cubic generators in the generating set of its vanishing ideal.

Similarly, in the second exceptional case, the 5 points of a g^2_5 are contained in a unique conic in the plane they span, and quadrics alone do not cut out the curve.

The first step of Petri's analysis builds upon a proof by Max Noether.

Theorem 2.1 [Noether 1880]. *A nonhyperelliptic canonical curve $C \subset \mathbb{P}^{g-1}$ is projectively normal, i.e., the maps*

$$H^0(\mathbb{P}^{g-1}, \mathcal{O}(n)) \to H^0(C, \omega_C^{\otimes n})$$

are surjective for every n.

Proof. Noether's proof is a clever application of the basepoint-free pencil trick. This is a method which, according to Mumford, Zariski taught to all of his students. Let $|D|$ be a basepoint-free pencil on a curve, and let $\mathcal{L}$ be a further line bundle on C. Then the Koszul complex

$$0 \to \Lambda^2 H^0(\mathcal{O}_C(D)) \otimes \mathcal{L}(-D) \to H^0(\mathcal{O}_C(D)) \otimes \mathcal{L} \to \mathcal{L}(D) \to 0$$

is an exact sequence. To see this, note that locally, at least one section of the line bundle $\mathcal{O}_C(D)$ does not vanish. Thus the kernel of the multiplication map

$$H^0(\mathcal{O}_C(D)) \otimes H^0(\mathcal{L}) \to H^0(\mathcal{L}(D))$$

is isomorphic with $H^0(\mathcal{L}(-D))$. (Note $\Lambda^2 H^0(\mathcal{O}_C(D)) \cong \mathbb{C}$, as $h^0(\mathcal{O}_C(D)) = 2$.)

Consider $p_1, \dots, p_g$ general points on C and the divisor $D = p_1 + \cdots + p_{g-2}$ built from the first $g-2$ points. Then the images of these points span $\mathbb{P}^{g-1}$ and the span of any subset of less than $g-1$ points intersects the curve in no further points. Choose a basis $\omega_1, \dots, \omega_g \in H^0(\omega_C)$ that is, up to scalars, dual to these points, i.e., $\omega_i(p_j) = 0$ for $i \neq j$ and $\omega_i(p_i) \neq 0$. Then $|\omega_C(-D)|$ is a basepoint-free pencil spanned by ω_{g-1}, ω_g. If we apply the basepoint-free pencil trick to this pencil and $\mathcal{L} = \omega_C$, then we obtain the sequence

$$0 \to \Lambda^2 H^0 \omega_C(-D) \otimes H^0 \mathcal{O}_C(D) \to H^0 \omega_C(-D) \otimes H^0 \omega_C \xrightarrow{\mu} H^0 \omega_C^{\otimes 2}(-D),$$

and the image of

$$\mu\colon H^0(\omega_C(-D)) \otimes H^0(\omega_C) \to H^0(\omega_C^{\otimes 2}(-D)) \tag{2-1}$$

is $2g-1$ dimensional because $h^0(\omega_C(-D)) = 2$ and $h^0(\mathcal{O}_C(D)) = 1$. Thus μ in (2-1) is surjective, since $h^0(\omega_C^{\otimes 2}(-D)) = 2g-1$ holds by Riemann–Roch. On the other hand,

$$\omega_1^{\otimes 2}, \dots, \omega_{g-2}^{\otimes 2} \in H^0(\omega_C^{\otimes 2})$$

represent linearly independent elements of $H^0(\omega_C^{\otimes 2})/H^0(\omega_C^{\otimes 2}(-D))$, hence represent a basis, and the map $H^0(\omega_C) \otimes H^0(\omega_C) \to H^0(\omega_C^{\otimes 2})$ is surjective as well. This proves quadratic normality.

The surjectivity of the multiplication maps

$$H^0(\omega_C^{\otimes n-1}) \otimes H^0(\omega_C) \to H^0(\omega_C^{\otimes n})$$

for $n \geq 3$ is similar, but easier: $\omega_1^{\otimes n}, \dots, \omega_{g-2}^{\otimes n} \in H^0(\omega_C^{\otimes n})$ are linearly independent modulo the codimension $g-2$ subspace $H^0(\omega_C^{\otimes n}(-D))$, and the map

$$H^0(\omega_C^{\otimes n-1}) \otimes H^0(\omega_C(-D)) \to H^0(\omega_C^{\otimes n}(-D))$$

is surjective simply because $H^1(\omega_C^{\otimes n-2}(D)) = 0$ for $n \geq 3$. □

Corollary 2.2. *The Hilbert function of the coordinate ring of a canonical curve takes the values*

$$\dim(S/I_C)_n = \begin{cases} 1 & \text{if } n = 0 \\ g & \text{if } n = 1 \\ (2n-1)(g-1) & \text{if } n \geq 2. \end{cases}$$ □

Proof of Petri's theorem. Petri's analysis begins with the map μ in (2-1) above. Choose homogeneous coordinates $x_1, \dots, x_g$ such that $x_i \mapsto \omega_i$. Since $\omega_i \otimes \omega_j \in H^0(\omega_C^{\otimes 2}(-D))$ for $1 \leq i < j \leq g-2$, we find the polynomials

$$f_{ij} := x_i x_j - \sum_{r=1}^{g-2} a_{ij}^r x_r - b_{ij} \in I_C, \tag{2-2}$$

where the a_{ij}^r and b_{ij} are linear and quadratic, respectively, in $\mathbb{C}[x_{g-1}, x_g]$. We may choose a monomial order such that $\mathbf{L}(f_{ij}) = x_i x_j$. Since $\binom{g-2}{2} = \binom{g+1}{2} - (3g-3)$, these quadrics span $(I_C)_2$. On the other hand, they do not form a Gröbner basis for I_C because the $(g-2)\binom{n+1}{2} + (n+1)$ monomials $x_i^k x_{g-1}^\ell x_g^m$ with $i = 1, \dots, g-2$ and $k + \ell + m = n$ represent a basis for $(S/\langle x_i x_j \mid 1 \leq i < j \leq g-2 \rangle)_n$, which is still larger. We therefore need $g-3$ further cubic Gröbner basis elements. To find these, Petri considers the basepoint-free pencil trick applied to $|\omega_C(-D)|$ and $\mathcal{L} = \omega_C^{\otimes 2}(-D)$. The cokernel of the map

$$H^0(\omega_C(-D)) \otimes H^0(\omega_C^{\otimes 2}(-D)) \to H^0(\omega_C^{\otimes 3}(-2D)) \tag{2-3}$$

has dimension $h^1(\omega_C) = 1$. To find the missing element in $H^0(\omega_C^{\otimes 3}(-2D))$, Petri considers the linear form $\alpha_i = \alpha_i(x_{g-1}, x_g)$ in the pencil spanned by x_{g-1}, x_g that defines a tangent hyperplane to C at p_i. Then $\alpha_i \omega_i^{\otimes 2} \in H^0(\omega_C^{\otimes 3}(-2D))$ because $\omega_i^{\otimes 2}$ vanishes quadratically at all points $p_j \neq p_i$, while α_i vanishes doubly at p_i. Not all of these elements can be contained in the image of (2-3), since otherwise we would find $g-2$ further cubic Gröbner basis elements of type

$$\alpha_i x_i^2 + \text{ lower order terms},$$

where a lower order term is a term that is at most linear in $x_1, \dots, x_{g-2}$. As this is too many, at least one of the $\alpha_i \omega_i^{\otimes 2}$ spans the cokernel of the map (2-3).

We now argue by uniform position. Since C is irreducible, the behavior of $\alpha_i \omega_i^{\otimes 2}$ with respect to spanning of the cokernel is the same for any general choice of points $p_1, \dots, p_g$. So for general choices, each of these elements span the cokernel, and after adjusting scalars, we find that

$$G_{k\ell} := \alpha_k x_k^2 - \alpha_\ell x_\ell^2 + \text{ lower order terms} \tag{2-4}$$

are in I_C. Note that $G_{k\ell} = -G_{\ell k}$ and $G_{k\ell} + G_{\ell m} = G_{km}$. So this gives only $g-3$ further equations with leading terms $x_k^2 x_{g-1}$ for $k = 1, \dots, g-3$ up to a scalar. The last Gröbner basis element is a quartic H with leading term $\mathbf{L}(H) = x_{g-2}^3 x_{g-1}$, which we can obtain as a remainder of the Buchberger test applied to $x_{g-2} G_{k,g-2}$. There are no further Gröbner basis elements, because the quotient S/J of S by

$$J := \langle x_i x_j, x_k^2 x_{g-1}, x_{g-2}^3 x_{g-1} \mid 1 \le i < j \le g-2, 1 \le k \le g-3 \rangle$$

has the same Hilbert function as S/I_C. Hence $\mathbf{L}(I_C) = J$.

We now apply Buchberger's test to $x_k f_{ij}$ for a triple of distinct indices $1 \le i, j, k \le g-2$. Division with remainder yields a syzygy

$$x_k f_{ij} - x_j f_{ik} + \sum_{r \neq k} a_{ij}^r f_{rk} - \sum_{r \neq j} a_{ik}^r f_{rj} + \rho_{ijk} G_{kj} = 0 \qquad (2\text{-}5)$$

for a suitable coefficient $\rho_{ijk} \in \mathbb{C}$. (Moreover, comparing coefficients, we find that $a_{ij}^k = \rho_{ijk}\alpha_k$ holds. In particular, Petri's coefficients ρ_{ijk} are symmetric in i, j, k, since a_{ij}^k is symmetric in i, j.) Since C is irreducible, we have that for a general choice of $p_1, \dots, p_g$, either all coefficients $\rho_{ijk} \neq 0$ or all $\rho_{ijk} = 0$. In the first case, the cubics lie in the ideal generated by the quadrics.

In the second case, the f_{ij} are a Gröbner basis by themselves. Thus the zero locus $V(f_{ij} | 1 \le i < j \le g-2)$ of the quadrics f_{ij} define an ideal of a scheme X of dimension 2 and degree $g-2$. Since C is irreducible and nondegenerate, the surface X is irreducible and nondegenerate as well. Thus $X \subset \mathbb{P}^{g-2}$ is a surface of minimal degree. These were classified by Bertini; see, for instance, [Eisenbud and Harris 1987b]. Either X is a rational normal surface scroll, or X is isomorphic to the Veronese surface $\mathbb{P}^2 \hookrightarrow \mathbb{P}^5$. In the case of a scroll, the ruling on X cuts out a g_3^1 on C by Riemann–Roch. In the case of the Veronese surface, the preimage of C in $\mathbb{P}^2$ is a plane quintic. □

Perhaps the most surprising part of Petri's theorem is this: either I_C is generated by quadrics or there are precisely $g-3$ minimal cubic generators. It is a consequence of the irreducibility of C that no value in between 0 and $g-3$ is possible for the number of cubic generators. If we drop the assumption of irreducibility, then there are canonical curves with $1, \dots, g-5$ or $g-3$ cubic generators. For example, if we take a stable curve $C = C_1 \cup C_2$ with two smooth components of genus $g_i \ge 1$ intersecting in three points, so that C has genus $g = g_1 + g_2 + 2$, then the dualizing sheaf ω_C is very ample and the three intersection points lie on a line by the residue theorem. For general curves C_1 and C_2 of genus $g_i \ge 3$ for $i \in \{1, 2\}$, the ideal I_C has precisely one cubic generator; see [Schreyer 1991]. However, we could not find such an example with precisely $g-4$ generators. For genus $g = 5$, one cubic generator is excluded by the structure theorem of Buchsbaum–Eisenbud, and obstructions for larger g are unclear to us.

Conjecture 2.3. *Let $A = S/I$ be a graded artinian Gorenstein algebra with Hilbert function $\{1, g-2, g-2, 1\}$. Then I has $0, 1, \dots, g-5$ or $g-3$ cubic minimal generators.*

The veracity of this conjecture would imply the corresponding statement for reducible canonical curves because the artinian reduction $A := S/(I_C + \langle \ell_1, \ell_2 \rangle)$ of S/I_C, for general linear forms ℓ_1, ℓ_2, has Hilbert function $\{1, g-2, g-2, 1\}$.

Petri's analysis has been treated by Mumford [1975], and also in [Arbarello et al. 1985; Saint-Donat 1973; Shokurov 1971]. From our point of view, Gröbner bases and the use of uniform position simplify and clarify the treatment quite a bit. Mumford [1975] remarks that we now have seen all curves at least once, following a claim made in [Petri 1923]. We disagree with him on this point. If we introduce indeterminates for all of the coefficients in Petri's equations, then the scheme defined by the condition on the coefficients that f_{ij}, G_{kl}, and H form a Gröbner basis can have many components [Schreyer 1991; Little 1998]. It is not clear to us how to find the component corresponding to smooth curves, much less how to find closed points on this component.

3. Finite length modules and space curves

In the remaining part of these lectures, we report on how to find all curves in a Zariski open subset of the moduli space $\mathfrak{M}_g$ of curves of genus g for small g. In Section 4, we report on the known *unirationality* results for these moduli spaces. But first, we must discuss a method to explicitly construct space curves.

In this section, a *space curve* $C \subset \mathbb{P}^3$ will be a Cohen–Macaulay subscheme of pure dimension 1; in particular, C has no embedded points. We denote by $\mathcal{I}_C$ the ideal sheaf of C and by $I_C = \sum_{n \in \mathbb{Z}} H^0(\mathbb{P}^3, \mathcal{I}_C(n))$ the homogeneous ideal of C. The goal of this section is to construct a curve C of genus g and degree d. To do so, we will use work of Rao, who showed that the construction of C is equivalent to the creation of its Hartshorne–Rao module (see Rao's theorem).

Definition 3.1. The *Hartshorne–Rao module* of C is the finite length module

$$M = M_C := \sum_{n \in \mathbb{Z}} H^1(\mathbb{P}^3, \mathcal{I}_C(n)) \subset \sum_{n \in \mathbb{Z}} H^0(\mathbb{P}^3, \mathcal{O}(n)) \cong S := \mathbb{K}[x_0, .., x_3].$$

The Hartshorne–Rao module measures the deviation of C from being projectively normal. Furthermore, M_C plays an important role in liaison theory of curves in $\mathbb{P}^3$, which we briefly recall now.

Let $S := \mathbb{K}[x_0, \ldots, x_3]$ and $S_C := S/I_C$ denote the homogeneous coordinate ring of $\mathbb{P}^3$ and $C \subset \mathbb{P}^3$, respectively. By the Auslander–Buchsbaum–Serre formula [Eisenbud 1995, Theorem 19.9], S_C has projective dimension $\operatorname{pd}_S S_C \leq 3$. Thus its minimal free resolution has the form

$$0 \leftarrow S_C \leftarrow S \leftarrow F_1 \leftarrow F_2 \leftarrow F_3 \leftarrow 0,$$

with free graded modules $F_i = \oplus S(-j)^{\beta_{ij}}$. By the same formula in the local case, we see that the sheafified $\mathcal{G} := \ker(\widetilde{F}_1 \to \mathcal{O}_{\mathbb{P}^3})$ is always a vector bundle, and

$$0 \leftarrow \mathcal{O}_C \leftarrow \mathcal{O}_{\mathbb{P}^3} \leftarrow \bigoplus_j \mathcal{O}_{\mathbb{P}^3}(-j)^{\beta_{1j}} \leftarrow \mathcal{G} \leftarrow 0 \tag{3-1}$$

is a resolution by locally free sheaves. If C is arithmetically Cohen–Macaulay, then $F_3 = 0$ and $\mathcal{G}$ splits into a direct sum of line bundles. In this case, the ideal I_C is generated by the maximal minors of $F_1 \leftarrow F_2$ by the Hilbert–Burch theorem [Hilbert 1890; Burch 1968; Eisenbud 1995]. In general, we have

$$M_C \cong \sum_{n\in\mathbb{Z}} H^2(\mathbb{P}^3, \mathcal{G}(n)) \quad \text{and} \quad \sum_{n\in\mathbb{Z}} H^1(\mathbb{P}^3, \mathcal{G}(n)) = 0. \tag{3-2}$$

We explain now why curves linked by an even number of liaison steps have, up to a twist, the same Hartshorne–Rao module, thus illustrating its connection to liaison theory. We will then mention Rao's theorem, which states that the converse also holds.

Suppose that $f, g \in I_C$ are homogeneous forms of degree d and e without common factors. Let $X := V(f, g)$ denote the corresponding complete intersection, and let C' be the residual scheme defined by the homogeneous ideal $I_{C'} := (f, g) : I_C$ [Peskine and Szpiro 1974]. The locally free resolutions of $\mathcal{O}_C$ and $\mathcal{O}_{C'}$ are closely related, as follows. Applying $\mathcal{E}xt^2(-, \omega_{\mathbb{P}^3})$ to the sequence

$$0 \to \mathcal{I}_{C/X} \to \mathcal{O}_X \to \mathcal{O}_C \to 0$$

gives

$$0 \leftarrow \mathcal{E}xt^2(\mathcal{I}_{C/X}, \omega_{\mathbb{P}^3}) \leftarrow \omega_X \leftarrow \omega_C \leftarrow 0.$$

From $\omega_X \cong \mathcal{O}_X(d+e-4)$, we conclude that $\mathcal{E}xt^2(\mathcal{I}_{C/X}, \mathcal{O}_{\mathbb{P}^3}(-d-e)) \cong \mathcal{O}_{C'}$, and hence $\mathcal{I}_{C'/X} \cong \omega_C(-d-e+4)$. Now the mapping cone of

$$\begin{array}{ccccccccc}
0 \longleftarrow & \mathcal{O}_C & \longleftarrow & \mathcal{O}_{\mathbb{P}^3} & \longleftarrow & \bigoplus_j \mathcal{O}_{\mathbb{P}^3}(-j)^{\beta_{1j}} & \longleftarrow & \mathcal{G} & \longleftarrow 0 \\
& \uparrow & & \uparrow \cong & & \uparrow & & \uparrow & \\
0 \longleftarrow & \mathcal{O}_X & \longleftarrow & \mathcal{O}_{\mathbb{P}^3} & \longleftarrow & \mathcal{O}_{\mathbb{P}^3}(-d) \oplus \mathcal{O}_{\mathbb{P}^3}(-e) & \longleftarrow & \mathcal{O}_{\mathbb{P}^3}(-d-e) & \longleftarrow 0
\end{array}$$

dualized with $\mathcal{H}om(-, \mathcal{O}_{\mathbb{P}^3}(-d-e))$ gives

$$\begin{array}{ccccccc}
0 \to \mathcal{O}_{\mathbb{P}^3}(-d-e) & \longrightarrow & \bigoplus_j \mathcal{O}_{\mathbb{P}^3}(j-d-e)^{\beta_{1j}} & \longrightarrow & \mathcal{G}^*(-d-e) & \longrightarrow\!\!\!\!\to & \mathcal{I}_{C'/X} \\
\cong \downarrow & & \downarrow & & \downarrow & & \downarrow \\
0 \to \mathcal{O}_{\mathbb{P}^3}(-d-e) & \longrightarrow & \mathcal{O}_{\mathbb{P}^3}(-e) \oplus \mathcal{O}_{\mathbb{P}^3}(-d) & \longrightarrow & \mathcal{O}_{\mathbb{P}^3} & \longrightarrow\!\!\!\!\to & \mathcal{O}_X,
\end{array}$$

which yields the following locally free resolution of $\mathcal{O}_{C'}$:

$$0 \to \bigoplus_j \mathcal{O}_{\mathbb{P}^3}(j-d-e)^{\beta_{1j}} \to \mathcal{G}^*(-d-e) \oplus \mathcal{O}_{\mathbb{P}^3}(-e) \oplus \mathcal{O}_{\mathbb{P}^3}(-d) \\ \to \mathcal{O}_{\mathbb{P}^3} \to \mathcal{O}_{C'} \to 0.$$

In particular, after truncating this complex to resolve $I_{C'}$, one sees that

$$\begin{aligned} M_{C'} &:= \sum_{n\in\mathbb{Z}} H^1(\mathbb{P}^3, \mathcal{I}_{C'}(n)) \cong \sum_{n\in\mathbb{Z}} H^1(\mathbb{P}^3, \mathcal{G}^*(n-d-e)) \\ &\cong \sum_{n\in\mathbb{Z}} H^2(\mathbb{P}^3, \mathcal{G}(d+e-4-n))^* \cong \mathrm{Hom}_{\mathbb{K}}(M_C, \mathbb{K})(4-d-e). \end{aligned}$$

Thus curves that are related via an even number of liaison steps have the same Hartshorne–Rao module up to a twist. Rao's famous result says that the converse is also true.

Rao's theorem (Theorem 3.2) [Rao 1978]. *The even liaison classes of curves in $\mathbb{P}^3$ are in bijection with finite length graded S- modules up to twist.* □

Therefore the difficulty in constructing the desired space curve C (of degree d and genus g) lies completely in the construction of the appropriate Hartshorne–Rao module $M = M_C$. Upon constructing M, we may then obtain the desired ideal sheaf $\mathcal{I}_C$ as follows. Assume that we have a free S-resolution of M_C,

$$0 \leftarrow M_C \leftarrow F_0 \leftarrow F_1 \leftarrow F_2 \leftarrow F_3 \leftarrow F_4 \leftarrow 0,$$

with $F_i = \bigoplus_j S(-j)^{\beta_{ij}}$. Let $\mathcal{F} := \widetilde{N}$ be the sheafification of $N := \ker(F_1 \to F_0)$, the second syzygy module of M. In this case, $\mathcal{F}$ will be a vector bundle without line bundle summands such that $H^1_*(\mathcal{F}) \cong H^1_*(\mathcal{I}_C)$ and $H^2_*(\mathcal{F}) = 0$. Here, we have used the notation $H^i_*(\mathcal{F}) := \bigoplus_n H^i(\mathcal{F}(n))$. If we constructed the correct Hartshorne–Rao module M, then taking $\mathcal{L}_1$ and $\mathcal{L}_2$ to be appropriate choices of direct sums of line bundles on $\mathbb{P}^3$, a general homomorphism $\varphi \in \mathrm{Hom}(\mathcal{L}_1, \mathcal{F}\oplus\mathcal{L}_2)$ will produce the desired curve C, as we will obtain $\mathcal{I}_C$ as the cokernel of a map φ of the bundles

$$0 \longrightarrow \mathcal{L}_1 \xrightarrow{\varphi} \mathcal{F} \oplus \mathcal{L}_2 \longrightarrow \mathcal{I}_C \longrightarrow 0.$$

To compute the rank of $\mathcal{F}$ and to choose the direct sums of line bundles $\mathcal{L}_1$ and $\mathcal{L}_2$, we now make plausible assumptions about the Hilbert function of M_C. We illustrate this approach in the example of the construction of a smooth linearly normal curve C of degree $d = 11$ and genus $g = 10$. Since $2d > 2g-2$, the line bundle $\mathcal{O}_C(2)$ is already nonspecial. Hence by Riemann–Roch, we have that $h^0(\mathcal{O}_C(2)) = 22+1-10 = 13$.

Remark 3.2. If we assume that C is a curve of maximal rank, i.e., that all maps $H^0(\mathcal{O}_{\mathbb{P}^3}(n)) \to H^0(\mathcal{O}_C(n))$ are either injective or surjective, then we can compute the Hilbert function of M_C and I_C. Note that being of maximal rank is an open condition, so among the curves in the union $\mathcal{H}_{d,g}$ of the component of the Hilbert scheme $\mathrm{Hilb}_{dt+1-g}(\mathbb{P}^3)$ containing smooth curves, maximal rank curves form an open (and hopefully nonempty) subset. There is a vast literature on the existence of maximal rank curves; see, for example, [Fløystad 1991].

To gain insight into the Betti numbers of $M = M_C$, we use Hilbert's formula for the Hilbert series:

$$h_M(t) = \sum_{n\in\mathbb{Z}} \dim M_n t^n = \frac{\sum_{i=0}^{3}(-1)^i \sum_j \beta_{ij} t^j}{(1-t)^4}.$$

Since $h_{M_C}(t) = 3t^2 + 4t^3$ by our maximal rank assumption (Remark 3.2), we have

$$(1-t)^4 h_M(t) = 3t^2 - 8t^3 + 2t^4 + 12t^5 - 13t^6 + 4t^7,$$

and thus the Betti table of M must be

$$\beta(M) = \begin{array}{c|ccccc} & 0 & 1 & 2 & 3 & 4 \\ \hline 2 & 3 & 8 & 2 & . & . \\ 3 & . & . & 12 & 13 & 4 \end{array}$$

if we assume that M has a so called *natural resolution*, which means that for each degree j at most one β_{ij} is nonzero. Note that having a natural resolution is an open condition in a family of modules with constant Hilbert function.

Table 1 provides a detailed look at the Hilbert functions relevant to our computation. From these we see that $H^0_*(\mathcal{O}_C)$ and $S_C = S/I_C$ will have the potential Betti tables at the top of the next page, if we assume that they also have natural resolutions.

n	$h^1(\mathcal{I}_C(n))$	$h^0(\mathcal{O}_C(n))$	$h^0(\mathcal{O}_{\mathbb{P}^3}(n))$	$h^0(\mathcal{I}_C(n))$
0	0	1	1	0
1	0	4	4	0
2	3	13	10	0
3	4	24	20	0
4	0	35	35	0
5	0	46	56	10
6	0	55	84	29

Table 1. With our maximal rank assumption of Remark 3.2, this table provides the relevant Hilbert functions in the case $d = 11$ and $g = 10$.

$$\beta(H^0_*(\mathcal{O}_C)) = \begin{array}{c|ccc} & 0 & 1 & 2 \\ \hline 0 & 1 & . & . \\ 1 & . & . & . \\ 2 & 3 & 8 & 2 \\ 3 & . & . & 2 \end{array} \quad \text{and} \quad \beta(S_C) = \begin{array}{c|cccc} & 0 & 1 & 2 & 3 \\ \hline 0 & 1 & . & . & . \\ 1 & . & . & . & . \\ 2 & . & . & . & . \\ 3 & . & . & . & . \\ 4 & . & 10 & 13 & 4 \end{array}.$$

Comparing these Betti tables, we find the following plausible choices of $\mathcal{F}$, $\mathcal{L}_1$, and $\mathcal{L}_2$:

- We choose $\mathcal{F} := \widetilde{N}$, where $N = \ker(\psi : S^8(-3) \to S^3(-2))$ is a sufficiently general 3×8 matrix of linear forms; in particular, $\operatorname{rank} \mathcal{F} = 5$.
- Let $\mathcal{L}_1 := \mathcal{O}^2(-4) \oplus \mathcal{O}^2(-5)$ and $\mathcal{L}_2 := 0$.
- The map $\varphi \in \operatorname{Hom}(\mathcal{L}_1, \mathcal{F})$ is a sufficiently general homomorphism. Since the map $F_2 \to H^0_*(\mathcal{F})$ is surjective, the choice of φ amounts to choosing an inclusion $\mathcal{O}^2(-5) \to \mathcal{O}^{12}(-5)$, i.e., a point in the Grassmannian $\mathbb{G}(2, 12)$.
- Finally, $\mathcal{I}_C = \operatorname{coker} \varphi$.

It is not clear that general choices as above will necessarily yield a smooth curve. If the sheaf $\mathcal{H}om(\mathcal{L}_1, \mathcal{F} \oplus \mathcal{L}_2)$ happens to be generated by its global sections $\operatorname{Hom}(\mathcal{L}_1, \mathcal{F} \oplus \mathcal{L}_2)$, then a Bertini-type theorem as in [Kleiman 1974] would apply. However, since we have to take all generators of $H^0_*(\mathcal{F})$ in degree 4, this is not the case. On the other hand, there is no obvious reason that $\operatorname{coker} \varphi$ should not define a smooth curve, and upon construction, it is easy to check the smoothness of such an example using a computer algebra system, such as Macaulay2 or Singular. Doing this, we find that general choices do lead to a smooth curve.

Exercise 3.3. Construct examples of curves of degree and genus as prescribed in [Hartshorne 1977, Figure 18 on page 354], including those which were open cases at the time of the book's publication.

4. Random curves

In this section, we explain how the ideas of Section 3 lead to a computer-aided proof of the unirationality of the moduli space $\mathfrak{M}_g$ of curves of genus g, when g is small. We will illustrate this approach by example, through the case of genus $g = 12$ and degree $d = 13$ in Theorem 4.5.

Definition 4.1. A variety X is called *unirational* if there exists a dominant rational map $\mathbb{A}^n \dashrightarrow X$. A variety X is called *uniruled* if there exists a dominant rational map $\mathbb{A}^1 \times Y \dashrightarrow X$ for some variety Y that does not factor through Y. A smooth projective variety X has *Kodaira dimension* κ if the section ring

$R_X := \sum_{n\geq 0} H^0(X, \omega_X^{\otimes n})$ of pluri-canonical forms on X has a Hilbert function with growth rate $h^0(\omega_X^{\otimes n}) \in O(n^\kappa)$. We say that X has *general type* if $\kappa = \dim X$, the maximal possible value.

Since the pluri-genera $h^0(\omega_X^{\otimes n})$ are birational invariants, being of general type does not depend on a choice of a smooth compactification. Thus we may also speak of general type for quasiprojective varieties.

Unirationality and general type are on opposite ends of birational geometry. If a variety is of general type, then there exists no rational curve through a general point of X [Kollár 1996, Corollary IV.1.11]. On the other hand, uniruled varieties have the pluri-canonical ring $R_X = (R_X)_0 = \mathbb{C}$ and thus (by convention) have Kodaira dimension $\kappa = -\infty$. In fact, even if X is unirational, then we can connect any two general points of X by a rational curve.

We now recall results concerning the unirationality of the moduli space $\mathfrak{M}_g$. There are positive results for small genus, followed by negative results for large genus.

Theorem 4.2 ([Severi 1921] for $g \leq 10$; [Sernesi 1981; Chang and Ran 1984] for $g = 12, 11, 13$; [Verra 2005] for $g = 14$.). *The moduli space $\mathfrak{M}_g$ of curves of genus g is unirational for $g \leq 14$.*

Theorem 4.3 [Harris and Mumford 1982; Eisenbud and Harris 1987a; Farkas 2006; 2009a; 2009b]. *The moduli space $\mathfrak{M}_g$ of curves of genus g is of general type for $g \geq 24$ or $g = 22$. The moduli space M_{23} has Kodaira dimension ≥ 2.* □

We call this beautiful theorem a negative result because it says that it will be very difficult to write down explicitly a general curve of large genus. Given a family of curves of genus $g \geq 24$ that pass through a general point of $\mathfrak{M}_g$, say via an explicit system of equations with varying coefficients, none of the essential coefficients is a free parameter. All of the coefficients will satisfy some complicated algebraic relations. On the other hand, in unirational cases, there exists a dominant family of curves whose parameters vary freely.

In principle, we can compute a dominating family explicitly along with a unirationality proof. In practice, this is often out of reach using current computer algebra systems; however, the following approach is feasible today in many cases. By replacing each free parameter in the construction of the family by a randomly chosen value in the ground field, the computation of an explicit example is possible. In particular, over a finite field $\mathbb{F}$, where it is natural to use the constant probability distribution on $\mathbb{F}$, a unirationality proof brings with it the possibility of choosing random points in $\mathfrak{M}_g(\mathbb{F})$, i.e., to compute a random curve. These curves can then be used for further investigations of the moduli space, as well as to considerably simplify the existing unirationality proofs. The advantage of using such random curves in the unirationality proof is that, with high probability, they will be smooth curves, while in a theoretical treatment, smoothness is always a delicate issue.

To begin this construction, we first need some information on the projective models of a general curve. This is the content of Brill– Noether theory. Let

$$W^r_d(C) := \{L \in \mathrm{Pic}^d(C) \mid h^0(C, L) \geq r+1\} \subset \mathrm{Pic}^d(C)$$

denote the space of line bundles of degree d on C that give rise to a morphism $C \to \mathbb{P}^r$.

Theorem 4.4. *Let C be a smooth projective curve of genus g.*

(1) [Brill and Noether 1874] *At every point*, $\dim W^r_d(C) \geq \rho := g-(r+1)(g-d+r)$.

(2) [Griffiths and Harris 1980; Fulton and Lazarsfeld 1981] *If $\rho \geq 0$, then $W^r_d(C) \neq 0$, and if $\rho > 0$, then $W^r_d(C)$ is connected. Further, the tangent space of $W^r_d(C)$ at a point $L \in W^r_d(C) \setminus W^{r+1}_d(C)$ is*

$$T_L W^r_d(C) = \mathrm{Im}\, \mu_L^\perp \subset H^1(\mathcal{O}_C) = T_L \mathrm{Pic}^d(C),$$

where $\mu_L\colon H^0(L) \otimes H^0(\omega_C \otimes L^{-1}) \to H^0(\omega_C) = H^1(\mathcal{O}_C)^*$ denotes the Petri map.

(3) [Gieseker 1982] *If $C \in \mathfrak{M}_g$ is a general curve, then $W^r_d(C)$ is smooth of dimension ρ away from $W^{r+1}_d(C)$. More precisely, the Petri map μ_L is injective for all $L \in W^r_d(C) \setminus W^{r+1}_d(C)$.* □

We now illustrate the computer-aided unirationality proof of $\mathfrak{M}_g$ by example, through the case $g = 12, d = 13$ [Schreyer and Tonoli 2002]. This case is not amongst those covered in [Sernesi 1981] or [Chang and Ran 1984] (which are $g = 11, d = 12$, $g = 12, d = 12$, and $g = 13, d = 13$). We are choosing the case $d = 12, g = 13$ because it illustrates well the difficulty of this construction. For $g = 14$, see [Verra 2005] and, for a computer aided unirationality proof, [Schreyer 2013]. For a related Macaulay2 package, see [von Bothmer et al. 2011].

Theorem 4.5. *Let $g = 12$ and $d = 13$. Then $\mathrm{Hilb}_{dt+1-g}(\mathbb{P}^3)$ has a component $\mathcal{H}_{d,g}$ that is unirational and dominates the moduli space $\mathfrak{M}_g$ of curves of genus g.*

Proof. This proof proceeds as follows. We first compute the Hilbert function and expected syzygies of the Hartshorne–Rao module $M = H^1_*(\mathcal{I}_C)$, the coordinate ring S_C, and the section ring $R := H^0_*(\mathcal{O}_C)$. We then use this information to choose generic matrices which realize the free resolution of M. Finally, we show that this construction leads to a family of curves that dominate $\mathfrak{M}_{12}$ and generically contains smooth curves.

We first choose r so that a general curve has a model of degree $d = 13$ in $\mathbb{P}^r$. In our case, we choose $r = 3$ so that $g - d + r = 2$. To compute the Hilbert function and expected syzygies of the Hartshorne–Rao module $M = H^1_*(\mathcal{I}_C)$, the coordinate ring $S_C = S/I_C$, and the section ring $R = H^0_*(\mathcal{O}_C)$, we assume the open condition that C has maximal rank, i.e.,

$$H^0(\mathbb{P}^3, \mathcal{O}(n)) \to H^0(C, L^n)$$

is of maximal rank for all n, as in Remark 3.2. In this case, $h_M(t) = 5t^2 + 8t^3 + 6t^4$, which has Hilbert numerator

$$h_M(t)(1-t)^4 = 5t^2 - 12t^3 + 4t^4 + 4t^5 + 9t^6 - 16t^{10} + 6t^{11}.$$

If M has a natural resolution, so that for each j at most one $\beta_{ij}(M)$ is nonzero, then M has the Betti table

$$\beta(M) \;=\; \begin{array}{c|ccccc} & 0 & 1 & 2 & 3 & 4 \\ \hline 2 & 5 & 12 & 4 & . & . \\ 3 & . & . & 4 & . & . \\ 4 & . & . & 9 & 16 & 6 \end{array}.$$

If we assume the open condition that S_C and R have natural syzygies as well, then their Betti tables are

$$\beta(S_C) \;=\; \begin{array}{c|ccc} & 0 & 1 & 2 \\ \hline 0 & 1 & . & . \\ 1 & . & . & . \\ 2 & 3 & 12 & 4 \\ 3 & . & . & 2 \end{array} \quad \text{and} \quad \beta(R) \;=\; \begin{array}{c|cccc} & 0 & 1 & 2 & 3 \\ \hline 0 & 1 & . & . & . \\ 1 & . & . & . & . \\ 2 & . & . & . & . \\ 3 & . & . & . & . \\ 4 & . & 2 & . & . \\ 5 & . & 9 & 16 & 6 \end{array}.$$

We conclude that once we have constructed the Hartshorne–Rao module $M = M_C$, say via its representation

$$0 \leftarrow M \leftarrow S^5(-2) \leftarrow S^{12}(-3),$$

we may choose $\mathcal{F}$ to be the kernel of

$$0 \leftarrow \mathcal{O}^5(-2) \leftarrow \mathcal{O}^{12}(-3) \leftarrow \mathcal{F} \leftarrow 0$$

and set $\mathcal{L}_1 := \mathcal{O}(-4)^4 \oplus \mathcal{O}^2(-5)$ and $\mathcal{L}_2 := 0$. Then C is determined by M and the choice of a point in $\mathbb{G}(2,4)$. In particular, as mentioned earlier, constructing C is equivalent to constructing the finite length module M with the desired syzygies.

If we choose the presentation matrix ϕ of M to be given by a general (or random) 5×12 matrix of linear forms, then its cokernel will be a module with

Hilbert series $5t^2 + 8t^3 + 2t^4$. In other words, to get the right Hilbert function for M, we must force 4 linear syzygies. To do this, choose a general (or random) 12×4 matrix ψ of linear forms. Then

$$\ker(\psi^t : S^{12}(1) \to S^4(2))$$

has at least $12 \cdot 4 - 4 \cdot 10 = 8$ generators in degree 0. In fact, there are precisely 8 and a general point in $\mathbb{G}(5, 8)$ gives rise to a 12×5 matrix φ^t of linear forms. This means that $M := \operatorname{coker}(\varphi\colon S^12(-3) \to S^5(-2))$ to have Hilbert series $5t^2 + 8t^3 + 6t^4$, due to the forced 4 linear syzygies.

Having constructed M, it remains to prove that this construction leads to a family of curves that dominates $\mathfrak{M}_{12}$. To this end, we compute a random example C, say over a finite prime field $\mathbb{F}_p$, and confirm its smoothness. Since we may regard our computation over $\mathbb{F}_p$ as the reduction modulo p of a construction defined over an open part of $\operatorname{Spec}\mathbb{Z}$, semicontinuity allows us to establish the existence of a smooth example defined over $\mathbb{Q}$ with the same syzygies.

We now consider the universal family $\mathfrak{W}^r_d \subset \mathfrak{Pic}^d$ over $\mathfrak{M}_g$ and a neighborhood of our example $(C, L) \in \mathfrak{Pic}^d$. Note that the codimension of $\mathfrak{W}^r_d$ is at most $(r+1)(g-d+r) = 4 \cdot 2 = 8$. On the other hand, we claim that the Petri map μ_L for (C, L) is injective. (Recall the definition of μ_L from Theorem 4.4.) To see this, note that the Betti numbers of $H^0_*(\omega_C)$ correspond to the dual of the resolution of $H^0_*(\mathcal{O}_C)$, so

$$\beta(H^0_*(\omega_C)) = \begin{array}{c|ccc} & 0 & 1 & 2 \\ \hline -1 & 2 & . & . \\ 0 & 4 & 12 & 3 \\ 1 & . & . & . \\ 2 & . & . & 1 \end{array}$$

Thus there are no linear relations among the two generators in $H^0(\omega_C \otimes L^{-1})$, which means that the $\mu_L\colon H^0(L) \otimes H^0(\omega_C \otimes L^{-1})$ is injective. From this we see that $\dim W^r_d(C)$ has dimension 4 at (C, L), and the constructed family dominates for dimension reasons. □

The unirationality of $\mathfrak{M}_{15}$ and $\mathfrak{M}_{16}$ are open; however, these moduli spaces are uniruled.

Theorem 4.6 (Chang and Ran [1986; 1991]; see also [Bruno and Verra 2005; Farkas 2009a]). *The moduli space $\mathfrak{M}_{15}$ is rationally connected, and $\mathfrak{M}_{16}$ is uniruled.* □

To explain why the unirationality in these cases is more difficult to approach using the method of Theorem 4.5, we conclude with a brief discussion on the space models of curves of genus $g = 16$. By Brill–Noether theory, a general curve C of genus 16 has finitely many models of degree $d = 15$ in $\mathbb{P}^3$. Again assuming the maximal rank condition of Remark 3.2, the Hartshorne–Rao module $M = H^1_*(\mathcal{I}_C)$ has Hilbert series

$$H_M(t) = 5t^2 + 10t^3 + 10t^4 + 4t^5$$

and expected syzygies

$$\beta(M) = \begin{array}{c|ccccc} & 0 & 1 & 2 & 3 & 4 \\ \hline 2 & 5 & 10 & . & & \\ 3 & . & . & 4 & . & . \\ 4 & . & . & 9 & 6 & . \\ 5 & . & . & . & 6 & 4 \end{array}.$$

The section ring $H^0_*(\mathcal{O}_C)$ and the coordinate ring S_C have expected syzygies

$$\beta(H^0_*(\mathcal{O}_C)) = \begin{array}{c|ccc} & 0 & 1 & 2 \\ \hline 0 & 1 & . & . \\ 1 & . & . & . \\ 2 & 5 & 10 & . \\ 3 & . & . & 4 \end{array} \quad \text{and} \quad \beta(S_C) = \begin{array}{c|cccc} & 0 & 1 & 2 & 3 \\ \hline 0 & 1 & . & . & . \\ 1 & . & . & . & . \\ 2 & . & . & . & . \\ 3 & . & . & . & . \\ 4 & . & . & . & . \\ 5 & . & 9 & 6 & . \\ 6 & . & . & 6 & 4 \end{array}.$$

Proposition 4.7. *A general curve C of genus $g = 16$ and degree $d = 15$ in $\mathbb{P}^3$ has syzygies as above. In particular, the Hartshorne–Rao module M_C uniquely determines C. Furthermore, the rational map from the component $\mathcal{H}_{d,g}$ of the Hilbert scheme* $\mathrm{Hilb}_{15t+1-16}(\mathbb{P}^3)$ *that dominates $\mathfrak{M}_{16}$ defined by*

$$\begin{aligned} \mathcal{H}_{d,g} &\dashrightarrow \{\text{ 20 determinantal points }\} \\ C &\mapsto \Gamma := \operatorname{supp}\operatorname{coker}(\varphi^t : \mathcal{O}^6(-1) \to \mathcal{O}^4) \end{aligned}$$

is dominant. Here $\varphi\colon S^4(-9) \to S^6(-8)$ denotes the linear part of the last syzygy matrix of M.

Proof. For the first statement, it suffices to find an example with the expected syzygies, since Betti numbers behave semicontinuously in a family of modules with constant Hilbert function. We may even take a reducible example, provided that it is smoothable. Consider the union $C := E_1 \cup E_2 \cup E_3$ of three smooth curves of genus 2 and degree 5, such that $E_i \cap E_j$ for $i \neq j$ consists of 4 nodes of C. Then C has degree $d = 3 \cdot 5 = 15$ and genus $g = 3 \cdot 2 + 4 \cdot 3 - 2 = 16$. Clearly, C is smoothable as an abstract curve. For general choices, it is smoothable as an embedded curve because the g^3_{15} on the reducible curve is an isolated smooth point in W^3_{15} (as we will see), so the smooth curves nearby have an isolated g^3_{15} as well.

It is easy to find such a union over a finite field $\mathbb{F}$. Start with the 12 intersection points $\{p_1, \dots, p_4\} \cup \{p_5, \dots, p_8\} \cup \{p_9, \dots, p_{12}\}$ randomly chosen in $\mathbb{P}^3(\mathbb{F})$. To construct E_1, pick at random a quadric Q_1 in the pencil of quadrics through $\{p_1, \dots, p_8\}$. Next, we must check if the tangent hyperplane of Q_1 in a point, say p_1, intersects Q_1 in a pair of lines individually defined over $\mathbb{F}$; this will happen about 50% of the time. Once this is true, choose one of the lines, call it L_1. Then $|\mathcal{O}_{Q_1}(3) \otimes \mathcal{O}_{Q_1}(-L_1)|$ is a linear system of class $(3, 2)$ on $Q_1 \cong \mathbb{P}^1 \times \mathbb{P}^1$. We may take E_1 as a general curve in this linear system that passes through $\{p_1, \dots, p_8\}$. Similarly, we choose E_2 using $\{p_1, \dots, p_4, p_9, \dots, p_{12}\}$ and E_3 starting with $\{p_5, \dots, p_{12}\}$. The union of the E_i yields the desired curve C, and it a straightforward computation to check that C has the expected Hartshorne–Rao module and syzygies.

The second statement can be proved by showing that the appropriate map between tangent spaces is surjective for this example. This involves computing appropriate Ext-groups. Define

$$\begin{aligned} \overline{M} &:= \operatorname{coker}(S^6(-2) \oplus S^6(-1) \to S^4) \\ &= \operatorname{Ext}^4_S(M, S(-9)) \\ &= \operatorname{Hom}_K(M, K)(-5) \\ \text{and} \quad N &:= \operatorname{coker}(\varphi^t : S^6(-1) \to S^4). \end{aligned}$$

Then there is a short exact sequence

$$0 \to P \to N \to \overline{M} \to 0$$

of modules with Hilbert series

$$\begin{aligned} h_{\overline{M}}(t) &= 4 + 10t + 10t^2 + 5t^3 \\ h_N(t) &= 4 + 10t + 16t^2 + 20t^3 + 20t^4 + 20t^5 + \cdots \\ h_P(t) &= \qquad\qquad\quad 6t^2 + 15t^3 + 20t^4 + 20t^5 + \cdots . \end{aligned}$$

The group $\operatorname{Ext}^1_S(\overline{M}, \overline{M})$ governs the deformation theory of $\overline{M}$ (and M). More details can be found in [Hartshorne 2010], for example, Theorem 2.7 applied in the affine case. More precisely, the degree 0 part of this Ext-group is the tangent space of homogeneous deformations of M, which in turn is isomorphic to the tangent space of the Hilbert scheme in C. Similarly, in the given example, $\operatorname{Ext}^1(N, N)_0$ can be identified with the tangent space to the space of twenty determinantal points. Note that we have the diagram

$$\begin{array}{ccccc} \operatorname{Ext}^1_S(\overline{M}, \overline{M}) & \longrightarrow & \operatorname{Ext}^1_S(N, \overline{M}) & \longrightarrow & \operatorname{Ext}^1_S(P, \overline{M}). \\ & & \uparrow & & \\ & & \operatorname{Ext}^1_S(N, N) & & \\ & & \uparrow & & \\ & & \operatorname{Ext}^1_S(N, P) & & \end{array}$$

In our example, computation shows that

$$\begin{aligned} \dim \operatorname{Ext}^1_S(\overline{M}, \overline{M})_0 &= 60, \\ \dim \operatorname{Ext}^1_S(N, \overline{M})_0 = \dim \operatorname{Ext}^1_S(N, N)_0 &= 45, \quad \text{and} \\ \dim \operatorname{Ext}^1_S(P, \overline{M})_0 = \dim \operatorname{Ext}^1_S(N, P)_0 &= 0. \end{aligned}$$

Thus the induced map $\operatorname{Ext}^1_S(\overline{M}, \overline{M})_0 \to \dim \operatorname{Ext}^1_S(N, N)_0$ is surjective with 15-dimensional kernel, as expected. □

Exercise 4.8. Fill in the computational details in of the proof of Proposition 4.7 and Theorem 4.5 using your favorite computer algebra system.

Remark 4.9. In the proof of Proposition 4.7, the module P has syzygies

$$\beta(P) = \begin{array}{c|cccc} & 0 & 1 & 2 & 3 \\ \hline 2 & 6 & 9 & . & . \\ 3 & . & 4 & 6 & . \\ 4 & . & . & 6 & 5 \end{array}.$$

The cokernel of $\psi^t : S^6(-1) \to S^5$ has support on a determinantal curve E of degree 15 and genus 26, which is smooth for general C. The points Γ form a divisor on E with $h^0(E, \mathcal{O}_E(\Gamma)) = 1$. The curves E and C do not intersect; in fact, we have no idea how the curve E is related to C, other than the fact that it can be constructed from the syzygies of M. It is possible that $\mathfrak{M}_{16}$ is not unirational, and, even if $\mathfrak{M}_{16}$ is unirational, it could be that the component of the Hilbert scheme containing C is itself not unirational.

It is not clear to us whether it is a good idea to start with the determinantal points Γ in Proposition 4.7. Perhaps entirely different purely algebraic methods might lead to a unirational construction of the modules M, and we invite the reader to discover such an approach.

References

[Aprodu and Farkas 2011] M. Aprodu and G. Farkas, "Green's conjecture for curves on arbitrary $K3$ surfaces", *Compos. Math.* **147**:3 (2011), 839–851.

[Aprodu and Nagel 2010] M. Aprodu and J. Nagel, *Koszul cohomology and algebraic geometry*, University Lecture Series **52**, Amer. Math. Soc., Providence, RI, 2010.

[Aprodu and Voisin 2003] M. Aprodu and C. Voisin, "Green–Lazarsfeld's conjecture for generic curves of large gonality", *C. R. Math. Acad. Sci. Paris* **336**:4 (2003), 335–339.

[Arbarello et al. 1985] E. Arbarello, M. Cornalba, P. A. Griffiths, and J. Harris, *Geometry of algebraic curves, I*, Grundlehren der Mathematischen Wissenschaften **267**, Springer, New York, 1985.

[Babbage 1939] D. W. Babbage, "A note on the quadrics through a canonical curve", *J. London Math. Soc.* **14** (1939), 310–315.

[von Bothmer et al. 2011] H. C. G. von Bothmer, F. Geiß, and F. O. Schreyer, "RandomCurves", Macaulay2 package, 2011, http://tinyurl.com/RandomCurves-tar-gz.

[Brill and Noether 1874] A. Brill and M. Noether, "Uber die algebraischen Funktionen und ihre Anwendung in der Geometrie", *Math. Ann.* **7** (1874), 269–310.

[Bruno and Verra 2005] A. Bruno and A. Verra, "$\mathcal{M}_{15}$ is rationally connected", pp. 51–65 in *Projective varieties with unexpected properties*, edited by C. Ciliberto et al., Walter de Gruyter, Berlin, 2005.

[Buchberger 1965] B. Buchberger, *Ein Algorithmus zum Auffinden der Basiselemente des Restklassenrings*, Dissertation, Universität Innsbruck, 1965, http://tinyurl.com/BuchbergerPhD-pdf.

[Buchberger 1970] B. Buchberger, "Ein algorithmisches Kriterium für die Lösbarkeit eines algebraischen Gleichungssystems", *Aequationes Math.* **4** (1970), 374–383.

[Burch 1968] L. Burch, "On ideals of finite homological dimension in local rings", *Proc. Cambridge Philos. Soc.* **64** (1968), 941–948.

[Chang and Ran 1984] M. C. Chang and Z. Ran, "Unirationality of the moduli spaces of curves of genus 11, 13 (and 12)", *Invent. Math.* **76**:1 (1984), 41–54.

[Chang and Ran 1986] M.-C. Chang and Z. Ran, "The Kodaira dimension of the moduli space of curves of genus 15", *J. Differential Geom.* **24**:2 (1986), 205–220.

[Chang and Ran 1991] M.-C. Chang and Z. Ran, "On the slope and Kodaira dimension of $\overline{\mathcal{M}}_g$ for small g", *J. Differential Geom.* **34**:1 (1991), 267–274.

[Decker et al. 2011] W. Decker, G.-M. Greuel, G. Pfister, and H. Schönemann, "Singular 3-1-3", 2011, http://www.singular.uni-kl.de. A computer algebra system for polynomial computations.

[Dickson 1913] L. E. Dickson, "Finiteness of the odd perfect and primitive abundant numbers with n prime factors", *Amer. J. Math.* **35**:4 (1913), 413–422.

[Eisenbud 1995] D. Eisenbud, *Commutative algebra: With a view toward algebraic geometry*, Graduate Texts in Mathematics **150**, Springer, New York, 1995.

[Eisenbud and Harris 1987a] D. Eisenbud and J. Harris, "The Kodaira dimension of the moduli space of curves of genus ≥ 23", *Invent. Math.* **90**:2 (1987), 359–387.

[Eisenbud and Harris 1987b] D. Eisenbud and J. Harris, "On varieties of minimal degree (a centennial account)", pp. 3–13 in *Algebraic geometry, Bowdoin, 1985* (Brunswick, ME, 1985), edited by S. J. Bloch, Proc. Sympos. Pure Math. **46**, Amer. Math. Soc., Providence, RI, 1987.

[Eliahou and Kervaire 1990] S. Eliahou and M. Kervaire, "Minimal resolutions of some monomial ideals", *J. Algebra* **129**:1 (1990), 1–25.

[Farkas 2006] G. Farkas, "Syzygies of curves and the effective cone of $\overline{\mathcal{M}}_g$", *Duke Math. J.* **135**:1 (2006), 53–98.

[Farkas 2009a] G. Farkas, "Birational aspects of the geometry of $\overline{\mathcal{M}}_g$", pp. 57–110 in *Surveys in differential geometry, XIV: Geometry of Riemann surfaces and their moduli spaces*, edited by L. Ji et al., Surv. Differ. Geom. **14**, Int. Press, Somerville, MA, 2009.

[Farkas 2009b] G. Farkas, "Koszul divisors on moduli spaces of curves", *Amer. J. Math.* **131**:3 (2009), 819–867.

[Fløystad 1991] G. Fløystad, "Construction of space curves with good properties", *Math. Ann.* **289**:1 (1991), 33–54.

[Fulton and Lazarsfeld 1981] W. Fulton and R. Lazarsfeld, "On the connectedness of degeneracy loci and special divisors", *Acta Math.* **146**:3-4 (1981), 271–283.

[Gieseker 1982] D. Gieseker, "Stable curves and special divisors: Petri's conjecture", *Invent. Math.* **66**:2 (1982), 251–275.

[Gordan 1899] P. Gordan, "Neuer Beweis des Hilbertschen Satzes über homogene Funktionen", *Nachrichten König. Ges. der Wiss. zu Gött.* **1899** (1899), 240–242.

[Grayson and Stillman] D. R. Grayson and M. E. Stillman, "Macaulay2", a software system for research in algebraic geometry, http://www.math.uiuc.edu/Macaulay2/.

[Green 1984] M. L. Green, "Koszul cohomology and the geometry of projective varieties", *J. Differential Geom.* **19**:1 (1984), 125–171.

[Green and Lazarsfeld 1986] M. Green and R. Lazarsfeld, "On the projective normality of complete linear series on an algebraic curve", *Invent. Math.* **83**:1 (1986), 73–90.

[Griffiths and Harris 1980] P. Griffiths and J. Harris, "On the variety of special linear systems on a general algebraic curve", *Duke Math. J.* **47**:1 (1980), 233–272.

[Harris and Mumford 1982] J. Harris and D. Mumford, "On the Kodaira dimension of the moduli space of curves", *Invent. Math.* **67**:1 (1982), 23–86.

[Hartshorne 1977] R. Hartshorne, *Algebraic geometry*, Graduate Texts in Mathematics **52**, Springer, New York-Heidelberg, 1977.

[Hartshorne 2010] R. Hartshorne, *Deformation theory*, Graduate Texts in Mathematics **257**, Springer, New York, 2010.

[Hilbert 1890] D. Hilbert, "Ueber die Theorie der algebraischen Formen", *Math. Ann.* **36**:4 (1890), 473–534.

[Hirschowitz and Ramanan 1998] A. Hirschowitz and S. Ramanan, "New evidence for Green's conjecture on syzygies of canonical curves", *Ann. Sci. École Norm. Sup.* (4) **31**:2 (1998), 145–152.

[Kleiman 1974] S. L. Kleiman, "The transversality of a general translate", *Compositio Math.* **28** (1974), 287–297.

[Kollár 1996] J. Kollár, *Rational curves on algebraic varieties*, Ergebnisse der Mathematik und ihrer Grenzgebiete. 3. Folge. A Series of Modern Surveys in Mathematics **32**, Springer, Berlin, 1996.

[Little 1998] J. B. Little, "Canonical curves and the Petri scheme", pp. 381–392 in *Gröbner bases and applications* (Linz, 1998), edited by B. Buchberger and F. Winkler, London Math. Soc. Lecture Note Ser. **251**, Cambridge Univ. Press, 1998.

[Macaulay 1916] F. S. Macaulay, *The algebraic theory of modular systems*, Cambridge Univ. Press, 1916.

[Mukai 1992] S. Mukai, "Curves and symmetric spaces", *Proc. Japan Acad. Ser. A Math. Sci.* **68**:1 (1992), 7–10.

[Mumford 1975] D. Mumford, *Curves and their Jacobians*, The University of Michigan Press, Ann Arbor, MI, 1975.

[Noether 1880] M. Noether, "Ueber die invariante Darstellung algebraischer Functionen", *Math. Ann.* **17**:2 (1880), 263–284.

[Peeva and Stillman 2008] I. Peeva and M. Stillman, "The minimal free resolution of a Borel ideal", *Expo. Math.* **26**:3 (2008), 237–247.

[Peskine and Szpiro 1974] C. Peskine and L. Szpiro, "Liaison des variétés algébriques, I", *Invent. Math.* **26** (1974), 271–302.

[Petri 1923] K. Petri, "Über die invariante Darstellung algebraischer Funktionen einer Veränderlichen", *Math. Ann.* **88**:3-4 (1923), 242–289.

[Rao 1978] A. P. Rao, "Liaison among curves in $\mathbf{P}^3$", *Invent. Math.* **50**:3 (1978), 205–217.

[Saint-Donat 1973] B. Saint-Donat, "On Petri's analysis of the linear system of quadrics through a canonical curve", *Math. Ann.* **206** (1973), 157–175.

[Schreyer 1980] F. O. Schreyer, *Die Berechnung von Syzygien mit dem verallgemeinerten Weierstraßschen Divisionssatz und eine Anwendung auf analytische Cohen–Macaulay Stellenalgebren minimaler Multiplizität*, Diplomarbeit, Universität Hamburg, 1980.

[Schreyer 1986] F.-O. Schreyer, "Syzygies of canonical curves and special linear series", *Math. Ann.* **275**:1 (1986), 105–137.

[Schreyer 1991] F.-O. Schreyer, "A standard basis approach to syzygies of canonical curves", *J. Reine Angew. Math.* **421** (1991), 83–123.

[Schreyer 2003] F.-O. Schreyer, "Some topics in computational algebraic geometry", pp. 263–278 in *Advances in algebra and geometry* (Hyderabad, 2001), edited by C. Musili et al., Hindustan Book Agency, New Delhi, 2003.

[Schreyer 2013] F.-O. Schreyer, "Computer aided unirationality proofs of moduli spaces", pp. 257–280 in *Handbook of moduli, III*, edited by G. Farkas and I. Morrison, Adv. Lect. Math. (ALM) **26**, Int. Press, Somerville, MA, 2013.

[Schreyer and Tonoli 2002] F.-O. Schreyer and F. Tonoli, "Needles in a haystack: special varieties via small fields", pp. 251–279 in *Computations in algebraic geometry with Macaulay 2*, edited by D. Eisenbud et al., Algorithms Comput. Math. **8**, Springer, Berlin, 2002.

[Sernesi 1981] E. Sernesi, "L'unirazionalità della varietà dei moduli delle curve di genere dodici", *Ann. Scuola Norm. Sup. Pisa Cl. Sci.* (4) **8**:3 (1981), 405–439.

[Severi 1921] F. Severi, *Vorlesungen über algebraische Geometrie*, Teubner, Leipzig, 1921.

[Shokurov 1971] V. V. Shokurov, "The Noether–Enriques theorem on canonical curves", *Mat. Sb. (N.S.)* **86(128)**:3 (1971), 367–408. In Russian; translated in *Math. USSR Sb.* **15**:3 (1971), 361–403.

[Verra 2005] A. Verra, "The unirationality of the moduli spaces of curves of genus 14 or lower", *Compos. Math.* **141**:6 (2005), 1425–1444.

[Voisin 1988] C. Voisin, "Courbes tétragonales et cohomologie de Koszul", *J. Reine Angew. Math.* **387** (1988), 111–121.

[Voisin 2002] C. Voisin, "Green's generic syzygy conjecture for curves of even genus lying on a $K3$ surface", *J. Eur. Math. Soc. (JEMS)* **4**:4 (2002), 363–404.

[Voisin 2005] C. Voisin, "Green's canonical syzygy conjecture for generic curves of odd genus", *Compos. Math.* **141**:5 (2005), 1163–1190.

cberkesc@math.duke.edu *Department of Mathematics, Duke University, Box 90320, Durham, 27708, United States*

schreyer@math.uni-sb.de *Mathematik und Informatik, Universität des Saarlandes, Campus E2 4, D-66123 Saarbrücken, Germany*

Commutative Algebra and Noncommutative Algebraic Geometry, I
MSRI Publications
Volume **67**, 2015

Vector bundles and ideal closure operations

HOLGER BRENNER

This is an introduction to the use of vector bundle techniques to ideal closure operations, in particular to tight closure and related closures like solid closure and plus closure. We also briefly introduce the theory of vector bundles in general, with an emphasis on smooth projective curves, and discuss the relationship between forcing algebras and closure operations.

Introduction

An ideal operation is an assignment which provides for every ideal I in a commutative ring R a further ideal I' fulfilling certain structural conditions such as $I \subseteq I'$, $I'' = I'$, and an inclusion $I \subseteq J$ should induce an inclusion $I' \subseteq J'$. The most important examples are the radical of an ideal, whose importance stems from Hilbert's Nullstellensatz, the integral closure $\bar{I}$, which plays a crucial role in the normalization of blow-up algebras, and tight closure I^*, which is a closure operation in positive characteristic invented by Hochster and Huneke. In the context of tight closure, many other closure operations were introduced such as plus closure I^+, Frobenius closure I^F, solid closure $I^\star$, dagger closure $I^\dagger$, parasolid closure.

MSC2010: primary 14H60; secondary 13A35.

Keywords: tight closure, plus closure, solid closure, local cohomology, Frobenius morphism, semistable vector bundle, torsor, affine variety, superheight.

In this survey article we want to describe how the concepts of forcing algebras, vector bundles and their torsors can help to understand closure operations. This approach can be best understood by looking at the fundamental question whether $f \in (f_1, \ldots, f_n)' = I'$. In his work on solid closure, Hochster considered the forcing algebra

$$A = R[T_1, \ldots, T_n]/(f_1 T_1 + \cdots + f_n T_n + f)$$

and put properties of this R-algebra in relation to the closure operation. Suppose that the ideal I is primary to the maximal ideal $\mathfrak{m}$ in a local normal domain R of dimension d. Then the containment $f \in I^\star$ is equivalent to the property that the local cohomology $H^d_{\mathfrak{m}}(A)$ does not vanish. This is still a difficult property, however the situation becomes somehow geometrically richer. Because of $H^d_{\mathfrak{m}}(A) \cong H^{d-1}(T, \mathcal{O}_A)$, where $T = D(\mathfrak{m}A)$ is the open subset of $\operatorname{Spec} A$ above the punctured spectrum $U = D(\mathfrak{m})$, we can study global cohomological properties of T. The first syzygy module $\operatorname{Syz}(f_1, \ldots, f_n)$ is a locally free sheaf on U and acts on T in a locally trivial way. The scheme T is therefore a torsor for the syzygy bundle, which are classified by $H^1(U, \operatorname{Syz}(f_1, \ldots, f_n))$, and the element f determines this class. So the ideal operation is reflected by global properties of T, which is locally just an affine space over the base.

If the ring and the ideal are graded, then these objects have their counterpart on the corresponding projective varieties. This allows to apply results and machinery from algebraic geometry to closure operations, like intersection theory, semistability conditions, ampleness, moduli spaces of vector bundles, deformations. As this translation works best in dimension two, where the corresponding projective varieties are curves, we will focus here on this case. This approach has led in dimension two over a finite field to a positive solution to the tantalizing question whether tight closure is plus closure, and to negative solutions to arithmetic variation and to the localization problem in tight closure theory. The aim of this article is to provide an introduction to this techniques and to show how they help to solve problems from tight closure theory.

Throughout we assume a basic knowledge of commutative algebra and algebraic geometry including local cohomology and sheaf cohomology; once in a while we will use some notions and results from tight closure theory.

Acknowledgements. This article grew out of lectures I gave on several occasions in Barcelona, Kolkatta, Tehran, Medellín, Ann Arbor and at MSRI during the Introductory Workshop for the Special Year on Commutative Algebra 2012–2013. I thank the organizers of the Special Year, David Eisenbud, Srikanth Iyengar, Ezra Miller, Anurag Singh and Karen Smith for the opportunity to be part in the Introductory Workshop. I thank MSRI for its hospitality during the workshop and in the spring semester 2013. I thank Matteo Varbaro who gave the tutorials for the lectures at MSRI.

For corrections and remarks I thank Mohsen Asgharzadeh, Daniel Brinkmann, Amartya Dutta, Neena Gupta, Axel Stäbler, Matteo Varbaro and Juan Vélez. I thank the referee for careful reading and several helpful remarks.

1. Linear equations, forcing algebras and closure operations

Systems of linear equations. We start with some linear algebra. Let K be a field. We consider a system of linear homogeneous equations over K,

$$\begin{aligned} f_{11}t_1 + \cdots + f_{1n}t_n &= 0, \\ f_{21}t_1 + \cdots + f_{2n}t_n &= 0, \\ &\vdots \\ f_{m1}t_1 + \cdots + f_{mn}t_n &= 0, \end{aligned}$$

where the f_{ij} are elements in K. The solution set to this system of homogeneous equations is a vector space V over K (a subvector space of K^n), its dimension is $n - \operatorname{rk}(A)$, where $A = (f_{ij})_{ij}$ is the matrix given by these elements. Additional elements $f_1, \dots, f_m \in K$ give rise to the system of inhomogeneous linear equations,

$$\begin{aligned} f_{11}t_1 + \cdots + f_{1n}t_n &= f_1, \\ f_{21}t_1 + \cdots + f_{2n}t_n &= f_2, \\ &\vdots \\ f_{m1}t_1 + \cdots + f_{mn}t_n &= f_m. \end{aligned}$$

The solution set T of this inhomogeneous system may be empty, but nevertheless it is tightly related to the solution space of the homogeneous system. First of all, there exists an action

$$V \times T \to T, \quad (v, t) \mapsto v + t,$$

because the sum of a solution of the homogeneous system and a solution of the inhomogeneous system is again a solution of the inhomogeneous system. Theis action is a group action of the group $(V, +, 0)$ on the set T. Moreover, if we fix one solution $t_0 \in T$ (supposing that at least one solution exists), then there exists a bijection

$$V \to T, \quad v \mapsto v + t_0.$$

This means that the group V acts simply transitive on T, and so T can be identified with the vector space V, however not in a canonical way.

Suppose now that X is a geometric object (a topological space, a manifold, a variety, a scheme, the spectrum of a ring) and that instead of elements in the field K we have functions

$$f_{ij} : X \to K$$

on X (which are continuous, or differentiable, or algebraic). We form the matrix of functions $A = (f_{ij})_{ij}$, which yields for every point $P \in X$ a matrix $A(P)$ over K. Then we get from these data the space

$$V = \left\{ (P; t_1, \ldots, t_n) \mid A(P) \begin{pmatrix} t_1 \\ \vdots \\ t_n \end{pmatrix} = 0 \right\} \subseteq X \times K^n$$

together with the projection to X. For a fixed point $P \in X$, the fiber V_P of V over P is the solution space to the corresponding system of homogeneous linear equations given by inserting P into f_{ij}. In particular, all fibers of the map

$$V \to X$$

are vector spaces (maybe of nonconstant dimension). These vector space structures yield an addition[1]

$$V \times_X V \to V, \quad (P; s_1, \ldots, s_n; t_1, \ldots, t_n) \mapsto (P; s_1 + t_1, \ldots, s_n + t_n)$$

(only points in the same fiber can be added). The mapping

$$X \to V, \quad P \mapsto (P; 0, \ldots, 0)$$

is called the *zero-section*.

Suppose now that additional functions

$$f_1, \ldots, f_m : X \to K$$

are given. Then we can form the set

$$T = \left\{ (P; t_1, \ldots, t_n) \mid A(P) \begin{pmatrix} t_1 \\ \vdots \\ t_n \end{pmatrix} = \begin{pmatrix} f_1(P) \\ \vdots \\ f_n(P) \end{pmatrix} \right\} \subseteq X \times K^n$$

with the projection to X. Again, every fiber T_P of T over a point $P \in X$ is the solution set to the system of inhomogeneous linear equations which arises by inserting P into f_{ij} and f_i. The actions of the fibers V_P on T_P (coming from linear algebra) extend to an action

$$V \times_X T \to T, \quad (P; t_1, \ldots, t_n; s_1, \ldots, s_n) \mapsto (P; t_1 + s_1, \ldots, t_n + s_n).$$

[1] $V \times_X V$ is the fiber product of $V \to X$ with itself.

Also, if a (continuous, differentiable, algebraic) map

$$s : X \to T$$

with $s(P) \in T_P$ exists, then we can construct a (continuous, differentiable, algebraic) isomorphism between V and T. However, different from the situation in linear algebra (which corresponds to the situation where X is just one point), such a section does rarely exist.

These objects T have new and sometimes difficult global properties which we try to understand in these lectures. We will work mainly in an algebraic setting and restrict to the situation where just one equation

$$f_1 T_1 + \cdots + f_n T_n = f$$

is given. Then in the homogeneous case ($f = 0$) the fibers are vector spaces of dimension $n - 1$ or n, and the later holds exactly for the points $P \in X$ where $f_1(P) = \cdots = f_n(P) = 0$. In the inhomogeneous case the fibers are either empty or of dimension (as a scheme) $n - 1$ or n. We give some typical examples.

Example 1.1. We consider the line $X = \mathbb{A}^1_K$ (or $X = K, \mathbb{R}, \mathbb{C}$ etc.) with the (identical) function x. For $f_1 = x$ and $f = 0$, i.e., for the homogeneous equation $xt = 0$, the geometric object V consists of a horizontal line (corresponding to the zero-solution) and a vertical line over $x = 0$. So all fibers except one are zero-dimensional vector spaces. For the inhomogeneous equation $xt = 1$, T is a hyperbola, and all fibers are zero-dimensional with the exception that the fiber over $x = 0$ is empty.

For the homogeneous equation $0t = 0$, V is just the affine cylinder over the base line. For the inhomogeneous equation $0t = x$, T consists of one vertical line, almost all fibers are empty.

Example 1.2. Let X denote a plane (K^2, $\mathbb{R}^2$, $\mathbb{A}^2_K$) with coordinate functions x and y. We consider an inhomogeneous linear equation of type

$$x^a t_1 + y^b t_2 = x^c y^d .$$

The fiber of the solution set T over a point $\neq (0, 0)$ is one-dimensional, whereas the fiber over $(0, 0)$ has dimension two (for $a, b, c, d \geq 1$). Many properties of T depend on these four exponents.

In (most of) these examples we can observe the following behavior. On an open subset, the dimension of the fibers is constant and equals $n - 1$, whereas the fiber over some special points degenerates to an n-dimensional solution set (or becomes empty).

Forcing algebras. We describe now the algebraic setting of systems of linear equations depending on a base space. For a commutative ring R, its spectrum $X = \operatorname{Spec}(R)$ is a topological space on which the ring elements can be considered as functions. The value of $f \in R$ at a prime ideal $P \in \operatorname{Spec}(R)$ is just the image of f under the ring homomorphism $R \to R/P \to \kappa(P) = Q(R/P)$. In this interpretation, a ring element is a function with values in different fields. Suppose that R contains a field K. Then an element $f \in R$ gives rise to the ring homomorphism

$$K[Y] \to R, \quad Y \mapsto f,$$

which gives rise to a scheme morphism

$$\operatorname{Spec}(R) \to \operatorname{Spec}(K[Y]) \cong \mathbb{A}_K^1.$$

This is another way to consider f as a function on $\operatorname{Spec}(R)$ with values in the affine line.

The following construction appeared first in [Hochster 1994] in the context of solid closure.

Definition 1.3. Let R be a commutative ring and let $f_1, \ldots, f_n$ and f be elements in R. Then the R-algebra

$$R[T_1, \ldots, T_n]/(f_1 T_1 + \cdots + f_n T_n - f)$$

is called the *forcing algebra* of these elements (or these data).

The forcing algebra B forces f to lie inside the extended ideal $(f_1, \ldots, f_n)B$ (hence the name). For every R-algebra S such that $f \in (f_1, \ldots, f_n)S$ there exists a (non unique) ring homomorphism $B \to S$ by sending T_i to the coefficient $s_i \in S$ in an expression $f = s_1 f_1 + \cdots + s_n f_n$.

The forcing algebra induces the spectrum morphism

$$\varphi : \operatorname{Spec}(B) \to \operatorname{Spec}(R).$$

Over a point $P \in X = \operatorname{Spec}(R)$, the fiber of this morphism is given by

$$\operatorname{Spec}\big(B \otimes_R \kappa(P)\big),$$

and we can write

$$B \otimes_R \kappa(P) = \kappa(P)[T_1, \ldots, T_n]/(f_1(P)T_1 + \cdots + f_n(P)T_n - f(P)),$$

where $f_i(P)$ means the evaluation of the f_i in the residue class field. Hence the $\kappa(P)$-points in the fiber are exactly the solutions to the inhomogeneous linear equation $f_1(P)T_1 + \cdots + f_n(P)T_n = f(P)$. In particular, all the fibers are (empty or) affine spaces.

Forcing algebras and closure operations. Let R denote a commutative ring and let $I = (f_1, \ldots, f_n)$ be an ideal. Let $f \in R$ and let

$$B = R[T_1, \ldots, T_n]/(f_1 T_1 + \cdots + f_n T_n - f)$$

be the corresponding forcing algebra and

$$\varphi : \mathrm{Spec}(B) \to \mathrm{Spec}(R)$$

the corresponding spectrum morphism. How are properties of φ (or of the R-algebra B) related to certain ideal closure operations?

We start with some examples. The element f belongs to the ideal I if and only if we can write $f = r_1 f_1 + \cdots + r_n f_n$ with $r_i \in R$. By the universal property of the forcing algebra this means that there exists an R-algebra homomorphism

$$B \to R,$$

hence $f \in I$ holds if and only if φ admits a scheme section. This is also equivalent to

$$R \to B$$

admitting an R-module section (R being a direct module summand of B) or B being a pure R-algebra (so for forcing algebras properties might be equivalent which are not equivalent for arbitrary algebras).

The radical of an ideal. Now we look at the *radical* of the ideal I,

$$\mathrm{rad}(I) = \left\{ f \in R \mid f^k \in I \text{ for some } k \right\}.$$

The importance of the radical comes mainly from Hilbert's Nullstellensatz, saying that for algebras of finite type over an algebraically closed field there is a natural bijection between radical ideals and closed algebraic zero-sets. So geometrically one can see from an ideal only its radical. As this is quite a coarse closure operation we should expect that this corresponds to a quite coarse property of the morphism φ as well. Indeed, it is true that $f \in \mathrm{rad}(I)$ if and only if φ is surjective. This is true since the radical of an ideal is the intersection of all prime ideals in which it is contained. Hence an element f belongs to the radical if and only if for all residue class homomorphisms

$$\theta : R \to \kappa(\mathfrak{p})$$

where I is sent to 0, also f is sent to 0. But this means for the forcing equation that whenever the equation degenerates to 0, then also the inhomogeneous part becomes zero, and so there will always be a solution to the inhomogeneous equation.

Exercise. Define the radical of a submodule inside a module.

Integral closure of an ideal. Another closure operation is *integral closure* (see [Huneke and Swanson 2006]). It is defined by

$$\bar{I} = \left\{ f \in R \mid f^k + a_1 f^{k-1} + \cdots + a_{k-1} f + a_k = 0 \ for some\ k \text{ and } a_i \in I^i \right\}.$$

This notion is important for describing the normalization of the blow up of the ideal I. Another characterization (assume that R is noetherian) is that there exists a $z \in R$, not contained in any minimal prime ideal of R, such that $zf^n \in I^n$ holds for all n. Another equivalent property — the valuative criterion — is that for all ring homomorphisms

$$\theta : R \to D$$

to a discrete valuation domain D the containment $\theta(f) \in \theta(I)D$ holds.

The characterization of the integral closure in terms of forcing algebras requires some notions from topology. A continuous map

$$\varphi : X \to Y$$

between topological spaces X and Y is called a *submersion*, if it is surjective and if Y carries the image topology (quotient topology) under this map. This means that a subset $W \subseteq Y$ is open if and only if its preimage $\varphi^{-1}(W)$ is open. Since the spectrum of a ring endowed with the Zarisiki topology is a topological space, this notion can be applied to the spectrum morphism of a ring homomorphism. With this notion we can state that $f \in \bar{I}$ if and only if the forcing morphism

$$\varphi : \operatorname{Spec}(B) \to \operatorname{Spec}(R)$$

is a universal submersion (universal means here that for any ring change $R \to R'$ to a noetherian ring R', the resulting homomorphism $R' \to B'$ still has this property). The relation between these two notions stems from the fact that also for universal submersions there exists a criterion in terms of discrete valuation domains: A morphism of finite type between two affine noetherian schemes is a universal submersion if and only if the base change to any discrete valuation domain yields a submersion (see [SGA 1 1971, Remarque 2.6]). For a morphism

$$Z \to \operatorname{Spec}(D)$$

(D a discrete valuation domain) to be a submersion means that above the only chain of prime ideals in $\operatorname{Spec}(D)$, namely $(0) \subset \mathfrak{m}_D$, there exists a chain of prime ideals $\mathfrak{p}' \subseteq \mathfrak{q}'$ in Z lying over this chain. This pair-lifting property holds for a universal submersion

$$\operatorname{Spec}(S) \longrightarrow \operatorname{Spec}(R)$$

for any pair of prime ideals $\mathfrak{p} \subseteq \mathfrak{q}$ in $\operatorname{Spec}(R)$. This property is stronger than lying over (which means surjective) but weaker than the going-down or the going-up property (in the presence of surjectivity).

If we are dealing only with algebras of finite type over the complex numbers $\mathbb{C}$, then we may also consider the corresponding complex spaces with their natural topology induced from the euclidean topology of $\mathbb{C}^n$. Then universal submersive with respect to the Zariski topology is the same as submersive in the complex topology (the target space needs to be normal).

Example 1.4. Let K be a field and consider $R = K[X]$. Since this is a principal ideal domain, the only interesting forcing algebras (if we are only interested in the local behavior around (X)) are of the form $K[X, T]/(X^n T - X^m)$. For $m \geq n$ this $K[X]$-algebra admits a section (corresponding to the fact that $X^m \in (X^n)$), and if $n \geq 1$ there exists an affine line over the maximal ideal (X). So now assume $m < n$. If $m = 0$, then we have a hyperbola mapping to an affine line, with the fiber over (X) being empty, corresponding to the fact that 1 does not belong to the radical of (X^n) for $n \geq 1$. So assume finally $1 \leq m < n$. Then X^m belongs to the radical of (X^n), but not to its integral closure (which is the identical closure on a one-dimensional regular ring). We can write the forcing equation as $X^n T - X^m = X^m(X^{n-m}T - 1)$. So the spectrum of the forcing algebra consists of a (thickened) line over (X) and of a hyperbola. The forcing morphism is surjective, but it is not a submersion. For example, the preimage of $V(X) = \{(X)\}$ is a connected component hence open, but this single point is not open.

Example 1.5. Let K be a field and let $R = K[X, Y]$ be the polynomial ring in two variables. We consider the ideal $I = (X^2, Y)$ and the element X. This element belongs to the radical of this ideal; hence the forcing morphism

$$\operatorname{Spec}\big(K[X, Y, T_1, T_2]/(X^2 T_1 + Y T_2 - X\big) \to \operatorname{Spec}\big(K[X, Y]\big)$$

is surjective. We claim that it is not a submersion. For this we look at the reduction modulo Y. In $K[X, Y]/(Y) \cong K[X]$ the ideal I becomes (X^2) which does not contain X. Hence by the valuative criterion for integral closure, X does not belong to the integral closure of the ideal. One can also say that the chain $V(X, Y) \subset V(Y)$ in the affine plane does not have a lift (as a chain) to the spectrum of the forcing algebra.

For the ideal $I = (X^2, Y^2)$ and the element XY the situation looks different. Let

$$\theta : K[X, Y] \to D$$

be a ring homomorphism to a discrete valuation domain D. If X or Y is mapped to 0, then also XY is mapped to 0 and hence belongs to the extended ideal. So assume that $\theta(X) = u\pi^r$ and $\theta(Y) = v\pi^s$, where π is a local parameter of D and u and v are units. Then $\theta(XY) = uv\pi^{r+s}$ and the exponent is at least the minimum of $2r$ and $2s$, hence

$$\theta(XY) \in (\pi^{2r}, \pi^{2s}) = (\theta(X^2), \theta(Y^2))D.$$

So XY belongs to the integral closure of (X^2, Y^2) and the forcing morphism

$$\operatorname{Spec}\bigl(K[X, Y, T_1, T_2]/(X^2T_1 + Y^2T_2 - XY)\bigr) \to \operatorname{Spec}\bigl(K[X, Y]\bigr)$$

is a universal submersion.

Continuous closure. Suppose now that $R = \mathbb{C}[X_1, \dots, X_k]$. Then every polynomial $f \in R$ can be considered as a continuous function

$$f : \mathbb{C}^k \to \mathbb{C}, \quad (x_1, \dots, x_k) \mapsto f(x_1, \dots, x_k),$$

in the complex topology. If $I = (f_1, \dots, f_n)$ is an ideal and $f \in R$ is an element, we say that f belongs to the *continuous closure* of I, if there exist continuous functions

$$g_1, \dots, g_n : \mathbb{C}^k \to \mathbb{C}$$

such that

$$f = \sum_{i=1}^{n} g_i f_i$$

(as an identity of functions). The same definition works for $\mathbb{C}$-algebras of finite type; see [Brenner 2006a; Epstein and Hochster 2011; Kollár 2012].

It is not at all clear at once that there may exist polynomials $f \notin I$ but inside the continuous closure of I. For $\mathbb{C}[X]$ it is easy to show that the continuous closure is (like the integral closure) just the ideal itself. We also remark that when we would only allow holomorphic functions $g_1, \dots, g_n$ then we could not get something larger. However, with continuous functions $g_1, g_2 : \mathbb{C}^2 \to \mathbb{C}$ we can for example write

$$X^2Y^2 = g_1X^3 + g_2Y^3.$$

Continuous closure is always inside the integral closure and hence also inside the radical. The element XY does not belong to the continuous closure of $I = (X^2, Y^2)$, though it belongs to the integral closure of I. In terms of forcing algebras, an element f belongs to the continuous closure if and only if the complex forcing mapping

$$\varphi_{\mathbb{C}} : \operatorname{Spec}(B)_{\mathbb{C}} \to \operatorname{Spec}(R)_{\mathbb{C}}$$

(between the corresponding complex spaces) admits a continuous section.

2. Vector bundles and torsors

Geometric vector bundles. We have seen that the fibers of the spectrum of a forcing algebra are (empty or) affine spaces. However, this is not only fiberwise true, but more general: If we localize the forcing algebra at f_i we get

$$(R[T_1, \dots, T_n]/(f_1T_1 + \cdots + f_nT_n - f))_{f_i} \cong R_{f_i}[T_1, \dots, T_{i-1}, T_{i+1}, \dots, T_n],$$

since we can write

$$T_i = -\sum_{j \neq i} \frac{f_j}{f_i} T_j + \frac{f}{f_i}.$$

So over every $D(f_i)$ the spectrum of the forcing algebra is an $(n-1)$-dimensional affine space over the base. So locally, restricted to $D(f_i)$, we have isomorphisms

$$T|_{D(f_i)} \cong D(f_i) \times \mathbb{A}^{n-1}.$$

On the intersections $D(f_i) \cap D(f_j)$ we get two identifications with affine space, and the transition morphisms are linear if $f = 0$, but only affine-linear in general (because of the translation with $\frac{f}{f_i}$).

So the forcing algebra has locally the form $R_{f_i}[T_1, \dots, T_{i-1}, T_{i+1}, \dots, T_n]$ and its spectrum $\operatorname{Spec}(B)$ has locally the form $D(f_i) \times \mathbb{A}_K^{n-1}$. This description holds on the union $U = \bigcup_{i=1}^n D(f_i)$. Moreover, in the homogeneous case ($f = 0$) the transition mappings are linear. Hence $V|_U$, where V is the spectrum of a homogeneous forcing algebra, is a geometric vector bundle according to the following definition.

Definition 2.1. Let X denote a scheme. A scheme V equipped with a morphism

$$p : V \to X$$

is called a *geometric vector bundle* of rank r over X if there exists an open covering $X = \bigcup_{i \in I} U_i$ and U_i-isomorphisms

$$\psi_i : U_i \times \mathbb{A}^r = \mathbb{A}^r_{U_i} \to V|_{U_i} = p^{-1}(U_i)$$

such that for every open affine subset $U \subseteq U_i \cap U_j$ the transition mappings

$$\psi_j^{-1} \circ \psi_i : \mathbb{A}^r_{U_i}|_U \to \mathbb{A}^r_{U_j}|_U$$

are linear automorphisms; that is, they are induced by an automorphism of the polynomial ring $\Gamma(U, \mathcal{O}_X)[T_1, \dots, T_r]$ given by $T_i \mapsto \sum_{j=1}^r a_{ij} T_j$.

Here we can restrict always to affine open coverings. If X is separated then the intersection of two affine open subschemes is again affine and then it is enough to check the condition on the intersections. The trivial bundle of rank r is the r-dimensional affine space $\mathbb{A}^r_X$ over X, and locally every vector bundle looks like this. Many properties of an affine space are enjoyed by general vector bundles. For example, in the affine space we have the natural addition

$$+ : \mathbb{A}_U^r \times_U \mathbb{A}_U^r \to \mathbb{A}_U^r, \quad (v_1, \ldots, v_r, w_1, \ldots, w_r) \mapsto (v_1 + w_1, \ldots, v_r + w_r),$$

and this carries over to a vector bundle, that is, we have an addition

$$\alpha : V \times_X V \to V.$$

The reason for this is that the isomorphisms occurring in the definition of a geometric vector bundle are linear, hence the addition on $V|_U$ coming from an isomorphism with some affine space over U is independent of the chosen isomorphism. For the same reason there is a unique closed subscheme of V called the *zero-section* which is locally defined to be $0 \times U \subseteq \mathbb{A}_U^r$. Also, multiplication by a scalar, i.e., the mapping

$$\cdot : \mathbb{A}_U \times_U \mathbb{A}_U^r \to \mathbb{A}_U^r, \quad (s, v_1, \ldots, v_r) \mapsto (sv_1, \ldots, sv_r),$$

carries over to a scalar multiplication

$$\cdot : \mathbb{A}_X \times_X V \to V.$$

In particular, for every point $P \in X$ the fiber $V_P = V \times_X P$ is an affine space over $\kappa(P)$.

For a geometric vector bundle $p : V \to X$ and an open subset $U \subseteq X$ one sets

$$\Gamma(U, V) = \{s : U \to V|_U \mid p \circ s = \mathrm{Id}_U\},$$

so this is the set of sections in V over U. This gives in fact for every scheme over X a set-valued sheaf. Because of the observations just mentioned, these sections can also be added and multiplied by elements in the structure sheaf, and so we get for every vector bundle a locally free sheaf, which is free on the open subsets where the vector bundle is trivial.

Definition 2.2. A coherent $\mathcal{O}_X$-module $\mathcal{F}$ on a scheme X is called *locally free* of rank r, if there exists an open covering $X = \bigcup_{i \in I} U_i$ and $\mathcal{O}_{U_i}$-module-isomorphisms $\mathcal{F}|_{U_i} \cong (\mathcal{O}_{U_i})^r$ for every $i \in I$.

Vector bundles and locally free sheaves are essentially the same objects.

Theorem 2.3. *Let X denote a scheme. Then the category of locally free sheaves on X and the category of geometric vector bundles on X are equivalent. A geometric vector bundle $V \to X$ corresponds to the sheaf of its sections, and a locally free sheaf $\mathcal{F}$ corresponds to the (relative) spectrum of the symmetric algebra of the dual module $\mathcal{F}^\vee$.*

The free sheaf of rank r corresponds to the affine space $\mathbb{A}_X^r$ over X.

Torsors of vector bundles. We have seen that

$$V = \operatorname{Spec}\big(R[T_1, \dots, T_n]/(f_1T_1 + \cdots + f_nT_n)\big)$$

acts on the spectrum of a forcing algebra

$$T = \operatorname{Spec}\big(R[T_1, \dots, T_n]/(f_1T_1 + \cdots + f_nT_n - f)\big)$$

by addition. The restriction of V to $U = D(f_1, \dots, f_n)$ is a vector bundle, and T restricted to U becomes a V-torsor.

Definition 2.4. Let V denote a geometric vector bundle over a scheme X. A scheme $T \to X$ together with an action

$$\beta : V \times_X T \to T$$

is called a geometric (Zariski) *torsor* for V (or a V-principal fiber bundle or a *principal homogeneous space*) if there exists an open covering $X = \bigcup_{i \in I} U_i$ and isomorphisms

$$\varphi_i : T|_{U_i} \to V|_{U_i}$$

such that the diagrams (we set $U = U_i$ and $\varphi = \varphi_i$)

$$\begin{array}{ccc} V|_U \times_U T|_U & \stackrel{\beta}{\to} & T|_U \\ \mathrm{Id} \times \varphi \downarrow & & \downarrow \varphi \\ V|_U \times_U V|_U & \stackrel{\alpha}{\to} & V|_U \end{array}$$

commute, where α is the addition on the vector bundle.

The torsors of vector bundles can be classified in the following way.

Proposition 2.5. *Let X denote a noetherian separated scheme and let*

$$p : V \to X$$

denote a geometric vector bundle on X with sheaf of sections $\mathcal{S}$. Then there exists a correspondence between first cohomology classes $c \in H^1(X, \mathcal{S})$ and geometric V-torsors.

Proof. We describe only the correspondence. Let T denote a V-torsor. There exists by definition an open covering $X = \bigcup_{i \in I} U_i$ such that there exist isomorphisms

$$\varphi_i : T|_{U_i} \to V|_{U_i}$$

which are compatible with the action of $V|_{U_i}$ on itself. The isomorphisms φ_i induce automorphisms

$$\psi_{ij} = \varphi_j \circ \varphi_i^{-1} : V|_{U_i \cap U_j} \to V|_{U_i \cap U_j}.$$

These automorphisms are compatible with the action of V on itself, and this means that they are of the form

$$\psi_{ij} = \mathrm{Id}_V |_{U_i \cap U_j} + s_{ij}$$

with suitable sections $s_{ij} \in \Gamma(U_i \cap U_j, \mathcal{S})$. This family defines a Čech cocycle for the covering and gives therefore a cohomology class in $H^1(X, \mathcal{S})$.

For the reverse direction, suppose that the cohomology class $c \in H^1(X, \mathcal{S})$ is represented by a Čech cocycle $s_{ij} \in \Gamma(U_i \cap U_j, \mathcal{S})$ for an open covering $X = \bigcup_{i \in I} U_i$. Set $T_i := V|_{U_i}$. We take the morphisms

$$\psi_{ij} : T_i|_{U_i \cap U_j} = V|_{U_i \cap U_j} \to V|_{U_i \cap U_j} = T_j|_{U_i \cap U_j}$$

given by $\psi_{ij} := \mathrm{Id}_V |_{U_i \cap U_j} + s_{ij}$ to glue the T_i together to a scheme T over X. This is possible since the cocycle condition guarantees the gluing condition for schemes (see [EGA I 1960, Chapter 0, §4.1.7]). The action of $T_i = V|_{U_i}$ on itself glues also together to give an action on T. □

It follows immediately that for an affine scheme (a scheme of type $\mathrm{Spec}(R)$) there is no nontrivial torsor for any vector bundle. There will however be in general many nontrivial torsors on the punctured spectrum (and on a projective variety).

Forcing algebras and induced torsors. As T_U is a V_U-torsor, and as every V-torsor is represented by a unique cohomology class, there should be a natural cohomology class coming from the forcing data. To see this, let R be a noetherian ring and $I = (f_1, \ldots, f_n)$ be an ideal. Then on $U = D(I)$ we have the short exact sequence

$$0 \to \mathrm{Syz}(f_1, \ldots, f_n) \to \mathcal{O}_U^n \to \mathcal{O}_U \to 0.$$

On the left we have a locally free sheaf of rank $n-1$ which we call the *syzygy sheaf* or *syzygy bundle*. It is the sheaf of sections in the geometric vector bundle

$$\mathrm{Spec}\big(R[T_1, \ldots, T_n]/(f_1T_1 + \cdots + f_nT_n)\big)|_U.$$

An element $f \in R$ defines an element $f \in \Gamma(U, \mathcal{O}_U)$ and hence a cohomology class $\delta(f) \in H^1(U, \mathrm{Syz}(f_1, \ldots, f_n))$. Hence f defines in fact a $\mathrm{Syz}(f_1, \ldots, f_n)$-torsor over U. We will see that this torsor is induced by the forcing algebra given by $f_1, \ldots, f_n$ and f.

Theorem 2.6. *Let R denote a noetherian ring, let $I = (f_1, \ldots, f_n)$ denote an ideal and let $f \in R$ be another element. Let $c \in H^1(D(I), \mathrm{Syz}(f_1, \ldots, f_n))$ be the corresponding cohomology class and let*

$$B = R[T_1, \ldots, T_n]/(f_1T_1 + \cdots + f_nT_n - f)$$

denote the forcing algebra for these data. Then the scheme $\operatorname{Spec}(B)|_{D(I)}$ *together with the natural action of the syzygy bundle on it is isomorphic to the torsor given by* c.

Proof. We compute the cohomology class $\delta(f) \in H^1(U, \operatorname{Syz}(f_1, \ldots, f_n))$ and the cohomology class given by the forcing algebra. For the first computation we look at the short exact sequence

$$0 \to \operatorname{Syz}(f_1, \ldots, f_n) \to \mathcal{O}_U^n \stackrel{f_1, \ldots, f_n}{\to} \mathcal{O}_U \to 0.$$

On $D(f_i)$, the element f is the image of $(0, \ldots, 0, f/f_i, 0, \ldots, 0)$ (the nonzero entry is at the i-th place). The cohomology class is therefore represented by the family of differences

$$\left(0, \ldots, 0, \frac{f}{f_i}, 0, \ldots, 0, -\frac{f}{f_j}, 0, \ldots, 0\right) \in \Gamma(D(f_i) \cap D(f_j), \operatorname{Syz}(f_1, \ldots, f_n)).$$

On the other hand, there are isomorphisms

$$V|_{D(f_i)} \to T|_{D(f_i)}, \quad (s_1, \ldots, s_n) \mapsto \left(s_1, \ldots, s_{i-1}, s_i + \frac{f}{f_i}, s_{i+1}, \ldots, s_n\right).$$

The composition of two such isomorphisms on $D(f_i f_j)$ is the identity plus the same section as before. □

Example 2.7. Let $(R, \mathfrak{m})$ denote a two-dimensional normal local noetherian domain and let f and g be two parameters in R. On $U = D(\mathfrak{m})$ we have the short exact sequence

$$0 \to \mathcal{O}_U \cong \operatorname{Syz}(f, g) \to \mathcal{O}_U^2 \stackrel{f, g}{\to} \mathcal{O}_U \to 0$$

and its corresponding long exact sequence of cohomology,

$$0 \to R \to R^2 \stackrel{f, g}{\to} R \stackrel{\delta}{\to} H^1(U, \mathcal{O}_X) \to \cdots.$$

The connecting homomorphism δ sends an element $h \in R$ to $\frac{h}{fg}$. The torsor given by such a cohomology class $c = \frac{h}{fg} \in H^1(U, \mathcal{O}_X)$ can be realized by the forcing algebra

$$R[T_1, T_2]/(fT_1 + gT_2 - h).$$

Note that different forcing algebras may give the same torsor, because the torsor depends only on the spectrum of the forcing algebra restricted to the punctured spectrum of R. For example, the cohomology class $\frac{1}{fg} = \frac{fg}{f^2g^2}$ defines one torsor, but the two fractions yield the two forcing algebras $R[T_1, T_2]/(fT_1 + gT_2 - 1)$ and $R[T_1, T_2]/(f^2T_1 + g^2T_2 - fg)$, which are quite different. The fiber over the maximal ideal of the first one is empty, whereas the fiber over the maximal ideal of the second one is a plane.

If R is regular, say $R = K[X, Y]$ (or the localization of this at (X, Y) or the corresponding power series ring) then the first cohomology classes are K-linear combinations of terms $1/(x^i y^j)$, for $i, j \geq 1$. They are realized by the forcing algebras

$$K[X, Y, T_1, T_2]/(X^i T_1 + Y^j T_2 - 1).$$

Since the fiber over the maximal ideal is empty, the spectrum of the forcing algebra equals the torsor. Or, the other way round, the torsor is itself an affine scheme.

3. Tight closure and cohomological properties of torsors

The closure operations we have considered so far can be characterized by some property of the forcing algebra. However, they can not be characterized by a property of the corresponding torsor alone. For example, for $R = K[X, Y]$, we may write

$$\frac{1}{XY} = \frac{X}{X^2Y} = \frac{XY}{X^2Y^2} = \frac{X^2Y^2}{X^3Y^3},$$

so the torsors given by the forcing algebras

$$R[T_1, T_2]/(XT_1 + YT_2 - 1), \quad R[T_1, T_2]/(X^2T_1 + YT_2 - X),$$
$$R[T_1, T_2]/(X^2T_1 + Y^2T_2 - XY) \quad \text{and} \quad R[T_1, T_2]/(X^3T_1 + Y^3T_2 - X^2Y^2)$$

are all the same (the restriction over $D(X, Y)$), but their global properties are quite different. We have a nonsurjection, a surjective nonsubmersion, a submersion which does not admit (for $K = \mathbb{C}$) a continuous section and a map which admits a continuous section.

We deal now with closure operations which depend only on the torsor which the forcing algebra defines, so they only depend on the cohomology class of the forcing data inside the syzygy bundle. Our main example is tight closure, a theory developed by Hochster and Huneke, and related closure operations like solid closure and plus closure. For background on tight closure see [Hochster and Huneke 1990; Huneke 1996; 1998].

Tight closure and solid closure. Let R be a noetherian domain of positive characteristic, let

$$F : R \to R, \quad f \mapsto f^p,$$

be the *Frobenius homomorphism*, and

$$F^e : R \to R, \quad f \mapsto f^q, \quad q = p^e,$$

its e-th iteration. Let I be an ideal and set

$$I^{[q]} = \text{extended ideal of } I \text{ under } F^e.$$

Then define the *tight closure* of I to be the ideal

$$I^* := \{f \in R \mid \text{there exists } z \neq 0 \text{ such that } zf^q \in I^{[q]} \text{ for all } q = p^e\}.$$

The element f defines the cohomology class $c \in H^1(D(I), \operatorname{Syz}(f_1, \ldots, f_n))$. Suppose that R is normal and that I has height at least 2 (think of a local normal domain of dimension at least 2 and an $\mathfrak{m}$-primary ideal I). Then the e-th Frobenius pull-back of the cohomology class is

$$F^{e*}(c) \in H^1\big(D(I), F^{e*}(\operatorname{Syz}(f_1, \ldots, f_n))\big) \cong H^1(D(I), \operatorname{Syz}(f_1^q, \ldots, f_n^q))$$

($q = p^e$) and this is the cohomology class corresponding to f^q. By the height assumption, $zF^{e*}(c) = 0$ if and only if $zf^q \in (f_1^q, \ldots, f_n^q)$, and if this holds for all e then $f \in I^*$ by definition. This shows already that tight closure under the given conditions does only depend on the cohomology class.

This is also a consequence of the following theorem, which gives a characterization of tight closure in terms of forcing algebra and local cohomology.

Theorem 3.1 [Hochster 1994, Theorem 8.6]. *Let R be a normal excellent local domain with maximal ideal $\mathfrak{m}$ over a field of positive characteristic. Let $f_1, \ldots, f_n$ generate an $\mathfrak{m}$-primary ideal I and let f be another element in R. Then $f \in I^*$ if and only if*

$$H_{\mathfrak{m}}^{\dim(R)}(B) \neq 0,$$

where $B = R[T_1, \ldots, T_n]/(f_1T_1 + \cdots + f_nT_n - f)$ denotes the forcing algebra of these elements.

If the dimension d is at least two, then

$$H_{\mathfrak{m}}^d(R) \to H_{\mathfrak{m}}^d(B) \cong H_{\mathfrak{m}B}^d(B) \cong H^{d-1}(D(\mathfrak{m}B), \mathcal{O}_B).$$

This means that we have to look at the cohomological properties of the complement of the exceptional fiber over the closed point, i.e., the torsor given by these data. If $H^{d-1}(D(\mathfrak{m}B), \mathcal{O}_B) = 0$ then this is true for all quasicoherent sheaves instead of the structure sheaf. This property can be expressed by saying that the *cohomological dimension* of $D(\mathfrak{m}B)$ is $\leq d-2$ and thus smaller than the cohomological dimension of the punctured spectrum $D(\mathfrak{m})$, which is exactly $d-1$. So belonging to tight closure can be rephrased by saying that the formation of the corresponding torsor does not change the cohomological dimension.

If the dimension is two, then we have to look at whether the first cohomology of the structure sheaf vanishes. This is true (by Serre's cohomological criterion for affineness, see below) if and only if the open subset $D(\mathfrak{m}B)$ is an *affine scheme* (the spectrum of a ring).

The right-hand side of the equivalence in Theorem 3.1 — the nonvanishing of the top-dimensional local cohomology — is independent of any characteristic assumption, and can be taken as the basis for the definition of another closure operation, called *solid closure*. So the theorem above says that in positive characteristic tight closure and solid closure coincide. There is also a definition of tight closure for algebras over a field of characteristic 0 by reduction to positive characteristic; see [Hochster 1996].

An important property of tight closure is that it is trivial for regular rings: $I^* = I$ for every ideal I. This rests upon Kunz's theorem [1969, 3.3] saying that the Frobenius homomorphism for regular rings is flat. This property implies the following cohomological property of torsors.

Corollary 3.2. *Let* $(R, \mathfrak{m})$ *denote a regular local ring of dimension* d *and of positive characteristic, let* $I = (f_1, \dots, f_n)$ *be an* $\mathfrak{m}$*-primary ideal and* $f \in R$ *an element with* $f \notin I$. *Let* $B = R[T_1, \dots, T_n]/(f_1T_1 + \cdots + f_nT_n - f)$ *be the corresponding forcing algebra. Then for the extended ideal* $\mathfrak{m}B$ *we have*

$$H^d_{\mathfrak{m}B}(B) = H^{d-1}(D(\mathfrak{m}B), \mathcal{O}_B) = 0.$$

Proof. This follows from Theorem 3.1 and $f \notin I^*$. □

In dimension two this is true in every (even mixed) characteristic.

Theorem 3.3. *Let* $(R, \mathfrak{m})$ *denote a two-dimensional regular local ring, let* $I = (f_1, \dots, f_n)$ *be an* $\mathfrak{m}$*-primary ideal and* $f \in R$ *an element with* $f \notin I$. *Let* $B = R[T_1, \dots, T_n]/(f_1T_1 + \cdots + f_nT_n - f)$ *be the corresponding forcing algebra. Then the extended ideal* $\mathfrak{m}B$ *satisfies*

$$H^2_{\mathfrak{m}B}(B) = H^1(D(\mathfrak{m}B), \mathcal{O}_B) = 0.$$

In particular, the open subset $T = D(\mathfrak{m}B)$ *is an affine scheme if and only if* $f \notin I$.

The main point for the proof of this result is that for $f \notin I$, the natural mapping

$$H^1(U, \mathcal{O}_X) \to H^1(T, \mathcal{O}_T)$$

is not injective by a Matlis duality argument. Since the local cohomology of a regular ring is explicitly known, this map annihilates some cohomology class of the form $1/(fg)$ where f, g are parameters. But then it annihilates the complete local cohomology module and then T is an affine scheme.

For nonregular two-dimensional rings it is a difficult question in general to decide whether a torsor is affine or not. A satisfactory answer is only known in the normal two-dimensional graded case over a field, which we will deal with in the final lectures.

In higher dimension in characteristic zero it is not true that a regular ring is *solidly closed* (meaning that every ideal equals its solid closure), as was shown by the following example of Paul Roberts.

Example 3.4 [Roberts 1994]. Let K be a field of characteristic 0 and let

$$B = K[X, Y, Z][U, V, W]/(X^3U + Y^3V + Z^3W - X^2Y^2Z^2).$$

Then the ideal $\mathfrak{a} = (X, Y, Z)B$ has the property that $H^3_{\mathfrak{a}}(B) \neq 0$. This means that in $R = K[X, Y, Z]$ the element $X^2Y^2Z^2$ belongs to the solid closure of the ideal (X^3, Y^3, Z^3); hence the three-dimensional polynomial ring is not solidly closed.

This example was the motivation for the introduction of parasolid closure [Brenner 2003a], which has all the good properties of solid closure but which is also trivial for regular rings.

If R is a normal local domain of dimension 2 and $I = (f_1, \dots, f_n)$ an $\mathfrak{m}$-primary ideal, then $f \in I^*$ (or inside the solid closure) if and only if $D(\mathfrak{m}) \subseteq \operatorname{Spec}(B)$ is not an affine scheme, where B denotes the forcing algebra. Here we will discuss in more detail, with this application in mind, when a scheme is affine.

Affine schemes. A scheme U is called *affine* if it is isomorphic to the spectrum of some commutative ring R. If the scheme is of finite type over a field (or a ring) K, then this is equivalent to saying that there exist global functions

$$g_1, \dots, g_m \in \Gamma(U, \mathcal{O}_U)$$

such that the mapping

$$U \to \mathbb{A}^m_K, \quad x \mapsto (g_1(x), \dots, g_m(x)),$$

is a closed embedding. The relation to cohomology is given by the following well-known theorem of Serre [Hartshorne 1977, Theorem III.3.7].

Theorem 3.5. *For U a noetherian scheme, the following properties are equivalent.*

(1) *U is an affine scheme.*

(2) *For every quasicoherent sheaf $\mathcal{F}$ on U and all $i \geq 1$ we have $H^i(U, \mathcal{F}) = 0$.*

(3) *For every coherent ideal sheaf $\mathcal{I}$ on U we have $H^1(U, \mathcal{I}) = 0$.*

It is in general a difficult question whether a given scheme U is affine. For example, suppose that $X = \operatorname{Spec}(R)$ is an affine scheme and

$$U = D(\mathfrak{a}) \subseteq X$$

is an open subset (such schemes are called *quasiaffine*) defined by an ideal $\mathfrak{a} \subseteq R$. When is U itself affine? The cohomological criterion above simplifies to the condition that $H^i(U, \mathcal{O}_X) = 0$ for $i \geq 1$.

Of course, if $\mathfrak{a} = (f)$ is a principal ideal (or up to radical a principal ideal), then $U = D(f) \cong \operatorname{Spec}(R_f)$ is affine. On the other hand, if $(R, \mathfrak{m})$ is a local ring of dimension ≥ 2, then

$$D(\mathfrak{m}) \subset \operatorname{Spec}(R)$$

is not affine, since

$$H^{d-1}(U, \mathcal{O}_X) = H^d_{\mathfrak{m}}(R) \neq 0$$

by the relation between sheaf cohomology and local cohomology and a theorem of Grothendieck [Bruns and Herzog 1993, Theorem 3.5.7].

Codimension condition. One can show that for an open affine subset $U \subseteq X$ the closed complement $Y = X \setminus U$ must be of pure codimension one (U must be the complement of the support of an effective divisor). In a regular or (locally $\mathbb{Q}$-) factorial domain the complement of every effective divisor is affine, since the divisor can be described (at least locally geometrically) by one equation. But it is easy to give examples to show that this is not true for normal three-dimensional domains. The following example is a standard example for this phenomenon and it is in fact given by a forcing algebra (we write here and in the following often small letters for the classes of the variables in the residue class ring).

Example 3.6. Let K be a field and consider the ring

$$R = K[X, Y, U, V]/(XU - YV).$$

The ideal $\mathfrak{p} = (x, y)$ is a prime ideal in R of height one. Hence the open subset $U = D(x, y)$ is the complement of an irreducible hypersurface. However, U is not affine. For this we consider the closed subscheme

$$\mathbb{A}^2_K \cong Z = V(u, v) \subseteq \operatorname{Spec}(R)$$

and

$$Z \cap U \subseteq U.$$

If U were affine, then also the closed subscheme $Z \cap U \cong \mathbb{A}^2_K \setminus \{(0, 0)\}$ would be affine, but this is not true, since the complement of the punctured plane has codimension 2.

Ring of global sections of affine schemes. For an open subset $U = D(\mathfrak{a}) \subseteq \operatorname{Spec}(R)$ its ring of global sections $\Gamma(U, \mathcal{O}_X)$ is difficult to compute in general. If R is a domain and $\mathfrak{a} = (f_1, \dots, f_n)$, then

$$\Gamma(U, \mathcal{O}_X) = R_{f_1} \cap R_{f_2} \cap \cdots \cap R_{f_n}.$$

This ring is not always of finite type over R, but it is if U is affine.

Lemma 3.7. *Let R be a noetherian ring and $U = D(\mathfrak{a}) \subseteq \operatorname{Spec}(R)$ an open subset.*

(1) *U is an affine scheme if and only if $\mathfrak{a}\Gamma(U, \mathcal{O}_X) = (1)$.*

(2) *If this holds, and $q_1 f_1 + \cdots + q_n f_n = 1$ with $f_1, \dots, f_n \in \mathfrak{a}$ and $q_i \in \Gamma(U, \mathcal{O}_X)$, then $\Gamma(U, \mathcal{O}_X) = R[q_1, \dots, q_n]$. In particular, the ring of global sections over U is finitely generated over R.*

Sketch of proof. (1) There always exists a natural scheme morphism

$$U \to \operatorname{Spec}\bigl(\Gamma(U, \mathcal{O}_X)\bigr),$$

and U is affine if and only if this morphism is an isomorphism. It is always an open embedding (because it is an isomorphism on the $D(f)$, $f \in \mathfrak{a}$), and the image is $D(\mathfrak{a}\Gamma(U, \mathcal{O}_X))$. This is everything if and only if the extended ideal is the unit ideal.

(2) We write $1 = q_1 f_1 + \cdots + q_n f_n$ and consider the natural morphism

$$U \to \operatorname{Spec}\bigl(R[q_1, \dots, q_n]\bigr)$$

corresponding to the ring inclusion $R[q_1, \dots, q_n] \subseteq \Gamma(U, \mathcal{O}_X)$. This morphism is again an open embedding and its image is everything. □

We give some examples of tight closure computations on the Fermat cubic $x^3 + y^3 + z^3 = 0$, a standard example in tight closure theory, with the methods we have developed so far.

Example 3.8. We consider the Fermat cubic $R = K[X, Y, Z]/(X^3 + Y^3 + Z^3)$, the ideal $I = (X, Y)$ and the element Z. We claim that for characteristic $\neq 3$ the element Z does not belong to the solid closure of I. Equivalently, the open subset

$$D(X, Y) \subseteq \operatorname{Spec}\bigl(R[S, T]/(XS + YT - Z)\bigr)$$

is affine. For this we show that the extended ideal inside the ring of global sections is the unit ideal. First of all we get the equation

$$X^3 + Y^3 = (XS + YT)^3 = X^3S^3 + 3X^2S^2YT + 3XSY^2T^2 + Y^3T^3$$

or, equivalently,

$$X^3(S^3-1)+3X^2YS^2T+3XY^2ST^2+Y^3(T^3-1)=0.$$

We write this as

$$\begin{aligned} X^3(S^3-1) &= -3X^2YS^2T-3XY^2ST^2-Y^3(T^3-1)\\ &= Y\left(-3X^2S^2T-3XYST^2-Y^2(T^3-1)\right), \end{aligned}$$

which yields on $D(X,Y)$ the rational function

$$Q=\frac{S^3-1}{Y}=\frac{-3X^2S^2T-3XYST^2-Y^2\left(T^3-1\right)}{X^3}.$$

This shows that $S^3-1=QY$ belongs to the extended ideal. Similarly, one can show that also the other coefficients $3S^2T$, $3ST^2$, T^3-1 belong to the extended ideal. Therefore in characteristic different from 3, the extended ideal is the unit ideal.

Example 3.9. We consider the Fermat cubic $R=K[X,Y,Z]/(X^3+Y^3+Z^3)$, the ideal $I=(X,Y)$ and the element Z^2. We claim that in positive characteristic $\neq 3$ the element Z^2 does belong to the tight closure of I. Equivalently, the open subset

$$D(X,Y)\subseteq \operatorname{Spec}\bigl(R[S,T]/(XS+YT-Z^2)\bigr)$$

is not affine. The element Z^2 defines the cohomology class

$$c=\frac{Z^2}{XY}\in H^1(D(X,Y),\mathcal{O}_X)$$

and its Frobenius pull-backs are $F^{e*}(c)=\frac{Z^{2q}}{X^qY^q}\in H^1(D(X,Y),\mathcal{O}_X)$. This cohomology module has a $\mathbb{Z}$-graded structure (the degree is given by the difference of the degree of the numerator and the degree of the denominator) and, moreover, it is 0 in positive degree (this is related to the fact that the corresponding projective curve is elliptic). Therefore for any homogeneous element $t\in R$ of positive degree we have $tF^{e*}(c)=0$ and so Z^2 belongs to the tight closure.

From this it follows also that in characteristic 0 the element Z^2 belongs to the solid closure, because affineness is an open property in an arithmetic (or any) family, which follows from Lemma 3.7 (1).

We give now a cohomological proof of a tight closure containment on the Fermat cubic for a nonparameter ideal. M. McDermott has raised the question whether

$$xyz\in(x^2,y^2,z^2)^* \text{ in } K[X,Y,Z]/(X^3+Y^3+Z^3).$$

This was answered positively by A. Singh [1998] by a long "equational" argument.

Example 3.10. Let $R = K[X, Y, Z]/(X^3 + Y^3 + Z^3)$, where K is a field of positive characteristic $p \neq 3$, $I = (x^2, y^2, z^2)$ and $f = xyz$. We consider the short exact sequence

$$0 \to \mathrm{Syz}(x^2, y^2, z^2) \to \mathcal{O}_U^3 \overset{x^2, y^2, z^2}{\to} \mathcal{O}_U \to 0$$

and the cohomology class

$$c = \delta(xyz) \in H^1\left(U, \mathrm{Syz}(x^2, y^2, z^2)\right).$$

We want to show that $zF^{e*}(c) = 0$ for all $e \geq 0$ (here the test element z equals the element z in the ring). It is helpful to work with the graded structure on this syzygy sheaf (or to work on the corresponding elliptic curve Proj R directly). Now the equation $x^3 + y^3 + z^3 = 0$ can be considered as a syzygy (of total degree 3) for x^2, y^2, z^2, yielding an inclusion

$$0 \to \mathcal{O}_U \to \mathrm{Syz}(x^2, y^2, z^2).$$

Since this syzygy does not vanish anywhere on U the quotient sheaf is invertible and in fact isomorphic to the structure sheaf. Hence we have

$$0 \to \mathcal{O}_U \to \mathrm{Syz}(x^2, y^2, z^2) \to \mathcal{O}_U \to 0$$

and the cohomology sequence

$$\to H^1(U, \mathcal{O}_U)_s \to H^1\big(U, \mathrm{Syz}(x^2, y^2, z^2)\big)_{s+3} \to H^1(U, \mathcal{O}_U)_s \to 0,$$

where s denotes the degree-s piece. The class c lives in $H^1\left(U, \mathrm{Syz}(x^2, y^2, z^2)\right)_3$, so its Frobenius pull-backs live in $H^1\left(U, F^{e*}\,\mathrm{Syz}(x^2, y^2, z^2)\right)_{3q}$, and we can have a look at the cohomology of the pull-backs of the sequence, i.e.,

$$\to H^1(U, \mathcal{O}_U)_0 \to H^1\big(U, F^{e*}\,\mathrm{Syz}(x^2, y^2, z^2)\big)_{3q} \to H^1(U, \mathcal{O}_U)_0 \to 0.$$

The class $zF^{e*}(c)$ lives in $H^1\big(U, F^{e*}\,\mathrm{Syz}(x^2, y^2, z^2)\big)_{3q+1}$. It is mapped on the right to $H^1(U, \mathcal{O}_U)_1$, which is 0 (because we are working over an elliptic curve), hence it comes from the left, which is $H^1(U, \mathcal{O}_U)_1 = 0$. So $zF^{e*}(c) = 0$ and $f \in \left(x^2, y^2, z^2\right)^*$.

Affineness and superheight. We have mentioned above that the complement of an affine open subset must have pure codimension 1. We have also seen in Example 3.6 that the nonaffineness can be established by looking at the behavior of the codimension when the situation is restricted to closed subschemes. The following definition and theorem is an algebraic version of this observation [Brenner 2002].

Definition 3.11. Let R be a noetherian commutative ring and let $I \subseteq R$ be an ideal. The (noetherian) *superheight* is the supremum

$$\sup\big(\operatorname{ht}(IS) : S \text{ is a notherian } R\text{-algebra}\big).$$

Theorem 3.12. *Let R be a noetherian commutative ring and let $I \subseteq R$ be an ideal and $U = D(I) \subseteq X = \operatorname{Spec}(R)$. Then the following are equivalent.*

(1) *U is an affine scheme.*

(2) *I has superheight ≤ 1 and $\Gamma(U, \mathcal{O}_X)$ is a finitely generated R-algebra.*

It is not true at all that the ring of global sections of an open subset U of the spectrum X of a noetherian ring is of finite type over this ring. This is not even true if X is an affine variety. This problem is directly related to Hilbert's fourteenth problem, which has a negative answer. We will later present examples (see Example 3.13) where U has superheight one, yet is not affine, hence its ring of global sections is not finitely generated.

Plus closure. For an ideal $I \subseteq R$ in a domain R define its *plus closure* by

$$I^+ = \{f \in R \mid \text{there exists a finite domain extension } R \subseteq T \text{ such that } f \in IT\}.$$

Equivalent: let R^+ be the *absolute integral closure* of R. This is the integral closure of R in an algebraic closure of the quotient field $Q(R)$ (first considered in [Artin 1971]). Then

$$f \in I^+ \text{ if and only if } f \in IR^+.$$

The plus closure commutes with localization; see [Huneke 1996, Exercise 12.2].

We also have the inclusion $I^+ \subseteq I^*$; see [Huneke 1996, Theorem 1.7].

Question. Is $I^+ = I^*$?

This is known as the *tantalizing question* in tight closure theory.

In terms of forcing algebras and their torsors, the containment inside the plus closure has the following geometric meaning (see [Brenner 2003c] for details): If R is a d-dimensional domain of finite type over a field, and $I = (f_1, \dots, f_n)$ is an $\mathfrak{m}$-primary ideal for some maximal ideal $\mathfrak{m}$ and $f \in R$, then $f \in I^+$ if and only if the spectrum of the forcing algebra contains a d-dimensional closed subscheme which meets the exceptional fiber (the fiber over the maximal ideal) in isolated points. This means that the superheight of the extended ideal to the forcing algebra is d or that the torsor contains a punctured d-dimensional closed subscheme. In this case the local cohomological dimension of the torsor must be d as well, since it contains a closed subscheme with this cohomological dimension. So also the plus closure depends only on the torsor.

In characteristic zero, the plus closure behaves very differently compared with positive characteristic. If R is a normal domain of characteristic 0, then the trace map shows that the plus closure is trivial, $I^+ = I$ for every ideal I.

Examples. In the following two examples we use results from tight closure theory to establish (non) -affineness properties of certain torsors.

Example 3.13. Let K be a field and consider the Fermat ring

$$R = K[X, Y, Z]/(X^d + Y^d + Z^d)$$

together with the ideal $I = (X, Y)$ and $f = Z^2$. For $d \geq 3$ we have $Z^2 \notin (X, Y)$. This element is however in the tight closure $(X, Y)^*$ of the ideal in positive characteristic (assume that the characteristic p does not divide d) and is therefore also in characteristic 0 inside the tight closure (in the sense of [Hochster 1996, Definition 3.1]) and inside the solid closure. Hence the open subset

$$D(X, Y) \subseteq \operatorname{Spec}\big(K[X, Y, Z, S, T]/(X^d + Y^d + Z^d, SX + TY - Z^2)\big)$$

is not an affine scheme. In positive characteristic, Z^2 is also contained in the plus closure $(X, Y)^+$ and therefore this open subset contains punctured surfaces (the spectrum of the forcing algebra contains two-dimensional closed subschemes which meet the exceptional fiber $V(X, Y)$ in only one point; the ideal (X, Y) has superheight two in the forcing algebra). In characteristic zero however, the superheight is one because plus closure is trivial for normal domains in characteristic 0, and therefore by Theorem 3.12 the algebra $\Gamma(D(X, Y), \mathcal{O}_B)$ is not finitely generated. For $K = \mathbb{C}$ and $d = 3$ one can also show that $D(X, Y)_{\mathbb{C}}$ is, considered as a complex space, a Stein space.

Example 3.14. Let K be a field of positive characteristic $p \geq 7$ and consider the ring

$$R = K[X, Y, Z]/(X^5 + Y^3 + Z^2)$$

together with the ideal $I = (X, Y)$ and $f = Z$. Since R has a rational singularity, it is F-regular, so all ideals are tightly closed. Therefore $Z \notin (X, Y)^*$ and so the torsor

$$D(X, Y) \subseteq \operatorname{Spec}\big(K[X, Y, Z, S, T]/(X^5 + Y^3 + Z^2, SX + TY - Z)\big)$$

is an affine scheme. In characteristic zero this can be proved by either using that R is a quotient singularity or by using the natural grading ($\deg X = 6$, $\deg Y = 10$, $\deg Z = 15$) where the corresponding cohomology class $Z/(XY)$ gets degree -1 and then applying the geometric criteria (see below) on the corresponding projective curve (rather the corresponding curve of the standard homogenization $U^{30} + V^{30} + W^{30} = 0$).

4. Cones over projective curves

We continue with the question when the torsors given by a forcing algebra over a two-dimensional ring are affine? We will look at the graded situation to be able to work on the corresponding projective curve.

In particular we want to address the following questions:

(1) Is there a procedure to decide whether the torsor is affine?

(2) Is it nonaffine if and only if there exists a geometric reason for it not to be affine (because the superheight is too large) ?

(3) How does the affineness vary in an arithmetic family, when we vary the prime characteristic?

(4) How does the affineness vary in a geometric family, when we vary the base ring?

In terms of tight closure, these questions are directly related to the tantalizing question of tight closure (is it the same as plus closure), the dependence of tight closure on the characteristic and the localization problem of tight closure.

Geometric interpretation in dimension two. We will restrict now to the two-dimensional homogeneous case in order to work on the corresponding projective curve. We want to find an object over the curve which corresponds to the forcing algebra or its induced torsor. The results of this part were developed in [Brenner 2003b; 2004; 2006c]; see also [Brenner 2008].

Let R be a two-dimensional standard-graded normal domain over an algebraically closed field K. Let $C = \operatorname{Proj}(R)$ be the corresponding smooth projective curve and let

$$I = (f_1, \ldots, f_n)$$

be an R_+-primary homogeneous ideal with generators of degrees $d_1, \ldots, d_n$. Then we get on C the short exact sequence

$$0 \to \operatorname{Syz}(f_1, \ldots, f_n)(m) \to \bigoplus_{i=1}^{n} \mathcal{O}_C(m - d_i) \stackrel{f_1, \ldots, f_n}{\to} \mathcal{O}_C(m) \to 0.$$

Here $\operatorname{Syz}(f_1, \ldots, f_n)(m)$ is a vector bundle, called the *syzygy bundle*, of rank $n-1$ and of degree

$$((n-1)m - \sum_{i=1}^{n} d_i) \deg(C).$$

Recall that the degree of a vector bundle $\mathcal{S}$ on a projective curve is defined as the degree of the invertible sheaf $\bigwedge^r \mathcal{S}$, where r is the rank of $\mathcal{S}$. The degree is additive on short exact sequences.

A homogeneous element f of degree m defines an element in $\Gamma(C, \mathcal{O}_C(m))$ and thus a cohomology class $\delta(f) \in H^1(C, \mathrm{Syz}(f_1, \ldots, f_n)(m))$, so this defines a torsor over the projective curve. We mention an alternative description of the torsor corresponding to a first cohomology class in a locally free sheaf which is better suited for the projective situation.

Remark 4.1. Let $\mathcal{S}$ denote a locally free sheaf on a scheme X. For a cohomology class $c \in H^1(X, \mathcal{S})$ one can construct a geometric object: Because of $H^1(X, \mathcal{S}) \cong \mathrm{Ext}^1(\mathcal{O}_X, \mathcal{S})$, the class defines an extension

$$0 \to \mathcal{S} \to \mathcal{S}' \to \mathcal{O}_X \to 0.$$

This extension is such that under the connecting homomorphism of cohomology, $1 \in \Gamma(X, \mathcal{O}_X)$ is sent to $c \in H^1(X, \mathcal{S})$. The extension yields a projective subbundle

$$\mathbb{P}(\mathcal{S}^\vee) \subset \mathbb{P}(\mathcal{S}'^\vee).$$

If V is the corresponding geometric vector bundle of $\mathcal{S}$, one may think of $\mathbb{P}(\mathcal{S}^\vee)$ as $\mathbb{P}(V)$ which consists for every base point $x \in X$ of all the lines in the fiber V_x passing through the origin. The projective subbundle $\mathbb{P}(V)$ has codimension one inside $\mathbb{P}(V')$, for every point it is a projective space lying (linearly) inside a projective space of one dimension higher. The complement is then over every point an affine space. One can show that the global complement

$$T = \mathbb{P}(\mathcal{S}'^\vee) \setminus \mathbb{P}(\mathcal{S}^\vee)$$

is another model for the torsor given by the cohomology class. The advantage of this viewpoint is that we may work, in particular when X is projective, in an entirely projective setting.

Semistability of vector bundles. In the situation of a forcing algebra of homogeneous elements, this torsor T can also be obtained as Proj B, where B is the (not necessarily positively) graded forcing algebra. In particular, it follows that the containment $f \in I^*$ is equivalent to the property that T is not an affine variety. For this properties, positivity (ampleness) properties of the syzygy bundle are crucial. We need the concept of (Mumford) - semistability.

Definition 4.2. Let $\mathcal{S}$ be a vector bundle on a smooth projective curve C. It is called *semistable* if

$$\mu(\mathcal{T}) = \frac{\deg(\mathcal{T})}{\mathrm{rk}(\mathcal{T})} \leq \frac{\deg(\mathcal{S})}{\mathrm{rk}(\mathcal{S})} = \mu(\mathcal{S})$$

for all subbundles $\mathcal{T}$.

Suppose that the base field has positive characteristic $p > 0$. Then $\mathcal{S}$ is called *strongly semistable*, if all (absolute) Frobenius pull-backs $F^{e*}(\mathcal{S})$ are semistable.

An important property of a semistable bundle of negative degree is that it can not have any global section $\neq 0$. Note that a semistable vector bundle need not be strongly semistable, the following is probably the simplest example.

Example 4.3. Let C be the smooth Fermat quartic given by $x^4 + y^4 + z^4$ and consider on it the syzygy bundle $\mathrm{Syz}(x, y, z)$ (which is also the restricted cotangent bundle from the projective plane). This bundle is semistable. Suppose that the characteristic is 3. Then its Frobenius pull-back is $\mathrm{Syz}(x^3, y^3, z^3)$. The curve equation gives a global nontrivial section of this bundle of total degree 4. But the degree of $\mathrm{Syz}(x^3, y^3, z^3)(4)$ is negative, hence it can not be semistable anymore.

The following example is related to Example 3.10.

Example 4.4. Let $R = K[X, Y, Z]/\left(X^3 + Y^3 + Z^3\right)$, where K is a field of positive characteristic $p \neq 3$, $I = \left(x^2, y^2, z^2\right)$, and $C = \mathrm{Proj}(R)$. The equation $x^3 + y^3 + z^3 = 0$ yields the short exact sequence

$$0 \to \mathcal{O}_C \to \mathrm{Syz}(x^2, y^2, z^2)(3) \to \mathcal{O}_C \to 0.$$

This shows that $\mathrm{Syz}(x^2, y^2, z^2)$ is strongly semistable, since the Frobenius pull-backs of this sequence show that all $F^{e*}\left(\mathrm{Syz}(x^2, y^2, z^2)\right)$ are semistable.

For a strongly semistable vector bundle $\mathcal{S}$ on C and a cohomology class $c \in H^1(C, \mathcal{S})$ with corresponding torsor we obtain the following affineness criterion (in characteristic zero we mean by strongly semistable just semistable).

Theorem 4.5. *Let C denote a smooth projective curve over an algebraically closed field K and let $\mathcal{S}$ be a strongly semistable vector bundle over C together with a cohomology class $c \in H^1(C, \mathcal{S})$. Then the torsor $T(c)$ is an affine scheme if and only if* $\deg(\mathcal{S}) < 0$ *and $c \neq 0$ ($F^e(c) \neq 0$ for all e in positive characteristic).*

This result rests on the ampleness of $\mathcal{S}'^{\vee}$ occurring in the dual exact sequence $0 \to \mathcal{O}_C \to \mathcal{S}'^{\vee} \to \mathcal{S}^{\vee} \to 0$ given by c (this rests on [Gieseker 1971] and [1971]). It implies for a strongly semistable syzygy bundle the following *degree formula* for tight closure.

Theorem 4.6. *Suppose that* $\mathrm{Syz}(f_1, \ldots, f_n)$ *is strongly semistable. Then*

$$R_m \subseteq I^* \quad \text{for } m \geq \frac{\sum d_i}{n-1}$$

and

$$R_m \cap I^* \subseteq I^F \quad \text{for } m < \frac{\sum d_i}{n-1},$$

where I^F is the Frobenius closure. In a relative setting, if R' is a finitely generated $\mathbb{Z}$-algebra and $R = R'/(p)$, then

$$R_m \cap I^* \subseteq I \text{ for } m < \frac{\sum d_i}{n-1}$$

for almost all prime numbers.

We indicate the proof of the inclusion result. The degree condition implies that $c = \delta(f) \in H^1(C, \mathcal{S})$ is such that $\mathcal{S} = \mathrm{Syz}(f_1, \ldots, f_n)(m)$ has nonnegative degree. Then also all Frobenius pull-backs $F^*(\mathcal{S})$ have nonnegative degree. Let $\mathcal{L} = \mathcal{O}(k)$ be a twist of the tautological line bundle on C such that its degree is larger than the degree of ω_C^{-1}, the dual of the canonical sheaf. Let $z \in H^0(Y, \mathcal{L})$ be a nonzero element. Then $zF^{e*}(c) \in H^1(C, F^{e*}(\mathcal{S}) \otimes \mathcal{L})$, and by Serre duality we have

$$H^1(C, F^{e*}(\mathcal{S}) \otimes \mathcal{L}) \cong H^0\big(F^{e*}(\mathcal{S}^\vee) \otimes \mathcal{L}^{-1} \otimes \omega_C\big)^\vee.$$

On the right we have a semistable sheaf of negative degree, which can not have a nontrivial section. Hence $zF^{e*}(c) = 0$ and therefore f belongs to the tight closure.

Harder–Narasimhan filtration. In general, there exists an exact criterion for the affineness of the torsor $T(c)$ depending on c and the *strong Harder–Narasimhan filtration* of $\mathcal{S}$. For this we give the definition of the Harder–Narasimhan filtration.

Definition 4.7. Let $\mathcal{S}$ be a vector bundle on a smooth projective curve C over an algebraically closed field K. Then the filtration

$$0 = \mathcal{S}_0 \subset \mathcal{S}_1 \subset \cdots \subset \mathcal{S}_{t-1} \subset \mathcal{S}_t = \mathcal{S}$$

of subbundles such that all quotient bundles $\mathcal{S}_k/\mathcal{S}_{k-1}$ are semistable with decreasing slopes $\mu_k = \mu(\mathcal{S}_k/\mathcal{S}_{k-1})$, is called the *Harder–Narasimhan filtration* of $\mathcal{S}$. This object exists uniquely by a theorem of Harder and Narasimhan [1975].

A Harder–Narasimhan filtration is called *strong* if all the quotients $\mathcal{S}_i/\mathcal{S}_{i-1}$ are strongly semistable. A Harder–Narasimhan filtration is not strong in general; however, by a theorem of A. Langer [2004, Theorem 2.7], there exists some Frobenius pull-back $F^{e*}(\mathcal{S})$ such that its Harder–Narasimhan filtration is strong.

Theorem 4.8. *Let C denote a smooth projective curve over an algebraically closed field K and let $\mathcal{S}$ be a vector bundle over C together with a cohomology class $c \in H^1(C, \mathcal{S})$. Let*

$$\mathcal{S}_1 \subset \mathcal{S}_2 \subset \cdots \subset \mathcal{S}_{t-1} \subset \mathcal{S}_t = F^{e*}(\mathcal{S})$$

be a strong Harder–Narasimhan filtration. We choose i such that $\mathcal{S}_i/\mathcal{S}_{i-1}$ has degree ≥ 0 and that $\mathcal{S}_{i+1}/\mathcal{S}_i$ has degree < 0. We set $\mathcal{Q} = F^{e*}(\mathcal{S})/\mathcal{S}_i$. Then the following are equivalent.

(1) *The torsor $T(c)$ is not an affine scheme.*

(2) *Some Frobenius power of the image of $F^{e*}(c)$ inside $H^1(X, \mathcal{Q})$ is 0.*

Plus closure in dimension two. Let K be a field and let R be a normal two-dimensional standard-graded domain over K with corresponding smooth projective curve C. A homogeneous $\mathfrak{m}$-primary ideal with homogeneous ideal generators $f_1, \ldots, f_n$ and another homogeneous element f of degree m yield a cohomology class

$$c = \delta(f) \in H^1(C, \operatorname{Syz}(f_1, \ldots, f_n)(m)).$$

Let $T(c)$ be the corresponding torsor. We have seen that the affineness of this torsor over C is equivalent to the affineness of the corresponding torsor over $D(\mathfrak{m}) \subseteq \operatorname{Spec}(R)$. Now we want to understand what the property $f \in I^+$ means for c and for $T(c)$. Instead of the plus closure we will work with the *graded plus closure* $I^{+\mathrm{gr}}$, where $f \in I^{+\mathrm{gr}}$ holds if and only if there exists a finite graded extension $R \subseteq S$ such that $f \in IS$. The existence of such an S translates into the existence of a finite morphism

$$\varphi : C' = \operatorname{Proj}(S) \to \operatorname{Proj}(R) = C$$

such that $\varphi^*(c) = 0$. Here we may assume that C' is also smooth. Therefore we discuss the more general question when a cohomology class $c \in H^1(C, \mathcal{S})$, where $\mathcal{S}$ is a locally free sheaf on C, can be annihilated by a finite morphism

$$C' \to C$$

of smooth projective curves. The advantage of this more general approach is that we may work with short exact sequences (in particular, the sequences coming from the Harder–Narasimhan filtration) in order to reduce the problem to semistable bundles which do not necessarily come from an ideal situation.

Lemma 4.9. *Let C denote a smooth projective curve over an algebraically closed field K, let $\mathcal{S}$ be a locally free sheaf on C and let $c \in H^1(C, \mathcal{S})$ be a cohomology class with corresponding torsor $T \to C$. Then the following conditions are equivalent.*

(1) *There exists a finite morphism*

$$\varphi : C' \to C$$

from a smooth projective curve C' such that $\varphi^*(c) = 0$.

(2) *There exists a projective curve $Z \subseteq T$.*

Proof. If (1) holds, then the pull-back $\varphi^*(T) = T \times_C C'$ is trivial (as a torsor), as it equals the torsor given by $\varphi^*(c) = 0$. Hence $\varphi^*(T)$ is isomorphic to a vector bundle and contains in particular a copy of C'. The image Z of this copy is a projective curve inside T. If (2) holds, then let C' be the normalization of Z. Since Z dominates C, the resulting morphism

$$\varphi : C' \to C$$

is finite. Since this morphism factors through T and since T annihilates the cohomology class by which it is defined, it follows that $\varphi^*(c) = 0$. □

We want to show that the cohomological criterion for (non) -affineness of a torsor along the Harder–Narasimhan filtration of the vector bundle also holds for the existence of projective curves inside the torsor, under the condition that the projective curve is defined over a finite field. This implies that tight closure is (graded) plus closure for graded $\mathfrak{m}$-primary ideals in a two-dimensional graded domain over a finite field.

Annihilation of cohomology classes of strongly semistable sheaves. We deal first with the situation of a strongly semistable sheaf $\mathcal{S}$ of degree 0. The following two results are from [Lange and Stuhler 1977]. We say that a locally free sheaf is *étale trivializable* if there exists a finite étale morphism $\varphi : C' \to C$ such that $\varphi^*(\mathcal{S}) \cong \mathcal{O}_{C'}^r$. Such bundles are directly related to linear representations of the étale fundamental group.

Lemma 4.10. *Let K denote a finite field (or the algebraic closure of a finite field) and let C be a smooth projective curve over K. Let $\mathcal{S}$ be a locally free sheaf over C. Then $\mathcal{S}$ is étale trivializable if and only if there exists some n such that $F^{n*}\mathcal{S} \cong \mathcal{S}$.*

Theorem 4.11. *Let K denote a finite field (or the algebraic closure of a finite field) and let C be a smooth projective curve over K. Let $\mathcal{S}$ be a strongly semistable locally free sheaf over C of degree 0. Then there exists a finite morphism*

$$\varphi : C' \to C$$

such that $\varphi^*(\mathcal{S})$ is trivial.

Proof. We consider the family of locally free sheaves $F^{e*}(\mathcal{S})$, $e \in \mathbb{N}$. Because these are all semistable of degree 0, and defined over the same finite field, we must have (by the existence of the moduli space for vector bundles) a repetition: $F^{e*}(\mathcal{S}) \cong F^{e'*}(\mathcal{S})$ for some $e' > e$. By Lemma 4.10 the bundle $F^{e*}(\mathcal{S})$ admits an étale trivialization $\varphi : C' \to C$. Hence the finite map $F^e \circ \varphi$ trivializes the bundle. □

Theorem 4.12. *Let K denote a finite field (or the algebraic closure of a finite field) and let C be a smooth projective curve over K. Let $\mathcal{S}$ be a strongly semistable locally free sheaf over C of nonnegative degree and let $c \in H^1(C, \mathcal{S})$ denote a cohomology class. Then there exists a finite morphism*

$$\varphi : C' \to C$$

such that $\varphi^*(c)$ is trivial.

Proof. If the degree of $\mathcal{S}$ is positive, then a Frobenius pull-back $F^{e*}(\mathcal{S})$ has arbitrary large degree and is still semistable. By Serre duality we get that $H^1(C, F^{e*}(\mathcal{S})) = 0$. So in this case we can annihilate the class by an iteration of the Frobenius alone. So suppose that the degree is 0. Then there exists by Theorem 4.11 a finite morphism which trivializes the bundle. So we may assume that $\mathcal{S} \cong \mathcal{O}_C^r$. Then the cohomology class has several components $c_i \in H^1(C, \mathcal{O}_C)$ and it is enough to annihilate them separately by finite morphisms. But this is possible by the parameter theorem of K. Smith [1994] (or directly using Frobenius and Artin–Schreier extensions). □

The general case. We look now at an arbitrary locally free sheaf $\mathcal{S}$ on C, a smooth projective curve over a finite field. We want to show that the same numerical criterion (formulated in terms of the Harder–Narasimhan filtration) for nonaffineness of a torsor holds also for the finite annihilation of the corresponding cohomology class (or the existence of a projective curve inside the torsor).

Theorem 4.13. *Let K denote a finite field (or the algebraic closure of a finite field) and let C be a smooth projective curve over K. Let $\mathcal{S}$ be a locally free sheaf over C and let $c \in H^1(C, \mathcal{S})$ denote a cohomology class. Let $\mathcal{S}_1 \subset \cdots \subset \mathcal{S}_t$ be a strong Harder–Narasimhan filtration of $F^{e*}(\mathcal{S})$. We choose i such that $\mathcal{S}_i/\mathcal{S}_{i-1}$ has degree ≥ 0 and that $\mathcal{S}_{i+1}/\mathcal{S}_i$ has degree < 0. We set $\mathcal{Q} = F^{e*}(\mathcal{S})/\mathcal{S}_i$. Then the following are equivalent.*

(1) *The class c can be annihilated by a finite morphism.*

(2) *Some Frobenius power of the image of $F^{e*}(c)$ inside $H^1(C, \mathcal{Q})$ is 0.*

Proof. Suppose that (1) holds. Then the torsor is not affine and hence by Theorem 4.8 also (2) holds. So suppose that (2) is true. By applying a certain power of the Frobenius we may assume that the image of the cohomology class in $\mathcal{Q}$ is 0. Hence the class stems from a cohomology class $c_i \in H^1(C, \mathcal{S}_i)$. We look at the short exact sequence

$$0 \to \mathcal{S}_{i-1} \to \mathcal{S}_i \to \mathcal{S}_i/\mathcal{S}_{i-1} \to 0,$$

where the sheaf of the right-hand side has a nonnegative degree. Therefore the image of c_i in $H^1(C, \mathcal{S}_i/\mathcal{S}_{i-1})$ can be annihilated by a finite morphism due to

Theorem 4.12. Hence after applying a finite morphism we may assume that c_i stems from a cohomology class $c_{i-1} \in H^1(C, \mathcal{S}_{i-1})$. Going on inductively we see that c can be annihilated by a finite morphism. □

Theorem 4.14. *Let C denote a smooth projective curve over the algebraic closure of a finite field K, let $\mathcal{S}$ be a locally free sheaf on C and let $c \in H^1(C, \mathcal{S})$ be a cohomology class with corresponding torsor $T \to C$. Then T is affine if and only if it does not contain any projective curve.*

Proof. Due to Theorem 4.8 and Theorem 4.13, for both properties the same numerical criterion does hold. □

These results imply the following theorem in the setting of a two-dimensional graded ring.

Theorem 4.15. *Let R be a standard-graded, two-dimensional normal domain over (the algebraic closure of) a finite field. Let I be an R_+-primary graded ideal. Then*

$$I^* = I^+.$$

This is also true for nonprimary graded ideals and also for submodules in finitely generated graded submodules. Moreover, G. Dietz [2006] has shown that one can get rid also of the graded assumption (of the ideal or module, but not of the ring).

5. Tight closure in families

After having understood tight closure and plus closure in the two-dimensional situation we proceed to a special three-dimensional situation, namely families of two-dimensional rings parametrized by a one-dimensional base scheme.

Affineness under deformations. We consider a base scheme B and a morphism

$$Z \to B$$

together with an open subscheme $W \subseteq Z$. For every base point $b \in B$ we get the open subset

$$W_b \subseteq Z_b$$

inside the fiber Z_b. It is a natural question to ask how properties of W_b vary with b. In particular we may ask how the cohomological dimension of W_b varies and how the affineness may vary.

In the algebraic setting we have a D-algebra S and an ideal $\mathfrak{a} \subseteq S$ (so $B = \operatorname{Spec}(D)$, $Z = \operatorname{Spec}(S)$ and $W = D(\mathfrak{a})$) which defines for every prime ideal $\mathfrak{p} \in \operatorname{Spec}(D)$ the extended ideal $\mathfrak{a}_\mathfrak{p}$ in $S \otimes_D \kappa(\mathfrak{p})$.

This question is already interesting when $B = \mathrm{Spec}(D)$ is an affine one-dimensional integral scheme, in particular in the following two situations.

(1) $B = \mathrm{Spec}(\mathbb{Z})$. Then we speak of an *arithmetic deformation* and want to know how affineness varies with the characteristic and what the relation is to characteristic zero.

(2) $B = \mathbb{A}^1_K = \mathrm{Spec}(K[t])$, where K is a field. Then we speak of a *geometric deformation* and want to know how affineness varies with the parameter t, in particular how the behavior over the special points where the residue class field is algebraic over K is related to the behavior over the generic point.

It is fairly easy using Lemma 3.7 (1) to show that if the open subset in the generic fiber is affine, then also the open subsets are affine for almost all special points.

We deal with this question where W is a torsor over a family of smooth projective curves (or a torsor over a punctured two-dimensional spectrum). The arithmetic as well as the geometric variant of this question are directly related to questions in tight closure theory. Because of the above mentioned degree criteria in the strongly semistable case, a weird behavior of the affineness property of torsors is only possible if we have a weird behavior of strong semistability.

Arithmetic deformations. We start with the arithmetic situation.

Example 5.1 [Brenner and Katzman 2006]. Consider $\mathbb{Z}[X, Y, Z]/(X^7+Y^7+Z^7)$ and take the ideal $I=(x^4, y^4, z^4)$ and the element $f=x^3y^3$. Consider reductions $\mathbb{Z} \to \mathbb{Z}/(p)$. Then

$$f \in I^* \text{ holds in } \mathbb{Z}/(p)[x, y, z]/(x^7+y^7+z^7) \text{ for } p \equiv 3 \bmod 7$$

and

$$f \notin I^* \text{ holds in } \mathbb{Z}/(p)[x, y, z]/(x^7+y^7+z^7) \text{ for } p \equiv 2 \bmod 7.$$

In particular, the bundle $\mathrm{Syz}(x^4, y^4, z^4)$ is semistable in the generic fiber, but not strongly semistable for any reduction $p \equiv 2 \bmod 7$. The corresponding torsor is an affine scheme for infinitely many prime reductions and not an affine scheme for infinitely many prime reductions.

In terms of affineness (or local cohomology) this example has the following properties: the ideal

$$(x, y, z) \subseteq \mathbb{Z}/(p)[x, y, z, s_1, s_2, s_3]/(x^7+y^7+z^7, s_1x^4+s_2y^4+s_3z^4+x^3y^3)$$

has cohomological dimension 1 if $p=3 \bmod 7$ and has cohomological dimension 0 (equivalently, $D(x, y, z)$ is an affine scheme) if $p=2 \bmod 7$.

Geometric deformations: a counterexample to the localization problem. Let $S \subseteq R$ be a multiplicative system and I an ideal in R. Then the *localization problem* of tight closure is the question whether the identity

$$(I^*)_S = (IR_S)^*$$

holds.

Here the inclusion $\subseteq$ is always true and $\supseteq$ is the problem. This means explicitly:

Question. If $f \in (IR_S)^*$, can we find an $h \in S$ such that $hf \in I^*$ holds in R?

Proposition 5.2. *Let $\mathbb{Z}/(p) \subset D$ be a one-dimensional domain and $D \subseteq R$ of finite type, and I an ideal in R. Suppose that localization holds and that*

$$f \in I^* \text{ holds in } R \otimes_D Q(D) = R_{D^*} = R_{Q(D)}$$

($S = D^* = D \setminus \{0\}$ is the multiplicative system). Then $f \in I^*$ holds in $R \otimes_D \kappa(\mathfrak{p})$ for almost all $\mathfrak{p}$ in Spec D.

Proof. By localization, there exists $h \in D$, $h \neq 0$, such that

$$hf \in I^* \text{ in } R.$$

By persistence of tight closure (under a ring homomorphism) we get

$$hf \in I^* \text{ in } R_{\kappa(\mathfrak{p})}.$$

The element h does not belong to $\mathfrak{p}$ for almost all $\mathfrak{p}$, so h is a unit in $R_{\kappa(\mathfrak{p})}$ and hence

$$f \in I^* \text{ in } R_{\kappa(\mathfrak{p})}$$

for almost all $\mathfrak{p}$. □

In order to get a counterexample to the localization property we will look now at geometric deformations:

$$D = \mathbb{F}_p[t] \subset \mathbb{F}_p[t][X, Y, Z]/(g) = S,$$

where t has degree 0 and X, Y, Z have degree 1 and g is homogeneous. Then (for every field $\mathbb{F}_p[t] \subseteq K$)

$$S \otimes_{\mathbb{F}_p[t]} K$$

is a two-dimensional standard-graded ring over K. For residue class fields of points of $\mathbb{A}^1_{\mathbb{F}_p} = \operatorname{Spec}(\mathbb{F}_p[t])$ we have basically two possibilities.

- $K = \mathbb{F}_p(t)$, the function field. This is the *generic* or *transcendental* case.
- $K = \mathbb{F}_q$, the *special* or *algebraic* or *finite* case.

How does $f \in I^*$ vary with K? To analyze the behavior of tight closure in such a family we can use what we know in the two-dimensional standard-graded situation.

In order to establish an example where tight closure does not behave uniformly under a geometric deformation we first need a situation where strong semistability does not behave uniformly. Such an example was given by Paul Monsky in terms of Hilbert–Kunz theory:

Example 5.3 [Monsky 1998]. Let

$$g = Z^4 + Z^2XY + Z(X^3 + Y^3) + (t + t^2)X^2Y^2.$$

Consider

$$S = \mathbb{F}_2[t, X, Y, Z]/(g).$$

Monsky proved the following results on the *Hilbert–Kunz multiplicity* of the maximal ideal (x, y, z) in $S \otimes_{\mathbb{F}_2[t]} L$, L a field:

$$e_{HK}(S \otimes_{\mathbb{F}_2[t]} L) = \begin{cases} 3 & \text{for } L = \mathbb{F}_2(t), \\ 3 + 4^{-d} & \text{for } L = \mathbb{F}_q = \mathbb{F}_2(\alpha)\ (t \mapsto \alpha,\ d = \deg(\alpha)). \end{cases}$$

By the geometric interpretation of Hilbert–Kunz theory (see [Brenner 2006b; 2007; Trivedi 2005]) this means that the restricted cotangent bundle

$$\mathrm{Syz}(x, y, z) = (\Omega_{\mathbb{P}^2})|_C$$

is strongly semistable in the transcendental case, but not strongly semistable in the algebraic case. In fact, for $d = \deg(\alpha)$, $t \mapsto \alpha$, where $L = \mathbb{F}_2(\alpha)$, the d-th Frobenius pull-back destabilizes (meaning that it is not semistable anymore).

The maximal ideal (x, y, z) can not be used directly. However, we look at the second Frobenius pull-back which is (characteristic two) just

$$I = (x^4, y^4, z^4).$$

By the degree formula we have to look for an element of degree 6. Let's take

$$f = y^3z^3.$$

This is our example (x^3y^3 does not work). First, by strong semistability in the transcendental case we have

$$f \in I^* \text{ in } S \otimes \mathbb{F}_2(t)$$

by the degree formula. If localization would hold, then f would also belong to the tight closure of I for almost all algebraic instances $\mathbb{F}_q = \mathbb{F}_2(\alpha)$, $t \mapsto \alpha$. Contrary to that we show that for all algebraic instances the element f belongs never to the tight closure of I.

Lemma 5.4. *Let* $\mathbb{F}_q = \mathbb{F}_p(\alpha)$, $t \mapsto \alpha$, $\deg(\alpha) = d$. *Set* $Q = 2^{d-1}$. *Then*

$$xyf^Q \notin I^{[Q]}.$$

Proof. This is an elementary but tedious computation [Brenner and Monsky 2010]. □

Theorem 5.5. *Tight closure does not commute with localization.*

Proof. One knows in our situation that xy is a so-called test element. Hence the previous lemma shows that $f \notin I^*$. □

In terms of affineness (or local cohomology) this example has the following properties: the ideal

$$(x, y, z) \subseteq \mathbb{F}_2(t)[x, y, z, s_1, s_2, s_3]/(g, s_1x^4 + s_2y^4 + s_3z^4 + y^3z^3)$$

has cohomological dimension 1 if t is transcendental and has cohomological dimension 0 (equivalently, $D(x, y, z)$ is an affine scheme) if t is algebraic.

Corollary 5.6. *Tight closure is not plus closure in graded dimension two for fields with transcendental elements.*

Proof. Consider

$$R = \mathbb{F}_2(t)[X, Y, Z]/(g).$$

In this ring $y^3z^3 \in I^*$, but it can not belong to the plus closure. Else there would be a curve morphism $Y \to C_{\mathbb{F}_2(t)}$ which annihilates the cohomology class c and this would extend to a morphism of relative curves almost everywhere. □

Corollary 5.7. *There is an example of a smooth projective* (*relatively over the affine line*) *variety* Z *and an effective divisor* $D \subset Z$ *and a morphism*

$$Z \to \mathbb{A}^1_{\mathbb{F}_2}$$

such that $(Z \setminus D)_\eta$ *is not an affine variety over the generic point* η, *but for every algebraic point* x *the fiber* $(Z \setminus D)_x$ *is an affine variety.*

Proof. Take $C \to \mathbb{A}^1_{\mathbb{F}_2}$ to be the Monsky quartic and consider the syzygy bundle

$$\mathcal{S} = \mathrm{Syz}(x^4, y^4, z^4)(6)$$

together with the cohomology class c determined by $f = y^3z^3$. This class defines an extension

$$0 \to \mathcal{S} \to \mathcal{S}' \to \mathcal{O}_C \to 0$$

and hence $\mathbb{P}(\mathcal{S}^*) \subset \mathbb{P}(\mathcal{S}'^*)$. Then $\mathbb{P}(\mathcal{S}'^*) \setminus \mathbb{P}(\mathcal{S}^*)$ is an example with the stated properties by the previous results. □

It is an open question whether such an example can exist in characteristic zero.

Generic results. Is it more difficult to decide whether an element belongs to the tight closure of an ideal or to the ideal itself? We discuss one situation where this is easier for tight closure.

Suppose that we are in a graded situation of a given ring (or a given ring dimension) and have fixed a number (at least the ring dimension) of homogeneous generators and their degrees. Suppose that we want to know the degree bound for (tight closure or ideal) inclusion for generic choice of the ideal generators. Generic means that we write the coefficients of the generators as indeterminates and consider the situation over the (large) affine space corresponding to these indeterminates or over its function field. This problem is already interesting and difficult for the polynomial ring: Suppose we are in $P = K[X, Y, Z]$ and want to study the generic inclusion bound for, say, $n \geq 4$ generic polynomials $F_1, \ldots, F_n$ all of degree a. What is the minimal degree number m such that $P_{\geq m} \subseteq (F_1, \ldots, F_n)$. The answer is

$$\left\lceil \frac{1}{2(n-1)}\left(3 - 3n + 2an + \sqrt{1 - 2n + n^2 + 4a^2 n}\right) \right\rceil .$$

This rests on the fact that the Fröberg conjecture has been solved in dimension 3, by D. Anick [1986]. (The Fröberg conjecture gives a precise description of the Hilbert function for an ideal in a polynomial ring which is generically generated. Here we only need to know in which degree the Hilbert function of the residue class ring becomes 0.)

The corresponding generic ideal inclusion bound for arbitrary graded rings depends heavily (already in the parameter case) on the ring itself. Surprisingly, the generic ideal inclusion bound for tight closure does not depend on the ring and is only slightly worse than the bound for the polynomial ring. The following theorem is due to Brenner and Fischbacher–Weitz [Brenner and Fischbacher-Weitz 2011].

Theorem 5.8. *Let $d \geq 1$ and $a_1, \ldots, a_n$ be natural numbers (the degree type), $n \geq d+1$. Let $K[x_0, x_1, \ldots, x_d] \subseteq R$ be a finite extension of standard-graded domains (a graded Noether normalization). Suppose that there exist n homogeneous polynomials $g_1, \ldots, g_n$ in $P = K[x_0, x_1, \ldots, x_d]$ with $\deg(g_i) = a_i$ such that $P_{\geq m} \subseteq (g_1, \ldots, g_n)$.*

(1) *$R_{m+d} \subseteq (f_1, \ldots, f_n)^*$ holds over the generic point of the parameter space (after the base change to the function field of this space) of homogeneous elements $f_1, \ldots, f_n$ in R of this degree type (the coefficients of the f_i are taken as indeterminates).*

(2) *If R is normal, then $R_{m+d+1} \subseteq (f_1, \ldots, f_n)^F \subseteq (f_1, \ldots, f_n)^*$ holds for (open) generic choice of homogeneous elements $f_1, \ldots, f_n$ in R of this degree type.*

Example 5.9. Suppose that we are in $K[x, y, z]$ and that $n = 4$ and $a = 10$. Then the generic degree bound for ideal inclusion in the polynomial ring is 19. Therefore by Theorem 5.8 the generic degree bound for tight closure inclusion in a three-dimensional graded ring is 21.

Example 5.10. Suppose that $n = d + 1$ in the situation of Theorem 5.8. Then the generic elements $f_1, \ldots, f_{d+1}$ are parameters. In the polynomial ring $P = K[x_0, x_1, \ldots, x_d]$ we have for parameters of degree $a_1, \ldots, a_{d+1}$ the inclusion

$$P_{\geq \sum_{i=0}^{d} a_i - d} \subseteq (f_1, \ldots, f_{d+1}),$$

because the graded Koszul resolution ends in $R(-\sum_{i=0}^{d} a_i)$ and

$$(H_{\mathfrak{m}}^{d+1}(P))_k = 0 \quad \text{for } k \geq -d.$$

So the theorem implies for a graded ring R finite over P that

$$P_{\geq \sum_{i=0}^{d} a_i} \subseteq (f_1, \ldots, f_{d+1})^*$$

holds for generic elements. But by the graded Briançon–Skoda theorem [Huneke 1998] this holds for parameters even without the generic assumption.

References

[Anick 1986] D. J. Anick, "Thin algebras of embedding dimension three", *J. Algebra* **100**:1 (1986), 235–259.

[Artin 1971] M. Artin, "On the joins of Hensel rings", *Advances in Math.* **7** (1971), 282–296.

[Brenner 2002] H. Brenner, "On superheight conditions for the affineness of open subsets", *J. Algebra* **247**:1 (2002), 37–56.

[Brenner 2003a] H. Brenner, "How to rescue solid closure", *J. Algebra* **265**:2 (2003), 579–605.

[Brenner 2003b] H. Brenner, *The theory of tight closure from the viewpoint of vector bundles*, Habilitationsschrift, Ruhr-Universität Bochum, 2003. arXiv math/0307161

[Brenner 2003c] H. Brenner, "Tight closure and projective bundles", *J. Algebra* **265**:1 (2003), 45–78.

[Brenner 2004] H. Brenner, "Slopes of vector bundles on projective curves and applications to tight closure problems", *Trans. Amer. Math. Soc.* **356**:1 (2004), 371–392.

[Brenner 2006a] H. Brenner, "Continuous solutions to algebraic forcing equations", preprint, 2006. arXiv math/0608611

[Brenner 2006b] H. Brenner, "The rationality of the Hilbert–Kunz multiplicity in graded dimension two", *Math. Ann.* **334**:1 (2006), 91–110.

[Brenner 2006c] H. Brenner, "Tight closure and plus closure in dimension two", *Amer. J. Math.* **128**:2 (2006), 531–539.

[Brenner 2007] H. Brenner, "The Hilbert–Kunz function in graded dimension two", *Comm. Algebra* **35**:10 (2007), 3199–3213.

[Brenner 2008] H. Brenner, "Tight closure and vector bundles", pp. 1–71 in *Three lectures on commutative algebra*, edited by G. Colomé-Nin et al., Univ. Lecture Ser. **42**, Amer. Math. Soc., Providence, RI, 2008.

[Brenner and Fischbacher-Weitz 2011] H. Brenner and H. Fischbacher-Weitz, "Generic bounds for Frobenius closure and tight closure", *Amer. J. Math.* **133**:4 (2011), 889–912.

[Brenner and Katzman 2006] H. Brenner and M. Katzman, "On the arithmetic of tight closure", *J. Amer. Math. Soc.* **19**:3 (2006), 659–672.

[Brenner and Monsky 2010] H. Brenner and P. Monsky, "Tight closure does not commute with localization", *Ann. of Math.* (2) **171**:1 (2010), 571–588.

[Bruns and Herzog 1993] W. Bruns and J. Herzog, *Cohen–Macaulay rings*, Cambridge Studies in Advanced Mathematics **39**, Cambridge University Press, 1993.

[Dietz 2006] G. D. Dietz, "Tight closure of finite length modules in graded rings", *J. Algebra* **306**:2 (2006), 520–534.

[EGA I 1960] A. Grothendieck, "Eléments de géométrie algébrique, I: Le langage des schémas", *Inst. Hautes Études Sci. Publ. Math.* **4** (1960), 5–228.

[Epstein and Hochster 2011] N. Epstein and M. Hochster, "Continuous closure, axes closure, and natural closure", preprint, 2011. arXiv 1106.3462

[Gieseker 1971] D. Gieseker, "p-ample bundles and their Chern classes", *Nagoya Math. J.* **43** (1971), 91–116.

[Harder and Narasimhan 1975] G. Harder and M. S. Narasimhan, "On the cohomology groups of moduli spaces of vector bundles on curves", *Math. Ann.* **212** (1975), 215–248.

[Hartshorne 1971] R. Hartshorne, "Ample vector bundles on curves", *Nagoya Math. J.* **43** (1971), 73–89.

[Hartshorne 1977] R. Hartshorne, *Algebraic geometry*, Graduate Texts in Mathematics **52**, Springer, New York, 1977.

[Hochster 1994] M. Hochster, "Solid closure", pp. 103–172 in *Commutative algebra: Syzygies, multiplicities, and birational algebra* (South Hadley, MA, 1992), edited by W. J. Heinzer et al., Contemp. Math. **159**, Amer. Math. Soc., Providence, RI, 1994.

[Hochster 1996] M. Hochster, "The notion of tight closure in equal characteristic zero", 1996. Appendix 1 to [Huneke 1996].

[Hochster and Huneke 1990] M. Hochster and C. Huneke, "Tight closure, invariant theory, and the Briançon–Skoda theorem", *J. Amer. Math. Soc.* **3**:1 (1990), 31–116.

[Huneke 1996] C. Huneke, *Tight closure and its applications*, CBMS Regional Conference Series in Mathematics **88**, Amer. Math. Soc., Providence, RI, 1996.

[Huneke 1998] C. Huneke, "Tight closure, parameter ideals, and geometry", pp. 187–239 in *Six lectures on commutative algebra* (Bellaterra, 1996), edited by J. Elias et al., Progr. Math. **166**, Birkhäuser, Basel, 1998.

[Huneke and Swanson 2006] C. Huneke and I. Swanson, *Integral closure of ideals, rings, and modules*, London Mathematical Society Lecture Note Series **336**, Cambridge University Press, 2006.

[Kollár 2012] J. Kollár, "Continuous closure of sheaves", *Michigan Math. J.* **61**:3 (2012), 475–491.

[Kunz 1969] E. Kunz, "Characterizations of regular local rings for characteristic p", *Amer. J. Math.* **91** (1969), 772–784.

[Lange and Stuhler 1977] H. Lange and U. Stuhler, "Vektorbündel auf Kurven und Darstellungen der algebraischen Fundamentalgruppe", *Math. Z.* **156**:1 (1977), 73–83.

[Langer 2004] A. Langer, "Semistable sheaves in positive characteristic", *Ann. of Math.* (2) **159**:1 (2004), 251–276.

[Monsky 1998] P. Monsky, "Hilbert–Kunz functions in a family: Point-S_4 quartics", *J. Algebra* **208**:1 (1998), 343–358.

[Roberts 1994] P. C. Roberts, "A computation of local cohomology", pp. 351–356 in *Commutative algebra: Syzygies, multiplicities, and birational algebra* (South Hadley, MA, 1992), edited by W. J. Heinzer et al., Contemp. Math. **159**, Amer. Math. Soc., Providence, RI, 1994.

[SGA 1 1971] A. Grothendieck, *Revêtements étales et groupe fondamental* (Séminaire de Géométrie Algébrique du Bois Marie 1960–1961), Lecture Notes in Math. **224**, Springer, Berlin, 1971.

[Singh 1998] A. K. Singh, "A computation of tight closure in diagonal hypersurfaces", *J. Algebra* **203**:2 (1998), 579–589.

[Smith 1994] K. E. Smith, "Tight closure of parameter ideals", *Invent. Math.* **115**:1 (1994), 41–60.

[Trivedi 2005] V. Trivedi, "Semistability and Hilbert–Kunz multiplicities for curves", *J. Algebra* **284**:2 (2005), 627–644.

hbrenner@uni-osnabrueck.de *Institut für Mathematik, Universität Osnabrück, Postfach 44 69, 49069 Osnabrück, Germany*

Commutative Algebra and Noncommutative Algebraic Geometry, I
MSRI Publications
Volume **67**, 2015

Hecke algebras and symplectic reflection algebras

MARIA CHLOUVERAKI

The current article is a short survey on the theory of Hecke algebras, and in particular Kazhdan–Lusztig theory, and on the theory of symplectic reflection algebras, and in particular rational Cherednik algebras. The emphasis is on the connections between Hecke algebras and rational Cherednik algebras that could allow us to obtain a generalised Kazhdan–Lusztig theory, or at least its applications, for all complex reflection groups.

1. Introduction

Finite Coxeter groups are finite groups of real matrices that are generated by reflections. They include the Weyl groups, which are fundamental in the classification of simple complex Lie algebras as well as simple algebraic groups. Iwahori–Hecke algebras associated to Weyl groups appear naturally as endomorphism algebras of induced representations in the study of finite reductive groups. They can also be defined independently as deformations of group algebras of finite Coxeter groups, where the deformation depends on an indeterminate q and a weight function L. For $q = 1$, we recover the group algebra. For a finite Coxeter group W, we will denote by $\mathcal{H}(W, L)$ the associated Iwahori–Hecke algebra.

When q is an indeterminate, the Iwahori–Hecke algebra $\mathcal{H}(W, L)$ is semisimple. By Tits's deformation theorem, there exists a bijection between the set of irreducible representations of $\mathcal{H}(W, L)$ and the set $\mathrm{Irr}(W)$ of irreducible representations of W. Using this bijection, Lusztig attaches to every irreducible representation of W an integer depending on L, thus defining the famous *a-function*. The a-function is used in his definition of *families of characters*, a partition of $\mathrm{Irr}(W)$ which plays a key role in the organisation of families of unipotent characters in the case of finite reductive groups.

Kazhdan–Lusztig theory is a key to understanding the representation theory of the Iwahori–Hecke algebra $\mathcal{H}(W, L)$. There exists a special basis of $\mathcal{H}(W, L)$, called the *Kazhdan–Lusztig basis*, which allows us to define the *Kazhdan–Lusztig cells* for $\mathcal{H}(W, L)$, a certain set of equivalence classes on W. The construction of Kazhdan–Lusztig cells yields the construction of representations for $\mathcal{H}(W, L)$. It also gives another, more combinatorial, definition for Lusztig's families of characters.

Now, when q specialises to a nonzero complex number η, and more specifically to a root of unity, the specialised Iwahori–Hecke algebra $\mathcal{H}_\eta(W, L)$ is not necessarily semisimple and we no longer have a bijection between its irreducible representations and $\mathrm{Irr}(W)$. We obtain then a decomposition matrix which records how the irreducible representations of the semisimple algebra split after the specialisation. A *canonical basic set* is a subset of $\mathrm{Irr}(W)$ in bijection with the irreducible representations of $\mathcal{H}_\eta(W, L)$ (and thus a labelling set for the columns of the decomposition matrix) with good properties. Its good properties ensure that the decomposition matrix has a lower unitriangular form while the a-function increases (roughly) down the columns. Canonical basic sets were defined by Geck and Rouquier [2001], who also proved their existence in certain cases with the use of Kazhdan–Lusztig theory. Thanks to the work of many people, canonical basic sets are now proved to exist and explicitly described for all finite Coxeter groups and for any choice of L.

Finite Coxeter groups are particular cases of complex reflection groups, that is, finite groups of complex matrices generated by "pseudoreflections". Their classification is due to Shephard and Todd [1954]: An irreducible complex reflection group either belongs to the infinite series $G(\ell, p, n)$ or is one of the 34 exceptional groups $G_4, \ldots, G_{37}$ (see Theorem 3.1). Important work in the last two decades has suggested that complex reflection groups will play a crucial, but not yet understood role in representation theory, and may even become as ubiquitous in the study of other mathematical structures. In fact, they behave so much like real reflection groups that Broué, Malle and Michel [Broué et al. 1999] conjectured that they could play the role of Weyl groups for, as yet mysterious, objects generalising finite reductive groups. These objects are called *spetses*.

Broué et al. [1998] defined Hecke algebras for complex reflection groups as deformations of their group algebras. A generalised Kazhdan–Lusztig cell theory for these algebras, known as *cyclotomic Hecke algebras*, is expected to help find spetses. Unfortunately, we do not have a Kazhdan–Lusztig basis for complex reflection groups. However, we can define families of characters using Rouquier's definition: Rouquier [1999] gave an alternative definition for Lusztig's families of characters by proving that, in the case of Weyl groups, they coincide with the blocks of the Iwahori–Hecke algebra over a certain ring, called the *Rouquier ring*. This definition generalises without problem to the case of complex reflection groups and their cyclotomic Hecke algebras, producing the so-called *Rouquier families*. These families have now been determined for all cyclotomic Hecke algebras of all complex reflection groups; see [Chlouveraki 2009].

We also have an a-function and can define canonical basic sets for cyclotomic Hecke algebras. Although there is no Kazhdan–Lusztig theory in the complex case, canonical basic sets are now known to exist for the groups of the infinite series $G(\ell, p, n)$ and for some exceptional ones. In order to obtain canonical basic sets for $G(\ell, 1, n)$, Geck and Jacon used Ariki's Theorem on the categorification of Hecke algebra representations and Uglov's work on canonical bases for higher level Fock spaces [Geck and Jacon 2006; Jacon 2004; 2007; Geck and Jacon 2011]. The result for $G(\ell, p, n)$ derives from that for $G(\ell, 1, n)$ with the use of Clifford theory [Genet and Jacon 2006; Chlouveraki and Jacon 2012].

In this paper we will see how we could use the representation theory of symplectic reflection algebras, and in particular rational Cherednik algebras, to obtain families of characters and canonical basic sets for cyclotomic Hecke algebras associated with complex reflection groups.

Symplectic reflection algebras are related to a large number of areas of mathematics such as combinatorics, integrable systems, real algebraic geometry, quiver varieties, symplectic resolutions of singularities and, of course, representation theory. They were introduced by Etingof and Ginzburg [2002] for the study of symplectic resolutions of the orbit space V/G, where V is a symplectic complex vector space and $G \subset \mathrm{Sp}(V)$ is a finite group acting on V. Verbitsky [2000] has shown that V/G admits a symplectic resolution only if (G, V) is a symplectic reflection group, that is, G is generated by symplectic reflections. Thanks to the insight by Etingof and Ginzburg, the study of the representation theory of symplectic reflection algebras has led to the (almost) complete classification of symplectic reflection groups (G, V) such that V/G admits a symplectic resolution.

Let (G, V) be a symplectic reflection group, and let TV^* denote the tensor algebra on the dual space V^* of V. The symplectic reflection algebra $\boldsymbol{H}_{t,\boldsymbol{c}}(G)$ associated to (G, V) is defined as the quotient of $TV^* \rtimes G$ by certain relations depending on a complex function $\boldsymbol{c}$ and a parameter t. The representation theory of $\boldsymbol{H}_{t,\boldsymbol{c}}(G)$ varies a lot according to whether t is zero or not. A complex reflection group $W \subset \mathrm{GL}(\mathfrak{h})$, where $\mathfrak{h}$ is a complex vector space, can be seen as a symplectic reflection group acting on $V = \mathfrak{h} \oplus \mathfrak{h}^*$. Symplectic reflection algebras associated with complex reflection groups are known as *rational Cherednik algebras*.

If $t \neq 0$, there exists an important category of representations of the rational Cherednik algebra, the *category* $\mathcal{O}$, and an exact functor, the KZ-*functor*, from $\mathcal{O}$ to the category of representations of a certain specialised cyclotomic Hecke algebra $\mathcal{H}_\eta(W)$ (the specialisation depends on the choice of parameters for the rational Cherednik algebra — every specialised Hecke algebra can arise this way). Category $\mathcal{O}$ is a highest weight category, and it comes equipped with a set of standard modules $\{\Delta(E) \mid E \in \mathrm{Irr}(W)\}$, a set of simple modules $\{\mathrm{L}(E) \mid E \in \mathrm{Irr}(W)\}$ and a decomposition matrix that records the number of times that $\mathrm{L}(E)$ appears in the composition series of $\Delta(E')$ for $E, E' \in \mathrm{Irr}(W)$. The exactness of KZ allows us to read off the decomposition matrix of $\mathcal{H}_\eta(W)$ from the decomposition matrix of category $\mathcal{O}$. Using this, we proved in [Chlouveraki et al. 2012] the existence of canonical basic sets for all finite Coxeter groups and for complex reflection groups of type $G(\ell, 1, n)$. In particular, we showed that E belongs to the canonical basic set for $\mathcal{H}_\eta(W)$ if and only if $\mathrm{KZ}(\mathrm{L}(E)) \neq 0$. Our proof of existence is quite general and it does not make use of Ariki's Theorem for type $G(\ell, 1, n)$. However, the explicit description of canonical basic sets in these cases by previous works answers simultaneously the question of which simple modules are killed by the KZ-functor; this appears to be new. We also proved that the images of the standard modules via the KZ-functor are isomorphic to the cell modules of Hecke algebras with cellular structure, but we will not go into that in this paper.

The case $t = 0$ yields the desired criterion for the space V/W to admit a symplectic resolution. It is a beautiful result due to Ginzburg and Kaledin [2004] and Namikawa [2011] that V/W admits a symplectic resolution if and only if the spectrum of the centre of $\boldsymbol{H}_{0,\boldsymbol{c}}(W)$ is smooth for generic $\boldsymbol{c}$. The space $X_{\boldsymbol{c}}(W) := \mathrm{Spec}(Z(\boldsymbol{H}_{0,\boldsymbol{c}}(W)))$ is called *generalised Calogero–Moser space*. Gordon [2003] introduced and studied extensively a finite-dimensional quotient of $\boldsymbol{H}_{0,\boldsymbol{c}}(W)$, called the *restricted rational Cherednik algebra*, whose simple modules are parametrised by $\mathrm{Irr}(W)$. The decomposition of this algebra into blocks induces a partition of $\mathrm{Irr}(W)$, known as *Calogero–Moser partition*. We have that $X_{\boldsymbol{c}}(W)$ is smooth if and only if the Calogero–Moser partition is trivial for all parabolic subgroups of W. Following the classification of irreducible complex reflection groups, and the works of Etingof and Ginzburg [2002], Gordon [2003] and Gordon and Martino [2009], Bellamy [2009] was able to prove that V/W admits a symplectic resolution if and only if $W = G(\ell, 1, n)$ or $W = G_4$.

A connection is conjectured between the Calogero–Moser partition and the families of characters, first suggested by Gordon and Martino [2009] for type B_n. In every case studied so far, the partition into Rouquier families (for a suitably chosen cyclotomic Hecke algebra) refines the Calogero–Moser partition

("Martino's conjecture"), while for finite Coxeter groups the two partitions coincide. The reasons for this connection are still unknown, since there is no apparent connection between Hecke algebras and rational Cherednik algebras at $t=0$. Inspired by this, and in an effort to construct a generalised Kazhdan–Lusztig cell theory, Bonnafé and Rouquier have used the Calogero–Moser partition to develop a "Calogero–Moser cell theory" which can be applied to all complex reflection groups [Bonnafé and Rouquier 2013]. The fruits of this very recent approach remain to be seen.

1A. ***Piece of notation and definition of blocks.*** Let R be a commutative integral domain and let F be the field of fractions of R. Let A be an R-algebra, free and finitely generated as an R-module. If R' is a commutative integral domain containing R, we will write $R'A$ for $R' \otimes_R A$.

Let now K be a field containing F such that the algebra KA is semisimple. The primitive idempotents of the centre $Z(KA)$ of KA are in bijection with the irreducible representations of KA. Let $\mathrm{Irr}(KA)$ denote the set of irreducible representations of KA. For $\chi \in \mathrm{Irr}(KA)$, let e_χ be the corresponding primitive idempotent of $Z(KA)$. There exists a unique partition $\mathrm{Bl}(A)$ of $\mathrm{Irr}(KA)$ that is the finest with respect to the property:

$$\sum_{\chi\in B} e_\chi \in A \quad \text{for all } B \in \mathrm{Bl}(A).$$

The elements $e_B := \sum_{\chi\in B} e_\chi$, for $B \in \mathrm{Bl}(A)$, are the primitive idempotents of $Z(A)$. We have

$$A \cong \prod_{B\in\mathrm{Bl}(A)} Ae_B.$$

The parts of $\mathrm{Bl}(A)$ are the *blocks* of A.

2. Iwahori–Hecke algebras

In this section we will focus on real reflection groups, while in the next section we will see what happens in the complex case.

2A. ***Kazhdan–Lusztig cells.*** Let (W, S) be a finite Coxeter system. By definition, W has a presentation of the form

$$W = \langle\, S \mid (st)^{m_{st}} = 1 \text{ for all } s, t \in S \,\rangle,$$

with $m_{ss} = 1$ and $m_{st} \geq 2$ for $s \neq t$. We have a *length function* $\ell : W \to \mathbb{Z}_{\geq 0}$ defined by $\ell(w) := \min\{r \mid w = s_{i_1} \dots s_{i_r} \text{ with } s_{i_j} \in S\}$ for all $w \in W$.

Let $L : W \to \mathbb{Z}_{\geq 0}$ be a weight function, that is, a map such that $L(ww') = L(w) + L(w')$ whenever $\ell(ww') = \ell(w) + \ell(w')$. For $s,\ t \in S$, we have $L(s) = L(t)$ whenever s and t are conjugate in W. Let q be an indeterminate. We define the *Iwahori–Hecke algebra* of W with parameter L , denoted by $\mathcal{H}(W, L)$, to be the $\mathbb{Z}[q, q^{-1}]$-algebra generated by elements $(T_s)_{s\in S}$ satisfying the relations:

$$(T_s - q^{L(s)})(T_s + q^{-L(s)}) = 0 \quad \text{and} \quad \underbrace{T_s T_t T_s T_t \ldots}_{m_{st}} = \underbrace{T_t T_s T_t T_s \ldots}_{m_{st}} \qquad \text{for } s \neq t.$$

If $L(s) = L(t)$ for all $s, t \in S$, we say that we are in the *equal parameter case*. Since L is a weight function, unequal parameters can only occur in irreducible types B_n, F_4 and dihedral groups $I_2(m)$ for m even.

Example 2.1. Let $W = \mathfrak{S}_3$. We have $W = \langle s, t \mid s^2 = t^2 = (st)^3 = 1\rangle$. Let $l := L(s) = L(t) \in \mathbb{Z}_{\geq 0}$. We have

$$\mathcal{H}(W, l) = \big\langle T_s, T_t \mid T_s T_t T_s = T_t T_s T_t,\ (T_s - q^l)(T_s + q^{-l}) = (T_t - q^l)(T_t + q^{-l}) = 0\big\rangle.$$

Let $w \in W$ and let $w = s_{i_1} \ldots s_{i_r}$ be a reduced expression for w, that is, $r = \ell(w)$. Set $T_w := T_{s_{i_1}} \ldots T_{s_{i_r}}$. As a $\mathbb{Z}[q, q^{-1}]$-module, $\mathcal{H}(W, L)$ is generated by the elements $(T_w)_{w\in W}$ satisfying the following multiplication formulas:

$$\begin{cases} T_s^2 = 1 + (q^{L(s)} - q^{-L(s)})\, T_s & \text{for } s \in S, \\ T_w T_{w'} = T_{ww'} & \text{if } \ell(ww') = \ell(w) + \ell(w'). \end{cases}$$

The elements $(T_w)_{w\in W}$ form a basis of $\mathcal{H}(W, L)$, the *standard basis*.

Let i be the algebra involution on $\mathcal{H}(W, L)$ given by $i(q) = q^{-1}$ and $i(T_s) = T_s^{-1}$ for $s \in S$ (as a consequence, $i(T_w) = T_{w^{-1}}^{-1}$ for all $w \in W$). By [Kazhdan and Lusztig 1979, Theorem 1.1] (see [Lusztig 1983, Proposition 2] for the unequal parameter case), for each $w \in W$, there exists an element $C_w \in \mathcal{H}(W, L)$ uniquely determined by the conditions

$$i(C_w) = C_w \quad \text{and} \quad \text{and } i(C_w) = T_w + \sum_{x\in W, x<w} P_{x,w}\, T_x,$$

where $<$ stands for the Chevalley–Bruhat order on W and $P_{x,w} \in q^{-1}\mathbb{Z}[q^{-1}]$. The elements $(C_w)_{w\in W}$ also form a basis of $\mathcal{H}(W, L)$, the *Kazhdan–Lusztig basis*.

Example 2.2. We have $C_1 = T_1 = 1$ and, for all $s \in S$,

$$C_s = \begin{cases} T_s & \text{if } L(s) = 0, \\ T_s + q^{-L(s)} T_1 & \text{if } L(s) > 0. \end{cases}$$

Using the Kazhdan–Lusztig basis, we can now define the following three preorders on W. For $x,\ y \in W$, we have

- $x \leq_{\mathcal{L}} y$ if C_x appears with nonzero coefficient in hC_y for some $h \in \mathcal{H}(W, L)$.
- $x \leq_{\mathcal{R}} y$ if C_x appears with nonzero coefficient in $C_y h'$ for some $h' \in \mathcal{H}(W, L)$.

- $x \leq_{\mathcal{LR}} y$ if C_x appears with nonzero coefficient in $hC_y h'$ for some $h, h' \in \mathcal{H}(W, L)$.

The preorder $\leq_{\mathcal{L}}$ defines an equivalence relation $\sim_{\mathcal{L}}$ on W as follows:

$$x \sim_{\mathcal{L}} y \iff x \leq_{\mathcal{L}} y \text{ and } y \leq_{\mathcal{L}} x.$$

The equivalence classes for $\sim_{\mathcal{L}}$ are called *left cells*. Similarly, one can define equivalence relations $\sim_{\mathcal{R}}$ and $\sim_{\mathcal{LR}}$ on W, whose equivalence classes are called, respectively, *right cells* and *two-sided cells*.

Example 2.3. For $W = \mathfrak{S}_3 = \{1, s, t, st, ts, sts = tst\}$ and $l > 0$,

- the left cells are $\{1\}$, $\{s, ts\}, \{t, st\}$ and $\{sts\}$;
- the right cells are $\{1\}$, $\{s, st\}, \{t, ts\}$ and $\{sts\}$;
- the two-sided cells are $\{1\}$, $\{s, t, st, ts\}$ and $\{sts\}$.

If $l = 0$, then all elements of W belong to the same cell (left, right or two-sided).

Let now $\mathfrak{C}$ be a left cell of W. The following two $\mathbb{Z}[q, q^{-1}]$-modules are left ideals of $\mathcal{H}(W, L)$:

$$\mathcal{H}_{\leq_{\mathcal{L}}\mathfrak{C}} = \langle C_y \mid y \leq_{\mathcal{L}} w,\, w \in \mathfrak{C}\rangle_{\mathbb{Z}[q,q^{-1}]},$$
$$\mathcal{H}_{<_{\mathcal{L}}\mathfrak{C}} = \langle C_y \mid y \leq_{\mathcal{L}} w,\, w \in \mathfrak{C},\, y \notin \mathfrak{C}\rangle_{\mathbb{Z}[q,q^{-1}]}.$$

Then

$$\mathcal{M}_{\mathfrak{C}} := \mathcal{H}_{\leq_{\mathcal{L}}\mathfrak{C}}/\mathcal{H}_{<_{\mathcal{L}}\mathfrak{C}}$$

is a free left $\mathcal{H}(W, L)$-module with basis indexed by the elements of $\mathfrak{C}$.

Let K be a field containing $\mathbb{Z}[q, q^{-1}]$ such that the algebra $K\mathcal{H}(W, L)$ is split semisimple (for example, take $K = \mathbb{C}(q)$). Then, since the left cells form a partition of W, we obtain a corresponding direct sum decomposition of $K\mathcal{H}(W, L)$:

$$K\mathcal{H}(W, L) \cong \bigoplus_{\mathfrak{C} \text{ left cell}} K\mathcal{M}_{\mathfrak{C}} \quad \text{(isomorphism of left } K\mathcal{H}(W, L)\text{-modules)}, \tag{2.4}$$

where $K\mathcal{M}_{\mathfrak{C}} := K \otimes_{\mathbb{Z}[q,q^{-1}]} \mathcal{M}_{\mathfrak{C}}$. We obtain analogous decompositions with respect to right and two-sided cells.

2B. ***Schur elements and the a-function.*** From now on, set $R := \mathbb{Z}[q, q^{-1}]$ and let K be a field containing R such that the algebra $K\mathcal{H}(W, L)$ is split semisimple.

Using the standard basis of the Iwahori–Hecke algebra, we define the linear map $\tau : \mathcal{H}(W, L) \to R$ by setting

$$\tau(T_w) := \begin{cases} 1 & \text{if } w = 1, \\ 0 & \text{otherwise.} \end{cases}$$

The map τ is a symmetrising trace on $\mathcal{H}(W, L)$, that is,

(a) $\tau(hh') = \tau(h'h)$ for all $h, h' \in \mathcal{H}(W, L)$, and

(b) the map $\widehat{\tau} : \mathcal{H}(W, L) \to \mathrm{Hom}_R(\mathcal{H}(W, L), R)$, $h \mapsto (x \mapsto \tau(hx))$ is an isomorphism of $\mathcal{H}(W, L)$-bimodules.

Moreover, the elements $(T_{w^{-1}})_{w\in W}$ form a basis of $\mathcal{H}(W, L)$ dual to the standard basis with respect to τ (that is, $\tau(T_{w^{-1}}T_{w'}) = \delta_{w,w'}$) [Geck and Pfeiffer 2000, Proposition 8.1.1]. The map τ is called the *canonical symmetrising trace* on $\mathcal{H}(W, L)$, because it specialises to the canonical symmetrising trace on the group algebra $\mathbb{Z}[W]$ when $q \mapsto 1$.

Now, the map τ can be extended to $K\mathcal{H}(W, L)$ by extension of scalars. By Tits's deformation theorem (see, for example, [Geck and Pfeiffer 2000, Theorem 7.4.6]), the specialisation $q \mapsto 1$ induces a bijection between the set of irreducible representations $\mathrm{Irr}(K\mathcal{H}(W, L))$ of $K\mathcal{H}(W, L)$ and the set of irreducible representations $\mathrm{Irr}(W)$ of W. For $E \in \mathrm{Irr}(W)$, let χ_E be the corresponding irreducible character of $K\mathcal{H}(W, L)$ and let ω_{χ_E} be the corresponding central character. We define

$$s_E := \chi_E(\hat{\tau}^{-1}(\chi_E))/\chi_E(1) = \omega_{\chi_E}(\hat{\tau}^{-1}(\chi_E))$$

to be the *Schur element* of $\mathcal{H}(W, L)$ associated with E. Geck has shown (see [Geck and Pfeiffer 2000, Proposition 7.3.9]) that $s_E \in \mathbb{Z}_K[q, q^{-1}]$ for all $E \in \mathrm{Irr}(W)$, where $\mathbb{Z}_K$ denotes the integral closure of $\mathbb{Z}$ in K. We have

$$\tau = \sum_{E\in\mathrm{Irr}(W)} \frac{1}{s_E}\chi_E \tag{2.5}$$

and

$$e_E = \frac{1}{s_E} \sum_{w\in W} \chi_E(T_w)\, T_{w^{-1}}, \tag{2.6}$$

where e_E is the primitive central idempotent of $K\mathcal{H}(W, L)$ corresponding to E. Both results are due to Curtis and Reiner [1962], but we follow the exposition in [Geck and Pfeiffer 2000, Theorem 7.2.6] and [loc. cit., Proposition 7.2.7], respectively.

Example 2.7. In the group algebra case ($L(s) = 0$ for all $s \in S$), we have $s_E = |W|/\chi_E(1)$ for all $E \in \mathrm{Irr}(W)$.

Example 2.8. The irreducible representations of the symmetric group $\mathfrak{S}_n$ are parametrised by the partitions of n. For $W = \mathfrak{S}_3$, there are three irreducible representations. Let $E^{(3)}$, $E^{(2,1)}$ and $E^{(1,1,1)}$ denote respectively the trivial, reflection and sign representation of $\mathfrak{S}_3$. We have

$$s_{E^{(3)}} = (q^{2l}+1)(q^{4l}+q^{2l}+1), \quad s_{E^{(2,1)}} = q^{2l}+1+q^{-2l},$$
$$s_{E^{(1,1,1)}} = (q^{-2l}+1)(q^{-4l}+q^{-2l}+1).$$

We can define the functions $a : \mathrm{Irr}(W) \to \mathbb{Z}$ and $A : \mathrm{Irr}(W) \to \mathbb{Z}$ by setting

$$a(E) := -\mathrm{val}_q(s_E) \quad \text{and} \quad A(E) := -\deg_q(s_E).$$

Note that both functions depend on L. For brevity, we will write a_E for $a(E)$ and A_E for $A(E)$.

Example 2.9. For $W = \mathfrak{S}_3$, we have

$$a_{E^{(3)}} = 0, \qquad a_{E^{(2,1)}} = 2l, \qquad a_{E^{(1,1,1)}} = 6l,$$
$$A_{E^{(3)}} = -6l, \quad A_{E^{(2,1)}} = -2l, \quad A_{E^{(1,1,1)}} = 0.$$

The Schur elements of $\mathcal{H}(W, L)$ have been explicitly calculated for all finite Coxeter groups:

- for type A_n by Steinberg [1951],
- for type B_n by Hoefsmit [1974],
- for type D_n by Benson and Gay [1977] (it derives from type B_n with the use of Clifford theory),
- for dihedral groups $I_2(m)$ by Kilmoyer and Solomon [1973],
- for F_4 by Lusztig [1979],
- for E_6 and E_7 by Surowski [1978],
- for E_8 by Benson [1979],
- for H_3 by Lusztig [1982],
- for H_4 by Alvis and Lusztig [1982].

There have been other subsequent proofs of the above results. For example, Iwahori–Hecke algebras of types A_n and B_n are special cases of Ariki–Koike algebras, whose Schur elements have been independently obtained by Geck, Iancu and Malle [2000] and Mathas [2004].

A case-by-case analysis shows that the Schur elements of $\mathcal{H}(W, L)$ can be written in the form

$$s_E = \xi_E\, q^{-a_E} \prod_{\Phi \in \mathrm{Cyc}_E} \Phi(q^{n_{E,\Phi}}), \tag{2.10}$$

where $\xi_E \in \mathbb{Z}_K$, $n_{E,\Phi} \in \mathbb{Z}_{>0}$ and Cyc_E is a family of K-cyclotomic polynomials (see [Geck and Pfeiffer 2000, Chapters 10 and 11; Chlouveraki 2009, Theorem 4.2.5]).

Example 2.11. For $W = \mathfrak{S}_3$, we have

$$s_{E^{(3)}} = \Phi_2(q^{2l})\Phi_3(q^{2l}), \quad s_{E^{(2,1)}} = q^{-2l}\Phi_3(q^{2l}),$$
$$s_{E^{(1,1,1)}} = q^{-6l}\Phi_2(q^{2l})\Phi_3(q^{2l}).$$

2C. ***Families of characters and Rouquier families.*** The *families of characters* are a special partition of the set of irreducible representations of W. In the case where W is a Weyl group, these families play an essential role in the definition of the families of unipotent characters for the corresponding finite reductive groups. Their original definition is due to Lusztig [1984, 4.2] and uses the a-function.

Let $I \subseteq S$ and consider the parabolic subgroup $W_I \subseteq W$ generated by I. Then we have a corresponding parabolic subalgebra $\mathcal{H}(W_I, L) \subseteq \mathcal{H}(W, L)$. By extension of scalars from R to K, we also have a subalgebra $K\mathcal{H}(W_I, L) \subseteq K\mathcal{H}(W, L)$, and a corresponding a-function on the set of irreducible representations of W_I. Denote by Ind_I^S the induction of representations from W_I to W. Let $E \in \operatorname{Irr}(W)$ and $M \in \operatorname{Irr}(W_I)$. We will write $M \rightsquigarrow_L E$ if E is a constituent of $\operatorname{Ind}_I^S(M)$ and $a_E = a_M$.

Definition 2.12. The partition of $\operatorname{Irr}(W)$ into *families* is defined inductively as follows: when $W = \{1\}$, there is only one family; it consists of the unit representation of W. Assume now that $W \neq \{1\}$ and that the families have already been defined for all proper parabolic subgroups of W. Then $E, E' \in \operatorname{Irr}(W)$ are in the same family of W if there exists a finite sequence $E = E_0, E_1, \ldots, E_r = E'$ in $\operatorname{Irr}(W)$ such that, for each $i \in \{0, 1, \ldots, r-1\}$, the following condition is satisfied: There exist a subset $I_i \subsetneq S$ and $M_i, M_i' \in \operatorname{Irr}(W_{I_i})$ such that M_i, M_i' belong to the same family of W_{I_i} and either

$$M_i \rightsquigarrow_L E_i \quad \text{and} \quad M_i' \rightsquigarrow_L E_{i+1}$$

or

$$M_i \rightsquigarrow_L E_i \otimes \varepsilon \quad \text{and} \quad M_i' \rightsquigarrow_L E_{i+1} \otimes \varepsilon,$$

where ε denotes the sign representation of W. We will also refer to these families as *Lusztig families*.

Lusztig [1987, 3.3 and 3.4] has shown that the functions a and A are both constant on the families of characters, that is, if E and E' belong to the same family, then $a_E = a_{E'}$ and $A_E = A_{E'}$.

The decomposition of W into two-sided cells can be used to facilitate the description of the partition of $\operatorname{Irr}(W)$ into families of characters. As we saw in the previous subsection, Tits's deformation theorem yields a bijection between $\operatorname{Irr}(K\mathcal{H}(W, L))$ and $\operatorname{Irr}(W)$. Let $E \in \operatorname{Irr}(W)$ and let V^E be the corresponding simple module of $K\mathcal{H}(W, L)$. Following the direct sum decomposition given by (2.4), there exists a left cell $\mathfrak{C}$ such that V^E is a constituent of $\mathcal{M}_{\mathfrak{C}}$; furthermore, all such left cells are contained in the same two-sided cell. This two-sided cell, therefore, only depends on E and will be denoted by $\mathcal{F}_E$. Thus, we obtain a natural surjective map

$$\mathrm{Irr}(W) \to \{\text{set of two-sided cells of } W\}, \qquad E \mapsto \mathcal{F}_E$$

(see [Lusztig 1984, 5.15] for the equal parameter case; the same argument works in general).

Definition 2.13. Let $E, E' \in \mathrm{Irr}(W)$. We will say that E and E' belong to the same *Kazhdan–Lusztig family* if $\mathcal{F}_E = \mathcal{F}_{E'}$.

The following remarkable result, relating Lusztig families and Kazhdan–Lusztig families, has been proved by Barbasch–Vogan and Lusztig for finite Weyl groups in the equal parameter case [Lusztig 1984, 5.25]. It was subsequently proved [Lusztig 2003, 23.3; Geck 2005] to hold for any finite Coxeter group and any weight function L, assuming that Lusztig's conjectures P1–P15 [Lusztig 2003, 14.2] are satisfied.

Theorem 2.14. *Assume that Lusztig's conjectures* P1–P15 *hold. The Lusztig families and the Kazhdan–Lusztig families coincide.*

Lusztig's conjectures P1–P15 concern properties of the Kazhdan–Lusztig basis which should hold for any Coxeter group and in the general multiparameter case. For the moment, Conjectures P1–P15 have been proved in the following cases:

- Equal parameter case for finite Weyl groups [Kazhdan and Lusztig 1980; Lusztig 2003; Springer 1982].
- Equal parameter case for H_3, H_4 and dihedral groups $I_2(m)$ [Alvis 1987; du Cloux 2006].
- Unequal parameter case for F_4 and dihedral groups $I_2(m)$ [Geck 2004; Geck 2011].
- Asymptotic case and some other cases for B_n [Bonnafé and Iancu 2003; Bonnafé 2006; Bonnafé et al. 2010].

Moreover, these are exactly the cases where we have a description of the Kazhdan–Lusztig cells and Kazhdan–Lusztig families. A conjectural combinatorial description of the Kazhdan–Lusztig cells for type B_n is given by Bonnafé et al. [2010].

Example 2.15. The group $\mathfrak{S}_3$ has three irreducible representations. For $l > 0$, each irreducible representation forms a family on its own. This is true in general for the symmetric group $\mathfrak{S}_n$. For $l = 0$, all irreducible representations belong to the same family. This is true in general for the group algebra ($L(s)=0$ for all $s \in S$) of every finite Coxeter group.

Rouquier [1999] gave an alternative definition for Lusztig's families. He showed that, for finite Weyl groups in the equal parameter case, the families of characters coincide with the blocks of the Iwahori–Hecke algebra $\mathcal{H}(W, L)$ over the *Rouquier ring*

$$\mathcal{R}_K(q) := \mathbb{Z}_K[q, q^{-1}, (q^n - 1)^{-1}_{n \geq 1}],$$

that is, following (2.6), the nonempty subsets B of $\mathrm{Irr}(W)$ which are minimal with respect to the property:

$$\sum_{E \in B} \frac{\chi_E(h)}{s_E} \in \mathcal{R}_K(q) \qquad \text{for all } h \in \mathcal{H}(W, L).$$

These are the *Rouquier families* of $\mathcal{H}(W, L)$. One advantage of this definition, as we will see in the next section, is that it can be also applied to complex reflection groups. This is important in the "Spetses project" [Broué et al. 1999; 2014].

Following the determination of Rouquier families for all complex reflection groups (see Section 3C for references), and thus for all finite Coxeter groups, one can check that Rouquier's result holds for all finite Coxeter groups for all choices of parameters (by comparing the Rouquier families with the already known Lusztig families [Lusztig 1984; 2003]); that is, we have the following:

Theorem 2.16. *Let (W, S) be a finite Coxeter system and let $\mathcal{H}(W, L)$ be an Iwahori–Hecke algebra associated to W. The Lusztig families and the Rouquier families of $\mathcal{H}(W, L)$ coincide.*

2D. *Canonical basic sets.* As we saw in Section 2C, the specialisation $q \mapsto 1$ yields a bijection between the set of irreducible representations of $K\mathcal{H}(W, L)$ and $\mathrm{Irr}(W)$. What happens though when q specialises to a complex number? The resulting Iwahori–Hecke algebra is not necessarily semisimple and the first questions that need to be answered are the following: What are the simple modules for the newly obtained algebra? Is there a good way to parametrise them? What are their dimensions? One major approach to answering these questions is through the existence of "canonical basic sets".

Let $\theta : \mathbb{Z}_K[q, q^{-1}] \to K(\eta)$, $q \mapsto \eta$ be a ring homomorphism such that η is a nonzero complex number. Let us denote by $\mathcal{H}_\eta(W, L)$ the algebra obtained as a specialisation of $\mathcal{H}(W, L)$ via θ. Set $\mathbb{K} := K(\eta)$. We have the following semisimplicity criterion [Geck and Pfeiffer 2000, Theorem 7.4.7]:

Theorem 2.17. *The algebra $\mathbb{K}\mathcal{H}_\eta(W, L)$ is semisimple if and only $\theta(s_E) \neq 0$ for all $E \in \mathrm{Irr}(W)$.*

Following (2.10), $\mathbb{K}\mathcal{H}_\eta(W, L)$ is semisimple unless η is a root of unity.

Example 2.18. The algebra $\mathbb{Q}(\eta)\mathcal{H}_\eta(\mathfrak{S}_3, l)$ is semisimple if and only if $\eta^{2l} \notin \{-1, \omega, \omega^2\}$, where $\omega := \exp(2\pi i/3)$.

If $\mathbb{K}\mathcal{H}_\eta(W, L)$ is semisimple, then, by Tits's deformation theorem, the specialisation θ yields a bijection between $\mathrm{Irr}(K\mathcal{H}(W, L))$ and $\mathrm{Irr}(\mathbb{K}\mathcal{H}_\eta(W, L))$. Thus, the irreducible representations of $\mathbb{K}\mathcal{H}_\eta(W, L)$ are parametrised by $\mathrm{Irr}(W)$. Hence, we need to see what happens when $\mathbb{K}\mathcal{H}_\eta(W, L)$ is not semisimple.

Let $R_0(K\mathcal{H}(W, L))$ and $R_0(\mathbb{K}\mathcal{H}_\eta(W, L))$ be respectively the Grothendieck groups of finitely generated $K\mathcal{H}(W, L)$-modules and $\mathbb{K}\mathcal{H}_\eta(W, L)$-modules. The classes $[U]$, where U ranges over simple $K\mathcal{H}(W, L)$-modules (respectively $\mathbb{K}\mathcal{H}_\eta(W, L)$-modules), generate $R_0(K\mathcal{H}(W, L))$ (respectively $R_0(\mathbb{K}\mathcal{H}_\eta(W, L))$). We obtain a well-defined decomposition map

$$d_\theta : R_0(K\mathcal{H}(W, L)) \to R_0(\mathbb{K}\mathcal{H}_\eta(W, L)),$$

such that, for all $E \in \mathrm{Irr}(W)$, we have

$$d_\theta([V^E]) = \sum_{M \in \mathrm{Irr}(\mathbb{K}\mathcal{H}_\eta(W, L))} [V^E : M][M].$$

The matrix

$$D_\theta = ([V^E : M])_{E \in \mathrm{Irr}(W),\, M \in \mathrm{Irr}(\mathbb{K}\mathcal{H}_\eta(W, L))}$$

is called the *decomposition matrix with respect to* θ. If $\mathbb{K}\mathcal{H}_\eta(W, L)$ is semisimple, then D_θ is a permutation matrix.

Definition 2.19. A *canonical basic set* with respect to θ is a subset $\mathcal{B}_\theta$ of $\mathrm{Irr}(W)$ such that

(a) there exists a bijection $\mathrm{Irr}(\mathbb{K}\mathcal{H}_\eta(W, L)) \to \mathcal{B}_\theta$, $M \mapsto E_M$;

(b) $[V^{E_M} : M] = 1$ for all $M \in \mathrm{Irr}(\mathbb{K}\mathcal{H}_\eta(W, L))$;

(c) if $[V^E : M] \neq 0$ for some $E \in \mathrm{Irr}(W)$, $M \in \mathrm{Irr}(\mathbb{K}\mathcal{H}_\eta(W, L))$, then either $E = E_M$ or $a_{E_M} < a_E$.

If a canonical basic set exists, the decomposition matrix has a lower unitriangular form (with an appropriate ordering of the rows). Thus, we can obtain a lot of information about the simple modules of $\mathbb{K}\mathcal{H}_\eta(W, L)$ from what we already know about the simple modules of $K\mathcal{H}(W, L)$.

A general existence result for canonical basic sets is proved by Geck [2007b, Theorem 6.6], following his earlier work [1998], and that of Geck and Rouquier [2001] and Geck and Jacon [2006]. Another proof is given in [Geck and Jacon 2011]. In every case canonical basic sets are known explicitly, thanks to the work of many people. For a complete survey on the topic, we refer the reader to [Geck and Jacon 2011].

Example 2.20. Let W be the symmetric group $\mathfrak{S}_n$. Then W is generated by the transpositions $s_i = (i, i+1)$ for all $i = 1, \ldots, n-1$, which are all conjugate in W. Set $l := L(s_1)$ and let η^{2l} be a primitive root of unity of order $e > 1$. By [Dipper and James 1986, Theorem 7.6], we have that, in this case, the canonical basic set $\mathcal{B}_\theta$ is the set of e-regular partitions (a partition is e-regular if it does not have e nonzero equal parts). For example, for $n = 3$, we have $\mathcal{B}_\theta = \{E^{(3)}, E^{(2,1)}\}$ for $e \in \{2, 3\}$, and $\mathcal{B}_\theta = \mathrm{Irr}(\mathfrak{S}_3)$ for $e > 3$.

3. Cyclotomic Hecke algebras

Cyclotomic Hecke algebras generalise the notion of Iwahori–Hecke algebras to the case of complex reflection groups. For any positive integer e we will write ζ_e for $\exp(2\pi i/e) \in \mathbb{C}$.

3A. ***Hecke algebras for complex reflection groups.*** Let $\mathfrak{h}$ be a finite dimensional complex vector space. A *pseudoreflection* is a nontrivial element $s \in \mathrm{GL}(\mathfrak{h})$ that fixes a hyperplane pointwise, that is, $\dim_{\mathbb{C}}\mathrm{Ker}(s - \mathrm{id}_{\mathfrak{h}}) = \dim_{\mathbb{C}}\mathfrak{h} - 1$. The hyperplane $\mathrm{Ker}(s - \mathrm{id}_{\mathfrak{h}})$ is the *reflecting hyperplane* of s. A *complex reflection group* is a finite subgroup of $\mathrm{GL}(\mathfrak{h})$ generated by pseudoreflections. The classification of (irreducible) complex reflection groups is due to Shephard and Todd [1954]:

Theorem 3.1. *Let $W \subset \mathrm{GL}(\mathfrak{h})$ be an irreducible complex reflection group (i.e., W acts irreducibly on $\mathfrak{h}$). Then one of the following assertions is true:*

- *There exist positive integers ℓ, p, n with $\ell/p \in \mathbb{Z}$ and $\ell > 1$ such that $(W, \mathfrak{h}) \cong (G(\ell, p, n), \mathbb{C}^n)$, where $G(\ell, p, n)$ is the group of all $n \times n$ monomial matrices whose nonzero entries are ℓ-th roots of unity, while the product of all nonzero entries is an (ℓ/p)-th root of unity.*
- *There exists a positive integer n such that $(W, \mathfrak{h}) \cong (\mathfrak{S}_n, \mathbb{C}^{n-1})$.*
- *$(W, \mathfrak{h})$ is isomorphic to one of the 34 exceptional groups G_n ($n = 4, \ldots, 37$).*

Remark 3.2. We have

$$G(1,1,n) \cong \mathfrak{S}_n, \quad G(2,1,n) \cong B_n, \quad G(2,2,n) \cong D_n, \quad G(m,m,2) \cong I_2(m),$$
$$G_{23} \cong H_3, \quad G_{28} \cong F_4, \quad G_{30} \cong H_4, G_{35} \cong E_6, \quad G_{36} \cong E_7, \quad G_{37} \cong E_8.$$

Let $W \subset \mathrm{GL}(\mathfrak{h})$ be a complex reflection group. Benard [1976] and Bessis [1997] have proved (using a case-by-case analysis) that the field K generated by the traces on $\mathfrak{h}$ of all the elements of W is a splitting field for W. The field K is called the *field of definition* of W. If $K \subseteq \mathbb{R}$, then W is a finite Coxeter group, and if $K = \mathbb{Q}$, then W is a Weyl group.

Let $\mathcal{A}$ be the set of reflecting hyperplanes of W. Let $\mathfrak{h}^{\text{reg}} := \mathfrak{h} \setminus \bigcup_{H \in \mathcal{A}} H$ and $B_W := \pi_1(\mathfrak{h}^{\text{reg}}/W, x_0)$, where x_0 is some fixed basepoint. The group B_W is the *braid group* of W. For every orbit $\mathcal{C}$ of W on $\mathcal{A}$, we set $e_{\mathcal{C}}$ the common order of the subgroups W_H, where H is any element of $\mathcal{C}$ and W_H is the pointwise stabiliser of H. Note that W_H is cyclic, for all $H \in \mathcal{A}$.

We choose a set of indeterminates $\boldsymbol{u} = (u_{\mathcal{C},j})_{(\mathcal{C} \in \mathcal{A}/W)(0 \leq j \leq e_{\mathcal{C}}-1)}$ and we denote by $\mathbb{Z}[\boldsymbol{u}, \boldsymbol{u}^{-1}]$ the Laurent polynomial ring in all the indeterminates $\boldsymbol{u}$. We define the *generic Hecke algebra* $\mathcal{H}(W)$ of W to be the quotient of the group algebra $\mathbb{Z}[\boldsymbol{u}, \boldsymbol{u}^{-1}]B_W$ by the ideal generated by the elements of the form

$$(\boldsymbol{s} - u_{\mathcal{C},0})(\boldsymbol{s} - u_{\mathcal{C},1}) \cdots (\boldsymbol{s} - u_{\mathcal{C},e_{\mathcal{C}}-1}),$$

where $\mathcal{C}$ runs over the set $\mathcal{A}/W$ and $\boldsymbol{s}$ runs over the set of monodromy generators around the images in $\mathfrak{h}^{\text{reg}}/W$ of the elements of $\mathcal{C}$ [Broué et al. 1998, Section 4].

From now on, we will make certain assumptions for $\mathcal{H}(W)$. These assumptions are known to hold for all finite Coxeter groups [Bourbaki 2002, IV, Section 2], $G(\ell, p, n)$ [Broué et al. 1999; Malle and Mathas 1998; Geck et al. 2000] and a few of the exceptional complex reflection groups [Marin 2012; 2014][1]; they are expected to be true for all complex reflection groups.

Hypothesis 3.3. (a) The algebra $\mathcal{H}(W)$ is a free $\mathbb{Z}[\boldsymbol{u}, \boldsymbol{u}^{-1}]$-module of rank equal to the order of W.

(b) There exists a symmetrising trace τ on $\mathcal{H}(W)$ that satisfies certain canonicality conditions [Broué et al. 1999, Sections 1 and 2]; the form τ specialises to the canonical symmetrising form on the group algebra when $u_{\mathcal{C},j} \mapsto \zeta_{e_{\mathcal{C}}}^{j}$.

Under these assumptions, Malle [1999, 5.2] has shown that there exists $N_W \in \mathbb{Z}_{>0}$ such that if we take

$$u_{\mathcal{C},j} = \zeta_{e_{\mathcal{C}}}^{j} v_{\mathcal{C},j}^{N_W}, \tag{3.4}$$

and set $\boldsymbol{v} := (v_{\mathcal{C},j})_{(\mathcal{C} \in \mathcal{A}/W)(0 \leq j \leq e_{\mathcal{C}}-1)}$, then the $K(\boldsymbol{v})$-algebra $K(\boldsymbol{v})\mathcal{H}(W)$ is split semisimple. By Tits's deformation theorem, it follows that the specialisation $v_{\mathcal{C},j} \mapsto 1$ induces a bijection between $\mathrm{Irr}(K(\boldsymbol{v})\mathcal{H}(W))$ and $\mathrm{Irr}(W)$. From now on, we will consider $\mathcal{H}(W)$ as an algebra over $\mathbb{Z}_K[\boldsymbol{v}, \boldsymbol{v}^{-1}]$, where $\mathbb{Z}_K$ denotes the integral closure of $\mathbb{Z}$ in K.

---

[1] Malle and Michel [2010] mention that these assumptions have been confirmed computationally by Müller in several exceptional cases, but this work is not published. Moreover, in [Broué and Malle 1993] the assumption (a) is proved for the groups G_4, G_5, G_{12} and G_{25}, but Marin [2014] pointed out that these proofs might contain a questionable argument.

Example 3.5. The group $W = G(\ell, 1, n)$ is isomorphic to the wreath product $(\mathbb{Z}/\ell\mathbb{Z}) \wr \mathfrak{S}_n$ and its splitting field is $K = \mathbb{Q}(\zeta_\ell)$. In this particular case, we can take $N_W = 1$. The algebra $K(\boldsymbol{v})\mathcal{H}(W)$ is generated by elements $\boldsymbol{s}, \boldsymbol{t}_1, \ldots, \boldsymbol{t}_{n-1}$ satisfying the braid relations of type B_n (given by

$$\boldsymbol{s}\boldsymbol{t}_1\boldsymbol{s}\boldsymbol{t}_1 = \boldsymbol{t}_1\boldsymbol{s}\boldsymbol{t}_1\boldsymbol{s}, \quad \boldsymbol{s}\boldsymbol{t}_i = \boldsymbol{t}_i\boldsymbol{s}, \quad \boldsymbol{t}_{i-1}\boldsymbol{t}_i\boldsymbol{t}_{i-1} = \boldsymbol{t}_i\boldsymbol{t}_{i-1}\boldsymbol{t}_i$$

for $i = 2, \ldots, n-1$ and $\boldsymbol{t}_i\boldsymbol{t}_j = \boldsymbol{t}_j\boldsymbol{t}_i$ for $|i-j| > 1$), together with the extra relations

$$(\boldsymbol{s} - v_{s,0})(\boldsymbol{s} - \zeta_\ell v_{s,1}) \cdots (\boldsymbol{s} - \zeta_\ell^{\ell-1} v_{s,\ell-1}) = 0, \quad (\boldsymbol{t}_i - v_{t,0})(\boldsymbol{t}_i + v_{t,1}) = 0,$$

for all $i = 1, \ldots, n-1$. The Hecke algebra of $G(\ell, 1, n)$ is also known as *Ariki–Koike algebra*, with the last quadratic relation usually looking like this:

$$(\boldsymbol{t}_i - q)(\boldsymbol{t}_i + 1) = 0,$$

where q is an indeterminate. The irreducible representations of $G(\ell, 1, n)$, and thus the irreducible representations of $K(\boldsymbol{v})\mathcal{H}(W)$, are parametrised by the ℓ-partitions of n.

Let now q be an indeterminate and let $\boldsymbol{m} = (m_{C,j})_{(C \in \mathcal{A}/W)(0 \le j \le e_C - 1)}$ be a family of integers. The $\mathbb{Z}_K$-algebra morphism

$$\varphi_{\boldsymbol{m}} : \mathbb{Z}_K[\boldsymbol{v}, \boldsymbol{v}^{-1}] \to \mathbb{Z}_K[q, q^{-1}], \quad v_{C,j} \mapsto q^{m_{C,j}}$$

is called a *cyclotomic specialisation*. The $\mathbb{Z}_K[q, q^{-1}]$-algebra $\mathcal{H}_{\varphi_{\boldsymbol{m}}}(W)$ obtained as the specialisation of $\mathcal{H}(W)$ via $\varphi_{\boldsymbol{m}}$ is called a *cyclotomic Hecke algebra* associated with W. The Iwahori–Hecke algebras defined in the previous section are cyclotomic Hecke algebras associated with real reflection groups. The algebra $K(q)\mathcal{H}_{\varphi_{\boldsymbol{m}}}(W)$ is split semisimple [Chlouveraki 2009, Proposition 4.3.4]. By Tits's deformation theorem, the specialisation $q \mapsto 1$ yields a bijection between $\mathrm{Irr}(K(q)\mathcal{H}_{\varphi_{\boldsymbol{m}}}(W))$ and $\mathrm{Irr}(W)$.

3B. ***Schur elements and the a-function.*** The symmetrising trace τ (see Hypothesis 3.3) can be extended to $K(\boldsymbol{v})\mathcal{H}(W)$ by extension of scalars, and can be used to define Schur elements $(s_E)_{E \in \mathrm{Irr}(W)}$ for $\mathcal{H}(W)$. The Schur elements of $\mathcal{H}(W)$ have been explicitly calculated for all complex reflection groups:

- for finite Coxeter groups see Section 2B ;
- for complex reflection groups of type $G(\ell, 1, n)$ by Geck et al. [2000] and Mathas [2004];
- for complex reflection groups of type $G(\ell, 2, 2)$ by Malle [1997];
- for the remaining exceptional complex reflection groups by Malle [1997; 2000].

With the use of Clifford theory, we obtain the Schur elements for type $G(\ell, p, n)$ from those of type $G(\ell, 1, n)$ when $n > 2$ or $n = 2$ and p is odd. The Schur elements for type $G(\ell, p, 2)$ when p is even derive from those of type $G(\ell, 2, 2)$. See [Malle 1995; Chlouveraki 2009, A.7].

Using a case-by-case analysis, we have been able to determine that the Schur elements of $\mathcal{H}(W)$ have the following form [Chlouveraki 2009, Theorem 4.2.5].

Theorem 3.6. *Let $E \in \mathrm{Irr}(W)$. The Schur element s_E is an element of $\mathbb{Z}_K[\boldsymbol{v}, \boldsymbol{v}^{-1}]$ of the form*

$$s_E = \xi_E N_E \prod_{i \in I_E} \Psi_{E,i}(M_{E,i}), \tag{3.7}$$

where

(a) *ξ_E is an element of $\mathbb{Z}_K$,*

(b) *$N_E = \prod_{\mathcal{C},j} v_{\mathcal{C},j}^{b_{\mathcal{C},j}}$ is a monomial in $\mathbb{Z}_K[\boldsymbol{v}, \boldsymbol{v}^{-1}]$ with $\sum_{j=0}^{e_{\mathcal{C}}-1} b_{\mathcal{C},j} = 0$ for all $\mathcal{C} \in \mathcal{A}/W$,*

(c) *I_E is an index set,*

(d) *$(\Psi_{E,i})_{i \in I_E}$ is a family of K-cyclotomic polynomials in one variable,*

(e) *$(M_{E,i})_{i \in I_E}$ is a family of monomials in $\mathbb{Z}_K[\boldsymbol{v}, \boldsymbol{v}^{-1}]$ such that if*

$$M_{E,i} = \prod_{\mathcal{C},j} v_{\mathcal{C},j}^{a_{\mathcal{C},j}},$$

then

$$\gcd(a_{\mathcal{C},j}) = 1 \quad \text{and} \quad \sum_{j=0}^{e_{\mathcal{C}}-1} a_{\mathcal{C},j} = 0 \quad \text{for all } \mathcal{C} \in \mathcal{A}/W.$$

Equation (3.7) gives the factorisation of s_E into irreducible factors. The monomials $(M_{E,i})_{i \in I_E}$ are unique up to inversion, and we will call them *potentially essential* for W.

Remark 3.8. Theorem 3.6 was independently obtained by Rouquier [2008, Theorem 3.5] using a general argument on rational Cherednik algebras.

Example 3.9. Consider $\mathfrak{S}_3$, which is isomorphic to $G(1, 1, 3)$. We have

$$s_{E^{(3)}} = \Phi_2(v_{t,0}v_{t,1}^{-1})\Phi_3(v_{t,0}v_{t,1}^{-1}), \qquad s_{E^{(2,1)}} = v_{t,0}^{-1}v_{t,1}\Phi_3(v_{t,0}v_{t,1}^{-1}),$$
$$s_{E^{(1,1,1)}} = v_{t,0}^{-3}v_{t,1}^{3}\Phi_2(v_{t,0}v_{t,1}^{-1})\Phi_3(v_{t,0}v_{t,1}^{-1}).$$

Let $\varphi_{\boldsymbol{m}} : v_{\mathcal{C},j} \mapsto q^{m_{\mathcal{C},j}}$ be a cyclotomic specialisation. The canonical symmetrising trace on $\mathcal{H}(W)$ specialises via $\varphi_{\boldsymbol{m}}$ to become the canonical symmetrising trace $\tau_{\varphi_{\boldsymbol{m}}}$ on $\mathcal{H}_{\varphi_{\boldsymbol{m}}}(W)$. The Schur elements of $\mathcal{H}_{\varphi_{\boldsymbol{m}}}(W)$ with respect to $\tau_{\varphi_{\boldsymbol{m}}}$ are

$(\varphi_{\boldsymbol{m}}(s_E))_{E\in\mathrm{Irr}(W)}$, hence they can be written in the form (2.10). We can again define functions $a^{\boldsymbol{m}} : \mathrm{Irr}(W) \to \mathbb{Z}$ and $A^{\boldsymbol{m}} : \mathrm{Irr}(W) \to \mathbb{Z}$ such that

$$a_E^{\boldsymbol{m}} := -\mathrm{val}_q(\varphi_{\boldsymbol{m}}(s_E)) \quad \text{and} \quad A_E^{\boldsymbol{m}} := -\deg_q(\varphi_{\boldsymbol{m}}(s_E)).$$

3C. ***Families of characters and Rouquier families.*** Let $\varphi_{\boldsymbol{m}} : v_{\mathcal{C},j} \mapsto q^{m_{\mathcal{C},j}}$ be a cyclotomic specialisation and let $\mathcal{H}_{\varphi_{\boldsymbol{m}}}(W)$ be the corresponding cyclotomic Hecke algebra associated with W. How can we define families of characters for $\mathcal{H}_{\varphi_{\boldsymbol{m}}}(W)$? We cannot apply Lusztig's original definition, because parabolic subgroups of complex reflection groups[2] do not have a nice presentation as in the real case, and certainly not a "corresponding" parabolic Hecke algebra. On the other hand, we do not have a Kazhdan–Lusztig basis for $\mathcal{H}_{\varphi_{\boldsymbol{m}}}(W)$, so we cannot construct Kazhdan–Lusztig cells and use them to define families of characters for complex reflection groups in the usual way. However, we can define the families of characters to be the Rouquier families of $\mathcal{H}_{\varphi_{\boldsymbol{m}}}(W)$, that is, the blocks of $\mathcal{H}_{\varphi_{\boldsymbol{m}}}(W)$ over the Rouquier ring $\mathcal{R}_K(q)$, where

$$\mathcal{R}_K(q) = \mathbb{Z}_K[q, q^{-1}, (q^n - 1)^{-1}_{n\geq 1}].$$

Similarly to the real case, the Rouquier families are the nonempty subsets B of $\mathrm{Irr}(W)$ that are minimal with respect to the property:

$$\sum_{E\in B} \frac{\varphi_{\boldsymbol{m}}(\chi_E(h))}{\varphi_{\boldsymbol{m}}(s_E)} \in \mathcal{R}_K(q) \qquad \text{for all } h \in \mathcal{H}(W).$$

Broué and Kim [2002] determined the Rouquier families for the complex reflection groups of type $G(\ell, 1, n)$, but their results are only true when ℓ is a power of a prime number or $\varphi_{\boldsymbol{m}}$ is a "good" cyclotomic specialisation. The same problem persists, and some new appear, in the determination of the Rouquier families for $G(\ell, p, n)$ by Kim [2005]. Malle and Rouquier [2003] calculated the Rouquier families for some exceptional complex reflection groups and the dihedral groups, for a certain choice of cyclotomic specialisation. More recently, we managed to determine the Rouquier families for all cyclotomic Hecke algebras of all complex reflection groups [Chlouveraki 2007; 2008b; 2009; 2010], thanks to their property of "semicontinuity" (the term is due to Cédric Bonnafé). In order to explain this property, we will need some definitions.

Let

$$M = \prod_{\mathcal{C},j} v_{\mathcal{C},j}^{a_{\mathcal{C},j}}$$

[2]The parabolic subgroups of a complex reflection group $W \subset \mathrm{GL}(\mathfrak{h})$ are the pointwise stabilisers of the subsets of $\mathfrak{h}$. It is a remarkable theorem by Steinberg [1964, Theorem 1.5] that all parabolic subgroups of W are again complex reflection groups.

be a potentially essential monomial for W. We say that the family of integers $\boldsymbol{m} = (m_{\mathcal{C},j})_{(\mathcal{C}\in\mathcal{A}/W)(0\le j\le e_{\mathcal{C}}-1)}$ *belongs to the potentially essential hyperplane* H_M (of $\mathbb{R}^{\sum_{\mathcal{C}} e_{\mathcal{C}}}$) if $\sum_{\mathcal{C},j} m_{\mathcal{C},j} a_{\mathcal{C},j} = 0$.

Suppose that $\boldsymbol{m}$ belongs to no potentially essential hyperplane. Then the Rouquier families of $\mathcal{H}_{\varphi_{\boldsymbol{m}}}(W)$ are called *Rouquier families associated with no essential hyperplane*. Now suppose that $\boldsymbol{m}$ belongs to a unique potentially essential hyperplane H. Then the Rouquier families of $\mathcal{H}_{\varphi_{\boldsymbol{m}}}(W)$ are called *Rouquier families associated with* H. If they do not coincide with the Rouquier families associated with no essential hyperplane, then H is called an *essential hyperplane* for W. All these notions are well-defined and they do not depend on the choice of $\boldsymbol{m}$ because of the following theorem [Chlouveraki 2009, Section 4.4].

Theorem 3.10 (semicontinuity property of Rouquier families). *Let*

$$\boldsymbol{m} = (m_{\mathcal{C},j})_{(\mathcal{C}\in\mathcal{A}/W)(0\le j\le e_{\mathcal{C}}-1)}$$

be a family of integers and let $\varphi_{\boldsymbol{m}} : v_{\mathcal{C},j} \mapsto q^{m_{\mathcal{C},j}}$ *be the corresponding cyclotomic specialisation. The Rouquier families of* $\mathcal{H}_{\varphi_{\boldsymbol{m}}}(W)$ *are unions of the Rouquier families associated with the essential hyperplanes that* $\boldsymbol{m}$ *belongs to and they are minimal with respect to that property.*

Thanks to the above result, it is enough to do calculations in a finite number of cases in order to obtain the families of characters for all cyclotomic Hecke algebras, whose number is infinite.

Example 3.11. For $W = \mathfrak{S}_3$, the Rouquier families associated with no essential hyperplane are trivial. The hyperplane H_M corresponding to the monomial $M = v_{t,0}v_{t,1}^{-1}$ is essential, and it is the unique essential hyperplane for $\mathfrak{S}_3$. Let $\varphi_{\boldsymbol{m}} : v_{t,j} \mapsto q^{m_j}$, $j = 0, 1$, be a cyclotomic specialisation. We have that $\boldsymbol{m} = (m_0, m_1)$ belongs to H_M if and only if $m_0 = m_1$. There is a single Rouquier family associated with H_M, which contains all irreducible representations of $\mathfrak{S}_3$.

We have also shown that the functions a and A are constant on the Rouquier families, for all cyclotomic Hecke algebras of all complex reflection groups [Chlouveraki 2008a; 2008b; 2010].

3D. ***Canonical basic sets.*** Given a cyclotomic Hecke algebra $\mathcal{H}_{\varphi_{\boldsymbol{m}}}(W)$ and a ring homomorphism $\theta : q \mapsto \eta \in \mathbb{C} \setminus \{0\}$, we obtain a semisimplicity criterion and a decomposition map exactly as in Section 2D. A canonical basic set with respect to θ is also defined in the same way.

In [Chlouveraki and Jacon 2011], we showed the existence of canonical basic sets with respect to any θ for all cyclotomic Hecke algebras associated with finite Coxeter groups, that is, when the weight function L in the definition of $\mathcal{H}(W, L)$ is also allowed to take negative values.

For nonreal complex reflection groups, things become more complicated. For $W = G(\ell, 1, n)$, consider the specialised Ariki–Koike algebra with relations

$$(s - \zeta_e^{s_0})(s - \zeta_e^{s_1}) \cdots (s - \zeta_e^{s_{\ell-1}}) = 0, \quad (t_i - \zeta_e)(t_i + 1) = 0 \quad \text{for } i = 1, \dots, n-1. \tag{3.12}$$

where $(s_0, \dots, s_{\ell-1}) \in \mathbb{Z}^\ell$ and $e \in \mathbb{Z}_{>0}$. With the use of Ariki's theorem [Ariki 1996] and Uglov's work [2000] on canonical bases for higher level Fock spaces, Geck and Jacon have shown that, for a suitable choice of $\boldsymbol{m}$, the corresponding function $a^{\boldsymbol{m}}$ yields a canonical basic set for the above specialised Ariki–Koike algebra [Geck and Jacon 2006; 2011; Jacon 2004; 2007]. This canonical basic set consists of the so-called Uglov ℓ-partitions [Jacon 2007, Definition 3.2]. However, this does not work the other way round: not all cyclotomic Ariki–Koike algebras admit canonical basic sets. For a study about which values of $\boldsymbol{m}$ yield canonical basic sets, see [Gerber 2014].

In [Chlouveraki and Jacon 2012], building on work by Genet and Jacon [2006], we generalised the above result to obtain canonical basic sets for all groups of type $G(\ell, p, n)$ with $n > 2$, or $n = 2$ and p odd.

Finally, for the exceptional complex reflection groups of rank 2 ($G_4, \dots, G_{22}$), we have shown the existence of canonical basic sets for the cyclotomic Hecke algebras appearing in [Broué and Malle 1993] with respect to any θ [Chlouveraki and Miyachi 2011].

4. Symplectic reflection algebras

Let V be a complex vector space of finite dimension n, and let $G \subset \mathrm{GL}(V)$ be a finite group. Let $\mathbb{C}[V]$ be the set of regular functions on V, which is the same thing as the symmetric algebra $\mathrm{Sym}(V^*)$ of the dual space of V. The group G acts on $\mathbb{C}[V]$ as follows:

$$^{g}f(v) := f(g^{-1}v) \quad \text{for all } g \in G,\ f \in \mathbb{C}[V],\ v \in V.$$

We set

$$\mathbb{C}[V]^G := \{f \in \mathbb{C}[V] \mid {}^{g}f = f \ \text{ for all } g \in G\},$$

the space of fixed points of $\mathbb{C}[V]$ under the action of G. It is a classical problem in algebraic geometry to try and understand as a variety the space

$$V/G = \mathrm{Spec}\, \mathbb{C}[V]^G.$$

Is the space V/G singular? How much? The first question is answered by the following result from [Shephard and Todd 1954; Chevalley 1955].

Theorem 4.1. *The following statements are equivalent*:

(1) *V/G is smooth.*

(2) *$\mathbb{C}[V]^G$ is a polynomial algebra, on n homogeneous generators.*

(3) *G is a complex reflection group.*

Example 4.2. Let $\mathfrak{S}_n$ act on $V = \mathbb{C}^n$ by permuting the coordinates. Let $\mathbb{C}[V] = \mathbb{C}[X_1, \dots, X_n]$ and let $\Sigma_1, \Sigma_2, \dots, \Sigma_n$ be the elementary symmetric polynomials in n variables. We have

$$\mathbb{C}[V]^{\mathfrak{S}_n} = \mathbb{C}[\Sigma_1(X_1, \dots, X_n), \Sigma_2(X_1, \dots, X_n), \dots, \Sigma_n(X_1, \dots, X_n)].$$

More generally, we have

$$\mathbb{C}[V]^{G(\ell,1,n)} = \mathbb{C}[\Sigma_1(X_1^\ell, \dots, X_n^\ell), \Sigma_2(X_1^\ell, \dots, X_n^\ell), \dots, \Sigma_n(X_1^\ell, \dots, X_n^\ell)],$$

where $G(\ell, 1, n) \cong (\mathbb{Z}/\ell\mathbb{Z})^n \rtimes \mathfrak{S}_n$ and $(\mathbb{Z}/\ell\mathbb{Z})^n$ acts on V by multiplying the coordinates by ℓ-th roots of unity. Note that $G(\ell, 1, n)$ acts irreducibly on V if and only if $\ell > 1$.

Example 4.3. Let $V = \mathbb{C}^2$ and let G be a finite subgroup of $\mathrm{SL}_2(\mathbb{C})$. Then G is not a complex reflection group (in fact, it contains no pseudoreflections at all). The singular space $\mathbb{C}^2/G$ is called a *Kleinian* (or *Du Val*) singularity. The simplest example we can take is

$$G = \left\{ \left(\begin{smallmatrix} 1 & 0 \\ 0 & 1 \end{smallmatrix}\right) \left(\begin{smallmatrix} -1 & 0 \\ 0 & -1 \end{smallmatrix}\right) \right\};$$

we will use it to illustrate further notions.

4A. *Symplectic reflection groups.* The group G in Example 4.3 might not be a complex reflection group, but it is a symplectic reflection group, which is quite close. Moreover, the space $\mathbb{C}^2/G$ is not smooth (following Theorem 4.1), but it admits a symplectic resolution.

Let (V, ω_V) be a symplectic vector space, let $\mathrm{Sp}(V)$ be the group of symplectic transformations on V and let $G \subset \mathrm{Sp}(V)$ be a finite group. The triple (G, V, ω_V) is called a *symplectic triple*. A symplectic triple is *indecomposable* if there is no G-equivariant splitting $V = V_1 \oplus V_2$ with $\omega_V(V_1, V_2) = 0$. Any symplectic triple is a direct sum of indecomposable symplectic triples.

Definition 4.4. Let (G, V, ω_V) be a symplectic triple and let $(V/G)_{\mathrm{sm}}$ denote the smooth part of V/G. A *symplectic resolution* of V/G is a resolution of singularities $\pi : X \to V/G$ such that there exists a complex symplectic form ω_X on X for which the isomorphism

$$\pi|_{\pi^{-1}((V/G)_{\mathrm{sm}})} : \pi^{-1}((V/G)_{\mathrm{sm}}) \to (V/G)_{\mathrm{sm}}$$

is a symplectic isomorphism.

The existence of a symplectic resolution for V/G is a very strong condition and implies that the map π has some very good properties; for example, π is "semismall" [Verbitsky 2000, Theorem 2.8]. Moreover, all crepant resolutions of V/G are symplectic [loc. cit., Theorem 2.5].

Verbitsky has shown that if V/G admits a symplectic resolution, then G is generated by symplectic reflections [loc. cit., Theorem 3.2].

Definition 4.5. A *symplectic reflection* is a nontrivial element $s \in \mathrm{Sp}(V)$ such that $\mathrm{rank}(s - \mathrm{id}_V) = 2$. The symplectic triple (G, V, ω_V) is a *symplectic reflection group* if G is generated by symplectic reflections.

Hence, if the space V/G admits a symplectic resolution, then (G, V, ω_V) is a symplectic reflection group; the converse is not true. The classification of such symplectic reflection groups is almost complete thanks to the representation theory of symplectic reflection algebras.

Example 4.6. Following Example 4.3, let G be the cyclic group of order 2, denoted by μ_2, acting on $V = \mathbb{C} \oplus \mathbb{C}^*$ by multiplication by -1. Let ω_V be the *standard symplectic form* on V, that is,

$$\omega_V(y_1 \oplus x_1, y_2 \oplus x_2) = x_2(y_1) - x_1(y_2). \tag{4.7}$$

Letting $\mathbb{C}[V] = \mathbb{C}[X, Y]$, we see that

$$\mathbb{C}[V]^G = \mathbb{C}[X^2, XY, Y^2] \cong \mathbb{C}[A, B, C]/(AC - B^2),$$

the quadratic cone. This has an isolated singularity at the origin, that is, at the zero orbit, which can be resolved by blowing up there. The resulting resolution $\pi : T^*\mathbb{P}^1 \to V/G$ is a symplectic resolution where $T^*\mathbb{P}^1$ has its canonical symplectic structure.

The classification of (indecomposable) symplectic reflection groups is due to Huffman and Wales [1976], Cohen [1980], and Guralnick and Saxl [2003]. Except for a finite list of explicit exceptions with $\dim_{\mathbb{C}}(V) \leq 10$, there are two classes of symplectic reflection groups:

- Wreath products. Let $\Gamma \subset \mathrm{SL}_2(\mathbb{C})$ be finite: such groups are called Kleinian subgroups and they preserve the canonical symplectic structure on $\mathbb{C}^2$. Set

$$V = \underbrace{\mathbb{C}^2 \oplus \mathbb{C}^2 \oplus \cdots \oplus \mathbb{C}^2}_{n \text{ summands}}$$

  with the symplectic form ω_V induced from that on $\mathbb{C}^2$ and let $G = \Gamma \wr \mathfrak{S}_n$ act in the obvious way on V.

- Complex reflection groups. Let $G \subset \mathrm{GL}(\mathfrak{h})$ be a complex reflection group. Set $V = \mathfrak{h} \oplus \mathfrak{h}^*$ with its standard symplectic form ω_V (see (4.7)) and with G acting diagonally.

In both these cases, (G, V, ω_V) is a symplectic reflection group.

Remark 4.8. Note that in the second case, where G is a complex reflection group, the space $\mathfrak{h}/G$ is smooth, but V/G is not. The symplectic reflections in (G, V, ω_V) are the pseudoreflections in $(G, \mathfrak{h})$.

Remark 4.9. There is a small overlap between the two main families of symplectic reflection groups, namely the complex reflection groups of type $G(\ell, 1, n)$.

Wang [1999, Sections 1.3 and 1.4] observes that if $G = \Gamma \wr \mathfrak{S}_n$ for some $\Gamma \subset \mathrm{SL}_2(\mathbb{C})$, then V/G has a symplectic resolution given by the Hilbert scheme of n points on the minimal resolution of the Kleinian singularity $\mathbb{C}^2/\Gamma$. In Section 6 we will see what happens in the case where G is a complex reflection group.

4B. *The symplectic reflection algebra $H_{t,c}(G)$.* From now on, let (G, V, ω_V) be a symplectic reflection group and let $\mathcal{S}$ be the set of all symplectic reflections in G.

Definition 4.10. The *skew-group ring* $\mathbb{C}[V] \rtimes G$ is, as a vector space, equal to $\mathbb{C}[V] \otimes \mathbb{C}G$ and the multiplication is given by

$$g \cdot f = {}^{g}f \cdot g \quad \text{for all } g \in G,\ f \in \mathbb{C}[V].$$

The centre $Z(\mathbb{C}[V] \rtimes G)$ of the skew-group ring is equal to $\mathbb{C}[V]^G$. It has been an insight of Etingof and Ginzburg [2002], which goes back to (at least) Crawley-Bovey and Holland [1998], that, in order to understand $\operatorname{Spec} \mathbb{C}[V]^G$, we could look at deformations of $\mathbb{C}[V] \rtimes G$, hoping that the centre of the deformed algebra is itself a deformation of $\mathbb{C}[V]^G$. These deformations are the symplectic reflection algebras.

Let $s \in \mathcal{S}$. The spaces $\operatorname{Im}(s - \mathrm{id}_V)$ and $\operatorname{Ker}(s - \mathrm{id}_V)$ are symplectic subspaces of V with $\dim_{\mathbb{C}} \operatorname{Im}(s - \mathrm{id}_V) = 2$ and $V = \operatorname{Im}(s - \mathrm{id}_V) \oplus \operatorname{Ker}(s - \mathrm{id}_V)$. Let ω_s be the 2-form on V whose restriction to $\operatorname{Im}(s - \mathrm{id}_V)$ is ω_V and whose restriction to $\operatorname{Ker}(s - \mathrm{id}_V)$ is zero. Let ω_{V^*} be the symplectic form on V^* corresponding to ω_V (under the identification of V and V^* induced by ω_V), and let TV^* denote the tensor algebra on V^*. Finally, let $\boldsymbol{c} : \mathcal{S} \to \mathbb{C}$ be a *conjugacy invariant function*, that is, a map such that

$$\boldsymbol{c}(gsg^{-1}) = \boldsymbol{c}(s) \quad \text{for all } s \in \mathcal{S},\ g \in G.$$

Definition 4.11. Let $t \in \mathbb{C}$. We define the *symplectic reflection algebra* $\boldsymbol{H}_{t,\boldsymbol{c}}(G)$ of G to be

$$\boldsymbol{H}_{t,\boldsymbol{c}}(G) := TV^* \rtimes G \Big/ \Big\langle [u, v] - \Big(t\,\omega_{V^*}(u, v) - 2 \sum_{s\in\mathcal{S}} \boldsymbol{c}(s)\,\omega_s(u, v)\,s\Big) \;\Big|\; u, v \in V^* \Big\rangle.$$

Note that the above definition simply describes how two vectors in V^* commute with each other in $\boldsymbol{H}_{t,\boldsymbol{c}}(G)$, and that we have $[u, v] \in \mathbb{C}G$ for all $u, v \in V^*$.

Remark 4.12. For all $\lambda \in \mathbb{C}^\times$, we have $\boldsymbol{H}_{\lambda t,\lambda\boldsymbol{c}}(G) \cong \boldsymbol{H}_{t,\boldsymbol{c}}(G)$. So we only need to consider the cases $t = 1$ and $t = 0$.

Remark 4.13. We have $\boldsymbol{H}_{0,0}(G) = \mathbb{C}[V] \rtimes G$.

Example 4.14. Let us consider the example of the cyclic group $\mu_2 = \langle s \rangle$ acting on $V = \mathbb{C}^2$, so that

$$sx = -x, \quad sy = -y \quad \text{and} \quad \omega_{V^*}(y, x) = 1,$$

where $\{x, y\}$ is a basis of $(\mathbb{C}^2)^*$. We have $\omega_s = \omega_{V^*}$, since $\mathrm{Im}(s - \mathrm{id}_V) = V$. Then $\boldsymbol{H}_{t,\boldsymbol{c}}(\mu_2)$ is the quotient of $\mathbb{C}\langle x, y, s\rangle$ by the relations:

$$s^2 = 1, \quad sx = -xs, \quad sy = -ys, \quad [y, x] = t - 2\boldsymbol{c}(s)s.$$

Example 4.15. Let $V = \mathbb{C}^2$. Then $\mathrm{Sp}(V) = \mathrm{SL}_2(\mathbb{C})$ and we can take G to be any finite subgroup of $\mathrm{SL}_2(\mathbb{C})$. Let $\{x, y\}$ be a basis of $(\mathbb{C}^2)^*$ such that $\omega_{V^*}(y, x) = 1$. Every $g \neq 1$ in G is a symplectic reflection and $\omega_g = \omega_{V^*}$. Then

$$\boldsymbol{H}_{t,\boldsymbol{c}}(G) = \mathbb{C}\langle x, y\rangle \rtimes G \Big/ \Big\langle [y, x] - \Big(t - 2 \sum_{g\in G\setminus\{1\}} \boldsymbol{c}(g)g\Big)\Big\rangle.$$

There is a natural filtration $\mathcal{F}$ on $\boldsymbol{H}_{t,\boldsymbol{c}}(G)$ given by putting V^* in degree one and G in degree zero. The crucial result by Etingof and Ginzburg is the Poincaré–Birkhoff–Witt (PBW) theorem [Etingof and Ginzburg 2002, Theorem 1.3].

Theorem 4.16. *There is an isomorphism of algebras*

$$\mathrm{gr}_{\mathcal{F}}(\boldsymbol{H}_{t,\boldsymbol{c}}(G)) \cong \mathbb{C}[V] \rtimes G,$$

given by $\sigma(v) \mapsto v$, $\sigma(g) \mapsto g$, where $\sigma(h)$ denotes the image of $h \in \boldsymbol{H}_{t,\boldsymbol{c}}(G)$ in $\mathrm{gr}_{\mathcal{F}}(\boldsymbol{H}_{t,\boldsymbol{c}}(G))$. In particular, there is an isomorphism of vector spaces

$$\boldsymbol{H}_{t,\boldsymbol{c}}(G) \cong \mathbb{C}[V] \otimes \mathbb{C}G.$$

Moreover, symplectic reflection algebras are the only deformations of $\mathbb{C}[V] \rtimes G$ with this property (PBW property).

The most important consequence of the PBW theorem is that it gives us an explicit basis of the symplectic reflection algebra. The proof of it is an application of a general result by Braverman and Gaitsgory: If I is a two-sided ideal of

$TV^* \rtimes G$ generated by a space U of elements of degree at most two, then [Braverman and Gaitsgory 1996, Theorem 0.5] gives necessary and sufficient conditions so that the quotient $TV^* \rtimes G/I$ has the PBW property. The PBW property also implies that $\boldsymbol{H}_{t,c}(G)$ has some good ring-theoretic properties, for example:

Corollary 4.17. (i) *The algebra* $\boldsymbol{H}_{t,c}(G)$ *is a Noetherian ring.*

(ii) $\boldsymbol{H}_{t,c}(G)$ *has finite global dimension.*

Remark 4.18. For general pairs (G, V) a description of PBW deformations of $\mathbb{C}[V] \rtimes G$ was originally given by Drinfeld [1986]. In the symplectic case this was rediscovered by Etingof and Ginzburg as above, and Drinfeld's general case was described in detail by Ram and Shepler [2003].

4C. *The spherical subalgebra.* We saw in the previous subsection that the skew-group ring $\mathbb{C}[V] \rtimes G$ is not commutative and that its centre $Z(\mathbb{C}[V] \rtimes G)$ is equal to $\mathbb{C}[V]^G$. We will now see that $\mathbb{C}[V] \rtimes G$ contains another subalgebra isomorphic to $\mathbb{C}[V]^G$.

Let $e := \frac{1}{|G|} \sum_{g \in G} g$ be the trivial idempotent in $\mathbb{C}G$. One can easily check that the map

$$\begin{aligned} \mathbb{C}[V]^G &\to e(\mathbb{C}[V] \rtimes G)e, \\ f &\mapsto efe, \end{aligned} \tag{4.19}$$

is an algebra isomorphism. We have $efe = fe$, for all $f \in \mathbb{C}[V]^G$.

Definition 4.20. We define the *spherical subalgebra* of $\boldsymbol{H}_{t,c}(G)$ to be the algebra

$$\boldsymbol{U}_{t,c}(G) := e\, \boldsymbol{H}_{t,c}(G)\, e.$$

The filtration $\mathcal{F}$ on $\boldsymbol{H}_{t,c}(G)$ induces, by restriction, a filtration on $\boldsymbol{U}_{t,c}(G)$. The PBW theorem, in combination with (4.19), implies that there is an isomorphism of algebras

$$\mathrm{gr}_{\mathcal{F}}(\boldsymbol{U}_{t,c}(G)) \cong e(\mathbb{C}[V] \rtimes G)e \cong \mathbb{C}[V]^G$$

and an isomorphism of vector spaces

$$\boldsymbol{U}_{t,c}(G) \cong \mathbb{C}[V]^G.$$

Thus, the spherical subalgebra provides a flat deformation of the coordinate ring of V/G, as desired.

Example 4.21. Let $G = \mu_2 = \langle s \rangle$ acting on $V = \mathbb{C}^2$ as in Example 4.14. Then $e = \frac{1}{2}(1+s)$. The spherical subalgebra $\boldsymbol{U}_{t,c}(\mu_2)$ is generated as a $\mathbb{C}$-algebra by

$$\boldsymbol{h} := -\tfrac{1}{2}e(xy+yx)e, \quad \boldsymbol{e} := \tfrac{1}{2}ex^2e \quad \text{and} \quad \boldsymbol{f} := \tfrac{1}{2}ey^2e.$$

There are relations

$$[\boldsymbol{e}, \boldsymbol{f}] = t\boldsymbol{h}, \quad [\boldsymbol{h}, \boldsymbol{e}] = -2t\boldsymbol{e}, \quad [\boldsymbol{h}, \boldsymbol{f}] = 2t\boldsymbol{f},$$
$$\boldsymbol{e}\boldsymbol{f} = (2\boldsymbol{c}(s) - \boldsymbol{h}/2)(t/2 - \boldsymbol{c}(s) - \boldsymbol{h}/2).$$

So if $t = 0$, $\boldsymbol{U}_{t,\boldsymbol{c}}(\mu_2)$ is commutative, while if $t = 1$, $\boldsymbol{U}_{t,\boldsymbol{c}}(\mu_2)$ is a central quotient of the enveloping algebra of $\mathfrak{sl}_2(\mathbb{C})$.

The space $\boldsymbol{H}_{t,\boldsymbol{c}}(G)e$ is a $(\boldsymbol{H}_{t,\boldsymbol{c}}(G), \boldsymbol{U}_{t,\boldsymbol{c}}(G))$-bimodule and it is called *the Etingof–Ginzburg sheaf.* The following result is known as the "double centraliser property" [Etingof and Ginzburg 2002, Theorem 1.5].

Proposition 4.22. (i) *The right $\boldsymbol{U}_{t,\boldsymbol{c}}(G)$-module $\boldsymbol{H}_{t,\boldsymbol{c}}(G)e$ is reflexive.*

(ii) $\mathrm{End}_{\boldsymbol{H}_{t,\boldsymbol{c}}(G)}(\boldsymbol{H}_{t,\boldsymbol{c}}(G)e)^{op} \cong \boldsymbol{U}_{t,\boldsymbol{c}}(G)$.

(iii) $\mathrm{End}_{\boldsymbol{U}_{t,\boldsymbol{c}}(G)^{op}}(\boldsymbol{H}_{t,\boldsymbol{c}}(G)e) \cong \boldsymbol{H}_{t,\boldsymbol{c}}(G)$.

This is important, because, in general, we have an explicit presentation of $\boldsymbol{H}_{t,\boldsymbol{c}}(G)$, but not of $\boldsymbol{U}_{t,\boldsymbol{c}}(G)$. The above result allows us to study $\boldsymbol{U}_{t,\boldsymbol{c}}(G)$ by studying $\boldsymbol{H}_{t,\boldsymbol{c}}(G)$ instead.

4D. *The centre of $\boldsymbol{H}_{t,\boldsymbol{c}}(G)$.* The behaviour of the centre of the spherical subalgebra observed in Example 4.21 is the same for all symplectic reflection groups [Etingof and Ginzburg 2002, Theorem 1.6].

Theorem 4.23. (i) *If $t = 0$, then $\boldsymbol{U}_{t,\boldsymbol{c}}(G)$ is commutative.*

(ii) *If $t \neq 0$, then $Z(\boldsymbol{U}_{t,\boldsymbol{c}}(G)) = \mathbb{C}$.*

Now the double centraliser property can be used to prove the following result relating the centres of $\boldsymbol{U}_{t,\boldsymbol{c}}(G)$ and $\boldsymbol{H}_{t,\boldsymbol{c}}(G)$.

Theorem 4.24 (the Satake isomorphism). *The map $z \mapsto ze$ defines an algebra isomorphism $Z(\boldsymbol{H}_{t,\boldsymbol{c}}(G)) \cong Z(\boldsymbol{U}_{t,\boldsymbol{c}}(G))$ for all parameters $(t, \boldsymbol{c})$.*

Corollary 4.25. (i) *If $t = 0$, then $Z(\boldsymbol{H}_{t,\boldsymbol{c}}(G)) \cong \boldsymbol{U}_{t,\boldsymbol{c}}(G)$.*

(ii) *If $t \neq 0$, then $Z(\boldsymbol{H}_{t,\boldsymbol{c}}(G)) = \mathbb{C}$.*

Thus, the symplectic reflection algebra $\boldsymbol{H}_{t,\boldsymbol{c}}(G)$ produces a commutative deformation of the space V/G when $t = 0$.

4E. *Symplectic resolutions.* In this subsection, we will focus on the case $t = 0$. Set $Z_{\boldsymbol{c}}(G) := Z(\boldsymbol{H}_{0,\boldsymbol{c}}(G))$. We have $Z_{\boldsymbol{c}}(G) \cong \boldsymbol{U}_{0,\boldsymbol{c}}(G)$, and so $\boldsymbol{H}_{0,\boldsymbol{c}}(G)$ is a finitely generated $Z_{\boldsymbol{c}}(G)$-module.

Definition 4.26. The *generalised Calogero–Moser space $X_{\boldsymbol{c}}(G)$* is defined to be the affine variety Spec $Z_{\boldsymbol{c}}(G)$.

Since the associated graded of $Z_{\boldsymbol{c}}(G)$ is $\mathbb{C}[V]^G$ (with respect to the filtration $\mathcal{F}$), $X_{\boldsymbol{c}}(G)$ is irreducible. The following result, due to Ginzburg and Kaledin [2004, Proposition 1.18 and Theorem 1.20] and Namikawa [2011, Corollary 2.10], gives us a criterion for V/G to admit a symplectic resolution, using the geometry of the generalised Calogero–Moser space.

Theorem 4.27. *Let (G, V, ω_V) be an (irreducible) symplectic reflection group. The space V/G admits a symplectic resolution if and only if $X_{\boldsymbol{c}}(G)$ is smooth for generic values of $\boldsymbol{c}$ (equivalently, there exists $\boldsymbol{c}$ such that $X_{\boldsymbol{c}}(G)$ is smooth).*

Example 4.28. Consider again the example of $\mu_2 = \langle s \rangle$ acting on $\mathbb{C}^2$. The centre of $\boldsymbol{H}_{0,\boldsymbol{c}}(\mu_2)$ is generated by $A := x^2$, $B := xy - \boldsymbol{c}(s)s$ and $C := y^2$. Thus,

$$X_{\boldsymbol{c}}(\mu_2) \cong \mathbb{C}[A, B, C]/(AC - (B + \boldsymbol{c}(s))(B - \boldsymbol{c}(s)))$$

is the affine cone over $\mathbb{P}^1 \subset \mathbb{P}^2$ when $\boldsymbol{c}(s) = 0$, but is a smooth affine surface for $\boldsymbol{c}(s) \neq 0$.

As we mentioned in Section 4A, if $G = \Gamma \wr \mathfrak{S}_n$ for some $\Gamma \subset \mathrm{SL}_2(\mathbb{C})$, then V/G always admits a symplectic resolution, that is, $X_{\boldsymbol{c}}(G)$ is smooth for generic $\boldsymbol{c}$. On the other hand, if $G \subset \mathrm{GL}(\mathfrak{h})$ is a complex reflection group acting on $V = \mathfrak{h} \oplus \mathfrak{h}^*$, this is not always the case. Etingof and Ginzburg proved that $X_{\boldsymbol{c}}(G)$ is smooth for generic $\boldsymbol{c}$ when $G = G(\ell, 1, n)$ [Etingof and Ginzburg 2002, Corollary 1.14]. However, Gordon showed that, for most finite Coxeter groups not of type A_n or B_n, $X_{\boldsymbol{c}}(G)$ is a singular variety for all choices of the parameter $\boldsymbol{c}$ [Gordon 2003, Proposition 7.3]. Finally, using the Calogero–Moser partition of $\mathrm{Irr}(G)$ described in [Gordon and Martino 2009], Bellamy [2009, Theorem 1.1] proved that $X_{\boldsymbol{c}}(G)$ is smooth for generic values of $\boldsymbol{c}$ if and only if $G = G(\ell, 1, n)$ or $G = G_4$. We will revisit this result in Section 6.

Following the classification of symplectic reflection groups, and all the works mentioned above, the classification of quotient singularities admitting symplectic resolutions is (almost) complete.

4F. *Rational Cherednik algebras.* From now on, let $W \subset \mathrm{GL}(\mathfrak{h})$ be a complex reflection group and let $V = \mathfrak{h} \oplus \mathfrak{h}^*$. There is a natural pairing $(\ ,\ ) : \mathfrak{h} \times \mathfrak{h}^* \to \mathbb{C}$ given by $(y, x) := x(y)$. Then the standard symplectic form ω_V on V is given by

$$\omega_V(y_1 \oplus x_1, y_2 \oplus x_2) = (y_1, x_2) - (y_2, x_1).$$

The triple (W, V, ω_V) is a symplectic reflection group. The set $\mathcal{S}$ of all symplectic reflections in (W, V, ω_V) coincides with the set of pseudoreflections in $(W, \mathfrak{h})$. Let $\boldsymbol{c} : \mathcal{S} \to \mathbb{C}$ be a conjugacy invariant function.

Definition 4.29. The *rational Cherednik algebra* of W is the symplectic reflection algebra $\boldsymbol{H}_{t,\boldsymbol{c}}(W)$ associated to (W, V, ω_V).

For $s \in \mathcal{S}$, fix $\alpha_s \in \mathfrak{h}^*$ to be a basis of the one-dimensional vector space $\mathrm{Im}(s - \mathrm{id}_V)|_{\mathfrak{h}^*}$ and $\alpha_s^\vee \in \mathfrak{h}$ to be a basis of the one-dimensional vector space $\mathrm{Im}(s - \mathrm{id}_V)|_{\mathfrak{h}}$. Then $\boldsymbol{H}_{t,\boldsymbol{c}}(W)$ is the quotient of $TV^* \rtimes W$ by the relations:

$$[x_1, x_2] = 0, \quad [y_1, y_2] = 0, \quad [y, x] = t(y, x) - 2\sum_{s\in\mathcal{S}} \boldsymbol{c}(s)\, \frac{(y, \alpha_s)(\alpha_s^\vee, x)}{(\alpha_s^\vee, \alpha_s)}\, s \tag{4.30}$$

for all $x_1, x_2, x \in \mathfrak{h}^*$ and $y_1, y_2, y \in \mathfrak{h}$.

Example 4.31. Let $W = \mathfrak{S}_n$ and $\mathfrak{h} = \mathbb{C}^n$. Choose a basis $x_1, \dots, x_n$ of $\mathfrak{h}^*$ and a dual basis $y_1, \dots, y_n$ of $\mathfrak{h}$ so that

$$\sigma x_i = x_{\sigma(i)}\sigma \quad \text{and} \quad \sigma(y_i) = y_{\sigma(i)}\sigma \qquad \text{for all } \sigma \in \mathfrak{S}_n,\ 1 \le i \le n.$$

The set $\mathcal{S}$ is the set of all transpositions in $\mathfrak{S}_n$. We denote by s_{ij} the transposition (i, j). Set

$$\alpha_{ij} := x_i - x_j \quad \text{and} \quad \alpha_{ij}^\vee = y_i - y_j \qquad \text{for all } 1 \le i < j \le n.$$

We have $(\alpha_{ij}^\vee, \alpha_{ij}) = 2$. There is a single conjugacy class in $\mathcal{S}$, so take $\boldsymbol{c} \in \mathbb{C}$. Then $\boldsymbol{H}_{t,\boldsymbol{c}}(\mathfrak{S}_n)$ is the quotient of $TV^* \rtimes \mathfrak{S}_n$ by the relations:

$$[x_i, x_j]=0, \quad [y_i, y_j]=0, \quad [y_i, x_i]=t-\boldsymbol{c}\sum_{j\neq i} s_{ij}, \quad [y_i, x_j]=\boldsymbol{c}\, s_{ij}, \qquad \text{for } i\neq j.$$

5. Rational Cherednik algebras at $t = 1$

The PBW theorem implies that the rational Cherednik algebra $\boldsymbol{H}_{1,\boldsymbol{c}}(W)$, as a vector space, has a "triangular decomposition"

$$\boldsymbol{H}_{1,\boldsymbol{c}}(W) \cong \mathbb{C}[\mathfrak{h}] \otimes \mathbb{C}W \otimes \mathbb{C}[\mathfrak{h}^*].$$

Another famous example of a triangular decomposition is the one of the enveloping algebra $U(\mathfrak{g})$ of a finite dimensional, semisimple complex Lie algebra $\mathfrak{g}$ (into the enveloping algebras of the Cartan subalgebra, the nilpotent radical of the Borel subalgebra and its opposite). In the representation theory of $\mathfrak{g}$, one of the categories of modules most studied, and best understood, is category $\mathcal{O}$, the abelian category generated by all highest weight modules. Therefore, it makes sense to want to construct and study an analogue of category $\mathcal{O}$ for rational Cherednik algebras.

5A. ***Category*** $\mathcal{O}$. Let $\boldsymbol{H}_{1,\boldsymbol{c}}(W)$-mod be the category of all finitely generated $\boldsymbol{H}_{1,\boldsymbol{c}}(W)$-modules. We say that a module $M \in \boldsymbol{H}_{1,\boldsymbol{c}}(W)$-mod is *locally nilpotent* for the action of $\mathfrak{h} \subset \mathbb{C}[\mathfrak{h}^*]$ if for each $m \in M$ there exists $N >> 0$ such that $\mathfrak{h}^N \cdot m = 0$.

Definition 5.1. We define $\mathcal{O}$ to be the category of all finitely generated $\boldsymbol{H}_{1,\boldsymbol{c}}(W)$-modules that are locally nilpotent for the action of $\mathfrak{h} \subset \mathbb{C}[\mathfrak{h}^*]$.

Remark 5.2. Each module in category $\mathcal{O}$ is finitely generated as a $\mathbb{C}[\mathfrak{h}]$-module.

Category $\mathcal{O}$ has been thoroughly studied in [Ginzburg et al. 2003]. Proofs of all its properties presented here can be found in this paper.

For all $E \in \mathrm{Irr}(W)$, we set

$$\Delta(E) := \boldsymbol{H}_{1,\boldsymbol{c}}(W) \otimes_{\mathbb{C}[\mathfrak{h}^*] \rtimes W} E,$$

where $\mathbb{C}[\mathfrak{h}^*]$ acts trivially on E (that is, the augmentation ideal $\mathbb{C}[\mathfrak{h}^*]_+$ acts on E as zero) and W acts naturally. The module $\Delta(E)$ belongs to $\mathcal{O}$ and is called a *standard module* (or *Verma module*). Each standard module $\Delta(E)$ has a simple head $\mathrm{L}(E)$ and the set

$$\{\mathrm{L}(E) \mid E \in \mathrm{Irr}(W)\}$$

is a complete set of pairwise nonisomorphic simple modules of the category $\mathcal{O}$. Every module in $\mathcal{O}$ has finite length, so we obtain a well-defined square decomposition matrix

$$\boldsymbol{D} = ([\Delta(E) : \mathrm{L}(E')])_{E, E' \in \mathrm{Irr}(W)},$$

where $[\Delta(E) : \mathrm{L}(E')]$ equals the multiplicity with which the simple module $\mathrm{L}(E')$ appears in the composition series of $\Delta(E)$. We have $[\Delta(E) : \mathrm{L}(E)] = 1$.

Proposition 5.3. *The following are equivalent*:

1. $\mathcal{O}$ *is semisimple.*
2. $\Delta(E) = \mathrm{L}(E)$ *for all* $E \in \mathrm{Irr}(W)$.
3. $\boldsymbol{D}$ *is the identity matrix.*

Now, there exist several orderings on the set of standard modules of $\mathcal{O}$ (and consequently on $\mathrm{Irr}(W)$) for which $\mathcal{O}$ is a highest weight category in the sense of [Cline et al. 1988] (see also [Rouquier 2008, Section 5.1]). If $<_{\mathcal{O}}$ is such an ordering on $\mathrm{Irr}(W)$, and if $[\Delta(E) : \mathrm{L}(E')] \neq 0$ for some $E, E' \in \mathrm{Irr}(W)$, then either $E = E'$ or $E' <_{\mathcal{O}} E$. Thus, we can arrange the rows of $\boldsymbol{D}$ so that the decomposition matrix is lower unitriangular. We will refer to these orderings on $\mathrm{Irr}(W)$ as *orderings on the category* $\mathcal{O}$. A famous example of such an ordering is the one given by the c-function.

5B. ***A change of parameters and the c-function.*** In order to relate rational Cherednik algebras with cyclotomic Hecke algebras via the KZ-functor in the next subsection, we need to change the parametrisation of $\boldsymbol{H}_{1,\boldsymbol{c}}(W)$. As in Section 3A, let $\mathcal{A}$ denote the set of reflecting hyperplanes of W. For $H \in \mathcal{A}$, let W_H be the pointwise stabiliser of H in W. The group W_H is cyclic and its order, denoted by $e_{\mathcal{C}}$, only depends on the orbit $\mathcal{C} \in \mathcal{A}/W$ that H belongs to. We have that

$$\mathcal{S} = \bigcup_{H \in \mathcal{A}} W_H \setminus \{1\}.$$

For each $s \in W_H \setminus \{1\}$, we have $\operatorname{Ker} \alpha_s = H$. Without loss of generality, we may assume that $\alpha_s = \alpha_{s'}$ and $\alpha_s^\vee = \alpha_{s'}^\vee$ for all $s, s' \in W_H \setminus \{1\}$. Set $\alpha_H := \alpha_s$ and $\alpha_H^\vee := \alpha_s^\vee$. Then the third relation in (4.30) becomes

$$[y, x] = (y, x) - 2 \sum_{H \in \mathcal{A}} \frac{(y, \alpha_H)(\alpha_H^\vee, x)}{(\alpha_H^\vee, \alpha_H)} \sum_{s \in W_H \setminus \{1\}} \boldsymbol{c}(s)\, s \quad \text{for all } x \in \mathfrak{h}^*, \ y \in \mathfrak{h}.$$

We define a family of complex numbers $\boldsymbol{k} = (k_{\mathcal{C},j})_{(\mathcal{C} \in \mathcal{A}/W)(0 \leq j \leq e_{\mathcal{C}}-1)}$ by

$$-2 \sum_{s \in W_H \setminus \{1\}} \boldsymbol{c}(s)\, s = \sum_{s \in W_H \setminus \{1\}} \left(\sum_{j=0}^{e_{\mathcal{C}}-1} \det(s)^{-j} (k_{\mathcal{C},j} - k_{\mathcal{C},j-1}) \right) s, \qquad \text{for } H \in \mathcal{C},$$

with $k_{\mathcal{C},-1} = 0$. This implies that

$$\boldsymbol{c}(s) = -\frac{1}{2} \sum_{j=0}^{e_{\mathcal{C}}-1} \det(s)^{-j} (k_{\mathcal{C},j} - k_{\mathcal{C},j-1}).$$

From now on, we will denote by $\boldsymbol{H_k}(W)$ the quotient of $TV^* \rtimes W$ by the relations

$$[x_1, x_2] = 0, \quad [y_1, y_2] = 0, \quad [y, x] = (y, x) + \sum_{H \in \mathcal{A}} \frac{(y, \alpha_H)(\alpha_H^\vee, x)}{(\alpha_H^\vee, \alpha_H)} \gamma_H,$$

where

$$\gamma_H = \sum_{w \in W_H \setminus \{1\}} \left(\sum_{j=0}^{e_{\mathcal{C}}-1} \det(w)^{-j} (k_{\mathcal{C},j} - k_{\mathcal{C},j-1}) \right) w,$$

for all $x_1, x_2, x \in \mathfrak{h}^*$ and $y_1, y_2, y \in \mathfrak{h}$. We have $\boldsymbol{H_k}(W) = \boldsymbol{H}_{1,\boldsymbol{c}}(W)$.

Let $E \in \operatorname{Irr}(W)$. We denote by c_E the scalar by which the element

$$-\sum_{H \in \mathcal{A}} \sum_{j=0}^{e_{\mathcal{C}}-1} \left(\sum_{w \in W_H} (\det w)^{-j} w \right) k_{\mathcal{C},j} \in Z(\mathbb{C}W)$$

acts on E. We obtain thus a function $c : \operatorname{Irr}(W) \to \mathbb{C},\ E \mapsto c_E$. The c-function defines an ordering $<_c$ on the category $\mathcal{O}$ as follows: For all $E, E' \in \operatorname{Irr}(W)$,

$$E' <_c E \ \text{ if and only if } \ c_E - c_{E'} \in \mathbb{Z}_{>0}.$$

Remark 5.4. If $c_E - c_{E'} \notin \mathbb{Z} \setminus \{0\}$ for all $E, E' \in \operatorname{Irr}(W)$, then $\boldsymbol{D}$ is the identity matrix, and thus $\mathcal{O}$ is semisimple.

Remark 5.5. In the rational Cherednik algebra literature the function c is usually taken to be the negative of the one defined here. In the context of this paper the above definition is more natural. In both cases we obtain an ordering on the category $\mathcal{O}$.

5C. ***The KZ-functor.*** Following [Ginzburg et al. 2003, 5.3], there exists an exact factor, known as the *Knizhnik–Zamalodchikov functor* or simply KZ, between the category $\mathcal{O}$ of $\boldsymbol{H_k}(W)$ and the category of representations of a certain specialised Hecke algebra $\mathcal{H}_{\boldsymbol{k}}(W)$. Using the notation of Section 3A, the specialised Hecke algebra $\mathcal{H}_{\boldsymbol{k}}(W)$ is a quotient of the group algebra $\mathbb{C}B_W$ by the ideal generated by the elements of the form

$$(\boldsymbol{s}-\exp(2\pi i k_{\mathcal{C},0}))(\boldsymbol{s}-\zeta_{e_\mathcal{C}}\exp(2\pi i k_{\mathcal{C},1}))\cdots(\boldsymbol{s}-\zeta_{e_\mathcal{C}}^{e_\mathcal{C}-1}\exp(2\pi i k_{\mathcal{C},e_\mathcal{C}-1})),$$

where $\mathcal{C}$ runs over the set $\mathcal{A}/W$ and $\boldsymbol{s}$ runs over the set of monodromy generators around the images in $\mathfrak{h}^{\text{reg}}/W$ of the elements of $\mathcal{C}$. The algebra $\mathcal{H}_{\boldsymbol{k}}(W)$ is obtained from the generic Hecke algebra $\mathbb{C}[\boldsymbol{v},\boldsymbol{v}^{-1}]\mathcal{H}(W)$ via the specialisation $\Theta: v_{\mathcal{C},j}^{N_W}\mapsto\exp(2\pi i k_{\mathcal{C},j})$ (recall that N_W is the power to which the indeterminates $v_{\mathcal{C},j}$ appear in the defining relations of the generic Hecke algebra so that the algebra $\mathbb{C}(\boldsymbol{v})\mathcal{H}(W)$ is split; see (3.4)). We always assume that Hypothesis 3.3 holds for $\mathcal{H}(W)$.

The functor KZ is represented by a projective object $P_{\text{KZ}}\in\mathcal{O}$, and we have $\mathcal{H}_{\boldsymbol{k}}(W)\cong\operatorname{End}_{\boldsymbol{H_k}(W)}(P_{\text{KZ}})^{op}$ [Ginzburg et al. 2003, 5.4]. Based on this, we have the following result due to Vale [2006, Theorem 2.1]:

Proposition 5.6. *The following are equivalent*:

(1) $\boldsymbol{H_k}(W)$ *is a simple ring.*

(2) $\mathcal{O}$ *is semisimple.*

(3) $\mathcal{H}_{\boldsymbol{k}}(W)$ *is semisimple.*

We can thus use the semisimplicity criterion for $\mathcal{H}_{\boldsymbol{k}}(W)$ given by Theorem 2.17 in order to determine for which values of $\boldsymbol{k}$ the category $\mathcal{O}$ is semisimple.

Now let $<_{\mathcal{O}}$ be any ordering on the category $\mathcal{O}$ as in Section 5A.

Proposition 5.7. *Set* $\boldsymbol{B}:=\{E\in\operatorname{Irr}(W)\mid \text{KZ}(\text{L}(E))\neq 0\}$.

(a) *The set* $\{\text{KZ}(\text{L}(E))\mid E\in\boldsymbol{B}\}$ *is a complete set of pairwise nonisomorphic simple* $\mathcal{H}_{\boldsymbol{k}}(W)$*-modules.*

(b) *For all* $E\in\operatorname{Irr}(W)$, $E'\in\boldsymbol{B}$, *we have*

$$[\Delta(E):\text{L}(E')]=[\text{KZ}(\Delta(E)):\text{KZ}(\text{L}(E'))].$$

(c) *If* $E\in\boldsymbol{B}$, *then* $[\text{KZ}(\Delta(E)):\text{KZ}(\text{L}(E))]=1$.

(d) *If* $[\mathrm{KZ}(\Delta(E)) : \mathrm{KZ}(\mathrm{L}(E'))] \neq 0$ *for some* $E \in \mathrm{Irr}(W)$ *and* $E' \in \boldsymbol{B}$, *then either* $E = E'$ *or* $E' <_{\mathcal{O}} E$.

Property (a) follows from [Ginzburg et al. 2003, Theorem 5.14]. For the proof of properties (b), (c) and (d), all of them deriving from the fact that KZ is exact, the reader may refer to [Chlouveraki et al. 2012, Proposition 3.1].

The simple modules killed by the KZ-functor are exactly the ones that do not have full support. Their determination, and thus the determination of the set $\boldsymbol{B}$, is a very difficult problem.

We also obtain a decomposition matrix $D_{\boldsymbol{k}}$ for the specialised Hecke algebra $\mathcal{H}_{\boldsymbol{k}}(W)$ with respect to the specialisation Θ. The rows of $D_{\boldsymbol{k}}$ are indexed by $\mathrm{Irr}(W)$ and its columns by $\mathrm{Irr}(\mathcal{H}_{\boldsymbol{k}}(W))$. Following Proposition 5.7, $D_{\boldsymbol{k}}$ can be obtained from the decomposition matrix $\boldsymbol{D}$ of the category $\mathcal{O}$ by removing the columns that correspond to the simple modules killed by the KZ-functor, that is, the columns labelled by $\mathrm{Irr}(W) \setminus \boldsymbol{B}$. This implies that $D_{\boldsymbol{k}}$ becomes lower unitriangular when its rows are ordered with respect to $<_{\mathcal{O}}$, in the same way that, in the cases where Θ factors through a cyclotomic Hecke algebra, the existence of a canonical basic set implies that $D_{\boldsymbol{k}}$ becomes lower unitriangular when its rows are ordered with respect to the a-function. If we could show that the a-function defines an ordering on the category $\mathcal{O}$, we would automatically obtain the existence of a canonical basic set for $\mathcal{H}_{\boldsymbol{k}}(W)$. At the same time, we would obtain the determination of $\boldsymbol{B}$ in the cases where canonical basic sets have already been explicitly described.

5D. *The $(a + A)$-function.* Let $\boldsymbol{m} = (m_{\mathcal{C},j})_{(\mathcal{C}\in\mathcal{A}/W)(0\leq j\leq e_{\mathcal{C}}-1)}$ be a family of integers and let $\varphi_{\boldsymbol{m}} : v_{\mathcal{C},j} \mapsto q^{m_{\mathcal{C},j}}$ be the corresponding cyclotomic specialisation for the Hecke algebra $\mathcal{H}(W)$. Let $\theta : q \mapsto \eta$ be a specialisation such that η is a nonzero complex number. If η is not a root of unity or $\eta = 1$, then, due to Theorem 2.17 and the form of the Schur elements of $\mathcal{H}_{\varphi_{\boldsymbol{m}}}(W)$, the specialised Hecke algebra $\mathcal{H}_\eta(W)$ is semisimple. So we may assume from now on that η is a root of unity of order $e > 1$, namely $\eta = \zeta_e^r$ for some $r \in \mathbb{Z}_{>0}$ such that $\gcd(e, r) = 1$.

Let $\boldsymbol{k} = (k_{\mathcal{C},j})_{(\mathcal{C}\in\mathcal{A}/W)(0\leq j\leq e_{\mathcal{C}}-1)}$ be the family of rational numbers defined by

$$k_{\mathcal{C},j} := \frac{rN_W}{e} m_{\mathcal{C},j} \quad \text{for all } \mathcal{C},\, j.$$

Then $\mathcal{H}_{\boldsymbol{k}}(W) = \mathcal{H}_\eta(W)$. Following [Chlouveraki et al. 2012, Section 3.3], we obtain the following equation which relates the functions $a^{\boldsymbol{m}}$ and $A^{\boldsymbol{m}}$ for $\mathcal{H}_{\varphi_{\boldsymbol{m}}}(W)$ with the c-function for $\boldsymbol{H}_{\boldsymbol{k}}(W)$:

$$a_E^{\boldsymbol{m}} + A_E^{\boldsymbol{m}} = \frac{e}{rN_W} c_E + \sum_{H\in\mathcal{A}} \sum_{j=0}^{e_{\mathcal{C}}-1} m_{\mathcal{C},j} \quad \text{for all } E \in \mathrm{Irr}(W), \tag{5.8}$$

where $\mathcal{C}$ denotes the orbit of $H \in \mathcal{A}$ under the action of W.

Remark 5.9. The above formula was also obtained in [Ginzburg et al. 2003, Section 6.2] for finite Weyl groups in the equal parameter case.

Equation (5.8) implies that $a^{\mathbf{m}} + A^{\mathbf{m}}$ yields the same ordering on $\mathrm{Irr}(W)$ as the c-function (note that in this case $c_E \in \mathbb{Q}$ for all $E \in \mathrm{Irr}(W)$). Thus, $a^{\mathbf{m}} + A^{\mathbf{m}}$ is also an ordering on the category $\mathcal{O}$, that is, if $[\Delta(E) : \mathrm{L}(E')] \neq 0$ for some $E, E' \in \mathrm{Irr}(W)$, then either $E = E'$ or $a^{\mathbf{m}}_{E'} + A^{\mathbf{m}}_{E'} < a^{\mathbf{m}}_{E} + A^{\mathbf{m}}_{E}$. If now the function $a^{\mathbf{m}}$ is compatible with $a^{\mathbf{m}} + A^{\mathbf{m}}$, that is, for all $E, E' \in \mathrm{Irr}(W)$,

$$a^{\mathbf{m}}_{E'} + A^{\mathbf{m}}_{E'} < a^{\mathbf{m}}_{E} + A^{\mathbf{m}}_{E} \Rightarrow a^{\mathbf{m}}_{E'} < a^{\mathbf{m}}_{E}, \tag{5.10}$$

then $a^{\mathbf{m}}$ is an ordering on the category $\mathcal{O}$ and we obtain the existence of a canonical basic set for $\mathcal{H}_{\varphi_{\mathbf{m}}}(W)$ with respect to θ by Proposition 5.7. This is true in several cases, but unfortunately not true in general. Some exceptional complex reflection groups where (5.10) holds and the above argument works are:

$$G_{23} = H_3, G_{24}, G_{27}, G_{29}, G_{30} = H_4.^3$$

This yields the existence of canonical basic sets for the groups G_{24}, G_{27} and G_{29}, which was not known before. To summarise, we have the following:

Proposition 5.11. *Let $W = G_n$, $n \in \{23, 24, 27, 29, 30\}$. Let $\mathbf{m}$ and $\mathbf{k}$ be defined as above, and let $E, E' \in \mathrm{Irr}(W)$. If $[\Delta(E) : \mathrm{L}(E')] \neq 0$, then either $E = E'$ or $a^{\mathbf{m}}_{E'} < a^{\mathbf{m}}_{E}$. In particular, we have $\mathrm{KZ}(\mathrm{L}(E)) \neq 0$ if and only if E belongs to the canonical basic set of $\mathcal{H}_{\varphi_{\mathbf{m}}}(W)$ with respect to $\theta : q \mapsto \zeta_e^r$.*

5E. *Canonical basic sets for Iwahori–Hecke algebras from rational Cherednik algebras.* Equation (5.8) has also allowed us to show that, in the case where W is a finite Coxeter group, and assuming that Lusztig's conjectures P1–P15 hold, the c-function is compatible with the ordering $\leq_{\mathcal{LR}}$ on two-sided cells, since a and A are (see [Geck 2009, Remark 5.4] for the a-function, [Lusztig 2003, Corollary 21.6] and [Chlouveraki and Jacon 2011, Proposition 2.8] for A). This in turn was crucial in showing [Chlouveraki et al. 2012, Corollary 4.7]:

Proposition 5.12. *Let (W, S) be a finite Coxeter group and let $\mathcal{H}(W, L)$ be the Iwahori–Hecke algebra of W with parameter L, as defined in Section 2A. For $H \in \mathcal{A}$, let $s_H \in W$ be the reflection with reflecting hyperplane H and let $\mathcal{C}$ be the orbit of H under the action of W. If $H' \in \mathcal{C}$, then we have $L(s_H) = L(s_{H'})$*

---

[3] The groups G_{23}, G_{24}, G_{27}, G_{29}, G_{30}, G_{31}, G_{33}, G_{34}, G_{35}, G_{36} and G_{37} are easy to check with a computer; they are all generated by pseudoreflections of order 2 whose reflecting hyperplanes belong to the same orbit.

and we can set $L_{\mathcal{C}} := L(s_H)$. *Let* $e, r \in \mathbb{Z}_{>0}$ *such that* $\gcd(e, r) = 1$, *and take, for all* $\mathcal{C} \in \mathcal{A}/W$,

$$k_{\mathcal{C},0} = \frac{rL_{\mathcal{C}}}{e} \quad \textit{and} \quad k_{\mathcal{C},1} = -\frac{rL_{\mathcal{C}}}{e}.$$

If $E \in \mathrm{Irr}(W)$, *then* $\mathrm{KZ}(\mathrm{L}(E)) \neq 0$ *if and only if* E *belongs to the canonical basic set of* $\mathcal{H}(W, L)$ *with respect to* $\theta : q \mapsto \zeta_e^r$.

The proof uses a connection, established in [Chlouveraki et al. 2012, Proposition 4.6], between category $\mathcal{O}$ and the cellular structure of the Iwahori–Hecke algebra. More specifically, if $E \in \mathrm{Irr}(W)$, then $\mathrm{KZ}(\Delta(E))$ is isomorphic to the cell module $W_\theta(E)$ defined in [Geck 2007a, Example 4.4]; we will not go into further details here. Note though that, in Proposition 5.12, we have not included the assumption that Lusztig's conjectures must hold. The reason is that the only case where they are not known to hold, the case of B_n, is covered by Corollary 5.18 below.

Remark 5.13. The above result can be generalised to the case where $k_{\mathcal{C},0} = \lambda L_{\mathcal{C}}$ and $k_{\mathcal{C},1} = -\lambda L_{\mathcal{C}}$for any complex number λ. If $\lambda \in \mathbb{Z}$ or $\lambda \in \mathbb{C} \setminus \mathbb{Q}$, then both category $\mathcal{O}$ and $\mathcal{H}_{\exp(2\pi i\lambda)}(W, L)$ are semisimple, so the statement trivially holds. If λ is a negative rational number, let us say $\lambda = -r/e$ for some $e, r \in \mathbb{Z}_{>0}$ with $\gcd(e, r) = 1$, and $E \in \mathrm{Irr}(W)$, then $\mathrm{KZ}(\mathrm{L}(E)) \neq 0$ if and only if E belongs to the canonical basic set of $\mathcal{H}(W, -L)$ with respect to $\theta : q \mapsto \zeta_e^r$. We recall now that the canonical basic sets for finite Coxeter groups where L can take negative values are described in [Chlouveraki and Jacon 2011]. In fact, E belongs to the canonical basic set of $\mathcal{H}(W, -L)$ with respect to $\theta : q \mapsto \zeta_e^r$ if and only if $E \otimes \varepsilon$ belongs to the canonical basic set of $\mathcal{H}(W, L)$ with respect to θ, where ε denotes the sign representation of W.

Proposition 5.12 yields the existence of canonical basic sets for all finite Coxeter groups in a uniform way. At the same time, it yields a description of the simple modules that are not killed by the KZ-functor, since canonical basic sets for finite Coxeter groups are explicitly known (see, for example, [Geck and Jacon 2011]). However, it does not imply that the a-function is an ordering on the category $\mathcal{O}$, because we do not know what happens with the simple modules killed by the KZ-functor. We do believe though that, for finite Coxeter groups, the a-function is an ordering on the category $\mathcal{O}$.

Example 5.14. Let W be the symmetric group $\mathfrak{S}_n$ and let $l := L(s)$ for every transposition $s \in \mathfrak{S}_n$ (there exists only one orbit $\mathcal{C}$ in $\mathcal{A}/W$). Let $\eta^{2l} := \zeta_e^r$ for some $e, r \in \mathbb{Z}_{>0}$ with $\gcd(e, r) = 1$. As we saw in Example 2.20, the canonical basic set $\mathcal{B}_\theta$ of $\mathcal{H}(W, l)$ with respect to $\theta : q \mapsto \eta$ consists of the e-regular partitions of n. Now take $k_{\mathcal{C},0} = r/2e$ and $k_{\mathcal{C},1} = -r/2e$. Let λ be a partition of

n and let E^λ be the corresponding irreducible representation of $\mathfrak{S}_n$. We have $\mathrm{KZ}(\mathrm{L}(E^\lambda)) \neq 0$ if and only if λ is e-regular.

5F. ***Canonical basic sets for Ariki–Koike algebras from rational Cherednik algebras.*** As we have said and seen earlier, there exist several orderings on the category $\mathcal{O}$. For $W = G(\ell, 1, n)$, where the irreducible representations are parametrised by the ℓ-partitions of n, one combinatorial ordering on the category $\mathcal{O}$ is given by Dunkl and Griffeth [2010, Theorem 4.1]. More precisely, in this case, there are two hyperplane orbits in $\mathcal{A}/W$; we will denote them by C_s and C_t. We have $e_{C_s} = \ell$ and $e_{C_t} = 2$. Let $(s_0, \ldots, s_{\ell-1}) \in \mathbb{Z}^\ell$ and $e \in \mathbb{Z}_{>0}$. We define $\boldsymbol{k} = (k_{C_s,0}, \ldots, k_{C_s,\ell-1}, k_{C_t,0}, k_{C_t,1})$ by

$$k_{C_s,j} = \frac{s_j}{e} - \frac{j}{\ell} \quad \text{for } j = 0, \ldots, \ell-1, \; k_{C_t,0} = \frac{1}{e}, \; k_{C_t,1} = 0. \tag{5.15}$$

Then the KZ-functor goes from the category $\mathcal{O}$ for $\boldsymbol{H}_{\boldsymbol{k}}(W)$ to the category of representations of the specialised Ariki–Koike algebra $\mathcal{H}_{\boldsymbol{k}}(W)$ with relations

$$(\boldsymbol{s}-\zeta_e^{s_0})(\boldsymbol{s}-\zeta_e^{s_1})\cdots(\boldsymbol{s}-\zeta_e^{s_{\ell-1}})=0, \quad (\boldsymbol{t}_i-\zeta_e)(\boldsymbol{t}_i+1)=0 \quad \text{for } i=1,\ldots,n-1,$$

as in (3.12).

Let $\lambda = (\lambda^{(0)}, \ldots, \lambda^{(\ell-1)})$ be an ℓ-partition of n. We will denote by E^λ the corresponding irreducible representation of $G(\ell, 1, n)$. We define the set of nodes of λ to be the set

$$[\lambda] = \{(a, b, c) : 0 \le c \le \ell - 1, \; a \ge 1, \; 1 \le b \le \lambda_a^{(c)}\}.$$

Let $\gamma = (a(\gamma), b(\gamma), c(\gamma)) \in [\lambda]$. We set $\vartheta(\gamma) := b(\gamma) - a(\gamma) + s_{c(\gamma)}$. We then have the following [Dunkl and Griffeth 2010, Proof of Theorem 4.1]:

Proposition 5.16. *Let λ, λ' be ℓ-partitions of n. If $[\Delta(E^\lambda) : \mathrm{L}(E^{\lambda'})] \neq 0$, then there exist orderings $\gamma_1, \gamma_2, \ldots, \gamma_n$ and $\gamma_1', \gamma_2', \ldots, \gamma_n'$ of the nodes of λ and λ' respectively, and nonnegative integers $\mu_1, \mu_2, \ldots, \mu_n$, such that, for all $1 \le i \le n$,*

$$\mu_i \equiv c(\gamma_i) - c(\gamma_i') \bmod \ell \quad \text{and} \quad \mu_i = c(\gamma_i) - c(\gamma_i') + \frac{\ell}{e}(\vartheta(\gamma_i') - \vartheta(\gamma_i)).$$

Now, there are several different cyclotomic Ariki–Koike algebras that produce the specialised Ariki–Koike algebra $\mathcal{H}_{\boldsymbol{k}}(W)$ defined above and they may have distinct a-functions attached to them. Using the combinatorial description of the a-function for $G(\ell, 1, n)$ given in [Geck and Jacon 2011, Section 5.5],[4] we

[4]This definition captures all a-functions for $G(\ell, 1, n)$ in the literature: the function $a^{\boldsymbol{m}}$ for $m_{C_s,j} = s_j\ell - ej$, $j = 0, \ldots, \ell-1$, given by Jacon [2007] and studied in the context of Uglov's work on canonical bases for higher level Fock spaces, and also the a-function for type B_n ($\ell = 2$) arising from the Kazhdan–Lusztig theory for Iwahori–Hecke algebras with unequal parameters (see [Geck and Jacon 2011, 6.7]).

showed in [Chlouveraki et al. 2012, Section 5] that it is compatible with the ordering on category $\mathcal{O}$ given by Proposition 5.16. Consequently, the a-function also defines a highest weight structure on $\mathcal{O}$, that is, we have the following:

Proposition 5.17. *Let λ, λ' be ℓ-partitions of n. If $[\Delta(E^{\lambda}) : \mathrm{L}(E^{\lambda'})] \neq 0$, then either $\lambda = \lambda'$ or $a_{E^{\lambda'}} < a_{E^{\lambda}}$.*

The above result, combined with Proposition 5.7, yields the following:

Corollary 5.18. *Let $W = G(\ell, 1, n)$. Let $(s_0, \dots, s_{\ell-1}) \in \mathbb{Z}^{\ell}$ and $e \in \mathbb{Z}_{>0}$. Let $\boldsymbol{k} = (k_{C_s,0}, \dots, k_{C_s,\ell-1}, k_{C_t,0}, k_{C_t,1})$ be defined as in (5.15). If λ is an ℓ-partition of n, then $\mathrm{KZ}(\mathrm{L}(E^{\lambda})) \neq 0$ if and only if E^{λ} belongs to the canonical basic set for $\mathcal{H}_{\boldsymbol{k}}(W)$ with respect to the a-function above.*

Thus, we obtain the existence of canonical basic sets for Ariki–Koike algebras without the use of Ariki's theorem. On the other hand, the description of the canonical basic sets for Ariki–Koike algebras by Jacon [2007, Main Theorem] yields a description of the set $\boldsymbol{B} = \{E^{\lambda} \in \mathrm{Irr}(W) \mid \mathrm{KZ}(\mathrm{L}(E^{\lambda})) \neq 0\}$: we have that $E^{\lambda} \in \boldsymbol{B}$ if and only if λ is an Uglov ℓ-partition.

Finally, we expect a result similar to Corollary 5.18 to hold in the case where $W = G(\ell, p, n)$ for $p > 1$.

6. Rational Cherednik algebras at $t = 0$

Let us now consider the rational Cherednik algebra $\boldsymbol{H}_{0,\boldsymbol{c}}(W)$. In this case, the centre of $\boldsymbol{H}_{0,\boldsymbol{c}}(W)$ is isomorphic to the spherical subalgebra of $\boldsymbol{H}_{0,\boldsymbol{c}}(W)$, that is, $Z(\boldsymbol{H}_{0,\boldsymbol{c}}(W)) \cong e\boldsymbol{H}_{0,\boldsymbol{c}}e$, where $e := \frac{1}{|W|}\sum_{w \in W} w$. So $\boldsymbol{H}_{0,\boldsymbol{c}}(W)$ is a finitely generated $Z(\boldsymbol{H}_{0,\boldsymbol{c}}(W))$-module. From now on, we set $Z_{\boldsymbol{c}}(W) := Z(\boldsymbol{H}_{0,\boldsymbol{c}}(W))$.

6A. *Restricted rational Cherednik algebras.* In the case of finite Coxeter groups the following was proved in [Etingof and Ginzburg 2002, Proposition 4.15], and the general case is due to Gordon [2003, Proposition 3.6].

Proposition 6.1. (a) *The subalgebra $\mathfrak{m} := \mathbb{C}[\mathfrak{h}]^W \otimes \mathbb{C}[\mathfrak{h}^*]^W$ of $\boldsymbol{H}_{0,\boldsymbol{c}}(W)$ is contained in $Z_{\boldsymbol{c}}(W)$.*

(b) *$Z_{\boldsymbol{c}}(W)$ is a free $\mathfrak{m}$-module of rank $|W|$.*

Let $\mathfrak{m}_+$ denote the ideal of $\mathfrak{m}$ consisting of elements with zero constant term.

Definition 6.2. We define the *restricted rational Cherednik algebra* to be

$$\overline{\boldsymbol{H}}_{0,\boldsymbol{c}}(W) := \boldsymbol{H}_{0,\boldsymbol{c}}(W)/\mathfrak{m}_+\boldsymbol{H}_{0,\boldsymbol{c}}(W).$$

This algebra was originally introduced, and extensively studied, in [Gordon 2003]. The PBW theorem implies that, as a vector space,

$$\overline{\boldsymbol{H}}_{0,\boldsymbol{c}}(W) \cong \mathbb{C}[\mathfrak{h}]^{\mathrm{co}W} \otimes \mathbb{C}W \otimes \mathbb{C}[\mathfrak{h}^*]^{\mathrm{co}W},$$

where $\mathbb{C}[\mathfrak{h}]^{\mathrm{co}W} = \mathbb{C}[\mathfrak{h}]/\langle \mathbb{C}[\mathfrak{h}]_+^W \rangle$ is the *coinvariant algebra*. Since W is a complex reflection group, $\mathbb{C}[\mathfrak{h}]^{\mathrm{co}W}$ has dimension $|W|$ and is isomorphic to the regular representation as a $\mathbb{C}W$-module. Thus, $\dim_{\mathbb{C}} \overline{\boldsymbol{H}}_{0,\boldsymbol{c}}(W) = |W|^3$.

Let $E \in \mathrm{Irr}(W)$. We set

$$\overline{\Delta}(E) := \overline{\boldsymbol{H}}_{0,\boldsymbol{c}}(W) \otimes_{\mathbb{C}[\mathfrak{h}^*]^{\mathrm{co}W} \rtimes W} E,$$

where $\mathbb{C}[\mathfrak{h}^*]^{\mathrm{co}W}$ acts trivially on E (that is, $\mathbb{C}[\mathfrak{h}^*]_+^{\mathrm{co}W}$ acts on E as zero) and W acts naturally. The module $\overline{\Delta}(E)$ is the *baby Verma module* of $\overline{\boldsymbol{H}}_{0,\boldsymbol{c}}(W)$ associated to E. We summarise, as is done in [Gordon 2003, Proposition 4.3], the results of Holmes and Nakano [1991] applied to this situation.

Proposition 6.3. *Let $E,\ E' \in \mathrm{Irr}(W)$.*

(i) *The baby Verma module $\overline{\Delta}(E)$ has a simple head, $\overline{\mathrm{L}}(E)$. Hence, $\overline{\Delta}(E)$ is indecomposable.*

(ii) $\overline{\Delta}(E) \cong \overline{\Delta}(E')$ *if and only if $E \cong E'$.*

(iii) *The set $\{\overline{\mathrm{L}}(E) \mid E \in \mathrm{Irr}(W)\}$ is a complete set of pairwise nonisomorphic simple $\overline{\boldsymbol{H}}_{0,\boldsymbol{c}}$-modules.*

6B. *The Calogero–Moser partition.* Recall that the generalised Calogero–Moser space $X_{\boldsymbol{c}}(W)$ is defined to be the affine variety Spec $Z_{\boldsymbol{c}}(W)$. By Theorem 4.27, $(\mathfrak{h} \oplus \mathfrak{h}^*)/W$ admits a symplectic resolution if and only if $X_{\boldsymbol{c}}(W)$ is smooth for generic values of $\boldsymbol{c}$. Etingof and Ginzburg proved that $X_{\boldsymbol{c}}(G)$ is smooth for generic $\boldsymbol{c}$ when $W = G(\ell, 1, n)$ [Etingof and Ginzburg 2002, Corollary 1.14]. Later, Gordon [2003, Proposition 7.3] showed that $X_{\boldsymbol{c}}(G)$ is a singular variety for all choices of the parameter $\boldsymbol{c}$ for the following finite Coxeter groups: D_{2n} $(n \geq 2)$, E_6, E_7, E_8, F_4, H_3, H_4 and $I_2(m)$ $(m \geq 5)$.

Now, since the algebra $\overline{\boldsymbol{H}}_{0,\boldsymbol{c}}$ is finite dimensional, we can define its blocks in the usual way (see Section 1A). Let $E,\ E' \in \mathrm{Irr}(W)$. Following [Gordon and Martino 2009], we define the *Calogero–Moser partition* of $\mathrm{Irr}(W)$ to be the set of equivalence classes of $\mathrm{Irr}(W)$ under the equivalence relation:

$$E \sim_{CM} E' \text{ if and only if } \overline{\mathrm{L}}(E) \text{ and } \overline{\mathrm{L}}(E') \text{ belong to the same block.}$$

We will simply write $CM_{\boldsymbol{c}}$-partition for the Calogero–Moser partition of $\mathrm{Irr}(W)$. The inclusion $\mathfrak{m} \subset Z_{\boldsymbol{c}}(W)$ defines a finite surjective morphism

$$\mathcal{Y} : X_{\boldsymbol{c}}(W) \longrightarrow \mathfrak{h}/W \times \mathfrak{h}^*/W,$$

where $\mathfrak{h}/W \times \mathfrak{h}^*/W = \mathrm{Spec}\,\mathfrak{m}$. Müller's theorem (see [Brown and Gordon 2001, Corollary 2.7]) implies that the natural map $\mathrm{Irr}(W) \to \mathcal{Y}^{-1}(0)$, $E \mapsto \mathrm{Supp}(\overline{\mathrm{L}}(E))$ factors through the $CM_{\boldsymbol{c}}$-partition. Using this fact, one can show that the geometry of $X_{\boldsymbol{c}}(W)$ is related to the $CM_{\boldsymbol{c}}$-partition in the following way.

Theorem 6.4. *The following are equivalent*:

(1) *The generalised Calogero–Moser space $X_{\mathbf{c}}(W)$ is smooth.*

(2) *The $CM_{\mathbf{c}}$-partition of* $\mathrm{Irr}(W')$ *is trivial for every parabolic subgroup W' of W.*

Using the above result and the classification of irreducible complex reflection groups (see Theorem 3.1), Bellamy [2009, Theorem 1.1]has shown the following:

Theorem 6.5. *Let W be an irreducible complex reflection group. The generalised Calogero–Moser space $X_{\mathbf{c}}(W)$ is smooth for generic values of $\mathbf{c}$ if and only if W is of type $G(\ell, 1, n)$ or G_4. In every other case, $X_{\mathbf{c}}(W)$ is singular for all choices of $\mathbf{c}$.*

Corollary 6.6. *Let W be an irreducible complex reflection group. The space $(\mathfrak{h} \oplus \mathfrak{h}^*)/W$ admits a symplectic resolution if and only if W is of type $G(\ell, 1, n)$ or G_4.*

6C. *The Calogero–Moser partition and Rouquier families.* It just so happens that the cases where $X_{\mathbf{c}}(W)$ is generically smooth, and the Calogero–Moser partition generically trivial, are exactly the cases where the Rouquier families are generically trivial (that is, the Rouquier families associated with no essential hyperplane are singletons). This, combined with the fact that the Calogero–Moser partition into blocks enjoys some property of semicontinuity, led to the question whether there is a connection between the two partitions.

The question was first asked by Gordon and Martino [2009] in terms of a connection between the Calogero–Moser partition and families of characters for type B_n. In their paper, they computed the $CM_{\mathbf{c}}$-partition, for all $\mathbf{c}$, for complex reflection groups of type $G(\ell, 1, n)$ and showed that for $\ell = 2$, using the conjectural combinatorial description of Kazhdan–Lusztig cells for type B_n by Bonnafé et al. [2010], the $CM_{\mathbf{c}}$-partition coincides with the partition into Kazhdan–Lusztig families. After that, Martino [2010] compared the combinatorial description of the $CM_{\mathbf{c}}$-partition for type $G(\ell, 1, n)$ given in [Gordon and Martino 2009] with the description of the partition into Rouquier families, given by Chlouveraki [2008b], for a suitable cyclotomic Hecke algebra $\mathcal{H}_{\mathbf{c}}$ of $G(\ell, 1, n)$ (different from the one defined in Section 5C). He showed that the two partitions coincide when ℓ is a power of a prime number (which includes the cases of type A_n and B_n), but not in general. In fact, he showed that the $CM_{\mathbf{c}}$-partition for $G(\ell, 1, n)$ is the same as the one obtained by Broué and Kim [2002]. He thus obtained the following two connections between the $CM_{\mathbf{c}}$-partition and the partition into Rouquier families for $G(\ell, 1, n)$, and he conjectured that they hold for every complex reflection group W [Martino 2010, 2.7]:

(a) The CM_c-partition for generic c coincides with the generic partition into Rouquier families (both being trivial for $W = G(\ell, 1, n)$);

(b) The partition into Rouquier families refines the CM_c-partition, for all choices of c; that is, if $E, E' \in \mathrm{Irr}(W)$ belong to the same Rouquier family of $\mathcal{H}_c$, then $E \sim_{CM} E'$.

Conditions (a) and (b) are known as "Martino's conjecture". Using the combinatorics of [Gordon and Martino 2009] and [Martino 2010], Bellamy [2012a] computed the CM_c-partition, for all c, and proved Martino's conjecture in the case where W is of type $G(\ell, p, n)$; note that when $p > 1$ the generic partitions in this case are not trivial. However, a counterexample for (a) was found recently by Thiel [2014] in the case where $W = G_{25}$. Thiel calculated the CM_c-partition for generic c for the exceptional complex reflection groups G_4, G_5, G_6, G_8, G_{10}, $G_{23} = H_3$, G_{24}, G_{25} and G_{26}. Comparing his results with the generic partition into Rouquier families for these groups, given by [Chlouveraki 2009], he showed that Part (a) of Martino's conjecture holds in every case[5] except for when $W = G_{25}$. In this particular case, the generic partition into Rouquier families simply refines the CM_c-partition for generic c. So we will state here as a conjecture only Part (b) of Martino's conjecture, which is still an open problem, and proved in all the above cases.

Conjecture 6.7 (Martino's conjecture). Let W be a complex reflection group. The partition into Rouquier families (for a suitably chosen cyclotomic Hecke algebra $\mathcal{H}_c$ of W) refines the CM_c-partition, for all choices of c; that is, if $E, E' \in \mathrm{Irr}(W)$ belong to the same Rouquier family of $\mathcal{H}_c$, then $E \sim_{CM} E'$.

Remark 6.8. Note that, in all the cases checked so far where W is a finite Coxeter group, the partition into Rouquier families and the CM_c-partition coincide. This covers the finite Coxeter groups of types A_n, B_n, D_n and the dihedral groups for all choices of c, and H_3 for generic c.

6D. *The Calogero–Moser partition and Kazhdan–Lusztig cells.* In an effort to develop a generalised Kazhdan–Lusztig cell theory, Bonnafé and Rouquier [2013] used the Calogero–Moser partition to define what they call *Calogero–Moser cells* for all complex reflection groups. An advantage of this, quite geometric, approach is that the Calogero–Moser partition exists naturally for all complex reflection groups. It also implies automatically the existence of a semicontinuity property for cells, a property that was conjectured and proved in some cases for Kazhdan–Lusztig cells by Bonnafé [2009]. However, Calogero–Moser cells are very hard to compute and their construction depends on an "uncontrollable" choice. After very long computations by Bonnafé and Rouquier, it is now confirmed that the

[5]For G_4 this was already known by Bellamy [2009].

Calogero–Moser cells coincide with the Kazhdan–Lusztig cells in the smallest possible cases (A_2, B_2, G_2); there is still a lot of work that needs to be done.

Acknowledgements

First of all, I would like to thank the Mathematical Sciences Research Institute (MSRI) for its hospitality during the programme "Noncommutative Algebraic Geometry and Representation Theory" and for its support. My thanks to the organisers of the Introductory Workshop, Michael Artin, Michel Van den Bergh and Toby Stafford, who invited me to give two talks on symplectic reflection algebras; the current paper is inspired by these talks. A special thanks to Toby Stafford for giving me more time to finish it. I feel a lot of gratitude towards Gwyn Bellamy and Iain Gordon for answering all my questions on symplectic reflection algebras (and they were a lot!), and for writing two excellent surveys [Bellamy 2012b; Gordon 2008] on the topic. Iain also offered to read this manuscript and suggested many corrections, for which I am grateful. I also thank Cédric Bonnafé for answering my questions in Luminy, reading this manuscript and making useful comments. My deepest gratitude towards Guillaume Pouchin for going through all the phases of this project with me. Finally, I would like to thank Gunter Malle and the referee of this paper for suggesting many corrections to the final version.

This material is based upon work supported by the National Science Foundation under Grant No. 0932078 000, while the author was in residence at the Mathematical Science Research Institute (MSRI) in Berkeley, California, during 2013. This research project is also implemented within the framework of the Action "Supporting Postdoctoral Researchers" of the Operational Program "Education and Lifelong Learning" (Action's Beneficiary: General Secretariat for Research and Technology), and is cofinanced by the European Social Fund (ESF) and the Greek State.

References

[Alvis 1987] D. Alvis, "The left cells of the Coxeter group of type H_4", *J. Algebra* **107**:1 (1987), 160–168.

[Alvis and Lusztig 1982] D. Alvis and G. Lusztig, "The representations and generic degrees of the Hecke algebra of type H_4", *J. Reine Angew. Math.* **336** (1982), 201–212.

[Ariki 1996] S. Ariki, "On the decomposition numbers of the Hecke algebra of $G(m, 1, n)$", *J. Math. Kyoto Univ.* **36**:4 (1996), 789–808.

[Bellamy 2009] G. Bellamy, "On singular Calogero-Moser spaces", *Bull. Lond. Math. Soc.* **41**:2 (2009), 315–326.

[Bellamy 2012a] G. Bellamy, "The Calogero-Moser partition for $G(m, d, n)$", *Nagoya Math. J.* **207** (2012), 47–77.

[Bellamy 2012b] G. Bellamy, "Symplectic reflection algebras", preprint, 2012. arXiv 1210.1239

[Benard 1976] M. Benard, "Schur indices and splitting fields of the unitary reflection groups", *J. Algebra* **38**:2 (1976), 318–342.

[Benson 1979] C. T. Benson, "The generic degrees of the irreducible characters of E_8", *Comm. Algebra* **7**:11 (1979), 1199–1209.

[Benson and Gay 1977] C. T. Benson and D. A. Gay, "On dimension functions of the generic algebra of type D_n", *J. Algebra* **45**:2 (1977), 435–438.

[Bessis 1997] D. Bessis, "Sur le corps de définition d'un groupe de réflexions complexe", *Comm. Algebra* **25**:8 (1997), 2703–2716.

[Bonnafé 2006] C. Bonnafé, "Two-sided cells in type B (asymptotic case)", *J. Algebra* **304**:1 (2006), 216–236.

[Bonnafé 2009] C. Bonnafé, "Semicontinuity properties of Kazhdan-Lusztig cells", *New Zealand J. Math.* **39** (2009), 171–192.

[Bonnafé and Iancu 2003] C. Bonnafé and L. Iancu, "Left cells in type B_n with unequal parameters", *Represent. Theory* **7** (2003), 587–609 (electronic).

[Bonnafé and Rouquier 2013] C. Bonnafé and R. Rouquier, "Cellules de Calogero–Moser", preprint, 2013. arXiv 1302.2720

[Bonnafé et al. 2010] C. Bonnafé, M. Geck, L. Iancu, and T. Lam, "On domino insertion and Kazhdan-Lusztig cells in type B_n", pp. 33–54 in *Representation theory of algebraic groups and quantum groups*, Progr. Math. **284**, Birkhäuser/Springer, New York, 2010.

[Bourbaki 2002] N. Bourbaki, *Lie groups and Lie algebras, Chapters 4–6*, Springer, Berlin, 2002.

[Braverman and Gaitsgory 1996] A. Braverman and D. Gaitsgory, "Poincaré-Birkhoff-Witt theorem for quadratic algebras of Koszul type", *J. Algebra* **181**:2 (1996), 315–328.

[Broué and Kim 2002] M. Broué and S. Kim, "Familles de caractères des algèbres de Hecke cyclotomiques", *Adv. Math.* **172**:1 (2002), 53–136.

[Broué and Malle 1993] M. Broué and G. Malle, *Zyklotomische Heckealgebren*, Astérisque **212**, Société Mathématique de France, Paris, 1993.

[Broué et al. 1998] M. Broué, G. Malle, and R. Rouquier, "Complex reflection groups, braid groups, Hecke algebras", *J. Reine Angew. Math.* **500** (1998), 127–190.

[Broué et al. 1999] M. Broué, G. Malle, and J. Michel, "Towards spetses, I", *Transform. Groups* **4**:2-3 (1999), 157–218. Dedicated to the memory of Claude Chevalley.

[Broué et al. 2014] M. Broué, G. Malle, and J. Michel, *Split spetses for primitive reflection groups*, Astérisque **359**, 2014.

[Brown and Gordon 2001] K. A. Brown and I. Gordon, "The ramification of centres: Lie algebras in positive characteristic and quantised enveloping algebras", *Math. Z.* **238**:4 (2001), 733–779.

[Chevalley 1955] C. Chevalley, "Invariants of finite groups generated by reflections", *Amer. J. Math.* **77** (1955), 778–782.

[Chlouveraki 2007] M. Chlouveraki, "Rouquier blocks of the cyclotomic Hecke algebras", *C. R. Math. Acad. Sci. Paris* **344**:10 (2007), 615–620.

[Chlouveraki 2008a] M. Chlouveraki, "Degree and valuation of the Schur elements of cyclotomic Hecke algebras", *J. Algebra* **320**:11 (2008), 3935–3949.

[Chlouveraki 2008b] M. Chlouveraki, "Rouquier blocks of the cyclotomic Ariki-Koike algebras", *Algebra Number Theory* **2**:6 (2008), 689–720.

[Chlouveraki 2009] M. Chlouveraki, *Blocks and families for cyclotomic Hecke algebras*, vol. 1981, Lecture Notes in Mathematics, Springer-Verlag, Berlin, 2009.

[Chlouveraki 2010] M. Chlouveraki, "Rouquier blocks of the cyclotomic Hecke algebras of $G(de, e, r)$", *Nagoya Math. J.* **197** (2010), 175–212.

[Chlouveraki and Jacon 2011] M. Chlouveraki and N. Jacon, "Schur elements and basic sets for cyclotomic Hecke algebras", *J. Algebra Appl.* **10**:5 (2011), 979–993.

[Chlouveraki and Jacon 2012] M. Chlouveraki and N. Jacon, "Schur elements for the Ariki-Koike algebra and applications", *J. Algebraic Combin.* **35**:2 (2012), 291–311.

[Chlouveraki and Miyachi 2011] M. Chlouveraki and H. Miyachi, "Decomposition matrices for d-Harish-Chandra series: the exceptional rank two cases", *LMS J. Comput. Math.* **14** (2011), 271–290.

[Chlouveraki et al. 2012] M. Chlouveraki, I. Gordon, and S. Griffeth, "Cell modules and canonical basic sets for Hecke algebras from Cherednik algebras", pp. 77–89 in *New trends in noncommutative algebra*, Contemp. Math. **562**, Amer. Math. Soc., Providence, RI, 2012.

[Cline et al. 1988] E. Cline, B. Parshall, and L. Scott, "Finite-dimensional algebras and highest weight categories", *J. Reine Angew. Math.* **391** (1988), 85–99.

[du Cloux 2006] F. du Cloux, "Positivity results for the Hecke algebras of noncrystallographic finite Coxeter groups", *J. Algebra* **303**:2 (2006), 731–741.

[Cohen 1980] A. M. Cohen, "Finite quaternionic reflection groups", *J. Algebra* **64**:2 (1980), 293–324.

[Crawley-Boevey and Holland 1998] W. Crawley-Boevey and M. P. Holland, "Noncommutative deformations of Kleinian singularities", *Duke Math. J.* **92**:3 (1998), 605–635.

[Curtis and Reiner 1962] C. W. Curtis and I. Reiner, *Representation theory of finite groups and associative algebras*, Pure and Applied Mathematics **11**, Interscience Publishers, New York, 1962.

[Dipper and James 1986] R. Dipper and G. James, "Representations of Hecke algebras of general linear groups", *Proc. London Math. Soc. (3)* **52**:1 (1986), 20–52.

[Drinfeld 1986] V. G. Drinfel′d, "Degenerate affine Hecke algebras and Yangians", *Funktsional. Anal. i Prilozhen.* **20**:1 (1986), 69–70. In Russian; translated in *Funct. Anal. Appl.* **20** (1986), 58–60.

[Dunkl and Griffeth 2010] C. Dunkl and S. Griffeth, "Generalized Jack polynomials and the representation theory of rational Cherednik algebras", *Selecta Math. (N.S.)* **16**:4 (2010), 791–818.

[Etingof and Ginzburg 2002] P. Etingof and V. Ginzburg, "Symplectic reflection algebras, Calogero-Moser space, and deformed Harish-Chandra homomorphism", *Invent. Math.* **147**:2 (2002), 243–348.

[Geck 1998] M. Geck, "Kazhdan-Lusztig cells and decomposition numbers", *Represent. Theory* **2** (1998), 264–277 (electronic).

[Geck 2004] M. Geck, "Computing Kazhdan-Lusztig cells for unequal parameters", *J. Algebra* **281**:1 (2004), 342–365.

[Geck 2005] M. Geck, "Left cells and constructible representations", *Represent. Theory* **9** (2005), 385–416 (electronic). Erratum, **11** (2007), 172–173.

[Geck 2007a] M. Geck, "Hecke algebras of finite type are cellular", *Invent. Math.* **169**:3 (2007), 501–517.

[Geck 2007b] M. Geck, "Modular representations of Hecke algebras", pp. 301–353 in *Group representation theory*, EPFL Press, Lausanne, 2007.

[Geck 2009] M. Geck, "Leading coefficients and cellular bases of Hecke algebras", *Proc. Edinb. Math. Soc. (2)* **52**:3 (2009), 653–677.

[Geck 2011] M. Geck, "On Iwahori-Hecke algebras with unequal parameters and Lusztig's isomorphism theorem", *Pure Appl. Math. Q.* **7**:3, Special Issue: In honor of Jacques Tits (2011), 587–620.

[Geck and Jacon 2006] M. Geck and N. Jacon, "Canonical basic sets in type B_n", *J. Algebra* **306**:1 (2006), 104–127.

[Geck and Jacon 2011] M. Geck and N. Jacon, *Representations of Hecke algebras at roots of unity*, vol. 15, Algebra and Applications, Springer, London, 2011.

[Geck and Pfeiffer 2000] M. Geck and G. Pfeiffer, *Characters of finite Coxeter groups and Iwahori-Hecke algebras*, vol. 21, London Mathematical Society Monographs. New Series, The Clarendon Press, Oxford University Press, New York, 2000.

[Geck and Rouquier 2001] M. Geck and R. Rouquier, "Filtrations on projective modules for Iwahori-Hecke algebras", pp. 211–221 in *Modular representation theory of finite groups (Charlottesville, VA, 1998)*, de Gruyter, Berlin, 2001.

[Geck et al. 2000] M. Geck, L. Iancu, and G. Malle, "Weights of Markov traces and generic degrees", *Indag. Math. (N.S.)* **11**:3 (2000), 379–397.

[Genet and Jacon 2006] G. Genet and N. Jacon, "Modular representations of cyclotomic Hecke algebras of type $G(r, p, n)$", *Int. Math. Res. Not.* **2006** (2006), Art. ID 93049, 18.

[Gerber 2014] T. Gerber, "Generalised canonical basic sets for Ariki-Koike algebras", *J. Algebra* **413** (2014), 364–401.

[Ginzburg and Kaledin 2004] V. Ginzburg and D. Kaledin, "Poisson deformations of symplectic quotient singularities", *Adv. Math.* **186**:1 (2004), 1–57.

[Ginzburg et al. 2003] V. Ginzburg, N. Guay, E. Opdam, and R. Rouquier, "On the category $\mathcal{O}$ for rational Cherednik algebras", *Invent. Math.* **154**:3 (2003), 617–651.

[Gordon 2003] I. Gordon, "Baby Verma modules for rational Cherednik algebras", *Bull. London Math. Soc.* **35**:3 (2003), 321–336.

[Gordon 2008] I. G. Gordon, "Symplectic reflection alegebras", pp. 285–347 in *Trends in representation theory of algebras and related topics*, EMS Ser. Congr. Rep., Eur. Math. Soc., Zürich, 2008.

[Gordon and Martino 2009] I. G. Gordon and M. Martino, "Calogero-Moser space, restricted rational Cherednik algebras and two-sided cells", *Math. Res. Lett.* **16**:2 (2009), 255–262.

[Guralnick and Saxl 2003] R. M. Guralnick and J. Saxl, "Generation of finite almost simple groups by conjugates", *J. Algebra* **268**:2 (2003), 519–571.

[Hoefsmit 1974] P. N. Hoefsmit, *Representations of Hecke algebras of finite groups with BN-pairs of classical type*, Ph.D. thesis, University of British Columbia, Ann Arbor, MI, 1974, http://search.proquest.com/docview/302757114.

[Holmes and Nakano 1991] R. R. Holmes and D. K. Nakano, "Brauer-type reciprocity for a class of graded associative algebras", *J. Algebra* **144**:1 (1991), 117–126.

[Huffman and Wales 1976] W. C. Huffman and D. B. Wales, "Linear groups containing an element with an eigenspace of codimension two", pp. 425–429 in *Proceedings of the Conference on Finite Groups (Univ. Utah, Park City, Utah, 1975)*, Academic Press, New York, 1976.

[Jacon 2004] N. Jacon, "On the parametrization of the simple modules for Ariki-Koike algebras at roots of unity", *J. Math. Kyoto Univ.* **44**:4 (2004), 729–767.

[Jacon 2007] N. Jacon, "Crystal graphs of higher level q-deformed Fock spaces, Lusztig a-values and Ariki-Koike algebras", *Algebr. Represent. Theory* **10**:6 (2007), 565–591.

[Kazhdan and Lusztig 1979] D. Kazhdan and G. Lusztig, "Representations of Coxeter groups and Hecke algebras", *Invent. Math.* **53**:2 (1979), 165–184.

[Kazhdan and Lusztig 1980] D. Kazhdan and G. Lusztig, "Schubert varieties and Poincaré duality", pp. 185–203 in *Geometry of the Laplace operator (Proc. Sympos. Pure Math., Univ. Hawaii, Honolulu, Hawaii, 1979)*, Proc. Sympos. Pure Math., XXXVI, Amer. Math. Soc., Providence, R.I., 1980.

[Kilmoyer and Solomon 1973] R. Kilmoyer and L. Solomon, "On the theorem of Feit-Higman", *J. Combinatorial Theory Ser. A* **15** (1973), 310–322.

[Kim 2005] S. Kim, "Families of the characters of the cyclotomic Hecke algebras of $G(de, e, r)$", *J. Algebra* **289**:2 (2005), 346–364.

[Lusztig 1979] G. Lusztig, "Unipotent representations of a finite Chevalley group of type E_8", *Quart. J. Math. Oxford Ser. (2)* **30**:119 (1979), 315–338.

[Lusztig 1982] G. Lusztig, "A class of irreducible representations of a Weyl group. II", *Nederl. Akad. Wetensch. Indag. Math.* **44**:2 (1982), 219–226.

[Lusztig 1983] G. Lusztig, "Left cells in Weyl groups", pp. 99–111 in *Lie group representations, I (College Park, Md., 1982/1983)*, Lecture Notes in Math. **1024**, Springer, Berlin, 1983.

[Lusztig 1984] G. Lusztig, *Characters of reductive groups over a finite field*, Annals of Mathematics Studies **107**, Princeton University Press, Princeton, NJ, 1984.

[Lusztig 1987] G. Lusztig, "Leading coefficients of character values of Hecke algebras", pp. 235–262 in *The Arcata Conference on Representations of Finite Groups (Arcata, Calif., 1986)*, Proc. Sympos. Pure Math. **47**, Amer. Math. Soc., Providence, RI, 1987.

[Lusztig 2003] G. Lusztig, *Hecke algebras with unequal parameters*, vol. 18, CRM Monograph Series, American Mathematical Society, Providence, RI, 2003.

[Malle 1995] G. Malle, "Unipotente Grade imprimitiver komplexer Spiegelungsgruppen", *J. Algebra* **177**:3 (1995), 768–826.

[Malle 1997] G. Malle, "Degrés relatifs des algèbres cyclotomiques associées aux groupes de réflexions complexes de dimension deux", pp. 311–332 in *Finite reductive groups (Luminy, 1994)*, Progr. Math. **141**, Birkhäuser, Boston, MA, 1997.

[Malle 1999] G. Malle, "On the rationality and fake degrees of characters of cyclotomic algebras", *J. Math. Sci. Univ. Tokyo* **6**:4 (1999), 647–677.

[Malle 2000] G. Malle, "On the generic degrees of cyclotomic algebras", *Represent. Theory* **4** (2000), 342–369 (electronic).

[Malle and Mathas 1998] G. Malle and A. Mathas, "Symmetric cyclotomic Hecke algebras", *J. Algebra* **205**:1 (1998), 275–293.

[Malle and Michel 2010] G. Malle and J. Michel, "Constructing representations of Hecke algebras for complex reflection groups", *LMS J. Comput. Math.* **13** (2010), 426–450.

[Malle and Rouquier 2003] G. Malle and R. Rouquier, "Familles de caractères de groupes de réflexions complexes", *Represent. Theory* **7** (2003), 610–640 (electronic).

[Marin 2012] I. Marin, "The cubic Hecke algebra on at most 5 strands", *J. Pure Appl. Algebra* **216**:12 (2012), 2754–2782.

[Marin 2014] I. Marin, "The freeness conjecture for Hecke algebras of complex reflection groups, and the case of the Hessian group G_{26}", *J. Pure Appl. Algebra* **218**:4 (2014), 704–720.

[Martino 2010] M. Martino, "The Calogero-Moser partition and Rouquier families for complex reflection groups", *J. Algebra* **323**:1 (2010), 193–205.

[Mathas 2004] A. Mathas, "Matrix units and generic degrees for the Ariki-Koike algebras", *J. Algebra* **281**:2 (2004), 695–730.

[Namikawa 2011] Y. Namikawa, "Poisson deformations of affine symplectic varieties", *Duke Math. J.* **156**:1 (2011), 51–85.

[Ram and Shepler 2003] A. Ram and A. V. Shepler, "Classification of graded Hecke algebras for complex reflection groups", *Comment. Math. Helv.* **78**:2 (2003), 308–334.

[Rouquier 1999] R. Rouquier, "Familles et blocs d'algèbres de Hecke", *C. R. Acad. Sci. Paris Sér. I Math.* **329**:12 (1999), 1037–1042.

[Rouquier 2008] R. Rouquier, "q-Schur algebras and complex reflection groups", *Mosc. Math. J.* **8**:1 (2008), 119–158, 184.

[Shephard and Todd 1954] G. C. Shephard and J. A. Todd, "Finite unitary reflection groups", *Canadian J. Math.* **6** (1954), 274–304.

[Springer 1982] T. A. Springer, "Quelques applications de la cohomologie d'intersection", pp. 249–273, exposé 589 in *Bourbaki Seminar, Vol. 1981/1982*, Astérisque **92**, Soc. Math. France, Paris, 1982.

[Steinberg 1951] R. Steinberg, "A geometric approach to the representations of the full linear group over a Galois field", *Trans. Amer. Math. Soc.* **71** (1951), 274–282.

[Steinberg 1964] R. Steinberg, "Differential equations invariant under finite reflection groups", *Trans. Amer. Math. Soc.* **112** (1964), 392–400.

[Surowski 1978] D. B. Surowski, "Degrees of irreducible characters of (B, N)-pairs of types E_6 and E_7", *Trans. Amer. Math. Soc.* **243** (1978), 235–249.

[Thiel 2014] U. Thiel, "A counter-example to Martino's conjecture about generic Calogero-Moser families", *Algebr. Represent. Theory* **17**:5 (2014), 1323–1348.

[Uglov 2000] D. Uglov, "Canonical bases of higher-level q-deformed Fock spaces and Kazhdan-Lusztig polynomials", pp. 249–299 in *Physical combinatorics (Kyoto, 1999)*, Progr. Math. **191**, Birkhäuser Boston, Boston, MA, 2000.

[Vale 2006] R. Vale, *On category $\mathcal{O}$ for the rational Cherednik algebra of the complex reflection group* $(\mathbb{Z}/\ell\mathbb{Z}) \wr S_n$, Ph.D. thesis, University of Glasgow, 2006.

[Verbitsky 2000] M. Verbitsky, "Holomorphic symplectic geometry and orbifold singularities", *Asian J. Math.* **4**:3 (2000), 553–563.

[Wang 1999] W. Wang, "Hilbert schemes, wreath products, and the McKay correspondence", preprint, 1999. arXiv 9912104

maria.chlouveraki@uvsq.fr *Laboratoire de Mathématiques UVSQ, Bâtiment Fermat, 45 Avenue des États-Unis, 78035 Versailles cedex, France*

Commutative Algebra and Noncommutative Algebraic Geometry, I
MSRI Publications
Volume **67**, 2015

Limits in commutative algebra and algebraic geometry

STEVEN DALE CUTKOSKY

In this survey article we explore various limits and asymptotic properties in commutative algebra and algebraic geometry. We show that several important invariants have good asymptotic behavior. We develop this and give some examples of pathological behavior for topics such as multiplicity of graded families of ideals, volumes of line bundles on schemes and regularity of powers of ideals.

1. Introduction

This article is on the general theme of limits arising in commutative algebra and algebraic geometry, in asymptotic multiplicity and rgularity. The four sections of this article are based on the four talks that I gave during the Spring semester of the special year on Commutative Algebra, held at MSRI during the 2012–2013 academic year. The first section is based on an Evans lecture I gave at Berkeley.

2. Asymptotic multiplicities

Multiplicity and projection from a point. We begin by discussing a formula involving the multiplicity of a point on a variety, which evolved classically. Proofs of the formula (1) can be found in Theorem 5.11 of [Mumford 1976] (over $k = \mathbb{C}$), in Section 11 of [Abhyankar 1998], Section 12 of [Lipman 1975] and Theorem 12.1 [Cutkosky 2009].

Suppose that $k = \bar{k}$ is an algebraically closed field and $X \subset \mathbb{P}_k^N$ is a d-dimensional projective variety. The degree of X is defined as

$$\#(X \cap L^{N-d})$$

where L^{N-d} is a generic linear subspace of $\mathbb{P}^N$ of dimension $N-d$. Suppose that $z \in \mathbb{P}^N$ is a closed point. Let $\pi_z : \mathbb{P}^N \dashrightarrow \mathbb{P}^{N-1}$ be the projection from z and $Y = \pi_z(X)$. We have the following formula relating $\deg(X)$ and $\deg(Y)$.

Partially supported by the National Science Foundation.

Theorem 2.1. *Suppose that $z \in X$ and X is not a cone over z. Then*

$$\deg(X) = [k(X):k(Y)]\deg(Y) + \text{mult}_z(X). \tag{1}$$

The index of function fields $[k(X):k(Y)]$ is equal to the number of points in X above a general point q of Y. In the case when $z \notin X$, $\text{mult}_z(X) = 0$. When $z \in X$, the correction term $\text{mult}_z(X)$ is the multiplicity $e_{m_R}(R)$ of the local ring $R = \mathcal{O}_{X,z}$, which is defined below. Formula (1), together with its role in resolution of singularities, was discussed by Zariski [1935, pp. 21–22]. At that time the correction term $\text{mult}_z(X)$ was not completely understood. In some cases, this number is a local intersection number, but not always. In fact, we have

$$e_{m_R}(R) \leq \dim_k(\mathcal{O}_{X,z})/I(L')\mathcal{O}_{X,z}) \tag{2}$$

where L' is a general linear space through z, and (2) is an equality if and only if $\mathcal{O}_{X,z}$ is Cohen–Macaulay [Zariski and Samuel 1960, Chapter VIII, Section 10, Theorem 23].

Multiplicity has a purely geometric construction (over $\mathbb{C}$), as explained in [Mumford 1976].

Multiplicity, graded families of ideals, filtrations and the Zariski subspace theorem. From now on in this section we suppose that (R, m_R) is a (Noetherian) local ring of dimension d.

A family of ideals $\{I_n\}_{n\in\mathbb{N}}$ of R is called a graded family of ideals if $I_0 = R$ and $I_m I_n \subseteq I_{m+n}$ for all m, n. $\{I_n\}$ is a filtration of R if, in addition, $I_{n+1} \subseteq I_n$ for all n.

The most basic example is $I_n = J^n$ where J is a fixed ideal of R.

Suppose that N is a finitely generated R-module, and I is an m_R-primary ideal. Let

$$t = \dim R/\text{ann}(N)$$

be the dimension of N. Let ℓ_R be the length of an R-module.

The theorem of Hilbert–Samuel is that the function $\ell_R(N/I^nN)$ is a polynomial in n of degree t for $n \gg 0$. This polynomial is called the Hilbert–Samuel polynomial.

The multiplicity of a finitely generated R-module N with respect to an m_R-primary ideal I is the leading coefficient of this polynomial times $t!$; that is,

$$e_I(N) = \lim_{n\to\infty} \frac{\ell_R(N/I^nN)}{n^r/t!}. \tag{3}$$

This multiplicity is always a natural number.

Multiplicity is the most basic invariant in resolution of singularities. Suppose X is an algebraic variety. A point $p \in X$ is nonsingular if $\mathcal{O}_{X,p}$ is a regular local ring [Zariski 1947]. The following theorem shows that $p \in X$ is a nonsingular point if and only if its multiplicity $e_{m_R}(R) = 1$ where $R = \mathcal{O}_{X,p}$.

Theorem 2.2. *If R is a regular local ring then $e_{m_R}(R) = 1$. If $e_{m_R}(R) = 1$ and R is formally equidimensional, then R is a regular local ring.*

The first statement follows since the associated graded ring of a regular local ring is a polynomial ring. The second statement is Theorem 40.6 of [Nagata 1959].

Suppose that $R \subset S$ are local rings with $m_R = m_S \cap R$. $I_n = m_S^n \cap R$ is a filtration of R. Let $\hat{R} = \lim_{\rightarrow} R/m_R^n$ be the m_R-adic completion of R and $\hat{S} = \lim_{\rightarrow} S/m_S^n$ be the m_S-adic completion of S.

The induced map $\hat{R} \to \hat{S}$ is an inclusion if and only if there exists a function $\sigma(n)$ such that $I_n \subset m_R^{\sigma(n)}$ and $\sigma(n) \to \infty$ as $n \to \infty$.

A fundamental theorem in algebraic geometry is the Zariski subspace theorem.

Theorem 2.3 [Zariski 1949; Abhyankar 1998, 10.6]. *Suppose that $R \subset S$ are local domains, essentially of finite type over a field k and R is analytically irreducible ($\hat{R}$ is a domain). Then the induced map $\hat{R} \to \hat{S}$ is an inclusion.*

While this is a theorem in algebraic geometry, the subspace theorem fails in complex analytic geometry.

Theorem 2.4 [Gabrielov 1971]. *There exist inclusions $R \subset S$ of convergent complex power series such that the induced map $\hat{R} \to \hat{S}$ of formal power series is not an inclusion.*

Even though the Zariski subspace theorem is true, the induced filtration $m_S^n \cap R$ is not the best behaved.

Theorem 2.5 [Cutkosky and Srinivas 1993]. *There exists an inclusion $R \to S$ of d-dimensional normal domains, essentially of finite type over the complex numbers, such that*

$$\lim_{n\to\infty} \frac{\ell_R(R/I_n)}{n^d}$$

exists but is an irrational number, where $I_n = m_S^n \cap R$.

The irrationality of the limit implies that $\bigoplus_{n\geq 0} I_n$ is not a finitely generated R-algebra.

Limits of lengths of graded families of ideals. Suppose that $\{I_n\}_{n\in\mathbb{N}}$ is a graded family of m_R-primary ideals (I_n is m_R-primary for $n \geq 1$) in a d-dimensional (Noetherian) local ring R. We pose the following question:

Question 2.6. When does

$$\lim_{n\to\infty} \frac{\ell_R(R/I_n)}{n^d} \tag{4}$$

exist?

This problem was considered in [Ein, Lazarsfeld and Smith 2003] and [Mustaţă 2002].

Lazarsfeld and Mustaţă [2009] showed that the limit exists for all graded families of m_R-primary ideals in R if R is a domain which is essentially of finite type over an algebraically closed field k with $R/m_R = k$. All of these assumptions are necessary in their proof. Their proof is by reducing the problem to one on graded linear series on a projective variety, and then using a method introduced by Okounkov [2003] to reduce the problem to one of counting points in an integral semigroup.

In [Cutkosky 2013a], it is shown that the limit exists for all graded families of m_R-primary ideals in R if R is analytically unramified ($\hat{R}$ is reduced), equicharacteristic and R/m_R is perfect.

In Example 5.3 of [Cutkosky 2014], an example is given of a nonreduced local ring R with a graded family of m_R-primary ideals $\{I_n\}$ such that the limit (4) does not exist. Dao and Smirnov communicated to me that they found this same example. They further showed that it was universal, proving the following theorem.

The nilradical $N(R)$ of a d-dimensional ring R is

$$N(R) = \{x \in R \mid x^n = 0 \text{ for some positive integer } n\}.$$

Recall that

$$\dim N(R) = \dim R/\text{ann}(N(R)),$$

so that $\dim N(R) = d$ if and only if there exists a minimal prime P of R such that $\dim R/P = d$ and R_P is not reduced.

Theorem 2.7 (Dao and Smirnov). *Suppose that R is a d-dimensional local ring such that* $\dim N(R) = d$. *Then there exists a graded family of m_R-primary ideals $\{I_n\}$ of R such that*

$$\lim_{n\to\infty} \frac{\ell_R(R/I_n)}{n^d}$$

does not exist.

We now state our general theorem, which gives necessary and sufficient conditions on a local ring R for all limits of graded families of m_R-primary ideals to exist.

Theorem 2.8 [Cutkosky 2014, Theorem 5.5]. *Suppose that R is a d-dimensional local ring. Then the limit*

$$\lim_{n\to\infty} \frac{\ell_R(R/I_n)}{n^d}$$

exists for all graded families of m_R-primary ideals $\{I_n\}$ of R if and only if $\dim N(\hat{R}) < d$.

If R is excellent, then $N(\hat{R}) = N(R)\hat{R}$, and the theorem is true with the condition $\dim N(\hat{R}) < d$ replaced with $\dim N(R) < d$. However, there exist Noetherian local domains R (so that $N(R) = 0$) such that $\dim N(\hat{R}) = \dim R$ [Nagata 1959, (E3.2)].

Before giving the proof of Theorem 2.8 in the next section, we turn to some of the theorem's applications. We start with some general "volume = multiplicity" formulas.

Theorem 2.9 [Cutkosky 2014, Theorem 6.5]. *Suppose that R is a d-dimensional, analytically unramified local ring, and $\{I_n\}$ is a graded family of m_R-primary ideals in R. Then*

$$\lim_{n\to\infty} \frac{\ell_R(R/I_n)}{n^d/d!} = \lim_{p\to\infty} \frac{e_{I_p}(R)}{p^d}$$

Related formulas have been proven in [Ein, Lazarsfeld and Smith 2003; Mustaţă 2002; Lazarsfeld and Mustaţă 2009]. This last paper proves the formula when R is essentially of finite type over an algebraically closed field k with $R/m_R = k$. All of these assumptions are necessary in their proof.

Suppose that R is a (Noetherian) local ring and I, J are ideals in R. The generalized symbolic power $I_n(J)$ is defined by

$$I_n(J) = I^n : J^\infty = \bigcup_{i=1}^{\infty} I^n : J^i.$$

Theorem 2.10 [Cutkosky 2014, Corollary 6.4]. *Suppose that R is an analytically unramified d-dimensional local ring. Let s be the constant limit dimension $s = \dim I_n(J)/I^n$ for $n \gg 0$. Suppose that $s < d$. Then*

$$\lim_{n\to\infty} \frac{e_{m_R}(I_n(J)/I^n)}{n^{d-s}}$$

exists.

This was proven in [Herzog, Puthenpurakal and Verma 2008] for ideals I and J in a d-dimensional local ring, with the assumption that $\bigoplus_{n\geq 0} I_n(J)$ is a finitely generated R-algebra.

3. The proof of Theorem 2.8

In this section we outline the proof of Theorem 2.8. See [Cutkosky 2013a; 2014] for details.

Proof that* $\dim N(\hat{R}) < d$ *implies limits exist. Suppose that R is a d-dimensional local ring with $\dim N(R) < d$ and $\{I_n\}$ is a graded family of m_R-primary ideals in R.

We have $\ell_{\hat{R}}(R/I_n\hat{R}) = \ell_R(R/I_n)$ for all n, so we may assume that $R = \hat{R}$ is complete; in particular, we may assume that R is excellent with $\dim N(R) < d$. There exists a positive integer c such that $m_R^c \subseteq I_1$, which implies that

$$m_R^{nc} \subseteq I_n \text{ for all positive } n. \tag{5}$$

Let $N = N(R)$ and $A = R/N$. We have short exact sequences

$$0 \to N/N \cap I_i R \to R/I_i R \to A/I_i A \to 0,$$

from which we deduce that there exists a constant $\alpha > 0$ such that

$$\ell_R(N/N \cap I_i R) \leq \ell_R(N/m_R^{ci} N) \leq \alpha i^{\dim N} \leq \alpha i^{d-1}.$$

Replacing R with A and I_n with $I_n A$, we thus reduce to the case that R is reduced. Using the following lemma, we then reduce to the case that R is a complete domain (so that it is analytically irreducible).

Lemma 3.1 [Cutkosky 2013a, Corollary 6.4]. *Suppose that R is a reduced local domain, and $\{I_n\}$ is a graded family of m_R-primary ideals in R. Let $\{P_1, \dots, P_s\}$ be the set of minimal primes of R and let $R_i = R/P_i$. Then there exists $\alpha > 0$ such that*

$$\left|\left(\sum_{i=1}^{s} \ell_{R_i}(R_i/I_n R_i)\right) - \ell_R(R/I_n)\right| \leq \alpha n^{d-1} \quad \textit{for all } n.$$

We now present a method introduced in [Okounkov 2003] to compute limits of multiplicities. The method has been refined in [Lazarsfeld and Mustaţă 2009; Kaveh and Khovanskii 2012].

Suppose that $\Gamma \subset \mathbb{N}^{d+1}$ is a semigroup. Let $\Sigma(\Gamma)$ be the closed convex cone generated by Γ in $\mathbb{R}^{d+1}$. Define $\Delta(\Gamma) = \Sigma(\Gamma) \cap (\mathbb{R}^d \times \{1\})$. For $i \in \mathbb{N}$, let $\Gamma_i = \Gamma \cap (\mathbb{N}^d \times \{i\})$.

Theorem 3.2 [Okounkov 2003; Lazarsfeld and Mustaţă 2009]. *Suppose that Γ satisfies these conditions:*

1) *There exist finitely many vectors $(v_i, 1) \in \mathbb{N}^{d+1}$ spanning a semigroup $B \subseteq \mathbb{N}^{d+1}$ such that $\Gamma \subseteq B$ (boundedness).*
2) *The subgroup generated by Γ is the full integral lattice $\mathbb{Z}^{d+1}$.*

Then

$$\lim_{i \to \infty} \frac{\#\Gamma_i}{i^d} = \mathrm{vol}(\Delta(\Gamma))$$

exists.

We now return to the proof that $\dim N(\hat{R}) < d$ implies limits exist. Recall that we have reduced to the case that R is a complete domain. Let $\pi : X \to \operatorname{spec}(R)$ be the normalization of the blowup of m_R. Being excellent, X is of finite type over R. X is regular in codimension 1, so there exists a closed point $p \in \pi^{-1}(m_R)$ such that $S = \mathcal{O}_{X,p}$ is regular and dominates R. We have an inclusion $R \to S$ of d-dimensional local rings such that $m_S \cap R = m_R$ with equality of quotient fields $Q(R) = Q(S)$. Let $k = R/m_R$, $k' = S/m_S$. Since S is essentially of finite type over R, we have that $[k' : k] < \infty$. Let $y_1, \ldots, y_d$ be regular parameters in S. Choose $\lambda_1, \ldots, \lambda_d \in \mathbb{R}_+$ which are rationally independent with $\lambda_i \geq 1$. Prescribe a rank 1 valuation ν on $Q(R)$ by

$$\nu(y_1^{i_1} \cdots y_d^{i_d}) = i_1\lambda_1 + \cdots + i_d\lambda_d$$

and $\nu(\gamma) = 0$ if $\gamma \in S$ is a unit. The value group of ν is

$$\Gamma_\nu = \lambda_1\mathbb{Z} + \cdots + \lambda_d\mathbb{Z} \subseteq \mathbb{R}.$$

Let V_ν be the valuation ring of ν. Then

$$k' = S/m_S \cong V_\nu/m_\nu.$$

For $\lambda \in \mathbb{R}_+$, define valuation ideals in V_ν by

$$K_\lambda = \{f \in Q(R) \mid \nu(f) \geq \lambda\}$$

and

$$K_\lambda^+ = \{f \in Q(R) \mid \nu(f) > \lambda\}.$$

Now suppose that $I \subset R$ is an ideal and $\lambda \in \Gamma_\nu$ is nonnegative. We have an inclusion

$$I \cap K_\lambda / I \cap K_\lambda^+ \subseteq K_\lambda / K_{\lambda^+} \cong k'.$$

Thus

$$\dim_k I \cap K_\lambda / I \cap K_\lambda^+ \leq [k' : k].$$

Lemma 3.3 [Cutkosky 2013a, Lemma 4.3]. *There exists $\alpha \in \mathbb{Z}_+$ such that $K_{\alpha n} \cap R \subset m_R^n$ for all $n \in \mathbb{Z}_+$.*

The proof uses Huebl's linear Zariski subspace theorem [Hübl 2001] or Rees' Izumi theorem [Rees 1989]. The assumption that R is analytically irreducible is necessary for the lemma. Recalling the constant c of (5), let $\beta = \alpha c$. Then

$$K_{\beta n} \cap R \subset m_R^{nc} \subseteq I_n \tag{6}$$

for all n. For $1 \le t \le [k':k]$, define

$$\Gamma^{(t)} = \big\{(n_1,\dots,n_d,i)\in\mathbb{N}^{d+1} \mid \dim_k I_i\cap K_{n_1\lambda_1+\cdots+n_d\lambda_d}/I_i\cap K^+_{n_1\lambda_1+\cdots+n_d\lambda_d}\ge t \text{ and } n_1+\cdots+n_d\le\beta i\big\},$$

$$\hat\Gamma^{(t)} = \big\{(n_1,\dots,n_d,i)\in\mathbb{N}^{d+1} \mid \dim_k R\cap K_{n_1\lambda_1+\cdots+n_d\lambda_d}/R\cap K^+_{n_1\lambda_1+\cdots+n_d\lambda_d}\ge t \text{ and } n_1+\cdots+n_d\le\beta i\big\}.$$

Lemma 3.4 [Cutkosky 2014, Lemma 4.4]. *Suppose that $t\ge 1$, $0\ne f\in I_i$, $0\ne g\in I_j$ and*

$$\dim_k I_i\cap K_{\nu(f)}/I_i\cap K^+_{\nu(f)}\ge t.$$

Then

$$\dim_k I_{i+j}\cap K_{\nu(fg)}/I_{i+j}\cap K^+_{\nu(fg)}\ge t.$$

Since $\nu(fg)=\nu(f)+\nu(g)$, we conclude that when they are nonempty, $\Gamma^{(t)}$ and $\hat\Gamma^{(t)}$ are subsemigroups of $\mathbb{N}^{d+1}$.

Given $\lambda=n_1\lambda_1+\cdots+n_d\lambda_d$ such that $n_1+\cdots+n_d\le\beta i$, we have

$$\dim_k K_\lambda\cap I_i/K^+_\lambda\cap I_i = \#\{t \mid (n_1,\dots,n_d,i)\in\Gamma^{(t)}\}.$$

Recalling (6), we obtain

$$\begin{aligned}\ell_R(R/I_i) &= \ell_R(R/K_{\beta i}\cap R)-\ell_R(I_i/K_{\beta i}\cap I_i)\\ &= \Big(\sum_{0\le\lambda<\beta i}\dim_k K_\lambda\cap R/K^+_\lambda\cap R\Big)-\Big(\sum_{0\le\lambda<\beta i}\dim_k K_\lambda\cap I_i/K^+_\lambda\cap I_i\Big)\\ &= \Big(\sum_{t=1}^{[k':k]}\#\hat\Gamma_i^{(t)}\Big)-\Big(\sum_{t=1}^{[k':k]}\#\Gamma_i^{(t)}\Big),\end{aligned}$$

where $\Gamma_i^{(t)}=\Gamma^{(t)}\cap(\mathbb{N}^d\times\{i\})$ and $\hat\Gamma_i^{(t)}=\hat\Gamma^{(t)}\cap(\mathbb{N}^d\times\{i\})$. The semigroups $\Gamma^{(t)}$ and $\hat\Gamma^{(t)}$ satisfy the hypotheses of Theorem 3.2. Thus

$$\lim_{i\to\infty}\frac{\#\Gamma_i^{(t)}}{i^d}=\mathrm{vol}(\Delta(\Gamma^{(t)})$$

and

$$\lim_{i\to\infty}\frac{\#\hat\Gamma_i^{(t)}}{i^d}=\mathrm{vol}(\Delta(\hat\Gamma^{(t)})$$

so that

$$\lim_{i\to\infty}\frac{\ell_R(R/I_i)}{i^d}$$

exists.

An example where limits do not exist. Let $i_1 = 2$ and $r_1 = i_1/2$. For $j \geq 1$, inductively define i_{j+1} so that i_{j+1} is even and $i_{j+1} > 2^j i_j$. Let $r_{j+1} = i_{j+1}/2$. For $n \in \mathbb{Z}_+$, define

$$\sigma(n) = \begin{cases} 1 & \text{if } n = 1, \\ i_j/2 & \text{if } i_j \leq n < i_{j+1}. \end{cases} \tag{7}$$

The limit

$$\lim_{n\to\infty} \frac{\sigma(n)}{n} \tag{8}$$

does not exist, even when n is constrained to lie in an arithmetic sequence [Cutkosky 2014, Lemmas 6.1 and 6.2]. The following example shows that limits do not always exist on nonreduced local rings.

Example 3.5 [Cutkosky 2014, Example 5.3]. Let k be a field, $d > 0$ and R be the nonreduced d-dimensional local ring $R = k[[x_1, \ldots, x_d, y]]/(y^2)$. There exists a graded family of m_R-primary ideals $\{I_n\}$ in R such that the limit

$$\lim_{n\to\infty} \frac{\ell_R(R/I_n)}{n^d} \tag{9}$$

does not exist, even when n is constrained to lie in an arithmetic sequence.

This example was also found by Dao and Smirnov. They further showed that it is universal (Theorem 2.7).

Proof. Let $\bar{x}_1, \ldots, \bar{x}_d, \bar{y}$ be the classes of $x_1, \ldots, x_d, y$ in R. Let N_i be the set of monomials of degree i in the variables $\bar{x}_1, \ldots, \bar{x}_d$. Let $\sigma(n)$ be the function defined in (7). Define M_R-primary ideals I_n in R by $I_n = (N_n, \bar{y}N_{n-\sigma(n)})$ for $n \geq 1$ (and $I_0 = R$).

We first verify that $\{I_n\}$ is a graded family of ideals, by showing that $I_m I_n \subset I_{m+n}$ for all $m, n > 0$. This follows since

$$I_m I_n = (N_{m+n}, \bar{y}N_{(m+n)-\sigma(m)}, \bar{y}N_{(m+n)-\sigma(n)})$$

and $\sigma(j) \leq \sigma(k)$ for $k \geq j$.

R/I_n has a k-basis consisting of

$$\{N_i \mid i < n\} \quad \text{and} \quad \{\bar{y}N_j \mid j < n - \sigma(n)\}.$$

Thus

$$\ell_R(R/I_n) = \binom{n}{d} + \binom{n-\sigma(n)}{d},$$

and the limit (9) does not exist, even when n is constrained to lie in an arithmetic sequence, by (8). □

4. Volumes on schemes

Unless stated otherwise, we will assume that X is a d-dimensional proper scheme over an arbitrary field k. The vector space $N(X)$ of $\mathbb{R}$-divisors modulo numerical equivalence is defined in [Kleiman 1966], Chapter 1 of [Lazarsfeld 2004a], and extended to this level of generality (proper schemes over a field) in [Cutkosky 2013b]. In the case of a nonsingular variety, $N(X)$ is defined in the last section of this article.

If $\mathcal{L}$ is a line bundle on X, then the volume of $\mathcal{L}$ is

$$\operatorname{vol}(\mathcal{L}) = \limsup_{n\to\infty} \frac{\dim_k \Gamma(X, \mathcal{L}^n)}{n^d/d!}.$$

If $\mathcal{L}$ is ample, then $\operatorname{vol}(\mathcal{L}) = (\mathcal{L})^d$ (the self intersection number). $\operatorname{vol}(\mathcal{L})$ is well defined on $N(X)$. The volume can be irrational [Cutkosky and Srinivas 1993] even on a nonsingular projective variety over $\mathbb{C}$. The following is a fundamental result.

Theorem 4.1. *Suppose that X is a projective variety over an algebraically closed field k, and $\mathcal{L}$ is a line bundle on X. Then*

$$\operatorname{vol}(\mathcal{L}) = \lim_{n\to\infty} \frac{\dim_k \Gamma(X, \mathcal{L}^n)}{n^d/d!}$$

exists as a limit.

The function vol extends to a continuous, even continuously differentiable d-homogeneous function on $N(X)$ [Lazarsfeld 2004b; Boucksom, Favre and Jonsson 2009; Lazarsfeld and Mustaţă 2009]. This last result is true for $\operatorname{vol}(\mathcal{L})$ on $N(X)$ when X is a proper variety over an arbitrary field [Cutkosky 2013b].

There are several proofs of Theorem 4.1. Lazarsfeld [2004b] gave one when k is algebraically closed of characteristic zero, using Fujita approximation ([Fujita 1994], which requires resolution of singularities). Satoshi Takagi [2007] gave a proof when k is algebraically closed of characteristic $p > 0$ using de Jong's resolution [1996] after alterations and Fujita approximation.

There are also proofs by Okounkov [2003] for ample line bundles and by Lazarsfeld and Mustaţă [2009] using the cone method (using Theorem 3.2).

Definition 4.2. A graded linear series (for a line bundle $\mathcal{L}$) on a d-dimensional proper scheme X over a field k is a graded k-subalgebra

$$L = \bigoplus_{n\geq 0} L_n \subseteq \bigoplus_{n\geq 0} \Gamma(X, \mathcal{L}^n).$$

Of course L need not be a finitely generated k-algebra. For the definition of Iitaka–Kodaira dimension, we need the following definition:

$$\sigma(L) = \max\bigl\{s \mid \text{there exist } y_1, \dots, y_s \in L \text{ homogenenous of positive degree and algebraically independent over } k\bigr\}.$$

We define the Iitaka–Kodaira dimension of L to be

$$\varkappa(L) = \begin{cases} \sigma(L) - 1 & \text{if } \sigma(L) > 0, \\ -\infty & \text{if } \sigma(L) = 0. \end{cases}$$

The index of $\mathcal{L}$ is $m(L) = [\mathbb{Z} : G]$ where G is the subgroup generated by $\{n \mid L_n \neq 0\}$.

If X is reduced and $\varkappa(L) = -\infty$, then $L_m = 0$ for all $m > 0$ and if X is reduced and $\varkappa(L) \geq 0$, then there exist constants $0 < \alpha < \beta$ such that

$$\alpha n^{\varkappa(L)} < \dim_k L_{mn} < \beta n^{\varkappa(L)}$$

for $n \gg 0$. However, if X is not reduced, then we can have $\varkappa(L) = -\infty$ and $\dim_k L_n > n^d$ for all $n \gg 0$ [Cutkosky 2014, Section 12].

Theorem 4.3 ([Lazarsfeld and Mustaţă 2009] when $\dim X = \varkappa(L)$ and $m(L) = 1$; [Kaveh and Khovanskii 2012]). *Suppose that X is a projective variety over an algebraically closed field k, and L is a graded linear series on X with $\varkappa(L) \geq 0$, index $m = m(L)$. Then*

$$\lim_{n \to \infty} \frac{\dim_k L_{nm}}{n^{\varkappa(L)}}$$

exists.

In analogy with our Theorem 2.8 for existence of limits of graded families of ideals, we have the following necessary and sufficient conditions for the existence of limits on projective schemes.

Theorem 4.4 [Cutkosky 2014, Theorem 10.6]. *Suppose that X is a d-dimensional projective scheme over a field k with $d > 0$. Let*

$$\mathcal{N}_X = \{f \in \mathcal{O}_X \mid f^s = 0 \text{ for some positive integer } s\},$$

the nilradical of X. Let $\alpha \in \mathbb{N}$. Then the following are equivalent:

1) *For every graded linear series L on X with $\alpha \leq \varkappa(L)$, there exists a positive integer r such that*

$$\lim_{n \to \infty} \frac{\dim_k L_{a+nr}}{n^{\varkappa(L)}}$$

exists for every positive integer a.

2) *For every graded linear series L on X with $\alpha \le \varkappa(L)$, there exists an arithmetic sequence $a+nr$ (for fixed r and a depending on L) such that*

$$\lim_{n\to\infty} \frac{\dim_k L_{a+nr}}{n^{\varkappa(L)}}$$

exists.

3) $\dim \mathcal{N}_X < \alpha$.

The implication 3) $\Longrightarrow$ 1) holds if X is a proper scheme over a field k, or if X is a compact analytic space.

Corollary 4.5 [Cutkosky 2014, Theorem 10.7]. *Suppose that X is a proper d-dimensional scheme over a field k with $\dim \mathcal{N}_X < d$ and $\mathcal{L}$ is a line bundle on X. Then*

$$\operatorname{vol}(\mathcal{L}) = \lim_{n\to\infty} \frac{\dim_k \Gamma(X, \mathcal{L}^n)}{n^d/d!}.$$

5. Asymptotic regularity

We first give a comparison of local cohomology and sheaf cohomology. Let $R = k[x_0, \ldots, x_n]$, a polynomial ring over a field with the standard grading, and $\mathfrak{m} = (x_0, \ldots, x_n)$. Suppose that M is a graded module over R. Let $\tilde{M}$ be the sheafification of M on $\mathbb{P}^n$. We have an exact sequence of graded R-modules

$$0 \to H^0_{\mathfrak{m}}(M) \to M \to \bigoplus_{j\in\mathbb{Z}} H^0(\mathbb{P}^n, \tilde{M}(j)) \to H^1_{\mathfrak{m}}(M) \to 0$$

and isomorphisms

$$\bigoplus_{j\in\mathbb{Z}} H^i(\mathbb{P}^n, \tilde{M}(j)) \cong H^{i+1}_{\mathfrak{m}}(M) \text{ for } i \ge 1.$$

We have the interpretation

$$\bigoplus_{j\in\mathbb{Z}} H^0(\mathbb{P}^n, \tilde{M}(j)) \cong \varinjlim \operatorname{Hom}_R(\mathfrak{m}^n, M)$$

as an ideal transform.

We now define the regularity of a finitely generated graded R-module M:

$$a^i(M) = \begin{cases} \sup\{j \mid H^i_{\mathfrak{m}}(M)_j \ne 0\} & \text{if } H^i_{\mathfrak{m}}(M) \ne 0, \\ -\infty & \text{otherwise.} \end{cases}$$

The regularity of M is defined to be

$$\operatorname{reg}(M) = \max_i \{a^i(M) + i\}.$$

Interpreting R as the coordinate ring of $\mathbb{P}^n = \text{proj}(R)$, and considering the sheaf $\tilde{M}$ on $\mathbb{P}^n$ associate to M, we can define the regularity of $\tilde{M}$ to be

$$\begin{aligned}\text{reg}(\tilde{M}) &= \max\left\{m \mid H^i(\mathbb{P}^n, \tilde{M}(m-i-1)) \neq 0 \text{ for some } i \geq 1\right\}\\ &= \max_{i\geq 2}\{a^i(M)+i\}.\end{aligned}$$

Thus

$$\text{reg}(\tilde{M}) \leq \text{reg}(M).$$

We now give some interpretations of regularity of modules. Let

$$F_* : 0 \to \cdots \to F_j \to \cdots \to F_1 \to F_0 \to M \to 0$$

be a minimal free resolution of M as a graded R-module. Let b_j be the maximum degree of the generators of F_j. Then

$$\text{reg}(M) = \max\{b_j - j \mid j \geq 0\}.$$

In fact, we have (see [Eisenbud and Goto 1984])

$$\begin{aligned}\text{reg}(M) &= \max\{b_j - j \mid j \geq 0\}\\ &= \max\left\{n \mid \exists j \text{ such that } \text{Tor}_j^R(k, M)_{n+j} \neq 0\right\}\\ &= \max\left\{n \mid \exists j \text{ such that } H_{\mathfrak{m}}^j(M)_{n-j} \neq 0\right\}.\end{aligned}$$

We define

$$\text{reg}_i(M) = \max\left\{n \mid \text{Tor}_i^R(k, M)_n \neq 0\right\} - i.$$

Then

$$\text{reg}(M) = \max\{\text{reg}_i(M) \mid i \geq 0\}.$$

Further, $\text{reg}_0(M)$ is the maximum degree of a homogeneous generator of M.

We now discuss the regularity of powers of ideals. We outline the proof of [Cutkosky, Herzog and Trung 1999] showing that $\text{reg}(I^n)$ is a linear polynomial for large n.

Let $F_1, \ldots, F_s$ be homogeneous generators of $I \subset R = k[x_0, \ldots, x_n]$, with $\deg(F_i) = d_i$. The map $y_i \mapsto F_i$ induces a surjection of bigraded R-algebras

$$S = R[y_1, \ldots, y_s] \to R(I) = \bigoplus_{m\geq 0} I^m$$

where we have $\text{bideg}(x_i) = (1, 0)$ for $0 \leq i \leq n$, and $\text{bideg}(y_j) = (d_j, 1)$ for $1 \leq j \leq s$. We have

$$\text{Tor}_i^R(k, I^m)_a \cong \text{Tor}_i^S(S/\mathfrak{m}S, R(I))_{(a,m)}.$$

Theorem 5.1. *Let E be a finitely generated bigraded module over $k[y_1, \ldots, y_s]$. Then the function*

$$\rho_E(m) = \max\{a \mid E_{(a,m)} \neq 0\}$$

is a linear polynomial for $m \gg 0$.

Since $\mathrm{Tor}_i^S(S/\mathfrak{m}S, R(I))$ is a finitely generated bigraded $S/\mathfrak{m}S$ module, we have:

Theorem 5.2 [Cutkosky, Herzog and Trung 1999]. *For all $i \geq 0$, the function $reg_i(I^n)$ is a linear polynomial for $n \gg 0$.*

Theorem 5.3 [Cutkosky, Herzog and Trung 1999; Kodiyalam 2000]. *$reg(I^n)$ is a linear polynomial for $n \gg 0$.*

In the expression

$$\mathrm{reg}(I^n) = an + b$$

for $n \gg 0$, we have

$$a = \lim \frac{\mathrm{reg}(I^n)}{n} = \lim \frac{d(I^n)}{n} = \rho(I)$$

where $d(I^n) = \mathrm{reg}_0(I^n)$ is the maximal degree of a homogeneous generator of I^n, and (see [Kodiyalam 2000])

$$\rho(I) = \min\{\max\{d(J) \mid J \text{ is a graded reduction of } I\}.$$

Theorem 5.4 [Tài Hà 2011]. *Suppose that R is standard graded over a commutative Noetherian ring with unity, I is a graded ideal of R and M is a finitely generated graded R-module. Then there exists a constant e such that for $n \gg 0$,*

$$reg(I^n M) = \rho_M(I)n + e$$

where

$$\rho_M(I) = \min\{d(J) \mid J \textit{ is a } M\textit{-reduction of } I\}.$$

J is an M-reduction of I if $I^{n+1}M = JI^nM$ for some $n \geq 0$.

Theorem 5.5 [Eisenbud and Harris 2010]. *Let $R = k[x_1, \ldots, x_m]$. If I is R_+-primary and generated in a single degree, then the constant term of $reg(I^n)$ (for $n \gg 0$) is the maximum of the regularity of the fibers of the morphism defined by a minimal set of generators.*

Theorem 5.6 [Tài Hà 2011; Chardin 2013]. *The constant term of the regularity $reg(I^n)$, for I homogeneous in $R = k[x_0, \ldots, x_m]$ with generators all of the same degree, can be computed as the maximum of regularities of the localization of the structure sheaf of the graph of a rational map of $\mathbb{P}^m$ determined by I above points in the projection of the graph onto its second factor (the image of the rational map).*

We now give a comparison of $\operatorname{reg}_i(I^m)$, $a^i(I^m)$ and $\operatorname{reg}(I^m)$. We continue to study the graded polynomial ring $R = k[x_0, \ldots, x_n]$, and assume that I is a homogeneous ideal. Recall that

$$a^i(I^m) = \begin{cases} \sup\{j \mid H^i_{\mathfrak{m}}(I^m)_j \neq 0\} & \text{if } H^i_{\mathfrak{m}}(I^m) \neq 0, \\ -\infty & \text{otherwise,} \end{cases}$$

$$\operatorname{reg}_i(I^m) = \max\{n \mid \operatorname{Tor}_i^R(k, I^m)_n \neq 0\} - i,$$

and the regularity is

$$\operatorname{reg}(I^m) = \max_i\{a^i(I^m) + i\} = \max\{\operatorname{reg}_i(I^m) \mid i \geq 0\}.$$

We have shown that all of the functions $\operatorname{reg}_i(I^m)$ are eventually linear polynomials, so $\operatorname{reg}(I^m)$ is eventually a linear polynomial.

We first discuss the behavior of $a^i(I^m)$.

Theorem 5.7 [Cutkosky 2000]. *There is a homogeneous height two prime ideal I in $k[x_0, x_1, x_2, x_3]$ of a nonsingular space curve, such that*

$$a^2(I^m) = \lfloor m(9 + \sqrt{2}) \rfloor + 1 + \sigma(m)$$

for $m > 0$, where $\lfloor x \rfloor$ is the greatest integer in a real number x and

$$\sigma(m) = \begin{cases} 0 & \text{if } m = q_{2n} \text{ for some } n \in \mathbb{N}, \\ 1 & \text{otherwise,} \end{cases}$$

where q_n is defined recursively by

$$q_0 = 1, q_1 = 2, q_n = 2q_{n-1} + q_{2n-2},$$

computed from the convergents $\frac{p_n}{q_n}$ of the continued fraction expansion of $\sqrt{2}$.

The m such that $\sigma(m) = 0$ are very sparse, as $q_{2n} \geq 3^n$.

We also compute that

$$a^3(I^m) = \lfloor m(9 - \sqrt{2}) \rfloor - \tau(m)$$

where $0 \leq \tau(m) \leq$ constant is a bounded function, and

$$a^4(I^m) = -4.$$

Since $9 + \sqrt{2} \leq \lim_{m\to\infty} \dfrac{\operatorname{reg}(I^m)}{m} \in \mathbb{Z}_+$, we have

$$a^1(I^m) = \operatorname{reg}(I^m) = \text{ linear function for } m \gg 0.$$

We now discuss the proof of this theorem. We first need to review numerical equivalence, from [Kleiman 1966] and [Lazarsfeld 2004a, Chapter 1]. Let k be an algebraically closed field, and X be a nonsingular projective variety over k. Define

$$\begin{aligned}\mathrm{Div}(X) &= \text{divisors on } X\\ &:= \text{formal sums of codimension-1 subvarieties of } X.\end{aligned}$$

Numerical equivalence is defined by

$$D_1 \equiv D_2 \iff (D_1 \cdot C) = (D_2 \cdot C) \text{ for all curves } C \text{ on } X.$$

The $\mathbb{R}$-vector space

$$N(X) = (\mathrm{Div}(X)/\equiv) \otimes_{\mathbb{Z}} \mathbb{R}$$

is finite-dimensional (this is proved in [Lazarsfeld 2004a, Proposition 1.1.16], for instance).

A divisor D on X is ample if $H^0(\mathcal{O}_X(mD))$ gives a projective embedding of X for some $m \gg 0$.

Theorem 5.8. *A divisor D is ample if and only if $(D^d \cdot V) > 0$ for all d-dimensional irreducible subvarieties V of X.*

Taking $V = X$ this condition is $(D^{\dim X}) > 0$.

Set

$$\begin{aligned}\mathrm{A}(X) &= \text{ ample cone} = \text{convex cone in } N(X) \text{ generated by ample divisors}\\ \mathrm{Nef}(X) &= \text{nef cone}\\ &= \text{convex cone generated by numerically effective divisors}\\ &\qquad ((D \cdot C) \geq 0 \text{ for all curves } C \text{ on } X,)\\ \mathrm{NE}(X) &= \text{ convex cone generated by effective divisors}\\ &\qquad (h^0(\mathcal{O}_X(nD)) > 0 \text{ for some } n > 0.).\end{aligned}$$

Theorem 5.9.

$$\mathrm{A}(X) \subseteq \overline{\mathrm{A}(X)} = \mathrm{Nef}(X) \subseteq \overline{\mathrm{NE}(X)}$$

Here $\overline{T}$ denotes closure of T in the euclidean topology.

Suppose that S is a nonsingular projective surface. Then $N(S)$ has an intersection form $q(D) = (D^2)$ for D a divisor on S.

A K3 surface is a nonsingular projective surface with $H^1(\mathcal{O}_S) = 0$ and such that K_S is trivial. From the theory of K3 surfaces (as reviewed in Section 2 of [Cutkosky 2000] and using a theorem of Morrison [Morrison 1984]) it follows that there exists a K3 surface S such that $N(S) \cong \mathbb{R}^3$ and

$$q(D) = 4x^2 - 4y^2 - 4z^2$$

for $D = (x, y, z) \in \mathbb{R}^3$.

Lemma 5.10. *Suppose that C is an integral curve on S. Then $(C^2) \geq 0$.*

Proof. Suppose otherwise. Then $(C^2) = -2$, since S is a K3 surface. But 4 divides $q(C) = (C^2)$, a contradiction. □

Corollary 5.11. $\overline{\mathrm{NE}(S)} = \overline{\mathrm{A}(S)}$, *and*

$$\overline{\mathrm{NE}(S)} = \left\{(x, y, z) \mid 4x^2 - 4y^2 - 4z^2 \geq 0, x \geq 0\right\}.$$

Let $H = (1, 0, 0)$ (that is, let H be a divisor whose class is $(1, 0, 0)$). Then $H^0(\mathcal{O}_S(H))$ gives an embedding of S as a quartic surface in $\mathbb{P}^3$. Choose $(a, b, c) \in \mathbb{Z}^3$ such that $a > 0$, $a^2 - b^2 - c^2 > 0$ and $\sqrt{b^2 + c^2} \notin \mathbb{Q}$. (a, b, c) is in the interior of $\overline{\mathrm{NE}(S)}$, which is equal to $\mathrm{A}(S)$. There exists a nonsingular curve C on S such that $C = (a, b, c)$ in $N(X)$. Let

$$\lambda_2 = a + \sqrt{b^2 + c^2} \text{ and } \lambda_1 = a - \sqrt{b^2 + c^2}.$$

Suppose that $m, r \in \mathbb{N}$. Then

$$\begin{aligned} mH - rC &\in \overline{\mathrm{NE}(S)} && \text{and is ample if } r\lambda_2 < m, \\ mH - rC,\ rC - mH &\notin \overline{\mathrm{NE}(S)} && \text{if } r\lambda_1 < m < r\lambda_2, \\ rC - mH &\in \overline{\mathrm{NE}(S)} && \text{and is ample if } m < r\lambda_1. \end{aligned}$$

Choose $C = (a, b, c)$ so that $7 < \lambda_1 < \lambda_2$ and $\lambda_2 - \lambda_1 > 2$. Then by Riemann–Roch,

$$\begin{aligned} \chi(mH - rC) &= h^0(mH - rC) - h^1(mH - rC) + h^2(mH - rC) \\ &= \tfrac{1}{2}(mH - rC)^2 + 2. \end{aligned}$$

Theorem 5.12.

$$h^1(mH - rC) = \begin{cases} 0 & \textit{if } r\lambda_2 < m, \\ -\frac{1}{2}(mH - rC)^2 - 2 & \textit{if } r\lambda_1 < m < r\lambda_2, \end{cases}$$

$$h^2(mH - rC) = 0 \qquad \textit{if } r\lambda_1 < m.$$

Let $\overline{H}$ be a linear hyperplane on $\mathbb{P}^3$ such that $\overline{H} \cdot S = H$. Let $\mathcal{I}_C = \tilde{I}_C$, where I_C is the homogeneous ideal of C in the coordinate ring R of $\mathbb{P}^3$.

Let $\pi : X \to \mathbb{P}^3$ be the blowup of C. Let $E = \pi^*(C)$, the exceptional surface, $\bar{S}$ be the strict transform of S on X. $\bar{S} \cong S$ and $E \cdot \bar{S} = C$. For $m, r \in \mathbb{N}$ and $i \geq 0$,

$$H^i(\mathbb{P}^3, \mathcal{I}_C^r(m)) \cong H^i(X, \mathcal{O}_X(m\overline{H} - rE)).$$

In particular,

$$I_C^{(r)} = (I_C^r)^{\mathrm{sat}} \cong \bigoplus_{m \geq 0} H^0(\mathbb{P}^3, \mathcal{I}_C^r(m)) = \bigoplus_{m \geq 0} H^0(X, \mathcal{O}_X(m\overline{H} - rE)).$$

In the exact sequence

$$0 \to \mathcal{O}_X(-\bar{S}) \to \mathcal{O}_X \to \mathcal{O}_{\bar{S}} \to 0$$

we have $\bar{S} \sim 4\overline{H} - E$. Tensor with $\mathcal{O}_X((m+4)\overline{H} - (r+1)E)$ to get

$$0 \longrightarrow \mathcal{O}_X(m\overline{H} - rE) \longrightarrow \mathcal{O}_X((m+4)\overline{H} - (r+1)E) \longrightarrow \mathcal{O}_S((m+4)H - (r+1)C) \longrightarrow 0.$$

Since $h^i(\mathcal{O}_X(m\overline{H})) = 0$ for $i > 0$ and $m \geq 0$, our calculation of cohomology on S and induction gives

$$h^1(m\overline{H} - rE) = \begin{cases} 0 & \text{if } m > r\lambda_2, \\ h^1(mH - rE) & \text{if } m = \lfloor r\lambda_2 \rfloor \text{ or } m = \lfloor r\lambda_2 \rfloor - 1, \end{cases}$$

$$h^2(mH - rE) = 0 \qquad \text{if } m > \lambda_1 r,$$

$$h^3(mH - rE) = 0 \qquad \text{if } m > 4r.$$

By our calculation of cohomology on S, we have for $r, t \in \mathbb{N}$,

$$h^1(\mathcal{I}_C^r(t-1)) = h^1((t-1)\overline{H} - rE) \begin{cases} = 0 & \text{if } t \geq \lfloor r\lambda_2 \rfloor + 1 + \sigma(r), \\ \neq 0 & \text{if } t = \lfloor r\lambda_2 \rfloor + \sigma(r), \end{cases}$$

where

$$\sigma(r) = \begin{cases} 0 & \text{if } h^1(\lfloor r\lambda - 2 \rfloor H - rC) = 0, \\ 1 & \text{if } h^1(\lfloor r\lambda - 2 \rfloor H - rC) \neq 0. \end{cases}$$

We obtain, for $r \in \mathbb{N}$,

$$a^2(I_C^r) = \operatorname{reg}(I_C^{(r)}) = \operatorname{reg}((I_C^r)^{\text{sat}}) = \lfloor r\lambda_2 \rfloor + 1 + \sigma(r)$$

with

$$\lim_{r \to \infty} \frac{a^2(I_C^r)}{r} = \lim_{r \to \infty} \frac{\operatorname{reg}((I_C^r)^{\text{sat}})}{r} = \lambda_2 \notin \mathbb{Q}.$$

In this example, we have shown that the function

$$\operatorname{reg}((I_C^n)^{\text{sat}}) = \operatorname{reg}(I_C^{(n)})$$

has irrational behavior asymptotically. This is perhaps not so surprising, as its symbolic algebra

$$\bigoplus_{n \geq 0} I_C^{(n)}$$

is not a finitely generated R-algebra. An example of an ideal of a union of generic points in the plane whose symbolic algebras is not finitely generated was found and used by Nagata [1959] to give his counterexample to Hilbert's 14th problem. Roberts [1985] interpreted this example to give an example of a prime ideal of a space curve. Even for rational monomial curves this algebra may not be finitely generated, by an example in [Goto, Nishida and Watanabe 1994].

Holger Brenner [2013] has recently given a remarkable example showing that the Hilbert–Kunz multiplicity can be irrational.

We now discuss the regularity of coherent sheaves. Suppose that X is a projective variety, over a field k, and H is a very ample divisor on X. Suppose that $\mathcal{J} \subset \mathcal{O}_X$ is an ideal sheaf. Let $\pi : B(\mathcal{J}) \to X$ be the blowup of $\mathcal{J}$, with exceptional divisor F. The Seshadri constant of $\mathcal{J}$ is defined to be

$$s_H(\mathcal{J}) = \inf\{s \in \mathbb{R} \mid \pi^*(sH) - F \text{ is a very ample } \mathbb{R}\text{-divisor on } B(\mathcal{J}).\}$$

The regularity of $\mathcal{J}$ is defined to be

$$\operatorname{reg}_H(\mathcal{J}) = \max\{m \mid H^i(X, \mathcal{H} \otimes \mathcal{O}_X((m-i-1)H)) \neq 0.$$

Theorem 5.13 [Cutkosky, Ein and Lazarsfeld 2001]. *Suppose that $\mathcal{I} \subset \mathcal{O}_X$ is an ideal sheaf. Then*

$$\lim_{m\to\infty} \frac{reg_H(\mathcal{I}^m)}{m} = \lim_{m\to\infty} \frac{d_H(\mathcal{I}^m)}{m} = s_H(\mathcal{I}).$$

For an ideal sheaf J,

$$d_H(\mathcal{J}) = \text{ least integer } d \text{ such that } \mathcal{J}(dH) \text{ is globally generated.}$$

If H is a linear hyperplane on $\mathbb{P}^n$, and $\mathcal{I} = \tilde{I}$, we get the statement that the limit

$$\lim_{m\to\infty} \frac{\operatorname{reg}((I^m)^{\text{sat}})}{m} = \lim_{m\to\infty} \frac{d((I^m)^{\text{sat}})}{m}.$$

exists, where $d((I^m)^{\text{sat}})$ is the maximal degree of a generator of $(I^m)^{\text{sat}}$.

We now give an example of an irrational Seshadri constant. The ideal I of a nonsingular curve in $\mathbb{P}^3$ contained in a quartic which we considered earlier gives an example (see [Cutkosky 2000]):

$$s_H(\tilde{I}) = \lim_{m\to\infty} \frac{\operatorname{reg}_H(\tilde{I}^m)}{m} = \lim_{m\to\infty} \frac{\operatorname{reg}((I^m)^{\text{sat}})}{m} = 9 + \sqrt{2}.$$

We do have something like linear growth of regularity reg_H in the example. Recall that the example is of the homogeneous height two prime ideal I in $k[x_0, x_1, x_2, x_3]$ of a nonsingular projective space curve, such that

$$\begin{aligned}\operatorname{reg}_H(\tilde{I}^n) = \operatorname{reg}((I^n)^{\text{sat}}) &= \max\{a^i(I^n) + i \mid 2 \leq i \leq 4\} \\ &= \lfloor m(9+\sqrt{2}) \rfloor + 1 + \sigma(m)\end{aligned}$$

for $m > 0$, where $\lfloor x \rfloor$ is the greatest integer in a real number x and

$$\sigma(m) = \begin{cases} 0 & \text{if } m = q_{2n} \text{ for some } n \in \mathbb{N}, \\ 1 & \text{otherwise,} \end{cases}$$

where q_n is defined recursively by

$$q_0 = 1, q_1 = 2, q_n = 2q_{n-1} + q_{2n-2}.$$

Theorem 5.14 (Wenbo Niu [2011]). *Suppose that $\mathcal{I} = \tilde{I}$ is an ideal sheaf on $\mathbb{P}^n$. Then there is a bounded function $\tau(m)$, with $0 \le \tau(m) \le$ constant, such that*

$$reg_H(\mathcal{I}^m) = reg((I^n)^{\text{sat}}) = \lfloor s_H m \rfloor + \tau(m).$$

for all $m > 0$.

We conclude this section with some questions. Suppose that I is a homogeneous ideal in a polynomial ring S.

- Does
$$\lim_{n\to\infty} \frac{a^i(I^n)}{n}$$
exist for all i?
- Does
$$\lim_{n\to\infty} \frac{\text{reg}(I^{(n)})}{n}$$
exist? (The answer is yes if the singular locus of S/I has dimension ≤ 1; see [Herzog, Hoa and Trung 2002].)
- Does
$$\lim_{n\to\infty} \frac{\text{reg}(\text{in}(I^n))}{n}$$
exist?
- David Eisenbud has posed the following problem. Suppose that I is generated in a single degree. Explain (geometrically) the constant term b in the linear polynomial
$$\text{reg}(I^n) = an + b \quad \text{for } n \gg 0.$$

References

[Abhyankar 1998] S. S. Abhyankar, *Resolution of singularities of embedded algebraic surfaces*, 2nd ed., Springer, Berlin, 1998.

[Boucksom, Favre and Jonsson 2009] S. Boucksom, C. Favre, and M. Jonsson, "Differentiability of volumes of divisors and a problem of Teissier", *J. Algebraic Geom.* **18**:2 (2009), 279–308.

[Brenner 2013] H. Brenner, "Irrational Hilbert–Kunz multiplicities", preprint, 2013. arXiv 1305.5873

[Chardin 2013] M. Chardin, "Powers of ideals and the cohomology of stalks and fibers of morphisms", *Algebra Number Theory* **7**:1 (2013), 1–18.

[Cutkosky 2000] S. D. Cutkosky, "Irrational asymptotic behaviour of Castelnuovo–Mumford regularity", *J. Reine Angew. Math.* **522** (2000), 93–103.

[Cutkosky 2009] S. D. Cutkosky, "Resolution of singularities for 3-folds in positive characteristic", *Amer. J. Math.* **131**:1 (2009), 59–127.

[Cutkosky 2013a] S. D. Cutkosky, "Multiplicities associated to graded families of ideals", *Algebra Number Theory* **7**:9 (2013), 2059–2083.

[Cutkosky 2013b] S. D. Cutkosky, "Teissier's problem on inequalities of nef divisors over an arbitrary field", preprint, 2013. To appear in *J. Algebra Appl.* arXiv 1304.1218

[Cutkosky 2014] S. D. Cutkosky, "Asymptotic multiplicities of graded families of ideals and linear series", *Adv. Math.* **264** (2014), 55–113.

[Cutkosky and Srinivas 1993] S. D. Cutkosky and V. Srinivas, "On a problem of Zariski on dimensions of linear systems", *Ann. of Math.* (2) **137**:3 (1993), 531–559.

[Cutkosky, Ein and Lazarsfeld 2001] S. D. Cutkosky, L. Ein, and R. Lazarsfeld, "Positivity and complexity of ideal sheaves", *Math. Ann.* **321**:2 (2001), 213–234.

[Cutkosky, Herzog and Trung 1999] S. D. Cutkosky, J. Herzog, and N. V. Trung, "Asymptotic behaviour of the Castelnuovo–Mumford regularity", *Compositio Math.* **118**:3 (1999), 243–261.

[Ein, Lazarsfeld and Smith 2003] L. Ein, R. Lazarsfeld, and K. E. Smith, "Uniform approximation of Abhyankar valuation ideals in smooth function fields", *Amer. J. Math.* **125**:2 (2003), 409–440.

[Eisenbud and Goto 1984] D. Eisenbud and S. Goto, "Linear free resolutions and minimal multiplicity", *J. Algebra* **88**:1 (1984), 89–133.

[Eisenbud and Harris 2010] D. Eisenbud and J. Harris, "Powers of ideals and fibers of morphisms", *Math. Res. Lett.* **17**:2 (2010), 267–273.

[Fujita 1994] T. Fujita, "Approximating Zariski decomposition of big line bundles", *Kodai Math. J.* **17**:1 (1994), 1–3.

[Gabrielov 1971] A. M. Gabrielov, "The formal relations between analytic functions", *Funkts. Anal. i Priložen.* **5**:4 (1971), 64–65. In Russian; translated in *Funct. Anal. Appl.* **5** (1972), 318–319.

[Goto, Nishida and Watanabe 1994] S. Goto, K. Nishida, and K. Watanabe, "Non–Cohen–Macaulay symbolic blow-ups for space monomial curves and counterexamples to Cowsik's question", *Proc. Amer. Math. Soc.* **120**:2 (1994), 383–392.

[Herzog, Hoa and Trung 2002] J. Herzog, L. T. Hoa, and N. V. Trung, "Asymptotic linear bounds for the Castelnuovo–Mumford regularity", *Trans. Amer. Math. Soc.* **354**:5 (2002), 1793–1809.

[Herzog, Puthenpurakal and Verma 2008] J. Herzog, T. J. Puthenpurakal, and J. K. Verma, "Hilbert polynomials and powers of ideals", *Math. Proc. Cambridge Philos. Soc.* **145**:3 (2008), 623–642.

[Hübl 2001] R. Hübl, "Completions of local morphisms and valuations", *Math. Z.* **236**:1 (2001), 201–214.

[de Jong 1996] A. J. de Jong, "Smoothness, semi-stability and alterations", *Inst. Hautes Études Sci. Publ. Math.* 83 (1996), 51–93.

[Kaveh and Khovanskii 2012] K. Kaveh and A. G. Khovanskii, "Newton–Okounkov bodies, semigroups of integral points, graded algebras and intersection theory", *Ann. of Math.* (2) **176**:2 (2012), 925–978.

[Kleiman 1966] S. L. Kleiman, "Toward a numerical theory of ampleness", *Ann. of Math.* (2) **84** (1966), 293–344.

[Kodiyalam 2000] V. Kodiyalam, "Asymptotic behaviour of Castelnuovo–Mumford regularity", *Proc. Amer. Math. Soc.* **128**:2 (2000), 407–411.

[Lazarsfeld 2004a] R. Lazarsfeld, *Positivity in algebraic geometry, I: Classical setting: Line bundles and linear series*, Ergebnisse der Mathematik und ihrer Grenzgebiete (3) **48**, Springer, Berlin, 2004.

[Lazarsfeld 2004b] R. Lazarsfeld, *Positivity in algebraic geometry, II: Positivity for vector bundles, and multiplier ideals*, Ergebnisse der Mathematik und ihrer Grenzgebiete (3) **49**, Springer, Berlin, 2004.

[Lazarsfeld and Mustaţă 2009] R. Lazarsfeld and M. Mustaţă, "Convex bodies associated to linear series", *Ann. Sci. Éc. Norm. Supér.* (4) **42**:5 (2009), 783–835.

[Lipman 1975] J. Lipman, "Introduction to resolution of singularities", pp. 187–230 in *Algebraic geometry* (Arcata, CA, 1974), edited by R. Hartshorne, Proc. Symp. Pure Math. **29**, Amer. Math. Soc., Providence, RI, 1975.

[Morrison 1984] D. R. Morrison, "On $K3$ surfaces with large Picard number", *Invent. Math.* **75**:1 (1984), 105–121.

[Mumford 1976] D. Mumford, *Algebraic geometry, I: Complex projective varieties*, Grundlehren der Mathematischen Wissenschaften **221**, Springer, Berlin, 1976.

[Mustaţă 2002] M. Mustaţă, "On multiplicities of graded sequences of ideals", *J. Algebra* **256**:1 (2002), 229–249.

[Nagata 1959] M. Nagata, "On the 14-th problem of Hilbert", *Amer. J. Math.* **81** (1959), 766–772.

[Niu 2011] W. Niu, "A vanishing theorem and asymptotic regularity of powers of ideal sheaves", *J. Algebra* **344** (2011), 246–259.

[Okounkov 2003] A. Okounkov, "Why would multiplicities be log-concave?", pp. 329–347 in *The orbit method in geometry and physics* (Marseille, 2000), edited by C. Duval et al., Progr. Math. **213**, Birkhäuser, Boston, MA, 2003.

[Rees 1989] D. Rees, "Izumi's theorem", pp. 407–416 in *Commutative algebra* (Berkeley, CA, 1987), edited by M. Hochster et al., Math. Sci. Res. Inst. Publ. **15**, Springer, 1989.

[Roberts 1985] P. C. Roberts, "A prime ideal in a polynomial ring whose symbolic blow-up is not Noetherian", *Proc. Amer. Math. Soc.* **94**:4 (1985), 589–592.

[Tài Hà 2011] H. Tài Hà, "Asymptotic linearity of regularity and a^*-invariant of powers of ideals", *Math. Res. Lett.* **18**:1 (2011), 1–9.

[Takagi 2007] S. Takagi, "Fujita's approximation theorem in positive characteristics", *J. Math. Kyoto Univ.* **47**:1 (2007), 179–202.

[Zariski 1935] O. Zariski, *Algebraic surfaces*, Ergebnisse der Mathematik und ihrer Grenzgebiete (3) **5**, Springer, Berlin, 1935. Second supplemented edition, Springer, 1971.

[Zariski 1947] O. Zariski, "The concept of a simple point of an abstract algebraic variety", *Trans. Amer. Math. Soc.* **62** (1947), 1–52.

[Zariski 1949] O. Zariski, "A simple analytical proof of a fundamental property of birational transformations", *Proc. Nat. Acad. Sci. U. S. A.* **35** (1949), 62–66.

[Zariski and Samuel 1960] O. Zariski and P. Samuel, *Commutative algebra, II*, D. Van Nostrand, Princeton, NJ, 1960.

CutkoskyS@missouri.edu *Department of Mathematics, University of Missouri, Columbia, MO 65211, United States*

Commutative Algebra and Noncommutative Algebraic Geometry, I
MSRI Publications
Volume **67**, 2015

Introduction to uniformity in commutative algebra

CRAIG HUNEKE AND CLAUDIU RAICU

This article is based on three lectures given by the first author as part of an introductory workshop at MSRI for the program in Commutative Algebra, 2012–13. Additional comments and explanations are included, as well as a new section on the uniform Artin–Rees theorem. We deal with the theme of uniform bounds, both absolute and effective, as well as uniform annihilation of cohomology.

1. Introduction

The goal of these notes is to introduce the concept of *uniformity* in commutative algebra. Rather than giving a precise definition of what uniformity means, we will try to convey the idea of uniformity through a series of examples. As we'll soon see, uniformity is ubiquitous in commutative algebra: it may refer to absolute or effective bounds for certain natural invariants (ideal generators, regularity, projective dimension), or uniform annihilation of (co)homology functors (Tor, Ext, local cohomology). We will try to convince the reader that the simple exercise of thinking from a uniform perspective almost always leads to significant, interesting, and fundamental questions and theories. This theme has also been discussed by Schoutens [2000], who shows how uniform bounds can be useful in numerous contexts that we do not consider in this paper.

The first section of this paper, based on the first lecture in the workshop, is more elementary and introduces many basic concepts. The next three sections target specific topics and require more background in general, though an effort has been made to minimize the knowledge needed to read them. Each section has some exercises which the reader might solve to gain further understanding. The first section in particular has a great many exercises.

We begin to illustrate the theme of uniformity with what is probably the most basic theorem in commutative algebra:

Huneke was supported by NSF grant number 1259142.
Keywords: uniform bounds, symbolic powers, projective dimension, free resolution, regularity.

Theorem 1 (Hilbert's basis theorem [Eisenbud 1995, Theorem 1.2]). *If k is a field and n is a nonnegative integer, then any ideal in the polynomial ring $S = k[x_1, \dots, x_n]$ is finitely generated.*

As it stands, this theorem does give a type of uniformity, namely the property of being finitely generated. But this is quite general, and not absolute or effective. One first might try for absolute bounds:

Question 2. *Is there an absolute upper bound for the (minimal) number of generators of ideals in S?*

This has a positive answer in the case $n = 1$: $S = k[x_1]$ is a *principal ideal domain*, so any ideal in S can be generated by one element. However, for $n \geq 2$, it is easy to see that such an absolute bound cannot exist: the ideal $I = (x_1, x_2)^N$ can't be generated by fewer than $(N + 1)$ elements. One can then try to refine Question 2, which leads us to several interesting variations:

Question 3. *Is there an absolute upper bound for the number of generators of an ideal I in S, if*

(a) *we assume that I is prime?*

(b) *I is homogeneous and we impose bounds on the degrees of the generators of I?*

(c) *we are only interested in the generation of I up to radical?* (Recall that the *radical* $\sqrt{I}$ is the set $\{f \in S : f^r \in I \text{ for some } r\}$.)

For part (a) we have a positive answer in the case $n = 2$: any prime ideal $I \subset k[x_1, x_2]$ is either maximal, or has height one, or is zero, so it can be generated by at most two elements (because the ring is a UFD, height one primes are principal; for maximal ideals, see Exercise (1) on page 170). However, for $n \geq 3$ the assumption that I is prime is not sufficient to guarantee an absolute bound for its number of generators: in fact in [Moh 1974; 1979] a sequence of prime ideals $\mathfrak{p}_n \subset k[x, y, z]$ is constructed, where the minimal number of generators of $\mathfrak{p}_n$ is $n + 1$. For part (b), if we assume that I is generated in degree at most d, then the absolute bound for the number of generators of I is attained when $I = \mathfrak{m}^d$ is a power of the maximal ideal $\mathfrak{m}$, and is given by the binomial coefficient $\binom{n+d-1}{n-1}$; see Exercise (2). Part (c) is already quite subtle: every ideal $I \subset S$ is generated up to radical by n elements [Eisenbud and Evans 1973; Storch 1972].

Another variation of Question 2 is to ask whether one can find effective lower bounds for the number of generators of an ideal I of S. One such bound is obtained as a consequence of Krull's Hauptidealsatz (Principal ideal theorem), in terms of the *codimension* of the ideal I:

$$\operatorname{codim}(I) = \dim(S) - \dim(S/I) = n - \dim(S/I).$$

Theorem 4 [Matsumura 1980, Chapter 12; Eisenbud 1995, Chapter 10]. *The number of generators of I is at least as large as the codimension of I (and this inequality is sharp).*

In order to discuss further uniformity statements, we need to expand the set of invariants that we associate to ideals, and more generally to modules over the polynomial ring S. We start with the following:

Definition 5 (Hilbert function and Hilbert series). Let $M = \bigoplus_{i\in\mathbb{Z}} M_i$ denote a finitely generated graded S-module, written as the sum of its homogeneous components (so that $S_i \cdot M_j \subset M_{i+j}$). The *Hilbert function* $h_M : \mathbb{Z} \to \mathbb{Z}_{\geq 0}$ is defined by

$$h_M(i) = \dim_k(M_i).$$

We write $M(d)$ for the shifted module having $M(d)_i = M_{d+i}$. It follows that $h_{M(d)}(i) = h_M(i+d)$.

The *Hilbert series* $H_M(z)$ is the generating function associated to h_M:

$$H_M(z) = \sum_{i\in\mathbb{Z}} h_M(i) \cdot z^i.$$

In the case when $M = S$ is the polynomial ring itself, we have

$$h_S(d) = \begin{cases} \binom{n+d-1}{n-1} & \text{if } d \geq 0, \\ 0 & \text{if } d < 0. \end{cases} \tag{1-1}$$

The Hilbert series of S takes the simple form

$$H_S(z) = \frac{1}{(1-z)^n}.$$

It is a remarkable fact, which we explain next, that the Hilbert series of any finitely generated graded S-module is a rational function. An equivalent statement is contained in the following theorem of Hilbert.

Theorem 6 [Matsumura 1980, Chapter 10; Eisenbud 1995, Chapter 12]. *If M is a finitely generated graded S-module, then there exists a polynomial $p_M(t)$ with rational coefficients, such that $p_M(i) = h_M(i)$ for sufficiently large values of i.*

The polynomial p_M is called the *Hilbert polynomial* of M. Since the theorem is true for $M = S$ (as shown by (1-1)), it holds for free modules as well. To prove it in general, it is then enough to show that any M can be approximated by free modules in such a way that its Hilbert function is controlled by the Hilbert functions of the corresponding free modules.[1] Such approximations are realized via exact sequences. When working with graded modules, we will assume that every homomorphism $f : M \to N$ has degree 0, i.e., $f(M_i) \subset N_i$ for all $i \in \mathbb{Z}$. It follows that any short exact sequence of graded modules

$$0 \longrightarrow M \longrightarrow N \longrightarrow K \longrightarrow 0$$

restricts in degree i to an exact sequence

$$0 \longrightarrow M_i \longrightarrow N_i \longrightarrow K_i \longrightarrow 0,$$

yielding $h_N(i) = h_M(i) + h_K(i)$, and therefore

$$H_N(z) = H_M(z) + H_K(z).$$

Now if $f \in S_d$ is a form of degree d, then $K = S/(f)$ can be approximated by free modules via the exact sequence

$$0 \longrightarrow S(-d) \xrightarrow{f} S \longrightarrow K \longrightarrow 0.$$

It follows that

$$h_K(i) = h_S(i) - h_S(i-d) = \binom{i+n-1}{n-1} - \binom{i-d+n-1}{n-1},$$

which is a polynomial for $i \geq d$. For an arbitrary finitely generated graded module M, a similar approximation result holds, having Theorem 6 as a direct consequence:

Theorem 7 (Hilbert's syzygy theorem [Matsumura 1980, Chapter 18; Eisenbud 1995, Theorem 1.13]). *If M is a finitely generated graded S-module, then there exists a finite minimal graded free resolution*

$$0 \longrightarrow F_n \longrightarrow F_{n-1} \longrightarrow \cdots \longrightarrow F_1 \longrightarrow F_0 \longrightarrow M \longrightarrow 0. \qquad (1\text{-}2)$$

In this statement, *minimal* just means that the entries of the matrices defining the maps between the free modules in the resolution have entries in the homogeneous maximal ideal $\mathfrak{m}$ of S. Writing $F_i = \bigoplus_{j \in \mathbb{Z}} S(-j)^{\beta_{i,j}}$, we call the multiplicities $\beta_{i,j} = \beta_{i,j}(M)$ the *graded Betti numbers* of M. We say that M has a *pure resolution* if for each i there is at most one value of j for which $\beta_{i,j} \neq 0$. It has a *linear resolution* if $\beta_{i,j} = 0$ for $i \neq j$ (or more generally if there exists $c \in \mathbb{Z}$ such that $\beta_{i,j} = 0$ for $i - j \neq c$). It is not immediately obvious from (1-2)

[1] Although we are concentrating on the graded case, the Hilbert function of a local ring can be defined easily by passing to the associated graded ring. A remarkable uniform result about Hilbert functions was proved in [Srinivas and Trivedi 1997]: if the local ring is Cohen–Macaulay, and we fix its dimension and multiplicity, then there are only finitely many possible Hilbert functions.

that the Betti numbers are uniquely determined by M, but this follows from their alternative more functorial characterization [Eisenbud 2005, Prop. 1.7]:

$$\beta_{i,j} = \dim_k \operatorname{Tor}_i^S(M, k)_j.$$

Here we think of i as the *homological degree*, and of j as the *internal degree*.

Example 8 (The Koszul complex). If we take M equal to $k = S/(x_1, \ldots, x_n)$, the residue field, then its minimal graded free resolution is pure (even linear), given by the *Koszul complex* on $x_1, \ldots, x_n$:

$$0 \longrightarrow S(-n)^{\binom{n}{n}} \longrightarrow \cdots \longrightarrow S(-i)^{\binom{n}{i}} \longrightarrow \cdots \longrightarrow S(-1)^n \longrightarrow S \longrightarrow k \longrightarrow 0.$$

The graded Betti numbers are given by

$$\beta_{i,j}(k) = \begin{cases} 0 & \text{if } j \neq i, \\ \binom{n}{i} & \text{if } j = i. \end{cases}$$

The graded Betti numbers of a module are recorded into the *Betti table*, where the entry in row j and column i is $\beta_{i,i+j}$:

	0	1	2	$\cdots$
0	$\beta_{0,0}$	$\beta_{1,1}$	$\beta_{2,2}$	$\cdots$
1	$\beta_{0,1}$	$\beta_{1,2}$	$\beta_{2,3}$	$\cdots$
$\vdots$	$\vdots$	$\vdots$	$\vdots$	$\ddots$

The Koszul complex in Example 8 has Betti table

	0	1	2	$\cdots$	i	$\cdots$	$n-1$	n
0	1	n	$\binom{n}{2}$	$\cdots$	$\binom{n}{i}$	$\cdots$	n	1

The Hilbert function, the Hilbert series, and therefore the Hilbert polynomial of M can all be read off from the Betti table of M. We have

$$h_M(i) = \sum_{l=0}^{n} (-1)^l \cdot \sum_j \binom{i-j+n-1}{n-1} \beta_{l,j},$$

and

$$H_M(z) = \frac{\sum_j \Bigl(\sum_l (-1)^l \beta_{l,j} \Bigr) \cdot z^j}{(1-z)^n}.$$

The following basic invariants of M are also encoded by the Betti table:

(1) *Dimension*:

$$\dim(M) = \dim(S/\operatorname{ann}(M)) = \deg(p_M) + 1.$$

(2) *Multiplicity*:

$$e(M) = (\dim(M) - 1)! \cdot (\text{leading coefficient of } p_M).$$

(3) *Projective dimension*:

$$\text{pd}(M) = \text{length of the Betti table (the index of the last nonzero column)}.$$

(4) *Regularity*:

$$\text{reg}(M) = \text{width of the Betti table (the index of the last nonzero row)}.$$

Some of the most fundamental problems in commutative algebra involve understanding the (relative) uniform properties that these invariants exhibit. For example, it is true generally that

$$\text{codim}(I) \leq \text{pd}(S/I) \leq n, \tag{1-3}$$

where the first inequality is a consequence of the Auslander–Buchsbaum formula [Eisenbud 1995, Theorem 19.9], while the second follows from Theorem 7. If the equality $\text{codim}(I) = \text{pd}(S/I)$ holds in (1-3) then we say that S/I is *Cohen-Macaulay*.

How can one best use the idea of uniformity when considering the Hilbert syzygy theorem? There are several ways. One can relax the condition that S be a polynomial ring, and consider possibly infinite resolutions. But then what type of questions should be asked? We will consider this idea below. One could also ask for absolute bounds on the invariants (1)–(4) under certain restrictions on the degrees of the generators of I. One such question is due to Mike Stillman:

Question 9 [Peeva and Stillman 2009, Problem 3.8.1]. *Fix positive integers $d_1, \dots, d_m$. Is $pd(S/I)$ bounded when S is allowed to vary over the polynomial rings in any number of variables, and I over all the ideals $I = (f_1, \dots, f_m)$, where each f_i is a homogeneous polynomial of degree d_i? With the same assumptions, is $reg(S/I)$ bounded?*

It has been shown by Giulio Caviglia that the existence of a uniform bound for $\text{pd}(S/I)$ is equivalent to the existence of one for $\text{reg}(S/I)$ [Peeva 2011, Theorem 29.5].

There has been an increasing amount of work on Stillman's question. Recent contributions include [Ananyan and Hochster 2012], which gives a positive answer to Stillman's question when all the d_i equal 2, and the paper of Huneke, McCullough, Mantero, and Seceleanu [Huneke et al. 2013], which gives a sharp upper bound if all the d_i equal 2 and the codimension is also 2. See the exercises for more references.

We mentioned the idea of passing to infinite resolutions. What could possibly be an analogue of Hilbert's theorem in this case? One such analogue was proved by Hsin-Ju Wang [1994; 1999]. His theorem gives an effective Hilbert syzygy theorem in the following sense. If R is regular of dimension d, then the syzygy theorem implies that modules have projective dimension at most d, and in particular for every pair of R-modules M and N, $Ext_R^{d+1}(M, N) = 0$. In particular, if $R_{\mathfrak{p}}$ is regular for some prime ideal $\mathfrak{p}$, then there is an element not in $\mathfrak{p}$ which annihilates any given one of these higher Ext modules. If the singular locus is closed, this means that there is a power of the ideal defining it which annihilates any fixed $Ext_R^{d+1}(M, N)$. What one can now ask is a natural question from the point of view of uniformity: is there a *uniform* annihilator of these Ext modules as M and N vary? The result of Hsin-Ju Wang answers this question affirmatively:

Theorem 10. *Let $(R, \mathfrak{m})$ be a d-dimensional complete Noetherian local ring, and let I be the ideal defining the singular locus of R. Let M be a finitely generated R-module. There exists an integer k such that for all R-modules N,*

$$I^k Ext_R^{d+1}(M, N) = 0.$$

We can ask for more: can the uniform annihilators be determined effectively? The answer is yes. If $R = S/\mathfrak{p}$ is a quotient of a polynomial ring by a prime ideal, it is well-known (see [Eisenbud 1995, Section 16.6]) that the singular locus is inside the closed set determined by the appropriate size minors of the Jacobian matrix, and is equal to this closed set if the ground field is perfect. Wang proves that one can use elements in the Jacobian ideal to annihilate these Ext modules, so his result is also *effective* in the sense that specific elements of the ideal I can be constructed from a presentation of the algebra.

Before we leave this first section, let's look at yet another famous theorem of Hilbert, his Nullstellensatz, which identifies the radical of an ideal I in a polynomial ring S as the intersection of all maximal ideals that contain I. What can one ask in order to change this basic result into a uniform statement? One answer is that as it stands, the Nullstellensatz is a theoretical description of the radical of I, but it is not effective in the sense that information about I is not tied to information about its radical. There has been considerable work on making the Nullstellensatz "effective". We quote one result from [Kollár 1988], in a somewhat simplified form:

Theorem 11. *Let I be a homogeneous ideal in $S = k[x_1, \dots, x_n]$. Write $I = (f_1, \dots, f_m)$, where each f_i is homogeneous of degree $d_i \geq 3$. Let q be the minimum of m and n. If we let $D = d_1 d_2 \cdots d_q$, then*

$$(\sqrt{I})^D \subseteq I.$$

There are many variations, which include quadrics and non-homogeneous versions. In fact Kollár's theorem itself is more detailed and specific. Notice that the number of the degrees d_i in the product is absolutely bounded by the number of variables, even if the number of generators of I is extremely large compared to the dimension. The version we give is sharp, however, as the next example shows.

Example 12. Let $S = k[x_1, \dots, x_n]$ be a polynomial ring over a field k. Fix a degree d. Set

$$f_1 = x_1^d,\ f_2 = x_1 x_n^{d-1} - x_2^d, \dots,\ f_{n-1} = x_1 x_n^{d-1} - x_{n-1}^d.$$

If I is the ideal generated by these forms of degree d, then it is easy to see that the radical of I is the ideal $(x_1, \dots, x_{n-1})$. Moreover, $x_{n-1}^D \in I$ for $D = d^{n-1}$, but not for smaller values. We leave this fact to the reader to check.

Exercises.

Generators:

(1) Let k be a field. Prove that every maximal ideal in $k[x_1, \dots, x_n]$ is generated by n elements. In particular, every prime ideal in $k[x, y]$ is generated by at most two elements.

(2) Suppose that I is a homogeneous ideal in $S = k[x_1, \dots, x_n]$ generated by forms of degrees at most d, such that every variable is in the radical of I. Prove that I can be generated by at most the number of minimal generators of $\mathfrak{m}^d$, where $\mathfrak{m} = (x_1, \dots, x_n)$. Is the same statement true if one doesn't assume that the radical of I contains $\mathfrak{m}$?

(3) Let R be a standard graded ring over an infinite field, with homogeneous maximal ideal $\mathfrak{m}$. We say that an $\mathfrak{m}$-primary homogeneous ideal I is $\mathfrak{m}$-full if for every general linear form ℓ, $\mathfrak{m}I : \ell = I$. Prove that if I is $\mathfrak{m}$-full and J is homogeneous and contains I, then the minimal number of generators of J is at most the minimal number of generators of I.

(4) Let $\mathfrak{p}$ be a homogeneous prime ideal of a polynomial ring S such that $\mathfrak{p}$ contains no linear forms. It is not known whether or not $\mathfrak{p}$ is always generated by forms of degrees at most the multiplicity of $S/\mathfrak{p}$. Can you find examples where this estimate is sharp? What about if $\mathfrak{p}$ is not prime?

Radicals:

(5) Let M be an n by n matrix of indeterminates over the complex numbers $\mathbb{C}$, and let I be the ideal generated by the entries of the matrix M^n. Find n polynomials generating an ideal with the same radical as that of I. (Hint: use linear algebra.)

(6) It is an unsolved problem whether or not every non-maximal prime ideal in a polynomial ring $S = k[x_1, \ldots, x_n]$ can be generated up to radical by $n-1$ polynomials. Here is an explicit example in which the answer is not known, from Moh. Let $\mathfrak{p}$ be the defining ideal of the curve $k[t^6 + t^{31}, t^8, t^{10}]$ in a polynomial ring in 3 variables. Can you find a set of generators of $\mathfrak{p}$? It is conjectured that $\mathfrak{p}$ is generated up to radical by 2 polynomials. Why is this the least possible number of polynomials that could generate $\mathfrak{p}$ up to radical? If the characteristic of k is positive it is known that $\mathfrak{p}$ is generated up to radical by 2 polynomials. Assuming the characteristic is equal to 2, find such polynomials.

(7) Let $k = \mathbb{C}$ be the field of complex numbers, and let $\mathfrak{p}$ be the defining ideal of the surface $k[t^4, t^3s, ts^3, s^4]$. Find generators for $\mathfrak{p}$, and find three polynomials which generate $\mathfrak{p}$ up to radical. It is unknown whether or not there are 2 polynomials which generate $\mathfrak{p}$ up to radical, although this is known in positive characteristic.

(8) Let S be a polynomial ring, and let I be generated by forms of degrees $d_1, \ldots, d_s$. Suppose that f is in the radical of I, so that there is some N such that $f^N \in I$. Is there an effective bound for N? Take a guess. Find the best example you can to see that N must be large.

(9) Let R be a regular local ring, and let $I \subseteq R$ be an ideal such that R/I is Cohen–Macaulay. Let

$$F_\bullet := 0 \to F_n \xrightarrow{f_n} \ldots \xrightarrow{f_1} R \to R/I \to 0$$

be a minimal free resolution of R/I. Show that

$$\sqrt{I} = \sqrt{I(f_1)} = \cdots = \sqrt{I(f_n)},$$

where $I(f_i)$ is the ideal of R generated by the k_i–minors of f_i, where k_i is maximal with the property that the k_i–minors of f_i are not all zero.

Stillman's question:

(10) Let S be a polynomial ring and let I be an ideal generated by two forms. Show that the projective dimension of S/I is at most 2. What well-known statement is this equivalent to?

(11) Let S be a polynomial ring. It is known that if I is generated by three quadrics, then the projective dimension of S/I is at most 4. Find an example to see that 4 is attained, and try to prove this statement.

(12) The largest known projective dimension of a quotient S/I where I is generated by three cubics and S is a polynomial ring is 5. Can you find such an example?

(13) Prove the following strong form of Stillman's problem for monomial ideals: if I is generated by s monomials in a polynomial ring S, then the projective dimension of S/I is at most s.

(14) What about binomial ideals? Is there a bound similar to that in the previous question?

Infinite resolutions:

(15) If $R = S/I$ where S is a polynomial ring and I has a Gröbner basis of quadrics, then R is Koszul, i.e., the residue field has a linear resolution. Prove this.

(16) Suppose that $R = S/I$, where S is a polynomial ring, and I is homogeneous. If the regularity of the residue field of R is bounded, show that the regularity of every finitely generated graded R-module M is also bounded.

(17) Find an example of a resolution of the residue field of a standard graded algebra so that the degrees of the entries of the matrices in a minimal resolution (after choosing bases for the free modules) are at least any fixed number N. It is a conjecture of Eisenbud, Reeves and Totaro that one can always choose bases of the free modules in the resolution of a finitely generated graded module so that the entries in the whole (usually infinite) set of matrices are bounded.

(18) Let R be a Cohen–Macaulay standard graded algebra which is a domain of multiplicity e. Prove that the i-th total Betti number (the sum of all $\beta_{i,j}$ for $j \in \mathbb{Z}$) of any quotient R/I is at most e times the $(i-1)$-st total Betti number of R/I for large i. What sort of uniformity for total Betti numbers might one hope for?

Relations between invariants:

(19) Try to imagine a conjecture about effective bounds relating the multiplicity of $S/\mathfrak{p}$, where S is a polynomial ring and $\mathfrak{p}$ is a homogeneous prime not containing a linear form, and the regularity of $S/\mathfrak{p}$. Why should there be any relationship? Try the case in which $\mathfrak{p}$ is generated by a regular sequence of forms.

(20) Let S be a polynomial ring, and let I be an ideal generated by square-free monomials. The multiplicity of S/I is just the number of minimal primes $\mathfrak{p}$ over I such that the dimension of $S/\mathfrak{p}$ is maximal. Can you say anything about the regularity? For example, what if I is the edge ideal of a graph (see the discussion following Question 35)?

(21) Is there any relationship at all between the projective dimension (resp. regularity) of a quotient S/I (S a polynomial ring, I a homogeneous ideal) and

the projective dimension (resp. regularity) of $S/\sqrt{I}$? Try to give examples or formulate a problem.

(22) Answer the previous question when I is generated by monomials.

2. Reduction to characteristic p and integral closure

In this section we will discuss the solution to an uniformity question of John Mather, and illustrate in the process two important concepts in commutative algebra: reduction to characteristic p, and integral closure. Throughout this section, S will denote the power series ring $\mathbb{C}[[x_1, \ldots, x_n]]$. Given $f \in S$, we will write f_i for its i-th partial derivative $\partial f/\partial x_i$. We write $J(f)$ for the *Jacobian ideal* of f, $J(f) = (f_1, \ldots, f_n)$.

Question 13 (Mather [Huneke and Swanson 2006, Question 13.0.1]). *Consider an element $f \in S$ satisfying $f(0) = 0$. Does there exist an uniform integer N such that $f^N \in J(f)$?*

The answer to this turns out to be positive as we'll explain shortly, and in fact one can take $N = n$. Notice however that there is no a priori reason for such an N to even exist, that is, for f to be contained in $\sqrt{J(f)}$. Let's first look at some examples:

- If $f = x_1^2 + x_2^2$ then $J(f) = (2x_1, 2x_2)$, so $f \in J(f)$.
- If $f = x^2 - x$ then since $f_1 = 2x - 1$ is a unit, we get $f \in J(f)$.
- If f is a *homogeneous polynomial*, or more generally a *quasihomogeneous* one, then $f \in J(f)$ (recall that f is said to be *quasihomogeneous* if there exist weights d and $\omega_i \in \mathbb{Z}_{\geq 0}$ with the property that $f(t^{\omega_1}x_1, \ldots, t^{\omega_n}x_n) = t^d f(x_1, \ldots, x_n)$). The conclusion $f \in J(f)$ follows from the quasihomogeneous version of Euler's formula:
$$\sum_{i=1}^{n} \omega_i \cdot x_i \cdot f_i = d \cdot f.$$
- Even when f is not quasihomogeneous, there might exist an analytic change of coordinates which transforms it into a quasihomogeneous polynomial, so the conclusion $f \in J(f)$ still holds. For example, if $f = (x_1 - x_2^2) \cdot (x_1 - x_2^3)$, one can make the change of variable $y_1 = x_1 - x_2^2$, $y_2 = x_2 \cdot \sqrt{1 - x_2}$ and get $f = y_1 \cdot (y_1 + y_2^2)$ which is quasihomogeneous for $\omega_1 = 2$, $\omega_2 = 1$ and $d = 4$. The following theorem gives a partial converse to this observation:

Theorem 14 [Saito 1971]. *If the hypersurface $f = 0$ has an isolated singularity (or equivalently $\sqrt{J(f)} = (x_1, \ldots, x_n)$) then f is quasihomogeneous (with respect to some analytic change of coordinates) if and only if $f \in J(f)$.*

We begin answering Question 13 by looking first at the case $n = 1$: writing t for x_1, $S = \mathbb{C}[[t]]$ is a DVR, so any nonzero $f \in S$ can be written as $f = t^i \cdot u$, where $u(0) \neq 0$, i.e., u is a unit; we get

$$\frac{\partial f}{\partial t} = t^{i-1} \cdot \left(i \cdot u + t \cdot \frac{\partial u}{\partial t} \right) = t^{i-1} \cdot \text{unit},$$

so $f \in J(f)$. This calculation in fact shows that $f \in J(f)$ even for $n > 1$, provided that we first make a ring extension to a power series ring in one variable. More precisely, assume that $n > 1$ and consider an embedding $S \hookrightarrow K[[t]]$, given by $x_i \mapsto x_i(t)$, where K is any field extension of $\mathbb{C}$. The above calculation shows that f is contained in the ideal of $K[[t]]$ generated by the derivative df/dt. The Chain Rule yields

$$\frac{df}{dt} = \sum_{i=1}^{n} f_i \cdot \frac{dx_i}{dt} \in J(f) \cdot K[[t]],$$

so we conclude that $f \in J(f) \cdot K[[t]]$. This motivates the following

Definition 15 (Integral closure of ideals). Given an ideal $I \subset S$, the *integral closure* $\bar{I}$ of I is defined by

$$\bar{I} = \{g \in S : \varphi(g) \in \varphi(I) \ \text{ for every field extension } \mathbb{C} \subset K \text{ and every } \mathbb{C}\text{–algebra homomorphism } \varphi : S \to K[[t]]\}.$$

We have thus shown that $f \in \overline{J(f)}$, so our next goal is to understand better the relationship between an ideal and its integral closure. We have the following result.

Theorem 16 (Alternative characterizations of integral closure [Huneke and Swanson 2006, Theorem 6.8.3, Corollary 6.8.12]). *Given an ideal $I \subset S$ and an element $g \in S$, the following statements are equivalent*:

(a) $g \in \bar{I}$.

(b) *There exist $k \in \mathbb{Z}_{\geq 0}$ and $s_i \in I^i$ for $i = 1, \dots, k$, such that*

$$g^k = s_1 \cdot g^{k-1} + \cdots + s_i \cdot g^{k-i} + \cdots + s_k.$$

(c) *There exists $c \in S \setminus \{0\}$ such that*

$$c \cdot g^m \in I^m \text{ for every } m \in \mathbb{Z}_{\geq 0}.$$

It is worth thinking for a moment about the differences between these three characterizations. In fact, they are very different, and we shall need all three of them. The first characterizes integral closure in a non-constructive way, since the definition depends on arbitrary homomorphisms to discrete valuation rings. Nonetheless, we have seen the power of this definition by using it to show that power series are integral over the ideal generated by their partials. The second characterization shows that one needs only a finite set of data to determine integral closures. In particular, it is clear from this characterization that integral closure behaves well under numerous operations such as homomorphisms. Finally, the third characterization is the easiest to use in the sense that it is a weak condition, but the condition by its nature involves an infinite set of equations.

Since $f \in \overline{J(f)}$, part (b) of Theorem 16 implies that $f^k \in J(f)$ for some k, but in principle k could depend on f. The goal of the rest of this section is to show that we can choose $k = n$, independently of f. We will do so by passing to characteristic $p > 0$. A word of caution is in order here, which is that the conclusion $f \in \sqrt{J(f)}$ fails in positive characteristic: if $f = g^p$, then $J(f) = 0$. Nevertheless, we will prove the following:

Theorem 17 [Skoda and Briançon 1974; Lipman and Sathaye 1981; Huneke and Swanson 2006, Chapter 13]. *Assume that R is a regular local ring of characteristic $p > 0$, and consider an ideal $J = (g_1, \dots, g_t)$ in R. If $g \in \overline{J}$ then $g^t \in J$.*

The advantage of working in positive characteristic is the existence of the *Frobenius endomorphism* F sending every element x to x^p. In the case of a regular local ring, the Frobenius endomorphism is in fact flat [Kunz 1969], which yields the following:

Theorem 18 (Test for ideal membership). *Consider a regular local ring R of characteristic $p > 0$, an ideal $J = (g_1, \dots, g_t)$, and an element $g \in R$. We have that $g \in J$ if and only if there exists $c \in R \setminus \{0\}$ such that for all e, $c \cdot g^{p^e}$ is in the Frobenius power*

$$J^{[p^e]} := (g_1^{p^e}, \dots, g_t^{p^e}).$$

Remark. This test for ideal membership has been conceptualized into an important closure operation called tight closure (see [Hochster and Huneke 1990]). If R is a Noetherian ring of characteristic p, I is an ideal, and $x \in R$, we say that x is in the *tight closure* of I if there exists an element $c \in R$, not in any minimal prime of R, such that $cx^{p^e} \in I^{[p^e]}$ for all large e. The set of all elements in the tight closure of I forms a new ideal I^*, called the *tight closure of I*.

Proof of Theorem 18. Assume that $g \notin J$ so that $(J : g) = \{x \in R : x \cdot g \in J\}$ is a proper ideal, and consider the exact sequence

$$0 \longrightarrow R/(J : g) \xrightarrow{g} R/J \longrightarrow R/(J, g) \longrightarrow 0. \tag{2-1}$$

Since F is flat, pulling back (2-1) along F^e preserves exactness, yielding the sequence

$$0 \longrightarrow R/(J:g)^{[p^e]} \xrightarrow{g^{p^e}} R/J^{[p^e]} \longrightarrow R/(J,g)^{[p^e]} \longrightarrow 0. \tag{2-2}$$

Since $(J,g)^{[p^e]} = (J^{[p^e]}, g^{[p^e]})$, we get by comparing (2-2) with the analogue of (2-1), namely

$$0 \longrightarrow R/(J^{[p^e]}:g^{p^e}) \xrightarrow{g^{p^e}} R/J^{[p^e]} \longrightarrow R/(J^{[p^e]}, g^{p^e}) \longrightarrow 0,$$

that $(J:g)^{[p^e]} = (J^{[p^e]}:g^{p^e})$. The condition $c \cdot g^{p^e} \in J^{[p^e]}$ for all e then becomes

$$c \in \bigcap_{e \geq 0} (J:g)^{[p^e]} \subset \bigcap_{e \geq 0} (J:g)^{p^e} = 0,$$

where the last equality follows from the Krull intersection theorem [Eisenbud 1995, Corollary 5.4]. □

Proof of Theorem 17. By Theorem 18, it suffices to find $c \neq 0$ such that $c \cdot (g^t)^{p^e} \in J^{[p^e]}$. Since $g \in \bar{J}$, we know by Theorem 16(c) that there exists $c \neq 0$ such that $c \cdot g^m \in J^m$ for all m. Taking $m = t \cdot p^e$, we have in particular that $c \cdot (g^t)^{p^e} \in J^{t \cdot p^e}$. Since $J^{t \cdot p^e}$ is generated by monomials $g_1^{i_1} \cdots g_t^{i_t}$, with $i_1 + \cdots + i_t = t \cdot p^e$, for each such monomial at least one of the exponents i_j satisfies $i_j \geq p^e$. It follows that $J^{t \cdot p^e} \subset J^{[p^e]}$, so $c \cdot (g^t)^{p^e} \in J^{[p^e]}$, concluding the proof of the theorem. □

We now explain the last ingredient needed to answer Mather's question, which is reduction to characteristic p. We will use it to show that the statement of Theorem 17 holds in characteristic 0 for the power series ring S:

Theorem 19. *Let $S = \mathbb{C}[[x_1, \ldots, x_n]]$ and consider an ideal $I = (f_1, \ldots, f_t)$ in S. If $f \in \bar{I}$ then $f^t \in I$.*

Proof. Suppose that the conclusion of the theorem fails, so $f^t \notin I$. The idea is to produce a regular ring R in characteristic p, an ideal $J \subset R$ and an element $g \in R$ that fail the conclusion of Theorem 17, obtaining a contradiction. The point here is that the hypotheses of Theorem 17 depend only on finite amount of data, which can be carried over to positive characteristic: this is essential in any argument involving reduction to characteristic p.

We now need a major theorem: Néron desingularization [Artin and Rotthaus 1988] states that we can write S as a directed union of smooth $\mathbb{C}[x_1, \ldots, x_n]$-algebras. This amazing theorem allows one to descend from power series, which a priori have infinitely many coefficients, to a more finite situation. Given any finite subset, $\mathcal{S}$ say, of S, we can choose one such algebra T containing $\mathcal{S}$. According to the equivalent description of the condition $f \in \bar{I}$ in part (b) of Theorem 16, there exist k and elements $s_i \in I^i$ such that

$$f^k = \sum_i s_i \cdot f^{k-i}. \tag{2-3}$$

To express the containment $s_i \in I^i$, we choose coefficients $c^i_\alpha \in S$ such that

$$s_i = \sum_{\substack{\alpha=(\alpha_1,\dots,\alpha_t)\\ \alpha_1+\dots+\alpha_t=i}} c^i_\alpha \cdot f_1^{\alpha_1} \cdots f_t^{\alpha_t}. \tag{2-4}$$

We will then require the smooth subalgebra T of S to contain $\mathcal{S} = \{f, f_j, s_i, c^i_\alpha\}$ for all j, i and α. We write I_T for the ideal $(f_1, \dots, f_t)$ of T. Since $f^t \notin I = I_T S$, it must be that $f^t \notin I_T$. Since T contains $\mathcal{S}$, (2-4) can be interpreted as an equality in T, which yields $s_i \in I_T^i$. Furthermore, (2-3) is an equation in T, so Theorem 16 applies to show that $f \in \overline{I_T}$.

Since T is smooth over $\mathbb{C}[x_1, \dots, x_n]$, it is in particular a finite type algebra over $\mathbb{C}$, so it can be written as a quotient of a polynomial ring $\mathbb{C}[y_1, \dots, y_r]$ by some ideal $(h_1, \dots, h_s)$. Each of the elements of $\mathcal{S}$ is then represented by the class of some polynomial in $\mathbb{C}[y_1, \dots, y_r]$, so collecting the coefficients of all these polynomials, as well as the coefficients of $h_1, \dots, h_s$, we obtain a finite subset $\mathcal{A} \subset \mathbb{C}$. We define $A = \mathbb{Z}[\mathcal{A}]$ to be the smallest subring of $\mathbb{C}$ containing $\mathcal{A}$. A is a finitely generated $\mathbb{Z}$-algebra, and we can consider the ring $T_A = A[y_1, \dots, y_r]/(h_1, \dots, h_s)$. T_A is called a *model* of T, having the property that $T_A \otimes_A \mathbb{C} = T$. Moreover, T_A contains all the elements of $\mathcal{S}$ (we use here an abuse of language: what we mean is that if we think of $A[y_1, \dots, y_r]$ as a subring of $\mathbb{C}[y_1, \dots, y_r]$, then every element of $\mathcal{S}$ is represented by some polynomial in $A[y_1, \dots, y_r]$). We write I_{T_A} for the ideal of T_A generated by $f_1, \dots, f_t$, and conclude as before that $f^t \notin I_{T_A}$ and $f \in \overline{I_{T_A}}$.

We are now ready to pass to characteristic $p > 0$. We first need to observe that if we write $Q(A)$ for the quotient field of A, then $T_A \otimes_A Q(A)$ is smooth over $Q(A)$, i.e., the map $A \to T_A$ is generically smooth: this follows from the fact that T is smooth over $\mathbb{C}$, together with the fact that applying the Jacobian criterion to the map $\mathbb{C} \to T = \mathbb{C}[y_1, \dots, y_r]/(h_1, \dots, h_s)$ is the same as applying it to $Q(A) \to T_A \otimes_A Q(A) = Q(A)[y_1, \dots, y_r]/(h_1, \dots, h_s)$. It follows that for a generic choice of a maximal ideal $\mathfrak{n} \subset A$, the quotient $R = T_A/\mathfrak{n}T_A$ is smooth over the finite field $A/\mathfrak{n}$, so in particular it is a regular local ring. Writing g (resp. g_i) for the class of $f \in T_A$ (resp. $f_i \in T_A$) in the quotient ring R, letting $J = (g_1, \dots, g_t)$, and observing that the equations (2-3) and (2-4) descend to R, we get that $g \in \overline{J}$. The condition $g^t \notin J$ follows from generic flatness and the genericity assumption on $\mathfrak{n}$: T_A is a finite type algebra over A, and multiplication by f^t on $(T_A/I_A) \otimes_A Q(A)$ is nonzero (if it were zero, then it would also be zero on $(T_A/I_A) \otimes_A Q(A) \otimes_{Q(A)} \mathbb{C} = T/I$, but $f^t \notin I$), i.e., multiplication by f^t on T_A/I_A is generically nonzero. It follows that for a generic choice of $\mathfrak{n}$, the ring R is a regular local ring in characteristic $p > 0$, containing an ideal $J = (g_1, \dots, g_t)$ and an element $g \in \overline{J}$ with $g^t \notin J$. This is in contradiction with Theorem 17, concluding our proof. □

Exercises.

(23) Let $S = \mathbb{C}[\![x_1, \dots, x_n]\!]$, let $f \in S$ with $f(0) = 0$, and let $\mathfrak{m}$ be the maximal ideal of S. Prove that $f \in \overline{\mathfrak{m} \cdot J(f)}$. It is not known whether or not $f \in \mathfrak{m} \cdot \overline{J(f)}$. If true, this would give a positive solution to the Eisenbud–Mazur conjecture (see Exercise (40) on page 181).

(24) Let $f(t), g(t)$ be polynomials with coefficients in a ring R, say $f(t) = a_n t^n + \cdots + a_0$, and $g(t) = b_n t^n + \cdots + b_0$. Let c_i be the coefficient of t^i in the product fg. Prove that the ideal generated by $a_i b_j$ is integral over the ideal generated by $c_{2n}, \dots, c_0$.

(25) Let S be a polynomial ring in n variables, and let $g_1, \dots, g_n$ be a regular sequence of forms of degree d (equivalently assume that they are forms of degree d, and that the radical of the ideal they generate is the homogeneous maximal ideal). Prove that $\overline{(g_1, \dots, g_n)} = \mathfrak{m}^d$.

3. Uniform Artin–Rees

In the last section we saw how to use characteristic p techniques (in a power series ring over the field $\mathbb{C}$) in order to give a uniform bound on the power of an element in the integral closure of an ideal I to be contained in I. The following more general result was first proved in [Skoda and Briançon 1974] for convergent power series over the complex numbers, and generalized to arbitrary regular local rings in [Lipman and Sathaye 1981].

Theorem 20. *Let R be a regular local ring and let I be an ideal generated by ℓ elements. Then for all $n \geq \ell$,*

$$\overline{I^n} \subseteq I^{n-\ell+1}.$$

Although apparently ℓ depends on the number of generators of I, in fact it can be made uniform. This is because if the residue field of R is infinite, every ideal is integral over an ideal generated by d elements, where d is the dimension of R. If the residue field is not infinite, then one can make a flat base change to that case and still prove that one can always choose $\ell = d$.

What about for Noetherian local rings which are not regular? Is there a uniform integer k such that for all ideals I, if $n \geq k$ then

$$\overline{I^n} \subseteq I^{n-k+1}?$$

The following conjecture was made in [Huneke 1992]:

Conjecture 21. *Let R be a reduced excellent Noetherian ring of finite Krull dimension. There exists an integer k, depending only on R, such that $\overline{I^n} \subseteq I^{n-k}$ for every ideal $I \subseteq R$ and all $n \geq k$?*

With a little thought, it is easy to see that at least one cannot choose such a k equal to the dimension of the ambient ring. For example if the dimension is one, then the statement $\overline{I^n} \subseteq I^{n-1+1}$ forces powers of all ideals to be integrally closed. For a local one-dimensional ring, this in turn forces the ring to be regular. On the other hand there will often be a uniform k in this special case. For example, let's suppose that R is a one-dimensional complete local domain. Its integral closure will be a DVR, say V. The integral closure of any ideal J in R is given by $JV \cap R$. There is a conductor ideal which is primary to the maximal ideal $\mathfrak{m}$, so there is a fixed integer k such that for every ideal I of R, I^k is in the conductor. But then,

$$\overline{I^n} = I^n V \cap R \subset I^{n-k},$$

since $I^k V \subset R$. Thinking about this analysis, it is not totally surprising that this question is closely connected with another uniform question dealing with the classical lemma of Artin and Rees.

The usual Artin–Rees lemma states that if R is Noetherian, $N \subseteq M$ are finitely generated R-modules, and I is an ideal of R, then there exists a $k > 0$ (depending on I, M, N) such that for all $n > k$, $I^n M \cap N = I^{n-k}(I^k M \cap N)$. A weaker statement which is sometimes just as useful is that for all $n > k$,

$$I^n M \cap N \subseteq I^{n-k} N.$$

How dependent upon I, M and N is the least such k? It is very easy to see that k fully depends upon both N and M, so the only uniformity that might occur is in varying the ideal I. The usual proof of the Artin–Rees lemma passes to the module $\mathfrak{M} := M \oplus IM \oplus I^2 M \oplus \cdots$, which is finitely generated over the Rees algebra $R[It]$ of I. Since the Rees algebra is Noetherian, every submodule of $\mathfrak{M}$ is finitely generated. Applying this fact to the submodule $\mathfrak{N} := N \oplus IM \cap N \oplus I^2 M \cap N \oplus \cdots$ of $\mathfrak{M}$ then easily gives the Artin–Rees lemma. On the face of it, there is no way that the integer k could be chosen uniformly, since it depends on the degrees of the generators of the submodule $\mathfrak{N}$ over the Rees algebra of I. Nonetheless, one can still make a rather optimistic conjecture [Huneke 1992]:

Conjecture 22. *Let R be an excellent Noetherian ring of finite Krull dimension. Let $N \subseteq M$ be two finitely generated R-modules. There exists an integer $k = k(N, M)$ such that for all ideals $I \subseteq R$ and all $n \geq k$,*

$$I^n M \cap N \subseteq I^{n-k} N.$$

In this generality, the conjecture is open. However, there is considerable literature giving lots of information about this conjecture and related problems. See, for example, [Aberbach 1993; Duncan and O'Carroll 1989; Ein et al. 2004;

Giral and Planas-Vilanova 2008; O'Carroll 1987; O'Carroll and Planas-Vilanova 2008; Planas-Vilanova 2000; Striuli 2007; Trivedi 1997].

It turns out that there is a very close relationship between these two conjectures, which is not at all apparent. One way to see such a connection is through results related to tight closure theory. Suppose that R is a d-dimensional local complete Noetherian ring of characteristic p which is reduced. The so-called "tight closure Briançon–Skoda theorem" [Hochster and Huneke 1990] states that for every ideal I, $\overline{I^n} \subseteq (I^{n-d+1})^*$, where J^* denotes the tight closure of an ideal J. If R is regular, every ideal is tightly closed. The point here is that R will have a nonzero test element c, not in any minimal prime. This means that c multiplies the tight closure of any ideal back into the ideal. Such elements are *uniform* annihilators, and are one of the most important features in the theory of tight closure. Suppose that there is a uniform Artin–Rees number k for the pair of R-modules, $(c) \subset R$. Then for every ideal I,

$$c\overline{I^n} \subseteq c(I^{n-d+1})^* \subset (c) \cap I^{n-d+1} \subset cI^{n-d-k+1}.$$

Since c is not in any minimal prime and R is reduced, it follows that c is a nonzerodivisor. We can cancel it to obtain that

$$\overline{I^n} \subseteq I^{n-d-k+1}.$$

Thus in this case, uniform Artin–Rees implies uniform Briançon–Skoda. In fact these conjectures are more or less equivalent.

Both conjectures were proved in fairly great generality in [Huneke 1992]:

Theorem 23 (Uniform Artin–Rees). *Let S be a Noetherian ring. Let $N \subseteq M$ be two finitely generated S-modules. If S satisfies any of the conditions below, then there exists an integer k such that for all ideals I of S, and for all $n \geq k$*

$$I^n M \cap N \subseteq I^{n-k} N.$$

(i) *S is essentially of finite type over a Noetherian local ring.*

(ii) *S is a reduced ring of characteristic p, and $S^{1/p}$ is module-finite over S.*

(iii) *S is essentially of finite type over $\mathbb{Z}$.*

We also have the following:

Theorem 24 (Uniform Briançon–Skoda). *Let S be a Noetherian reduced ring. If S satisfies any of the following conditions, then there exists a positive integer k such that $\overline{I^n} \subseteq I^{n-k}$ for all ideals I of S and all $n \geq k$.*

(i) *S is essentially of finite type over an excellent Noetherian local ring.*

(ii) *S is of characteristic p, and $S^{1/p}$ is module-finite over S.*

(iii) *S is essentially of finite type over $\mathbb{Z}$.*

Exercises.

(26) If a Noetherian ring R has the uniform Artin–Rees property, show that R/J (for any ideal $J \subseteq R$) also has the uniform Artin–Rees property.

(27) If a Noetherian ring R has the uniform Artin–Rees property and W is any multiplicatively closed subset of R, show that the localization R_W also has the uniform Artin–Rees property.

(28) Suppose that a Noetherian ring R has the uniform Artin–Rees property. Given a finitely generated R-module M, and an integer $i \geq 1$, show that there exists an integer $k \geq 1$ such that for all ideals I of R and all n,

$$I^k \operatorname{Tor}_i^R(R/I^n, M) = 0.$$

(29) Let $R = k[\![x, y]\!]$, k a field. Set $I = (x^n, y^n, x^{n-1}y)$, $J = (x^n, y^n)$. Prove that if $k < n$ then $I^\ell \neq J^{\ell-k} I^k$ for some $\ell \geq k+1$.

(30) Let R be a Noetherian domain which satisfies the uniform Artin–Rees theorem for every pair of finitely generated modules $N \subset M$. Let f be a nonzero element of R. Prove that there exists an integer k such that for every maximal ideal $\mathfrak{m}$ of R, $f \notin \mathfrak{m}^k$.

4. Symbolic powers

In this section, S will denote either a polynomial ring $k[x_1, \ldots, x_n]$ over some field k, or a regular local ring. The guiding problem will be the comparison between regular and symbolic powers of ideals in S. From a uniform perspective, we would like to understand whether the equality between small regular and symbolic powers guarantees the equality of all regular and symbolic powers. As we'll see, this is a very difficult question, but it gives rise to many interesting variations. We begin with a discussion of multiplicities, which will motivate the introduction of symbolic powers.

Recall from Section 1 that if S is a polynomial ring, I is a homogeneous ideal, and $R = S/I$, then $\dim_k(R_i) = h_R(i)$ is a polynomial function for sufficiently large values of i, and the multiplicity $e(R)$ is defined by the property that

$$h_R(i) = \frac{e(R)}{(\dim R - 1)!} i^{\dim R - 1} + (\text{lower order terms}).$$

We can define the multiplicity of a local ring $(R, \mathfrak{m})$ by letting $e(R) = e(\operatorname{gr}_{\mathfrak{m}}(R))$, where

$$\operatorname{gr}_{\mathfrak{m}}(R) = R/\mathfrak{m} \oplus \mathfrak{m}/\mathfrak{m}^2 \oplus \mathfrak{m}^2/\mathfrak{m}^3 \oplus \cdots = \bigoplus_{i \geq 0} \mathfrak{m}^i/\mathfrak{m}^{i+1}$$

is the *associated graded ring* of R with respect to $\mathfrak{m}$. Let's look at some examples of multiplicities:

Example 25. (a) If $S = k[x_1, \dots, x_n]$ and f is a form of degree d then $e(S/(f)) = d$. If S is a regular local ring, then

$$e(S/(f)) = \operatorname{ord}(f) := \max\{n : f \in \mathfrak{m}^n\}.$$

(b) If $S = k[x_1, \dots, x_n]$, and S/I is Cohen–Macaulay, having a *pure* resolution

$$0 \longrightarrow S(-d_c)^{\beta_c} \longrightarrow \cdots \longrightarrow S(-d_2)^{\beta_2} \longrightarrow S(-d_1)^{\beta_1} \longrightarrow S \longrightarrow S/I \longrightarrow 0,$$

where $c = \operatorname{codim}(I)$, then

$$e(S/I) = \frac{\prod_{j=1}^{c} d_j}{c!}.$$

(c) If $X \subset \mathbb{P}^n$ is a set consisting of r points, and if we write R_X for the homogeneous coordinate ring of X, then

$$e(R_X) = r.$$

Note that for a projective variety X, $e(R_X)$ is also called the *degree* of X.

(d) If $X = \mathbb{G}(2, n)$ is the Grassmannian of 2-planes in n-space, in its Plücker embedding, then its degree is (see for example [Mukai 1993])

$$e(R_X) = \frac{1}{n-1}\binom{2n-4}{n-2},$$

the $(n-2)$-nd *Catalan number.*

The following is a natural question when studying multiplicity:

Question 26. *How does the multiplicity behave under flat maps?*

We are interested in two types of flat maps:

I. A local flat ring homomorphism $(R, \mathfrak{m}) \to (R', \mathfrak{m}')$.

II. A localization map $R \to R_{\mathfrak{p}}$, where $\mathfrak{p}$ is a prime ideal.

For flat maps of type I, the behavior of multiplicity is the subject of an old conjecture:

Conjecture 27 [Lech 1960]. *If* $(R, \mathfrak{m}) \to (R', \mathfrak{m}')$ *is a local flat homomorphism then*

$$e(R) \le e(R').$$

It is amazing that very little progress has been made on this conjecture, although it is easy to state, and was made about 50 years ago! To paraphrase a famous line of Mel Hochster, it is somewhat of an insult to our field that we cannot answer this conjecture, one way or the other.

For flat maps of type II, if S is a regular local ring and $R = S/(f)$, we have (see [Nagata 1962, (38.3); Zariski 1949] and Exercise (36) on page 187)

$$e(R_{\mathfrak{p}}) = \max\{n : f \in \mathfrak{p}^n S_{\mathfrak{p}} \cap S\} \leq \max\{n : f \in \mathfrak{m}^n\} = e(R).$$

If we denote by $\mathfrak{p}^{(n)}$ the intersection $\mathfrak{p}^n S_{\mathfrak{p}} \cap S$, also called the n-th *symbolic power* of $\mathfrak{p}$, then the above inequality is equivalent to the containment

$$\mathfrak{p}^{(n)} \subseteq \mathfrak{m}^n. \tag{4-1}$$

For a ring R, we write $\operatorname{Spec}(S)$ (resp. $\operatorname{Max}(S)$) for the collection of its prime (resp. maximal) ideals. In general, if $S = k[x_1, \ldots, x_n]$ is a polynomial ring and $\mathfrak{p} \in \operatorname{Spec}(S)$, then by the Nullstellensatz [Eisenbud 1995, Theorem 4.19],

$$\mathfrak{p} = \bigcap_{\substack{\mathfrak{m} \in \operatorname{Max}(S) \\ \mathfrak{p} \subset \mathfrak{m}}} \mathfrak{m}.$$

The symbolic powers of $\mathfrak{p}$ can then be described (see [Eisenbud and Hochster 1979]) as

$$\mathfrak{p}^{(n)} = \bigcap_{\substack{\mathfrak{m} \in \operatorname{Max}(S) \\ \mathfrak{p} \subset \mathfrak{m}}} \mathfrak{m}^n, \tag{4-2}$$

generalizing the inclusion (4-1). If we think of $\mathfrak{p}$ as defining an affine variety X, then (4-2) characterizes $\mathfrak{p}^{(n)}$ as the polynomial functions that vanish to order n at the points of X. Symbolic powers make sense in a more general context. If $I = \mathfrak{p}_1 \cap \cdots \cap \mathfrak{p}_r$ is an intersection of prime ideals, then

$$I^{(n)} = \mathfrak{p}_1^{(n)} \cap \cdots \cap \mathfrak{p}_r^{(n)}. \tag{4-3}$$

For an arbitrary ideal I, see [Bauer et al. 2009, Definition 8.1.1].

One of the main questions regarding symbolic powers is the following:

Question 28 Regular versus symbolic powers. *How do I^n and $I^{(n)}$ compare? In particular, when are they equal for all n?*

Example 29. (a) If I is a *complete intersection ideal*, i.e., if it is generated by a regular sequence, then $I^n = I^{(n)}$ for all $n \geq 1$ (see Exercise (33) for the case when I is a prime ideal).

(b) Let X denote a generic $n \times n$ matrix with $n \geq 3$, let $\Delta = \det(X)$ and $I = I_{n-1}(X)$, the ideal of $(n-1) \times (n-1)$ minors of X. We have on one hand that the adjoint matrix $\operatorname{adj}(X)$ has entries in I, so $\det(\operatorname{adj}(X)) \in I^n$, and on the other hand $\operatorname{adj}(X) \cdot X = \Delta \cdot \mathbb{I}_n$ (where $\mathbb{I}_n$ denotes the $n \times n$ identity matrix), so $\det(\operatorname{adj}(X)) = \Delta^{n-1}$. We get $\Delta^{n-1} \in I^n$, from which it can be shown that $\Delta \in I^{(2)} \setminus I^2$ (see Exercise (42)).

(c) If $S = k[x, y, z]$, $\mathfrak{p} \in \operatorname{Spec}(S)$ with $\dim(S/\mathfrak{p}) = 1$, then the following are equivalent:
- $\mathfrak{p}^{(n)} = \mathfrak{p}^n$ for all $n \geq 1$.
- $\mathfrak{p}^{(2)} = \mathfrak{p}^2$.
- $\mathfrak{p}$ is locally a complete intersection.

The following conjecture is related to Example 29(b), and in particular it is true in the said example:

Conjecture 30 [Hübl 1999, Conjecture 1.3]. *If R is a regular local ring, $\mathfrak{p} \in$ Spec(R) and $f \in R$, with the property that $f^{n-1} \in \mathfrak{p}^n$, then $f \in \mathfrak{m} \cdot \mathfrak{p}$.*

As a consequence of Exercise (42), it follows under the assumptions of the conjecture that $f \in \mathfrak{p}^{(2)}$, but the conclusion $f \in \mathfrak{m} \cdot \mathfrak{p}$ turns out to be significantly harder. In characteristic 0, Conjecture 30 is equivalent to the Eisenbud–Mazur conjecture on evolutions (see Exercise (40) and [Eisenbud and Mazur 1997; Boocher 2008]).

A natural uniformity problem is to determine if it is enough to test the equality in Question 28 for finitely many values of n, and moreover to determine a uniform bound for these values. A precise version of this is the following:

Question 31. *Assume that S is a regular local ring, or a polynomial ring, and that $\mathfrak{p} \in$ Spec(S). If $\mathfrak{p}^{(\dim S)} = \mathfrak{p}^{\dim S}$, does it follow that $\mathfrak{p}^{(n)} = \mathfrak{p}^n$ for all $n \geq 1$?*

One could ask the same question, replacing the condition $\mathfrak{p}^{(\dim S)} = \mathfrak{p}^{\dim S}$ with a stronger one, namely $\mathfrak{p}^{(i)} = \mathfrak{p}^i$ for all $i \leq \dim(S)$. The equivalence between the two formulations is unknown in general, but in characteristic zero it would follow from a positive answer to the following:

Question 32. *Let S be a regular local ring containing $\mathbb{C}$, and let $\mathfrak{p} \in$ Spec(S). Does it follow that there exists a nonzerodivisor of degree 1 in the associated graded ring $gr_{\mathfrak{p}}(S)$?*

There are two test cases where much is known about Question 31: points in $\mathbb{P}^2$ and square-free monomial ideals.

Case I: points in $\mathbb{P}^2_k$. Consider a set X of r points in $\mathbb{P}^2_k$, and let $I = I_X$ be its defining ideal. Since codim$(I) = 2$, it follows from [Ein et al. 2001; Hochster and Huneke 2002] that $I^{(2n)} \subset I^n$ for all n, so in particular $I^{(4)} \subset I^2$. It is then natural to ask:

Question 33 [Huneke 2006, Question 0.4]. *Is it true that $I^{(3)} \subset I^2$?*

Bocci and Harbourne gave a positive answer to this question when $I = I_X$ is the ideal of a generic set of points [Bocci and Harbourne 2010]. In characteristic 2, the inclusion follows from the techniques of [Hochster and Huneke 2002] (see [Bauer et al. 2009, Example 8.4.4]). Harbourne formulated some general conjectures for arbitrary homogeneous ideals which would imply a positive answer to Question 33 (see [Bauer et al. 2009, Conjecture 8.4.2, Conjecture 8.4.3] or [Harbourne and Huneke 2013, Conjecture 4.1.1]). Unfortunately, it turns out that the relation between symbolic powers and ordinary powers is much more subtle, even in the case of points in $\mathbb{P}^2$: Question 33 was recently given a negative answer in [Dumnicki et al. 2013].

One measure of how close the ordinary powers are to symbolic powers is given by comparing the least degrees of their generators. Given a homogeneous ideal J, we write $\alpha(J)$ for the smallest degree of a minimal generator of J. Since $I^n \subset I^{(n)}$, $\alpha(I^{(n)}) \leq \alpha(I^n) = n \cdot \alpha(I)$, or equivalently $\alpha(I) \geq \alpha(I^{(n)})/n$. The sequence $\alpha(I^{(n)})/n$ is always convergent (see Exercise (43)), and it has made a surprising appearance in the construction of Nagata's counter-example to Hilbert's 14th problem.

Theorem 34 [Nagata 1960]. *If X is a set of r generic points in $\mathbb{P}^2_{\mathbb{C}}$ and $I = I_X$ is its defining ideal, then*

(1) $\lim_{n\to\infty} \alpha(I^{(n)})/n \leq \sqrt{r}$.

(2) *If $r = s^2$ is a perfect square with $s \geq 4$, then $\alpha(I^{(n)})/n > s$.*

If we take $r = s^2$ with $s \geq 4$, then the ring $R = \bigoplus_{n\geq 0} I^{(n)}$ is a ring of invariants that is not finitely generated.

Case II: square-free monomial ideals. Consider the ideal

$$I = (xy, xz, yz) = (x, y) \cap (x, z) \cap (y, z)$$

defining a set of 3 non-collinear points in $\mathbb{P}^2$. We have

$$I^{(2)} = (x, y)^2 \cap (x, z)^2 \cap (y, z)^2,$$

and it is easily checked that $xyz \in I^{(2)} \setminus I^2$.

More generally, any square-free monomial ideal I can be written as an intersection $\mathfrak{p}_1 \cap \cdots \cap \mathfrak{p}_s$ of prime ideals, where each $\mathfrak{p}_i$ is generated by a subset of the variables. If $\operatorname{codim}(\mathfrak{p}_i) = c_i$ then $x_1 \cdots x_n \in \mathfrak{p}_i^{c_i}$. Taking $c = \operatorname{codim}(I) = \min(c_i)$ we get that

$$x_1 \cdots x_n \in \mathfrak{p}_1^c \cap \cdots \cap \mathfrak{p}_s^c = I^{(c)}.$$

It follows that if $I^c = I^{(c)}$, then $x_1 \cdots x_n$ must be contained in I^c; thus I contains c monomials with disjoint support, i.e.,

$$I \text{ contains a regular sequence consisting of } c \text{ monomials.} \tag{4-4}$$

If $I^{(n)} = I^n$ for all n then (4-4) holds for all ideals J obtained from I by setting variables equal to 0 or 1 (such an ideal J is called a *minor* of I). This raises the following question, posed by Gitler, Valencia and Villarreal.

Question 35 [Gitler et al. 2007]. *If* (4-4) *holds for all minors of* I, *does it follow that* $I^{(n)} = I^n$ *for all* n?

This question is equivalent to a max-flow-min-cut conjecture due to Conforti and Cornuéjols [Cornuéjols 2001, Conjecture 1.6], and it is open except in the case when I is generated by quadrics. Note that to any square-free monomial ideal I generated by quadrics one can associate a graph G as follows: the vertices of G correspond to the variables in the ring, and two vertices x_i and x_j are joined by an edge if $x_i x_j \in I$. Conversely, starting with a graph G one can reverse the preceding construction to get a monomial ideal I generated by quadrics. I is called the *edge ideal* of the graph G. With this terminology, we have:

Theorem 36 [Gitler et al. 2007]. *If* I *is the edge ideal of a graph* G, *then the following are equivalent*:

(1) $I^{(n)} = I^n$ *for all* $n \geq 1$.

(2) I *is packed, i.e.*, (4-4) *holds after setting any subset of the variables to be equal to* 0 *or* 1.

(3) G *is bipartite.*

Exercises.

(31) A famous theorem of Rees says that if R is a Noetherian local ring which is formally equidimensional (i.e., its completion is equidimensional), and I is primary to the maximal ideal $\mathfrak{m}$, then $f \in \bar{I}$ if and only if $e(I) = e(I + (f))$. Prove the easy direction of this theorem.

(32) Let $S = k[[x_1, \dots, x_n]]$ and let $f_i = x_1^{a_{i1}} + \cdots + x_n^{a_{in}}$. Assume that the ideal I generated by the f_i's has the property that S/I is a finite dimensional vector space. Give a formula, in terms of the exponents a_{ij}, for the dimension of this vector space (which is also the length of S/I, or the multiplicity of the ideal I).

(33) Let $\mathfrak{p}$ be a prime ideal generated by a regular sequence in a regular local ring (or polynomial ring). Prove that $\mathfrak{p}^{(n)} = \mathfrak{p}^n$ for all $n \geq 1$.

(34) Prove that if I is a reduced ideal in a polynomial ring, then $I^{(n)} \cdot I^{(m)} \subset I^{(n+m)}$.

(35) With the notation from the previous exercise, prove that the graded algebra

$$T := \bigoplus_{n \geq 0} I^{(n)}$$

is Noetherian if and only if there exists an integer k such that for all n,

$$(I^{(k)})^n = I^{(kn)}.$$

(36) Let S be a regular local ring and let $f \in S$ be a nonzero, nonunit element in S. Prove that the multiplicity of $R := S/(f)$ is equal to the order of f.

(37) Prove the following result of Chudnovsky [1981], which he showed using transcendental methods: if I is the ideal of a set of points in the projective plane over the complex numbers, then

$$\alpha(I^{(N)}) \geq \frac{N\alpha(I)}{2},$$

where (as before) $\alpha(\quad)$ denotes the least degree of a minimal generator of a homogeneous ideal.

(38) Let I be the ideal of at most five points in the projective plane. Prove that $I^{(3)} \subset I^2$.

(39) Let I be an ideal of points in the projective plane over a field of characteristic 2. Prove that $I^{(3)} \subset I^2$.

(40) The Eisenbud–Mazur conjecture states that if S is a power series ring over a field of characteristic 0, then for every prime ideal $\mathfrak{p}$,

$$\mathfrak{p}^{(2)} \subset \mathfrak{m}\mathfrak{p},$$

where $\mathfrak{m}$ denotes the maximal ideal of S. Prove this when $\mathfrak{p}$ is homogeneous.

(41) Prove the Eisenbud–Mazur conjecture assuming that for every $f \in S$ (S as in Exercise (40) and f not a unit) f is not a minimal generator of the integral closure of its partial derivatives.

(42) Let R be a regular local ring, and let $\mathfrak{p}$ be a prime ideal. Set $G = \operatorname{gr}_{\mathfrak{p}}(R)$, the associated graded ring of $\mathfrak{p}$. If $f \in R$, write f^* for the leading form of f in G. Show that if f^* is nilpotent in G, then $f \in \mathfrak{p}^{(n)}$ but $f \notin \mathfrak{p}^n$ for some n.

(43) Let I be a homogeneous ideal in a polynomial ring S, satisfying $I = \sqrt{I}$. Prove that the limit of $\alpha(I^{(m)})/m$ exists as m goes to infinity.

References

[Aberbach 1993] I. M. Aberbach, "Test elements in excellent rings with an application to the uniform Artin–Rees property", *Proc. Amer. Math. Soc.* **118**:2 (1993), 355–363.

[Ananyan and Hochster 2012] T. Ananyan and M. Hochster, "Ideals generated by quadratic polynomials", *Math. Res. Lett.* **19**:1 (2012), 233–244.

[Artin and Rotthaus 1988] M. Artin and C. Rotthaus, "A structure theorem for power series rings", pp. 35–44 in *Algebraic geometry and commutative algebra, I*, edited by H. Hijikata et al., Kinokuniya, Tokyo, 1988.

[Bauer et al. 2009] T. Bauer, S. Di Rocco, B. Harbourne, M. Kapustka, A. Knutsen, W. Syzdek, and T. Szemberg, "A primer on Seshadri constants", pp. 33–70 in *Interactions of classical and numerical algebraic geometry*, edited by D. J. Bates et al., Contemp. Math. **496**, Amer. Math. Soc., Providence, RI, 2009.

[Bocci and Harbourne 2010] C. Bocci and B. Harbourne, "Comparing powers and symbolic powers of ideals", *J. Algebraic Geom.* **19**:3 (2010), 399–417.

[Boocher 2008] A. Boocher, *The module of differentials, evolutions, and the Eisenbud–Mazur conjecture*, Senior thesis, University of Notre Dame, 2008, http://www.maths.ed.ac.uk/~aboocher/writings/seniorthesis.pdf.

[Chudnovsky 1981] G. V. Chudnovsky, "Singular points on complex hypersurfaces and multidimensional Schwarz lemma", pp. 29–69 in *Seminar on number theory* (Paris, 1979–80), edited by M.-J. Bertin, Progr. Math. **12**, Birkhäuser, Boston, 1981.

[Cornuéjols 2001] G. Cornuéjols, *Combinatorial optimization: packing and covering*, CBMS-NSF Regional Conference Series in Applied Mathematics **74**, Society for Industrial and Applied Mathematics (SIAM), Philadelphia, PA, 2001.

[Dumnicki et al. 2013] M. Dumnicki, T. Szemberg, and H. Tutaj-Gasińska, "Counterexamples to the $I^{(3)} \subset I^2$ containment", *J. Algebra* **393** (2013), 24–29.

[Duncan and O'Carroll 1989] A. J. Duncan and L. O'Carroll, "A full uniform Artin–Rees theorem", *J. Reine Angew. Math.* **394** (1989), 203–207.

[Ein et al. 2001] L. Ein, R. Lazarsfeld, and K. E. Smith, "Uniform bounds and symbolic powers on smooth varieties", *Invent. Math.* **144**:2 (2001), 241–252.

[Ein et al. 2004] L. Ein, R. Lazarsfeld, K. E. Smith, and D. Varolin, "Jumping coefficients of multiplier ideals", *Duke Math. J.* **123**:3 (2004), 469–506.

[Eisenbud 1995] D. Eisenbud, *Commutative algebra: with a view toward algebraic geometry*, Graduate Texts in Mathematics **150**, Springer, New York, 1995.

[Eisenbud 2005] D. Eisenbud, *The geometry of syzygies: a second course in commutative algebra and algebraic geometry*, Graduate Texts in Mathematics **229**, Springer, New York, 2005.

[Eisenbud and Evans 1973] D. Eisenbud and E. G. Evans, Jr., "Every algebraic set in n-space is the intersection of n hypersurfaces", *Invent. Math.* **19** (1973), 107–112.

[Eisenbud and Hochster 1979] D. Eisenbud and M. Hochster, "A Nullstellensatz with nilpotents and Zariski's main lemma on holomorphic functions", *J. Algebra* **58**:1 (1979), 157–161.

[Eisenbud and Mazur 1997] D. Eisenbud and B. Mazur, "Evolutions, symbolic squares, and Fitting ideals", *J. Reine Angew. Math.* **488** (1997), 189–201.

[Giral and Planas-Vilanova 2008] J. M. Giral and F. Planas-Vilanova, "Integral degree of a ring, reduction numbers and uniform Artin–Rees numbers", *J. Algebra* **319**:8 (2008), 3398–3418.

[Gitler et al. 2007] I. Gitler, C. E. Valencia, and R. H. Villarreal, "A note on Rees algebras and the MFMC property", *Beiträge Algebra Geom.* **48**:1 (2007), 141–150.

[Harbourne and Huneke 2013] B. Harbourne and C. Huneke, "Are symbolic powers highly evolved?", *J. Ramanujan Math. Soc.* **28A** (2013), 247–266.

[Hochster and Huneke 1990] M. Hochster and C. Huneke, "Tight closure, invariant theory, and the Briançon–Skoda theorem", *J. Amer. Math. Soc.* **3**:1 (1990), 31–116.

[Hochster and Huneke 2002] M. Hochster and C. Huneke, "Comparison of symbolic and ordinary powers of ideals", *Invent. Math.* **147**:2 (2002), 349–369.

[Hübl 1999] R. Hübl, "Evolutions and valuations associated to an ideal", *J. Reine Angew. Math.* **517** (1999), 81–101.

[Huneke 1992] C. Huneke, "Uniform bounds in Noetherian rings", *Invent. Math.* **107**:1 (1992), 203–223.

[Huneke 2006] C. Huneke, "Open problems on powers of ideals", Lecture notes, 2006, http://www.aimath.org/WWN/integralclosure/Huneke.pdf. Notes from an AIM workshop on Integral Closure, Multiplier Ideals, and Cores.

[Huneke and Swanson 2006] C. Huneke and I. Swanson, *Integral closure of ideals, rings, and modules*, London Mathematical Society Lecture Note Series **336**, Cambridge University Press, 2006.

[Huneke et al. 2013] C. Huneke, P. Mantero, J. McCullough, and A. Seceleanu, "The projective dimension of codimension two algebras presented by quadrics", *J. Algebra* **393** (2013), 170–186.

[Kollár 1988] J. Kollár, "Sharp effective Nullstellensatz", *J. Amer. Math. Soc.* **1**:4 (1988), 963–975.

[Kunz 1969] E. Kunz, "Characterizations of regular local rings for characteristic p", *Amer. J. Math.* **91** (1969), 772–784.

[Lech 1960] C. Lech, "Note on multiplicities of ideals", *Ark. Mat.* **4** (1960), 63–86.

[Lipman and Sathaye 1981] J. Lipman and A. Sathaye, "Jacobian ideals and a theorem of Briançon–Skoda", *Michigan Math. J.* **28**:2 (1981), 199–222.

[Matsumura 1980] H. Matsumura, *Commutative algebra*, 2nd ed., Mathematics Lecture Note Series **56**, Benjamin/Cummings, Reading, MA, 1980.

[Moh 1974] T. T. Moh, "On the unboundedness of generators of prime ideals in power series rings of three variables", *J. Math. Soc. Japan* **26** (1974), 722–734.

[Moh 1979] T. T. Moh, "On generators of ideals", *Proc. Amer. Math. Soc.* **77**:3 (1979), 309–312.

[Mukai 1993] S. Mukai, "Curves and Grassmannians", pp. 19–40 in *Algebraic geometry and related topics* (Inchon, 1992), edited by J.-H. Yang and Y. Namikawa, Int. Press, Cambridge, MA, 1993.

[Nagata 1960] M. Nagata, "On the fourteenth problem of Hilbert", pp. 459–462 in *Proc. Internat. Congress Math. 1958*, Cambridge Univ. Press, 1960.

[Nagata 1962] M. Nagata, *Local rings*, Interscience Tracts in Pure and Applied Mathematics **13**, Interscience, New York-London, 1962.

[O'Carroll 1987] L. O'Carroll, "A uniform Artin–Rees theorem and Zariski's main lemma on holomorphic functions", *Invent. Math.* **90**:3 (1987), 647–652.

[O'Carroll and Planas-Vilanova 2008] L. O'Carroll and F. Planas-Vilanova, "Irreducible affine space curves and the uniform Artin–Rees property on the prime spectrum", *J. Algebra* **320**:8 (2008), 3339–3344.

[Peeva 2011] I. Peeva, *Graded syzygies*, Algebra and Applications **14**, Springer, London, 2011.

[Peeva and Stillman 2009] I. Peeva and M. Stillman, "Open problems on syzygies and Hilbert functions", *J. Commut. Algebra* **1**:1 (2009), 159–195.

[Planas-Vilanova 2000] F. Planas-Vilanova, "The strong uniform Artin–Rees property in codimension one", *J. Reine Angew. Math.* **527** (2000), 185–201.

[Saito 1971] K. Saito, "Quasihomogene isolierte Singularitäten von Hyperflächen", *Invent. Math.* **14** (1971), 123–142.

[Schoutens 2000] H. Schoutens, "Uniform bounds in algebraic geometry and commutative algebra", pp. 43–93 in *Connections between model theory and algebraic and analytic geometry*, edited by A. Macintyre, Quad. Mat. **6**, Dept. Math., Seconda Univ. Napoli, Caserta, 2000.

[Skoda and Briançon 1974] H. Skoda and J. Briançon, "Sur la clôture intégrale d'un idéal de germes de fonctions holomorphes en un point de $\mathbf{C}^n$", *C. R. Acad. Sci. Paris Sér. A* **278** (1974), 949–951.

[Srinivas and Trivedi 1997] V. Srinivas and V. Trivedi, "On the Hilbert function of a Cohen–Macaulay local ring", *J. Algebraic Geom.* **6**:4 (1997), 733–751.

[Storch 1972] U. Storch, "Bemerkung zu einem Satz von M. Kneser", *Arch. Math. (Basel)* **23** (1972), 403–404.

[Striuli 2007] J. Striuli, "A uniform Artin–Rees property for syzygies in rings of dimension one and two", *J. Pure Appl. Algebra* **210**:2 (2007), 577–588.

[Trivedi 1997] V. Trivedi, "Hilbert functions, Castelnuovo–Mumford regularity and uniform Artin–Rees numbers", *Manuscripta Math.* **94**:4 (1997), 485–499.

[Wang 1994] H.-J. Wang, "On the Fitting ideals in free resolutions", *Michigan Math. J.* **41**:3 (1994), 587–608.

[Wang 1999] H.-J. Wang, "A uniform property of affine domains", *J. Algebra* **215**:2 (1999), 500–508.

[Zariski 1949] O. Zariski, "A fundamental lemma from the theory of holomorphic functions on an algebraic variety", *Ann. Mat. Pura Appl.* (4) **29** (1949), 187–198.

huneke@virginia.edu
Department of Mathematics, University of Virginia, 141 Cabell Drive, Kerchof Hall, Charlottesville, VA 22904-4137, United States

craicu@nd.edu
Department of Mathematics, University of Notre Dame, 255 Hurley, Notre Dame, IN 46556, United States

Institute of Mathematics "Simion Stoilow" of the Romanian Academy

Commutative Algebra and Noncommutative Algebraic Geometry, I
MSRI Publications
Volume **67**, 2015

Noncommutative motives and their applications

MATILDE MARCOLLI AND GONÇALO TABUADA

This survey is based on lectures given by the authors during the program "Noncommutative algebraic geometry and representation theory" at MSRI in the Spring 2013. It covers recent work by the authors on noncommutative motives and their applications, and is intended for a broad mathematical audience. In Section 1 we recall the main features of Grothendieck's theory of motives. In Sections 2 and 3 we introduce several categories of noncommutative motives and describe their relation with the classical commutative counterparts. In Section 4 we formulate the noncommutative analogues of Grothendieck's standard conjectures of type C and D, of Voevodsky's smash-nilpotence conjecture, and of Kimura–O'Sullivan finite-dimensionality conjecture. Section 5 is devoted to recollections of the (super-)Tannakian formalism. In Section 6 we introduce the noncommutative motivic Galois (super-)groups and their unconditional versions. In Section 7 we explain how the classical theory of (intermediate) Jacobians can be extended to the noncommutative world. Finally, in Section 8 we present some applications to motivic decompositions and to Dubrovin's conjecture.

Marcolli was partially supported by the grants DMS-0901221, DMS-1007207, DMS-1201512, and PHY-1205440. Tabuada was partially supported by the National Science Foundation CAREER Award #1350472 and by the Fundação para a Ciência e a Tecnologia (Portuguese Foundation for Science and Technology) through the project UID/MAT/00297/2013 (Centro de Matemática e Aplicações).

The authors are very grateful to the organizers Michael Artin, Victor Ginzburg, Catharina Stroppel, Toby Stafford, Michel Van den Bergh, and Efim Zelmanov for kindly giving us the opportunity to present our recent work. They would like also to thank the anonymous referee for comments and corrections.

1. Grothendieck's theory of motives

We recall here the main features of Grothendieck's classical theory of pure motives, which will be useful when passing to the noncommutative world. These facts are well-known and we refer the reader to [André 2004; Jannsen et al. 1994; Manin 1968] for more detailed treatments. Let k be a base field and F a field of coefficients.

Let $\mathcal{V}(k)$ be the category of smooth projective k-schemes. The category of pure motives is obtained from $\mathcal{V}(k)$ by linearization, idempotent completion, and inversion of the Lefschetz motive.

1.1. *Correspondences.* The linearization of $\mathcal{V}(k)$ is obtained by replacing the morphisms of schemes with correspondences. Concretely, the correspondences $\mathrm{Corr}_{\sim,F}(X,Y) := \mathcal{Z}^{\dim(X)}_{\sim,F}(X \times Y)$ from X to Y are the F-linear combinations of algebraic cycles in $X \times Y$ of codimension equal to $\dim(X)$. This includes the case of ordinary morphisms by viewing their graphs as correspondences. The composition of correspondences is obtained by pulling back the cycles to the product $X \times Y \times Z$, taking their intersection product there, and pushing forward the result to the product $X \times Z$:

$$\begin{aligned} &\mathrm{Corr}_{\sim,F}(X,Y) \times \mathrm{Corr}_{\sim,F}(Y,Z) \to \mathrm{Corr}_{\sim,F}(X,Z), \\ &\qquad\qquad (\alpha,\beta) \mapsto (\pi_{XZ})_*(\pi_{XY}^*(\alpha) \bullet \pi_{YZ}^*(\beta)). \end{aligned} \tag{1-1}$$

1.2. *Equivalence relations on algebraic cycles.* One of the important steps in the construction of the category of pure motives is the choice of an equivalence relation on algebraic cycles. The usual choices are rational equivalence, homological equivalence, and numerical equivalence. Rational equivalence depends upon the moving lemma and gives rise to the category of Chow motives. Homological equivalence depends on the choice of a "good" cohomology theory (Weil cohomology theory) and gives rise to the category of homological motives. Numerical equivalence depends only on the intersection product between algebraic cycles and gives rise to the category of numerical motives. These three equivalence relations are summarized as follows:

- A correspondence α from X to Y is rationally trivial, $\alpha \sim_{\mathrm{rat}} 0$, if there exists a $\beta \in \mathcal{Z}^*_{\sim,F}(X \times Y \times \mathbb{P}^1)$ such that $\alpha = \beta(0) - \beta(\infty)$.
- A correspondence α from X to Y is homologically trivial, $\alpha \sim_{\mathrm{hom}} 0$, if it vanishes under a chosen Weil cohomology theory.

- A correspondence α from X to Y is numerically trivial, $\alpha \sim_{\text{num}} 0$, if it has a trivial intersection number with every other algebraic cycle.

It is well-known that $\sim_{\text{rat}} \neq \sim_{\text{num}}$. The question of whether $\sim_{\text{hom}} = \sim_{\text{num}}$ remains open and is part of an important set of conjectures about motives which will be described below (see §1.10).

The category of pure motives has different properties depending on the equivalence relation.

1.3. ***Pure motives.*** The symmetric monoidal category of effective pure motives $\text{Mot}^{\text{eff}}_{\sim,F}(k)$ is defined as follows: the objects are the pairs (X, p) (with $X \in \mathcal{V}(k)$ and p an idempotent of $\text{Corr}_{\sim,F}(X, X)$), the morphisms are the correspondences

$$\text{Hom}_{\text{Mot}^{\text{eff}}_{\sim,F}(k)}((X, p), (Y, q)) = p\text{Corr}_{\sim,F}(X, Y)q,$$

the composition is induced from (1-1), and the symmetric monoidal structure is given by $(X, p) \otimes (Y, q) = (X \times Y, p \times q)$. In what follows we will write $h_F(X)$ instead of (X, Δ_X).

The effective pure motive $h_F(\mathbb{P}^1)$ decomposes as $h^0_F(\mathbb{P}^1) \oplus h^2_F(\mathbb{P}^1) \simeq 1 \oplus \mathbf{L}$, where $1 = h_F(\text{Spec}(k))$ and $\mathbf{L}$ is called the Lefschetz motive.

The symmetric monoidal category of pure motives $\text{Mot}_{\sim,F}(k)$ is obtained from $\text{Mot}^{\text{eff}}_{\sim,F}(k)$ by formally inverting the Lefschetz motive. The formal inverse $\mathbf{L}^{-1}$ is called the Tate motive $\mathbb{Q}(1)$ (one writes $\mathbb{Q}(n) := \mathbb{Q}(1)^{\otimes n}$). Concretely, in the category of pure motives the objects are triples $(X, p, m) := (X, p) \otimes (\mathbf{L}^{-1})^{\otimes m} = (X, p) \otimes \mathbb{Q}(m)$ and the morphisms are given by

$$\text{Hom}_{\text{Mot}_{\sim,F}(k)}((X, p, m), (Y, q, n)) = p\text{Corr}^{m-n}_{\sim,F}(X, Y)q.$$

The category $\text{Mot}_{\sim,F}(k)$ is additive. In the case where $m = n$, the direct sum $(X, p, m) \oplus (Y, q, n)$ is defined as $(X \amalg Y, p \oplus q, m)$. The general case reduces to this one using the Lefschetz motive.

Inverting the Lefschetz motive has therefore the effect of introducing arbitrary shifts in the codimension of the algebraic cycles, instead of using only algebraic cycles of codimension equal to $\dim(X)$. One has a canonical (contravariant) symmetric monoidal functor

$$h_F : \mathcal{V}(k)^{\text{op}} \to \text{Mot}_{\sim,F}(k), \quad X \mapsto h_F(X),$$

which sends a morphism $f : X \to Y$ to the transpose of its graph $\Gamma(f) \subset X \times Y$.

1.4. ***Chow and homological motives.*** The category $\text{Mot}_{\sim,F}(k)$ with $\sim = \sim_{\text{rat}}$ (resp. $\sim = \sim_{\text{hom}}$) is called the category of Chow motives (resp. homological motives) and is denoted by $\mathsf{Chow}_F(k)$ (resp. by $\mathsf{Hom}_F(k)$).

1.5. ***Numerical motives.*** One of the most important results in the theory of pure motives was obtained in [Jannsen 1992]. It asserts that the numerical equivalence relation is the "best one" from the point of view of the resulting properties of the category. More precisely, Jannsen proved that the following conditions are equivalent:

- $\mathrm{Mot}_{\sim,F}(k)$ is a semisimple abelian category;
- $\mathrm{Corr}_{\sim,F}(X,X)$ is a finite-dimensional semisimple F-algebra for every X;
- The equivalence relation $\sim$ is equal to $\sim_{\mathrm{num}}$.

The category $\mathrm{Mot}_{\sim,F}(k)$ with $\sim=\sim_{\mathrm{num}}$ is called the category of numerical motives $\mathsf{Num}_F(k)$

1.6. ***Smash-nilpotence.*** Voevodsky [1995] introduced the equivalence relation of smash-nilpotence on algebraic cycles, $\sim_{\otimes\mathrm{nil}}$, and conjectured the following:

Conjecture $V(X)$. $\qquad \mathcal{Z}^*_{\sim_{\otimes\mathrm{nil}},F}(X)=\mathcal{Z}^*_{\sim_{\mathrm{num}},F}(X)$.

The $\otimes_{\mathrm{nil}}$-ideal of an F-linear, additive, symmetric monoidal category $\mathcal{C}$ is defined as

$$\otimes_{\mathrm{nil}}(a,b)=\{g\in\mathrm{Hom}_{\mathcal{C}}(a,b)\mid g^{\otimes n}=0 \text{ for } n\gg 0\}.$$

The quotient functor $\mathcal{C}\to\mathcal{C}/\otimes_{\mathrm{nil}}$ is F-linear, additive, symmetric monoidal, and conservative. If $\mathcal{C}$ is idempotent complete, then $\mathcal{C}/\otimes_{\mathrm{nil}}$ is also idempotent complete. One denotes by $\mathsf{Voev}_F(k):=\mathsf{Chow}_F(k)/\otimes_{\mathrm{nil}}$ the category of Chow motives up to smash-nilpotence.

1.7. ***All together.*** The different categories of pure motives are related by a sequence of F-linear, additive, full, symmetric monoidal functors $\mathsf{Chow}_F(k)\to\mathsf{Voev}_F(k)\to\mathsf{Hom}_F(k)\to\mathsf{Num}_F(k)$.

1.8. ***Tate motives.*** The additive full subcategory of $\mathrm{Mot}_{\sim,F}(k)$ generated by $\mathbb{Q}(1)$ is called the category of (pure) Tate motives. This category is independent of the equivalence relation.

1.9. ***Weil cohomology theories.*** A Weil cohomology theory axiomatizes the properties of a "good" cohomology theory. It consists of a contravariant functor $H^*:\mathcal{V}(k)^{\mathrm{op}}\to\mathrm{GrVect}(F)$ to $\mathbb{Z}$-graded F-vector spaces equipped with the following data:

- Künneth isomorphisms $H^*(X\times Y)\simeq H^*(X)\otimes H^*(Y)$;
- trace maps $tr:H^{2\dim(X)}(X)(\dim(X))\to F$;
- cycle maps $\gamma_n:\mathcal{Z}^n_{\sim_{\mathrm{rat}},F}(X)\to H^{2n}(X)(n)$.

One assumes that $\dim H^2(\mathbb{P}^1) = 1$ and some natural compatibility conditions. Examples of Weil cohomology theories include de Rham, Betti, étale, and crystalline cohomology.

A great deal of difficulty in the theory of pure motives comes from the poor understanding of the cycle maps. The question of which cohomology classes are in the range of the γ_n is a notoriously difficult problem (which includes the Hodge conjecture below).

The idea of motives can be traced back to Grothendieck's quest for a universal cohomology lying behind all the different Weil cohomology theories.

1.10. *Grothendieck's standard conjectures.* Important conjectures relate the properties of the categories of pure motives with the geometry of schemes. The standard conjectures are traditionally labeled as type C, D, B, and I. They are summarized as follows:

- *Type C* (*Künneth*): the Künneth components of the diagonal Δ_X are algebraic cycles.
- *Type D* (Hom = Num): the homological and the numerical equivalence relations coincide.
- *Type B* (*Lefschetz*): the Lefschetz involution $\star_{L,X}$ is algebraic (with $\mathbb{Q}$-coefficients).
- *Type I* (*Hodge*): the quadratic form defined by the Hodge involution $\star_H$ is positive definite.

There are relations between these conjectures: type B and I imply type D and in characteristic zero type B implies all others. For our purposes, we will focus on types C and D.

2. From motives to noncommutative motives

As mentioned above, the origin of pure motives was Grothendieck's quest for a universal cohomology lying behind all the different Weil cohomology theories. The origin of noncommutative motives is similar. In the noncommutative world the basic objects are not schemes but rather dg categories. Instead of cohomology theories, one has homological type invariants such as algebraic K-theory, cyclic homology (and all its variants), topological Hochschild homology, etc. In analogy with the commutative world, one can try to identify a suitable universal invariant lying behind all these different invariants.

2.1. *Dg categories.* A differential graded (=dg) category $\mathcal{A}$ is a category whose morphism sets $\mathcal{A}(x, y)$ are cochain complexes of k-modules (k can more generally be a commutative ring) and whose composition law satisfies the Leibniz rule. A dg functor $F : \mathcal{A} \to \mathcal{B}$ is a functor which preserves this extra structure. For further details, we refer the reader to the pioneering work of Bondal and Kapranov [1990; 1989] and to the ICM survey [Keller 2006]. In what follows, we will denote by $\mathsf{dgcat}(k)$ the category of all (small) dg categories and dg functors.

Perfect complexes. Dg categories should be understood as "noncommutative schemes". The reason for this is that one can canonically associate to every scheme X a dg category, namely the dg category of perfect complexes $\mathrm{perf}_{\mathsf{dg}}(X)$ of $\mathcal{O}_X$-modules. This dg category enhances the classical derived category of perfect complexes $\mathrm{perf}(X)$ in the sense that the latter is obtained from the former by passing to degree zero cohomology. When k is a field and X is quasiprojective, Lunts and Orlov [2010] proved that this dg enhancement is in fact "unique". This construction gives rise to a well-defined (contravariant) symmetric monoidal functor

$$\mathcal{V}(k)^{\mathrm{op}} \to \mathsf{dgcat}(k), \qquad X \mapsto \mathrm{perf}_{\mathsf{dg}}(X).$$

An arbitrary dg category should be then considered as the dg category of perfect complexes of an hypothetical "noncommutative scheme".

Saturated dg categories. Kontsevich [2005; 2010; 2009] introduced a class of dg categories whose properties closely resemble those of perfect complexes on smooth proper schemes. These are called saturated dg categories. Concretely, a dg category $\mathcal{A}$ is saturated if it is perfect as a bimodule over itself and if for any two objects x, y we have $\sum_i \mathrm{rank}\, H^i\mathcal{A}(x, y) < \infty$. A k-scheme X is smooth and proper if and only if the associated dg category $\mathrm{perf}_{\mathsf{dg}}(X)$ is saturated.

As mentioned in [Kontsevich 2005], other examples of saturated dg categories arise from representation theory of (finite) quivers and from deformation by quantization.

2.2. *Morita equivalences.* A dg functor $F : \mathcal{A} \to \mathcal{B}$ is called a Morita equivalence if the restriction of scalars between derived categories $\mathcal{D}(\mathcal{B}) \xrightarrow{\sim} \mathcal{D}(\mathcal{A})$ is an equivalence of (triangulated) categories.

All the classical invariants, like algebraic K-theory, cyclic homology (and all its variants), topological Hochschild homology, etc, are Morita invariant in the sense that they send Morita equivalences to isomorphisms. It is then natural to consider dg categories up to Morita equivalence.

As proved in [Tabuada 2005], the category $\mathsf{dgcat}(k)$ carries a Quillen model structure whose weak equivalences are the Morita equivalences. Let us denote by $\mathsf{Hmo}(k)$ the homotopy category hence obtained. This category is symmetric monoidal and, as shown in [Cisinski and Tabuada 2012], the saturated dg categories can be characterized as the dualizable (or rigid) objects of $\mathsf{Hmo}(k)$.

Bondal and Van den Bergh [2003] proved that for every quasicompact quasiseparated k-scheme X the dg category $\mathrm{perf}_{\mathrm{dg}}(X)$ is isomorphic in $\mathsf{Hmo}(k)$ to a dg k-algebra with bounded cohomology.

Bondal and Kapranov's pretriangulated envelope. Using *one-sided twisted complexes*, Bondal and Kapranov [1990] constructed in the pretriangulated envelope of every dg category $\mathcal{A}$. Intuitively speaking, one formally adds to $\mathcal{A}$ (de-)suspensions, cones, cones of morphisms between cones, etc. Making use of the Morita model structure, this construction can be conceptually understood as a functorial fibrant resolution functor; consult [Tabuada 2005] for details.

Drinfeld's DG quotient. A very useful operation on triangulated categories is the Verdier quotient. Via a very elegant construction (reminiscent of Dwyer–Kan localization), Drinfeld [2004] lifted this operation to the setting of dg categories. Although very elegant, this construction didn't seem to satisfy any obvious universal property. The Morita model structure changed this state of affairs by allowing the characterizing of Drinfeld's DG quotient as an homotopy cofiber construction; consult [Tabuada 2010b] for details.

2.3. *Additive invariants.* Given a dg category $\mathcal{A}$, let $T(\mathcal{A})$ be the dg category of pairs (i, x), where $i \in \{1, 2\}$ and $x \in \mathcal{A}$. The complex of morphisms in $T(\mathcal{A})$ from (i, x) to (i', x') is given by $\mathcal{A}(x, x')$ if $i \leq i'$ and is zero otherwise. Composition is induced by $\mathcal{A}$. Intuitively speaking, $T(\mathcal{A})$ "dg categorifies" the notion of upper triangular matrix. Note that we have two inclusion dg functors $i_1 : \mathcal{A} \hookrightarrow T(\mathcal{A})$ and $i_2 : \mathcal{A} \hookrightarrow T(\mathcal{A})$. Let $E : \mathsf{dgcat}(k) \to \mathsf{A}$ be a functor with values in an additive category. We say that E is an additive invariant if it satisfies the following two conditions:

- It sends Morita equivalences to isomorphisms.
- Given any dg category $\mathcal{A}$, the inclusion dg functors induce an isomorphism

$$[E(i_1)\ E(i_2)] : E(\mathcal{A}) \oplus E(\mathcal{A}) \xrightarrow{\sim} E(T(\mathcal{A})).$$

Thanks to [Blumberg and Mandell 2012; Keller 1999; Quillen 1973; Schlichting 2006; Tabuada 2010a; Thomason and Trobaugh 1990; Waldhausen 1985], among other works, we know that all the invariants above are additive.

The universal additive invariant was constructed in [Tabuada 2005] as follows: consider the additive symmetric monoidal category $\mathsf{Hmo}_0(k)$ whose objects are the dg categories and whose morphisms are given by $\mathrm{Hom}_{\mathsf{Hmo}_0(k)}(\mathcal{A}, \mathcal{B}) := K_0\mathrm{rep}(\mathcal{A}, \mathcal{B})$, where $\mathrm{rep}(\mathcal{A}, \mathcal{B}) \subset \mathcal{D}(\mathcal{A}^{\mathrm{op}} \otimes^{\mathbf{L}} \mathcal{B})$ stands for the full triangulated subcategory of those $\mathcal{A}$-$\mathcal{B}$ bimodules which are perfect as right $\mathcal{B}$-modules. The composition law and the symmetric monoidal structure are induced from $\mathsf{Hmo}(k)$.

As explained in loc. cit., the canonical composed symmetric monoidal functor

$$U : \mathsf{dgcat}(k) \longrightarrow \mathsf{Hmo}(k) \longrightarrow \mathsf{Hmo}_0(k) \tag{2-1}$$

is the universal additive invariant, i.e., precomposition with U induces a bijection between additive functors $\overline{E} : \mathsf{Hmo}_0(k) \to \mathsf{A}$ and additive invariants $E : \mathsf{dgcat}(k) \to \mathsf{A}$. This suggests that $\mathsf{Hmo}_0(k)$ is the correct place where noncommutative motives should reside. Let us denote by $\mathsf{Hmo}_0(k)_F$ its F-linearization (F can more generally be a commutative ring).

2.4. *Computations.* In order to gain some sensibility with (2-1), we recall some computations:

- [Marcolli and Tabuada 2015] Let X be a smooth projective k-scheme whose derived category perf(X) admits a full exceptional collection of length n (see §8.2). In this case, $U(\mathrm{perf}_{\mathsf{dg}}(X))$ identifies with the direct sum of n copies of $U(k)$.
- [Tabuada and Van den Bergh 2015] Let X be a quasicompact quasiseparated k-scheme, A a sheaf of Azumaya algebras over X of rank r, and $\mathrm{perf}_{\mathsf{dg}}(A)$ the associated dg category. When $1/r$ belongs to the commutative ring F, $U(\mathrm{perf}_{\mathsf{dg}}(A))_F$ identifies with $U(\mathrm{perf}_{\mathsf{dg}}(X))_F$.
- [Bernardara and Tabuada 2014b] Let A be a central simple k-algebra of degree $d := \sqrt{\dim(A)}$ and $SB(A)$ the associated Severi–Brauer variety. In this case, $U(\mathrm{perf}_{\mathsf{dg}}(SB(A)))$ identifies with the following direct sum $U(k) \oplus U(A) \oplus U(A)^{\otimes 2} \oplus \cdots \oplus U(A)^{\otimes (d-1)}$.
- [Tabuada and Van den Bergh 2015] Let k be a perfect field, A a finite-dimensional k-algebra of finite global dimension, and $J(A)$ its Jacobson radical. In this case, we have $U(A) \simeq U(A/J(A))$.
- [Tabuada 2014] Let A and B be two central simple k-algebras and $[A]$ and $[B]$ their Brauer classes. In this case, $U(A)$ and $U(B)$ are isomorphic if and only if $[A] = [B]$.

2.5. *Noncommutative Chow motives.* Kontsevich [2005] introduced the symmetric monoidal category of noncommutative Chow motives $\mathsf{NChow}(k)_F$. It can be described as the idempotent completion $(-)^\natural$ of the full subcategory of $\mathsf{Hmo}_0(k)_F$ given by the saturated dg categories. Concretely, the objects are of the pairs $(\mathcal{A}, e)$ (with $\mathcal{A}$ a saturated dg category and e an idempotent), the morphisms are given by the noncommutative correspondences

$$\mathrm{Hom}_{\mathsf{NChow}(k)_F}((\mathcal{A}, e), (\mathcal{B}, e')) := eK_0\mathrm{rep}(\mathcal{A}, \mathcal{B})e' \simeq eK_0(\mathcal{A}^{\mathrm{op}} \otimes^{\mathbf{L}} \mathcal{B})_F e',$$

the composition law is induced by the derived tensor product of bimodules, and the symmetric monoidal structure is induced by the derived tensor product of dg categories.

Fundamental theorems. The fundamental theorems in algebraic K-theory and periodic cyclic homology, proved respectively in [Weibel 1989] and [Kassel 1987], are of major importance. Their proofs are not only very different but also quite involved. The category $\mathsf{NChow}(k)_F$ allowed a simple, unified and conceptual proof of these fundamental theorems; consult [Tabuada 2012] for details.

2.6. *A bridge from Chow to noncommutative Chow motives.* Noncommutative motives should, in a suitable sense, contain the category of motives. This idea was made precise in [Tabuada 2013] (following the original insight in [Kontsevich 2009]). The precise statement is the existence of a $\mathbb{Q}$-linear additive fully-faithful symmetric monoidal functor R making the following diagram commute

$$\begin{array}{ccc} \mathcal{V}(k)^{\mathrm{op}} & \xrightarrow{X \mapsto \mathrm{perf}_{\mathrm{dg}}(X)} & \mathsf{dgcat}(k) \\ \downarrow{\scriptstyle h_{\mathbb{Q}}} & & \downarrow{\scriptstyle U} \\ \mathsf{Chow}_{\mathbb{Q}}(k) & & \mathsf{Hmo}_0(k) \\ \downarrow{\scriptstyle \pi} & & \downarrow{\scriptstyle (-)_{\mathbb{Q}}} \\ \mathsf{Chow}_{\mathbb{Q}}(k)/_{-\otimes\mathbb{Q}(1)} & \xrightarrow[R]{} & \mathsf{NChow}(k)_{\mathbb{Q}} \subset \mathsf{Hmo}_0(k)^{\natural}_{\mathbb{Q}} \end{array} \tag{2-2}$$

where $\mathsf{Chow}_{\mathbb{Q}}(k)/_{-\otimes\mathbb{Q}(1)}$ stands for the orbit category. This bridge opens new horizons and opportunities of research by enabling the interchange of results between the commutative and the noncommutative world. This yoga was developed in [Tabuada 2013] in what regards Schur and Kimura–O'Sullivan finite-dimensionality (see §4.3 below), motivic measures, and motivic zeta functions.

The above diagram (2-2) holds more generally with $\mathbb{Q}$ replaced by any field F of characteristic zero.

Orbit categories. Given an F-linear, additive, symmetric monoidal category $\mathcal{C}$ and a $\otimes$-invertible object $\mathcal{O} \in \mathcal{C}$, the orbit category $\mathcal{C}/_{-\otimes\mathcal{O}}$ has the same objects as $\mathcal{C}$ and morphisms

$$\mathrm{Hom}_{\mathcal{C}/_{-\otimes\mathcal{O}}}(a,b) := \bigoplus_{i \in \mathbb{Z}} \mathrm{Hom}_{\mathcal{C}}(a, b \otimes \mathcal{O}^{\otimes i}).$$

The composition law and the symmetric monoidal structure are induced from $\mathcal{C}$. By construction, the orbit category comes equipped with a canonical symmetric monoidal projection functor $\pi : \mathcal{C} \to \mathcal{C}/_{-\otimes\mathcal{O}}$. Moreover, π is endowed with a 2-isomorphism $\pi \circ (- \otimes \mathcal{O}) \overset{\sim}{\Rightarrow} \pi$ and is 2-universal among all such functors.

3. Categories of noncommutative motives

3.1. ***Periodic cyclic homology as "noncommutative de Rham cohomology".*** Connes' periodic cyclic homology extends naturally from k-algebras to dg categories giving thus rise to a functor $HP^{\pm} : \mathsf{dgcat}(k) \to \mathrm{sVect}(k)$ to super k-vector spaces. In the case of a smooth k-scheme X (with k of characteristic zero), the Hochschild–Konstant–Rosenberg theorem show us that

$$HP^{\pm}(\mathrm{perf}_{\mathsf{dg}}(X)) \simeq HP^{\pm}(X) \simeq \Big(\bigoplus_{n \text{ even}} H^n_{dR}(X), \bigoplus_{n \text{ odd}} H^n_{dR}(X) \Big). \tag{3-1}$$

For this reason $HP^{\pm}$ is considered the noncommutative analogue of de Rham cohomology. For further details on this viewpoint, we invite the reader to consult the ICM address [Kaledin 2010].

As proved in [Marcolli and Tabuada 2011, Theorem 7.2], $HP^{\pm}$ induces symmetric monoidal functors

$$\overline{HP^{\pm}} : \mathsf{NChow}(k)_F \to \mathrm{sVect}(F), \quad \overline{HP^{\pm}} : \mathsf{NChow}(k)_F \to \mathrm{sVect}(k) \tag{3-2}$$

under the assumption that F is a field extension of k (left-hand-side) or the assumption that k is a field extension of F (right-hand-side).

3.2. ***Noncommutative homological motives.*** Making use of the above "noncommutative de Rham cohomology", one defines the symmetric monoidal category of noncommutative homological motives $\mathsf{NHom}(k)_F$ as the idempotent completion of the quotient category $\mathsf{NChow}(k)_F/\mathrm{Ker}(\overline{HP^{\pm}})$.

3.3. ***Noncommutative numerical motives.*** In order to define a category of noncommutative numerical motives one needs to extend to the noncommutative world the notion of intersection number. This can be done as follows. Let $(\mathcal{A}, e)$ and $(\mathcal{B}, e')$ be two noncommutative Chow motives. Given a noncommutative correspondence $\underline{\mathsf{B}} = e[\sum_i a_i \mathsf{B}_i]e'$ from $(\mathcal{A}, e)$ to $(\mathcal{B}, e')$ (recall that the B_i's are $\mathcal{A}$-$\mathcal{B}$-bimodules), and a noncommutative correspondence $\underline{\mathsf{B}'} = e'[\sum_j b_j \mathsf{B}'_j]e$ from $(\mathcal{B}, e')$ to $(\mathcal{A}, e)$, one defines their intersection number as the following sum

$$\langle \underline{\mathsf{B}}, \underline{\mathsf{B}'} \rangle := \sum_{i,j,n} (-1)^n a_i \cdot b_j \cdot \mathrm{rank}\, HH_n(\mathcal{A}; \mathsf{B}_i \otimes^{\mathbf{L}}_{\mathcal{B}} \mathsf{B}'_j),$$

where $HH_n(\mathcal{A}; \mathsf{B}_i \otimes^{\mathbf{L}}_{\mathcal{B}} \mathsf{B}'_j)$ stands for the n-th Hochschild homology group of $\mathcal{A}$ with coefficients in the $\mathcal{A}$-$\mathcal{A}$ bimodule $\mathsf{B}_i \otimes^{\mathbf{L}}_{\mathcal{B}} \mathsf{B}'_j$. The numerical equivalence relation on noncommutative Chow motives is obtained by declaring a noncommutative correspondence $\underline{\mathsf{B}}$ to be numerically trivial if $\langle \underline{\mathsf{B}}, \underline{\mathsf{B}}' \rangle = 0$ for all $\underline{\mathsf{B}}'$. This equivalence relation gives rise to the largest $\otimes$-ideal $\mathcal{N}$ strictly contained in $\mathsf{NChow}(k)_F$. The symmetric monoidal category of noncommutative numerical motives $\mathsf{NNum}(k)_F$ is then defined as the idempotent completion of the quotient category $\mathsf{NChow}(k)_F/\mathcal{N}$.

As proved in [Marcolli and Tabuada 2014c, Theorem 1.12], the functor R of diagram (2-2) descends to a $\mathbb{Q}$-linear additive fully-faithful symmetric monoidal functor $R_{\mathsf{N}} : \mathsf{Num}_{\mathbb{Q}}(k)/_{-\otimes\mathbb{Q}(1)} \to \mathsf{NNum}(k)_{\mathbb{Q}}$.

A different numerical equivalence relation on noncommutative Chow motives (based on the bilinear pairing $\sum_i (-1)^i \dim \mathrm{Ext}^i(-,-)$), was proposed in [Kontsevich 2005]. As proved in [Marcolli and Tabuada 2012, Theorem 1.1], Kontsevich's notion is equivalent to the above one.

3.4. ***Semisimplicity.*** As proved in [Marcolli and Tabuada 2014c, Theorem 1.1; 2011, Theorem 4.6], Jannsen's result (see §1.5) holds also in the noncommutative world. Concretely, under the assumption that k and F have same characteristic, $\mathsf{NNum}(k)_F$ is abelian semisimple. This was conjectured in [Kontsevich 2005]. Moreover, if $\mathcal{J}$ is a $\otimes$-ideal in $\mathsf{NChow}(k)_F$ for which the idempotent completion of $\mathsf{NChow}(k)_F/\mathcal{J}$ is abelian semisimple, then $\mathcal{J}$ agrees with $\mathcal{N}$.

As explained in [Marcolli and Tabuada 2014c, Corollary 1.1], the semisimplicity of $\mathsf{NNum}(k)_F$ (with k of characteristic zero) combined with the functor R of diagram (2-2) gives rise to an alternative proof of Jannsen's result.

3.5. ***Smash-nilpotence in the noncommutative world.*** Recall from §1.6 the definition of the $\otimes_{\mathrm{nil}}$-ideal. One denotes by $\mathsf{NVoev}(k)_F := \mathsf{NChow}(k)_F/\otimes_{\mathrm{nil}}$ the category of noncommutative Chow motives up to smash-nilpotence. As proved in [Marcolli and Tabuada 2014d, Proposition 3.1], the functor R of diagram (2-2) descends also to a symmetric monoidal fully-faithful functor $R_{\mathsf{V}} : \mathsf{Voev}_{\mathbb{Q}}(k)/_{-\otimes\mathbb{Q}(1)} \to \mathsf{NVoev}(k)_{\mathbb{Q}}$.

3.6. ***All together.*** The categories of noncommutative motives are related by F-linear, additive, full, symmetric monoidal functors $\mathsf{NChow}(k)_F \to \mathsf{NVoev}(k)_F \to \mathsf{NHom}(k)_F \to \mathsf{NNum}(k)_F$.

Given a saturated dg category $\mathcal{A}$, we will denote by $\sim_{\otimes\mathrm{nil}}, \sim_{\mathrm{hom}}, \sim_{\mathrm{num}}$ the equivalence relations on $\mathrm{Hom}_{\mathsf{NChow}(k)_F}(U(k)_F, U(\mathcal{A})_F) \simeq K_0(\mathcal{A})_F$ induced by the above functors.

3.7. ***Noncommutative Artin motives.*** The category of Artin motives $\mathsf{AM}_F(k)$ is by definition the smallest additive rigid idempotent complete full subcategory of $\mathsf{Chow}_F(k)$ containing the finite étale k-schemes. One defines the category of noncommutative Artin motives $\mathsf{NAM}(k)_{\mathbb{Q}}$ as the image of $\mathsf{AM}_{\mathbb{Q}}(k)$ under the functors $R \circ \pi$ of diagram (2-2). As proved in [Marcolli and Tabuada 2014b, Theorem 1.7], this latter category is independent of the equivalence relation.

4. Conjectures in the noncommutative world

Let $\mathcal{A}$ be a saturated dg category.

4.1. *Standard conjectures.* In [Marcolli and Tabuada 2011] we introduced the noncommutative analogues of Grothendieck's standard conjectures of type C and D.

Conjecture $C_{NC}(\mathcal{A})$. *The Künneth projectors*

$$\pi_{\mathcal{A}}^{+} : HP^{\pm}(\mathcal{A}) \to HP^{+}(\mathcal{A}) \to HP^{\pm}(\mathcal{A}),$$
$$\pi_{\mathcal{A}}^{-} : HP^{\pm}(\mathcal{A}) \to HP^{-}(\mathcal{A}) \to HP^{\pm}(\mathcal{A})$$

are algebraic, i.e., $\pi_{\mathcal{A}}^{\pm} = \overline{HP^{\pm}(\underline{\pi}_{\mathcal{A}}^{\pm})}$ *for noncommutative correspondences* $\underline{\pi}_{\mathcal{A}}^{\pm}$.

A weaker version of the standard conjecture of type C (Künneth) is the sign conjecture $C^{+}(X)$: *The Künneth projectors* $\pi_X^{+} = \sum_i \pi_X^{2i}$ *and* $\pi_X^{-} = \sum_i \pi_X^{2i+1}$ *are algebraic*. The restriction of C_{NC} to the commutative world is weaker than the sign conjecture in the sense that $C^{+}(X) \Rightarrow C_{NC}(\mathrm{perf}_{\mathrm{dg}}(X))$.

Conjecture $D_{NC}(\mathcal{A})$. $\qquad K_0(\mathcal{A})_F/{\sim_{\mathrm{hom}}} = K_0(\mathcal{A})_F/{\sim_{\mathrm{num}}}$.

Similarly, the restriction of D_{NC} to the commutative world is weaker than the standard conjecture of type D (Hom=Num) in the sense that $D(X) \Rightarrow D_{NC}(\mathrm{perf}_{\mathrm{dg}}(X))$.

4.2. *Smash-nilpotence conjecture.* Voevodsky's nilpotence conjecture (see §1.6) was extended in [Marcolli and Tabuada 2014d] to the noncommutative world as follows:

Conjecture $V_{NC}(\mathcal{A})$. $\qquad K_0(\mathcal{A})_F/{\sim_{\otimes_{\mathrm{nil}}}} = K_0(\mathcal{A})_F/{\sim_{\mathrm{num}}}$.

As proved in Theorem 4.1 of the same reference, the restriction of V_{NC} to the commutative world is equivalent to Voevodsky's smash-nilpotence conjecture in the sense that $V(X) \Leftrightarrow V_{NC}(\mathrm{perf}_{\mathrm{dg}}(X))$. This suggests that instead of attacking Voevodsky's conjecture V, one should alternatively attack conjecture V_{NC} (using noncommutative tools). The authors are currently developing this approach.

4.3. *Finite-dimensionality conjecture.* Let F be a field of characteristic zero and $\mathcal{C}$ an F-linear, idempotent complete, symmetric monoidal category. An object $a \in \mathcal{C}$ is called even (resp. odd) dimensional if $\wedge^n(a) = 0$ (resp. $\mathrm{Sym}^n(a) = 0$) for some $n > 0$. An object $a \in \mathcal{C}$ is called finite-dimensional if $a = a_{+} \oplus a_{-}$, with a_{+} (resp. a_{-}) even (resp. odd) dimensional. Kimura [2005] and O'Sullivan [2005] conjectured the following:

Conjecture $KS(X)$. *The Chow motive $h_F(X)$ is finite-dimensional.*

This conjecture was extended in [Marcolli and Tabuada 2014d] to the noncommutative world as follows:

Conjecture $KS_{NC}(\mathcal{A})$. *The noncommutative Chow motive $U(\mathcal{A})_F$ is finite-dimensional.*

The restriction of KS_{NC} to the noncommutative world is weaker than the Kimura–O'Sullivan finite-dimensionality conjecture in the sense that $KS(X) \Rightarrow KS_{NC}(\mathrm{perf}_{\mathrm{dg}}(X))$.

Under the assumption that k is a field extension of F (or vice-versa), it was proved in [Marcolli and Tabuada 2014d, Theorem 4.1] that conjectures $V_{NC}((\mathcal{A}^{\mathrm{op}})^{\otimes n} \otimes \mathcal{A}^{\otimes n})$, $n \geq 1$, combined with conjecture $C_{NC}(\mathcal{A})$, imply conjecture $KS_{NC}(\mathcal{A})$. Moreover, if conjecture KS_{NC} holds for every saturated dg category and all symmetric monoidal functors $\mathsf{NChow}(k)_F \to \mathrm{sVect}(K)$ (with K a field extension of F) factor through $\mathsf{NNum}(k)_F$, then conjecture V_{NC} also holds for every saturated dg category.

5. (Super-)Tannakian formalism

5.1. ***Tannakian categories.*** Let $(\mathcal{C}, \otimes, \mathbf{1})$ be an F-linear, abelian, symmetric monoidal category. In particular, we have commutativity and $\otimes$-unit constraints

$$\tau_{a,b} : a \otimes b \xrightarrow{\sim} b \otimes a, \quad \ell_a : a \xrightarrow{\sim} a \otimes \mathbf{1}, \quad r_a : \mathbf{1} \otimes a \xrightarrow{\sim} a,$$

and the following equality holds $\tau_{b,a} \circ \tau_{a,b} = \mathrm{id}_{a \otimes b}$. The category $\mathcal{C}$ is called rigid if there exists a duality functor $(-)^\vee : \mathcal{C} \to \mathcal{C}^{\mathrm{op}}$, evaluation maps $\epsilon : a \otimes a^\vee \to \mathbf{1}$, and coevaluation maps $\eta : \mathbf{1} \to a^\vee \otimes a$, for which the following composition is equal to the identity

$$a \xrightarrow{\ell_a} a \otimes \mathbf{1} \xrightarrow{\mathrm{id}_a \otimes \eta} a \otimes a^\vee \otimes a \xrightarrow{\epsilon \otimes \mathrm{id}_a} \mathbf{1} \otimes a \xrightarrow{r_a} a.$$

The categorical trace of an endomorphism $g : a \to a$ is defined as $tr(g) = \epsilon \circ \tau_{a^\vee \otimes a} \circ (\mathrm{id}_{a^\vee} \otimes g) \circ \eta$. The number $\dim(a) := tr(\mathrm{id}_a)$ is called the dimension or Euler characteristic of a.

A category $\mathcal{C}$ with the above properties, and with $\mathrm{End}(\mathbf{1}) = F$, is called Tannakian if there exists an exact faithful symmetric monoidal functor $\omega : \mathcal{C} \to \mathrm{Vect}(K)$ with values in a category of K-vector spaces (with K a field extension of F). The functor ω is called a fiber functor. When this holds with $K = F$, $\mathcal{C}$ is called a neutral Tannakian category.

If $\mathcal{C}$ is a neutral Tannakian category, then there is a $\otimes$-equivalence of categories $\mathcal{C} \simeq \mathrm{Rep}_F(\mathrm{Gal}(\mathcal{C}))$. The affine group scheme $\mathrm{Gal}(\mathcal{C})$ is given by the $\otimes$-automorphisms $\underline{\mathrm{Aut}}^{\otimes}(\omega)$ of the fiber functor ω.

Tannakian categories in characteristic zero were characterized in [Deligne 1990] as follows: an F-linear, abelian, rigid symmetric monoidal category $\mathcal{C}$, with $\mathrm{End}(\mathbf{1}) = F$, is Tannakian if and only if $\dim(a) \geq 0$ for all objects $a \in \mathcal{C}$.

5.2. ***Super-Tannakian categories.*** An F-linear, abelian, rigid symmetric monoidal category $\mathcal{C}$, with $\mathrm{End}(\mathbf{1}) = F$, is called super-Tannakian if there exists a super-fiber functor $\omega : \mathcal{C} \to \mathrm{sVect}(K)$ with values in a category of super K-vector spaces (with K a field extension of F). When this holds with $K = F$, $\mathcal{C}$ is called a neutral super-Tannakian category.

If $\mathcal{C}$ is a neutral super-Tannakian category, ω induces a $\otimes$-equivalence between $\mathcal{C}$ and the category $\mathrm{Rep}_F(\mathrm{sGal}(\mathcal{C}), \epsilon)$ of super-representations of the affine supergroup scheme $\mathrm{sGal}(\mathcal{C}) := \underline{\mathrm{Aut}}^{\otimes}(\omega)$ (the super-structure is given by the parity automorphism ϵ).

Super-Tannakian categories were also characterized in [Deligne 2002] as follows: an F-linear, abelian, rigid symmetric monoidal category $\mathcal{C}$, with $\mathrm{End}(\mathbf{1}) = F$, is super-Tannakian if and only if it is Schur-finite. When F is algebraically closed, $\mathcal{C}$ is neutral super-Tannakian if and only if it is Schur-finite.

Schur-finiteness. Let $\mathcal{C}$ be a category as above, S_n the symmetric group on n symbols, and $\mathbb{Q}[S_n]$ the associated group ring. Every partition λ of n gives rise to an idempotent $e_\lambda \in \mathbb{Q}[S_n]$ and to a Schur functor $S_\lambda : \mathcal{C} \to \mathcal{C}$, $a \mapsto e_\lambda(a^{\otimes n})$. The category $\mathcal{C}$ is called Schur-finite if all its objects are annihilated by some Schur functor.

5.3. ***Motivic Galois groups.*** Deligne's characterization of Tannakian categories is not satisfied in the case of $\mathsf{Num}_F(k)$ because $\dim(h_F(X))$ is equal to the Euler characteristic $\chi(X)$ of the k-scheme X which can be negative. Jannsen [1992] proved that if the standard conjecture of type C holds, then one can modify the commutativity constraints $\tau_{X,Y}$ using the algebraic cycles coming from the Künneth components of the diagonal. This has the effect of correcting the negative signs of the Euler characteristic. Let $\mathsf{Num}_F^{\dagger}(k)$ be the Tannakian category hence obtained. If the standard conjecture of type D also holds, then $\mathsf{Num}_F^{\dagger}(k)$ is a neutral Tannakian category and every Weil cohomology theory H^* is a fiber functor. Under these assumptions, one obtains a group scheme $\mathrm{Gal}(\mathsf{Num}_F^{\dagger}(k))$ called the motivic Galois group.

5.4. ***Motivic Galois supergroups.*** In contrast with §5.3, Deligne's characterization of super-Tannakian categories is satisfied in the case of $\mathsf{Num}_F(k)$. When F is algebraically closed, $\mathsf{Num}_F(k)$ is then a neutral Tannakian category. As a consequence, one obtains a supergroup scheme $\mathrm{sGal}(\mathsf{Num}_F(k))$ called the motivic Galois supergroup.

6. Noncommutative motivic Galois (super-)groups

Let k be a field of characteristic zero and F a field extension of k.

Assuming conjectures C_{NC} and D_{NC}, it was proved in [Marcolli and Tabuada 2011, Theorem 1.6] that the category $\mathsf{NNum}^{\dagger}(k)_F$ (obtained from $\mathsf{NNum}(k)_F$ by modifying the commutativity constrains) is neutral Tannakian. An explicit fiber functor is given by periodic cyclic homology. The associated group scheme $\mathrm{Gal}(\mathsf{NNum}^{\dagger}(k)_F)$ is called the noncommutative motivic Galois group.

By Theorem 1.2 of [Marcolli and Tabuada 2011], the category $\mathsf{NNum}(k)_F$ is super-Tannakian. When F is algebraically closed, $\mathsf{NNum}(k)_F$ is neutral super-Tannakian. Under these assumptions, one obtains a supergroup scheme $\mathrm{sGal}(\mathsf{NNum}(k)_F)$ called the noncommutative motivic Galois supergroup.

6.1. ***Comparison morphisms.*** Assuming conjectures C, D, C_{NC}, D_{NC}, we have well-defined (noncommutative) motivic Galois (super-)groups. As proved in [Marcolli and Tabuada 2011, Theorem 1.7], the composed functor

$$\mathsf{Num}_k(k) \xrightarrow{\pi} \mathsf{Num}_k(k)/_{-\otimes\mathbb{Q}(1)} \xrightarrow{R_{\mathsf{N}}} \mathsf{NNum}(k)_k \tag{6-1}$$

gives rise to faithfully-flat comparison morphisms

$$\mathrm{Gal}(\mathsf{NNum}^{\dagger}(k)_k) \twoheadrightarrow \mathrm{Ker}(t : \mathrm{Gal}(\mathsf{Num}_k^{\dagger}(k)) \twoheadrightarrow \mathbb{G}_m) \tag{6-2}$$

$$\mathrm{sGal}(\mathsf{NNum}(k)_k) \twoheadrightarrow \mathrm{Ker}(t : \mathrm{sGal}(\mathsf{Num}_k(k)) \twoheadrightarrow \mathbb{G}_m), \tag{6-3}$$

where $\mathbb{G}_m$ is the multiplicative group scheme and t is induced by the inclusion of Tate motives in $\mathsf{Num}_k(k)$. These comparison morphisms were suggested in [Kontsevich 2005]. Intuitively speaking, they show us that the $\otimes$-symmetries of the commutative world which can be lifted to the noncommutative world are precisely those that become trivial when restricted to Tate motives.

The proof of (6-2)–(6-3) makes use of the theory of Tate triples developed in [Deligne and Milne 1982], of a suitable extension of this theory to the super-Tannakian setting, and of Milne's work [2007] on quotients of Tannakian categories. The key step is the description of the right-hand-side of (6-2) (resp. of (6-3)) as the Galois group (resp. supergroup) of the orbit category of $\mathsf{Num}_k^{\dagger}(k)$ (resp. of $\mathsf{Num}_k(k)$).

It is unclear at the moment if the kernel of these comparison morphisms is nontrivial. This problem is related to the existence of "truly noncommutative numerical motives", i.e., objects of $\mathsf{NNum}(k)_k$ that are not in the essential image of (6-1).

6.2. ***Unconditional version.*** The functors (3-2) descend to symmetric monoidal functors

$$\overline{HP^{\pm}} : \mathsf{NHow}(k)_F \longrightarrow \mathrm{sVect}(K). \tag{6-4}$$

Here, $K = F$ when F is a field extension of k and $K = k$ when k is a field extension of F. Let $\mathsf{NHom}(k)_F^{\pm}$ be the full subcategory of $\mathsf{NHom}(k)_F$ consisting of those noncommutative homological motives for which the Künneth projectors are algebraic. By changing the commutativity constraints of this latter category, one obtains a rigid symmetric monoidal category $\mathsf{NHom}^{\dagger}(k)_F^{\pm}$ and an F-linear symmetric monoidal functor $\mathsf{NHom}^{\dagger}(k)_F^{\pm} \to \mathrm{Vect}(K)$. Making use of techniques from [André and Kahn 2002a; 2002b], we showed in [Marcolli and Tabuada 2014d, §1] that the associated category $\mathsf{NNum}^{\dagger}(k)_F^{\pm}$ is Tannakian, semisimple, and that the canonical functor $\mathsf{NHom}^{\dagger}(k)_F^{\pm} \to \mathsf{NNum}^{\dagger}(k)_F^{\pm}$ admits a $\otimes$-section s^{NC} (unique up to conjugation by a $\otimes$-isomorphism). One obtains in this way a fiber functor

$$\omega : \mathsf{NNum}^{\dagger}(k)_F^{\pm} \xrightarrow{s^{NC}} \mathsf{NHom}^{\dagger}(k)_F^{\pm} \longrightarrow \mathrm{Vect}(K).$$

The associated group scheme $\mathrm{Gal}(\mathsf{NNum}^{\dagger}(k)_F^{\pm})$ is called the unconditional noncommutative motivic Galois group. As proved in [Marcolli and Tabuada 2014d, Theorem 1.7], we have a faithfully-flat comparison morphism

$$\mathrm{Gal}(\mathsf{NNum}^{\dagger}(k)_k^{\pm}) \twoheadrightarrow \quad \mathrm{Ker}(t : \mathrm{Gal}(\mathsf{Num}_k^{\dagger}(k)^{\pm}) \twoheadrightarrow \mathbb{G}_m). \tag{6-5}$$

Assuming conjectures C, D and C_{NC}, D_{NC}, the unconditional noncommutative motivic Galois group agree with the conditional one $\mathrm{Gal}(\mathsf{NNum}^{\dagger}(k)_k)$ and (6-5) identifies with (6-3).

6.3. *Base change.* As proved in [Marcolli and Tabuada 2014b, Theorem 1.9], one has the following short exact sequence

$$1 \longrightarrow \mathrm{Gal}(\mathsf{NNum}^{\dagger}(\bar{k})_F) \xrightarrow{I} \mathrm{Gal}(\mathsf{NNum}^{\dagger}(k)_F) \xrightarrow{P} \mathrm{Gal}(\bar{k}/k) \longrightarrow 1. \tag{6-6}$$

Here I is induced by the base change $- \otimes_k \bar{k} : \mathsf{NNum}^{\dagger}(k)_F \to \mathsf{NNum}^{\dagger}(\bar{k})_F$; the absolute Galois group $\mathrm{Gal}(\bar{k}/k)$ is obtained from the Tannakian formalism applied to the category of noncommutative Artin motives $\mathsf{NAM}(k)_F$; and finally P is induced by the inclusion of the latter category in $\mathsf{NNum}^{\dagger}(k)_F$.

The proof in [Deligne and Milne 1982] of the classical commutative counterpart of (6-6) makes full use of "commutative arguments" which don't seem to admit noncommutative analogues. The proof of (6-6) is not only very different but moreover much more conceptual from a categorical viewpoint. By extracting the key ingredients of this latter proof we have established in [Marcolli and Tabuada 2014b, Appendix A] a general result about short exact sequences of Galois groups. This led to a new proof of Deligne–Milne's short exact sequence which circumvents their "commutative arguments".

7. From noncommutative motives to motives via Jacobians

We have described in §2.6 a bridge from motives to noncommutative motives. One can ask is there a bridge in the opposite direction, associating a "commutative shadow" to every noncommutative motive? This (vague) idea can be implemented using the theory of Jacobians, suitably extended to the noncommutative world. Let $k \subseteq \mathbb{C}$ be an algebraically closed field.

7.1. *Jacobians.* Jacobians $J(C)$ of curves C are geometric models for the cohomology $H^1(C)$. The study of Jacobians is in fact one of the historic precursors of the theory of motives. Given an arbitrary smooth projective k-scheme X, the Picard $\mathrm{Pic}^0(X)$ and the Albanese $\mathrm{Alb}(X)$ varieties provide, in a similar way, geometric models for the pieces $H^1(X)$ and $H^{2\dim(X)-1}(X)$. In what concerns the remaining pieces of the cohomology, Griffiths' intermediate Jacobians

$$J_i(X) := \frac{H_B^{2i+1}(X,\mathbb{C})}{F^{i+1}H_B^{2i+1}(X,\mathbb{C}) + H_B^{2i+1}(X,\mathbb{Z})},\ 0 < i < \dim(X),$$

where H_B stands for Betti cohomology and F for the Hodge filtration, are not algebraic. Nevertheless, they contain an algebraic part $J_i^a(X) \subseteq J_i(X)$ defined by the image of the Abel–Jacobi map $AJ_i : CH_{\mathbb{Z}}^{i+1}(X)^{\mathrm{alg}} \to J_i(X)$, where $CH_{\mathbb{Z}}^{i+1}(X)^{\mathrm{alg}}$ is the group of algebraically trivial cycles of codimension $i+1$.

7.2. *Pairings.* Given a smooth projective k-scheme X, consider the following k-vector spaces

$$NH_{dR}^{2i+1}(X) := \sum_{C,\gamma_i} \mathrm{Image}\big(H_{dR}^1(C) \xrightarrow{H_{dR}^1(\gamma_i)} H_{dR}^{2i+1}(X)\big), \quad 0 \le i \le \dim(X)-1, \tag{7-1}$$

where C is a smooth projective curve and $\gamma_i : h_{\mathbb{Q}}(C) \to h_{\mathbb{Q}}(X)(i)$ is a morphism in $\mathsf{Chow}_{\mathbb{Q}}(k)$. Intuitively speaking, (7-1) are the odd pieces of de Rham cohomology that are generated by curves. By restricting the classical intersection pairings on de Rham cohomology to (7-1) one obtains

$$\langle -,-\rangle : NH_{dR}^{2\dim(X)-2i-1}(X) \times NH_{dR}^{2i+1}(X) \longrightarrow k, \quad 0 \le i \le \dim(X)-1. \tag{7-2}$$

7.3. *Jacobians of noncommutative motives.* In [Marcolli and Tabuada 2014a, Theorem 1.3] we constructed a $\mathbb{Q}$-linear additive "Jacobian" functor

$$\mathbf{J}(-) : \mathsf{NChow}(k)_{\mathbb{Q}} \longrightarrow \mathsf{Ab}_{\mathbb{Q}}(k) \tag{7-3}$$

with values in the category of abelian varieties up to isogeny. Among other properties, one has an isomorphism $\mathbf{J}(\mathrm{perf}_{\mathrm{dg}}(X)) \simeq \prod_{i=0}^{\dim(X)-1} J_i^a(X)$ whenever the above pairings (7-2) are nondegenerate. As explained in loc. cit., this is always the case for $i=0$ and $i=\dim(X)-1$ and the remaining cases follow from Grothendieck's standard conjecture of type B. Hence, the pairings (7-2) are nondegenerate for curves, surfaces, abelian varieties, complete intersections, uniruled threefolds, rationally connected fourfolds, and for any smooth hypersurface section, product, or finite quotient thereof (and if one trusts Grothendieck for all smooth projective k-schemes).

Given a noncommutative Chow motive N, the abelian variety $\mathbf{J}(N)$ was constructed as follows:

- First, via $h^1_{\mathbb{Q}}(-)$ and the fully-faithful functor R_{N} (see §3.3), one observes that $\mathrm{Ab}_{\mathbb{Q}}(k)$ identifies with a semisimple abelian full subcategory of $\mathsf{NNum}(k)_{\mathbb{Q}}$.
- Secondly, the semisimplicity of $\mathsf{NNum}(k)_{\mathbb{Q}}$ implies that N admits a unique direct sum decomposition $S_1\oplus\cdots\oplus S_n$ into simple objects.
- Finally, one defines $\mathbf{J}(N)$ as the smallest piece of the noncommutative numerical motive $N\simeq S_1\oplus\cdots\oplus S_n$ which contains the simple objects belonging to $\mathrm{Ab}_{\mathbb{Q}}(k)$.

Roughly speaking, (7-3) show us that the classical theory of Jacobians can be extended to the noncommutative world as long as one works with all the intermediate Jacobians simultaneously. Note that this restriction is an intrinsic feature of the noncommutative world which cannot be avoided because as soon as one passes from X to $\mathrm{perf}_{\mathrm{dg}}(X)$ one loses track of the individual pieces of $H^*_{dR}(X)$ (see (3-1)).

7.4. ***Some applications.*** The above theory of Jacobians of noncommutative motives allowed categorical Torelli theorems, a new proof of a classical theorem of Clemens and Griffiths concerning blow-ups of threefolds, and several new results on quadric fibrations and intersections of quadrics; see [Bernardara and Tabuada 2014a]. Recently, this theory allowed also the proof of a conjecture of Paranjape [1994] in the case of a complete intersection of either two quadrics or three odd-dimensional quadrics; see [Bernardara and Tabuada 2015].

8. Applications to motivic decompositions and to Dubrovin's conjecture

8.1. ***Motivic decompositions.*** It is well-know that $h(\mathbb{P}^n)=1\oplus\mathbf{L}\oplus\cdots\oplus\mathbf{L}^{\otimes n}$; see [Manin 1968]. Other examples of motivic decompositions containing only $\otimes$-powers of the Lefschetz motive arise from quadrics. Given a nondegenerate quadratic form (V,q) of dimension $n\geq 3$ defined over an algebraically closed field k of characteristic zero, let $Q_q\subset\mathbb{P}(V)$ be the associated smooth projective quadric of dimension $d:=n-2$. The motivic decomposition of Q_q, proved in

[Rost 1990], is

$$h_{\mathbb{Q}}(Q_q) \simeq \begin{cases} 1 \oplus \mathbf{L} \oplus \cdots \oplus \mathbf{L}^{\otimes d} & \text{for } d \text{ odd}, \\ 1 \oplus \mathbf{L} \oplus \cdots \oplus \mathbf{L}^{\otimes d} \oplus \mathbf{L}^{\otimes (d/2)} & \text{for } d \text{ even}. \end{cases}$$

Fano 3-folds also fit in this pattern. In this case, thanks to [Gorchinskiy and Guletskiĭ 2012], we have

$$h_{\mathbb{Q}}(X) \simeq 1 \oplus h^1_{\mathbb{Q}}(X) \oplus \mathbf{L}^{\oplus b} \oplus (h^1_{\mathbb{Q}}(J) \otimes \mathbf{L}) \oplus (\mathbf{L}^{\otimes 2})^{\oplus b} \oplus h^5_{\mathbb{Q}}(X) \oplus \mathbf{L}^{\otimes 3},$$

where $h^1_{\mathbb{Q}}(X)$ and $h^5_{\mathbb{Q}}(X)$ are the Picard and Albanese motives, $b = b_2(X) = b_4(X)$, and J is an abelian variety which is isogenous to the intermediate Jacobian when $k = \mathbb{C}$. Whenever the odd cohomology of X vanishes, this motivic decomposition reduces to a direct sum of $\otimes$-powers of the Lefschetz motive. Further examples include toric varieties and certain homogeneous spaces (see [Brosnan 2005]), and moduli spaces of pointed curves of genus zero (see [Chen et al. 2009]).

8.2. *Full exceptional collections.* A collection of objects $\{\mathcal{E}_1, \ldots, \mathcal{E}_n\}$ in a k-linear triangulated category $\mathcal{C}$ is called exceptional if $\mathrm{RHom}(\mathcal{E}_i, \mathcal{E}_i) = k$, for all i, and $\mathrm{RHom}(\mathcal{E}_i, \mathcal{E}_j) = 0$ for all $i > j$. It is called full if the objects $\mathcal{E}_1, \cdots, \mathcal{E}_n$ generate the triangulated category $\mathcal{C}$.

The derived category $\mathrm{Perf}(X) \simeq \mathcal{D}^b(\mathrm{Coh}(X))$ of a smooth projective k-scheme X admits a full exceptional collection in several cases: projective spaces (see [Beĭlinson 1978]), quadrics (see [Kapranov 1988]), toric varieties (see [Kawamata 2006]), certain homogeneous spaces (see [Kuznetsov and Polishchuk 2011]), moduli spaces of pointed curves of genus zero (see [Manin and Smirnov 2013]), and Fano 3-folds with vanishing odd cohomology (see [Ciolli 2005]). In all these examples the corresponding motivic decomposition contain only $\otimes$-powers of the Lefschetz motive. It is therefore natural to ask if there is a relation between these two notions. The answer is "yes", as we now explain.

8.3. *From exceptional collections to motivic decompositions.* Let X be a smooth projective k-scheme for which $\mathrm{perf}(X)$ admits a full exceptional collection $\{\mathcal{E}_1, \ldots, \mathcal{E}_n\}$. Then, as proved in [Marcolli and Tabuada 2015, Theorem 1.3], there is a choice of integers $\ell_1, \ldots, \ell_n \in \{0, \ldots, \dim(X)\}$ such that

$$h_{\mathbb{Q}}(X) \simeq \mathbf{L}^{\otimes \ell_1} \oplus \cdots \oplus \mathbf{L}^{\otimes \ell_n}. \tag{8-1}$$

The motivic decomposition (8-1) was obtained as follows:

- First, as mentioned in §2.4, the noncommutative Chow motive $U(\mathrm{perf}_{\mathrm{dg}}(X))_{\mathbb{Q}}$ decomposes into the direct sum of n copies of $U(k)_{\mathbb{Q}}$.

- Secondly, the commutativity of diagram (2-2) implies that $h_{\mathbb{Q}}(X)$ (considered as an object of the orbit category $\mathsf{NChow}_{\mathbb{Q}}(k)/_{-\otimes\mathbb{Q}(1)})$ decomposes into the direct sum of n copies of $\mathbf{L}$.
- Finally, one observes that the "fiber" of $\mathbf{L}$ under the projection functor from noncommutative Chow motives to its orbit category consists solely of $\otimes$-powers of the Lefschetz motive.

The decomposition (8-1) holds more generally with X a smooth proper Deligne–Mumford stack.

The decomposition (8-1) has recently greatly refined: instead of working with $\mathbb{Q}$-coefficients it suffices to invert the prime factors of the integer $(2\dim(X))!$; consult [Bernardara and Tabuada 2014b] for details.

8.4. *Dubrovin's conjecture.* At his ICM address, Dubrovin [$\geq$ 2015] conjectured a striking connection between Gromov–Witten invariants and derived categories of coherent sheaves. The most recent formulation, due to Hertling, Manin and Teleman [Hertling et al. 2009], is the following:

Conjecture. *Given a smooth projective $\mathbb{C}$-scheme X, the following holds*:

(i) *The quantum cohomology of X is semisimple if and only if X is Hodge–Tate (i.e the Hodge numbers $h^{p,q}(X)$ are zero for $p \neq q$) and* $\mathrm{perf}(X)$ *admits a full exceptional collection*;

(ii) *The Stokes matrix of the structure connection of the quantum cohomology identifies with the Gram matrix of the full exceptional collection.*

Thanks to [Bayer 2004; Golyshev 2009; Guzzetti 1999; Ueda 2005] and other works, both statements are known to be true in the case of projective spaces (and its blow-ups) and Grassmannians. Item (i) also holds for minimal Fano threefolds. Moreover, it is proved in [Hertling et al. 2009] that the Hodge–Tate property follows from the semisimplicity of quantum cohomology.

Making use of the above motivic decomposition (8-1), we proved in [Marcolli and Tabuada 2015, Proposition 1.9] that the Hodge–Tate property follows also from the existence of a full exceptional collection. As a consequence, this assumption can be removed from item (i) of Dubrovin's conjecture.

References

[André 2004] Y. André, *Une introduction aux motifs (motifs purs, motifs mixtes, périodes)*, Panoramas et Synthèses [Panoramas and Syntheses] **17**, Société Mathématique de France, Paris, 2004.

[André and Kahn 2002a] Y. André and B. Kahn, "Construction inconditionnelle de groupes de Galois motiviques", *C. R. Math. Acad. Sci. Paris* **334**:11 (2002), 989–994.

[André and Kahn 2002b] Y. André and B. Kahn, "Nilpotence, radicaux et structures monoïdales", *Rend. Sem. Mat. Univ. Padova* **108** (2002), 107–291.

[Bayer 2004] A. Bayer, "Semisimple quantum cohomology and blowups", *Int. Math. Res. Not.* **2004**:40 (2004), 2069–2083.

[Beĭlinson 1978] A. A. Beĭlinson, "Coherent sheaves on $\mathbf{P}^n$ and problems in linear algebra", *Funktsional. Anal. i Prilozhen.* **12**:3 (1978), 68–69.

[Bernardara and Tabuada 2014a] M. Bernardara and G. Tabuada, "From semi-orthogonal decompositions to polarized intermediate Jacobians via Jacobians of noncommutative motives", preprint, 2014. arXiv 1305.4687

[Bernardara and Tabuada 2014b] M. Bernardara and G. Tabuada, "Relations between the Chow motive and the noncommutative motive of a smooth projective variety", preprint, 2014. To appear in *Journal of Pure and Applied Algebra*. arXiv 1303.3172

[Bernardara and Tabuada 2015] M. Bernardara and G. Tabuada, "Chow groups of intersections of quadrics via homological projective duality and (Jacobians of) noncommutative motives", preprint, 2015. arXiv 1310.6020

[Blumberg and Mandell 2012] A. J. Blumberg and M. A. Mandell, "Localization theorems in topological Hochschild homology and topological cyclic homology", *Geom. Topol.* **16**:2 (2012), 1053–1120.

[Bondal and Kapranov 1989] A. I. Bondal and M. M. Kapranov, "Representable functors, Serre functors, and reconstructions", *Izv. Akad. Nauk SSSR Ser. Mat.* **53**:6 (1989), 1183–1205, 1337. In Russian; translated in *Math. USSR-Izv.* **35**:3 (1990), 519–541.

[Bondal and Kapranov 1990] A. I. Bondal and M. M. Kapranov, "Framed triangulated categories", *Mat. Sb. (N.S.)* **181**:5 (1990), 669–683. In Russian; translated in *Math. USSR-Sb.* **70**:1 (1991), 93–107.

[Bondal and van den Bergh 2003] A. Bondal and M. van den Bergh, "Generators and representability of functors in commutative and noncommutative geometry", *Mosc. Math. J.* **3**:1 (2003), 1–36.

[Brosnan 2005] P. Brosnan, "On motivic decompositions arising from the method of Białynicki-Birula", *Invent. Math.* **161**:1 (2005), 91–111.

[Chen et al. 2009] L. Chen, A. Gibney, and D. Krashen, "Pointed trees of projective spaces", *J. Algebraic Geom.* **18**:3 (2009), 477–509.

[Ciolli 2005] G. Ciolli, "On the quantum cohomology of some Fano threefolds and a conjecture of Dubrovin", *Internat. J. Math.* **16**:8 (2005), 823–839.

[Cisinski and Tabuada 2012] D.-C. Cisinski and G. Tabuada, "Symmetric monoidal structure on non-commutative motives", *J. K-Theory* **9**:2 (2012), 201–268.

[Deligne 1990] P. Deligne, "Catégories tannakiennes", pp. 111–195 in *The Grothendieck Festschrift*, vol. II, edited by P. Cartier et al., Progr. Math. **87**, Birkhäuser, Boston, 1990.

[Deligne 2002] P. Deligne, "Catégories tensorielles", *Mosc. Math. J.* **2**:2 (2002), 227–248.

[Deligne and Milne 1982] P. Deligne and J. Milne, "Tannakian categories", pp. 101–228 in *Hodge cycles, motives, and Shimura varieties*, edited by P. Deligne et al., Lecture Notes in Mathematics **900**, Springer, Berlin, 1982.

[Drinfeld 2004] V. Drinfeld, "DG quotients of DG categories", *J. Algebra* **272**:2 (2004), 643–691.

[Dubrovin $\geq$ 2015] B. Dubrovin, "Geometry and analytic theory of Frobenius manifolds", pp. 315–326 in *Proceedings of the International Congress of Mathematicians* (Berlin, 1998), vol. II.

[Golyshev 2009] V. V. Golyshev, "Minimal Fano threefolds: Exceptional sets and vanishing cycles", *Dokl. Akad. Nauk* **424**:2 (2009), 151–155. In Russian; translated in *Dokl. Math.* **79**:1 (2009), 16–20.

[Gorchinskiy and Guletskiĭ 2012] S. Gorchinskiy and V. Guletskiĭ, "Motives and representability of algebraic cycles on threefolds over a field", *J. Algebraic Geom.* **21**:2 (2012), 347–373.

[Guzzetti 1999] D. Guzzetti, "Stokes matrices and monodromy of the quantum cohomology of projective spaces", *Comm. Math. Phys.* **207**:2 (1999), 341–383.

[Hertling et al. 2009] C. Hertling, Y. I. Manin, and C. Teleman, "An update on semisimple quantum cohomology and F-manifolds", *Tr. Mat. Inst. Steklova* **264**:Mnogomernaya Algebraicheskaya Geometriya (2009), 69–76.

[Jannsen 1992] U. Jannsen, "Motives, numerical equivalence, and semi-simplicity", *Invent. Math.* **107**:3 (1992), 447–452.

[Jannsen et al. 1994] U. Jannsen, S. Kleiman, and J.-P. Serre (editors), *Motives*, Proceedings of Symposia in Pure Mathematics **55**, Amer. Math. Soc., Providence, RI, 1994.

[Kaledin 2010] D. Kaledin, "Motivic structures in non-commutative geometry", pp. 461–496 in *Proceedings of the International Congress of Mathematicians, II* (Hyderabad, India, 2010), edited by R. Bhatia et al., Hindustan Book Agency, New Delhi, 2010.

[Kapranov 1988] M. M. Kapranov, "On the derived categories of coherent sheaves on some homogeneous spaces", *Invent. Math.* **92**:3 (1988), 479–508.

[Kassel 1987] C. Kassel, "Cyclic homology, comodules, and mixed complexes", *J. Algebra* **107**:1 (1987), 195–216.

[Kawamata 2006] Y. Kawamata, "Derived categories of toric varieties", *Michigan Math. J.* **54**:3 (2006), 517–535.

[Keller 1999] B. Keller, "On the cyclic homology of exact categories", *J. Pure Appl. Algebra* **136**:1 (1999), 1–56.

[Keller 2006] B. Keller, "On differential graded categories", pp. 151–190 in *International Congress of Mathematicians, II* (Madrid, Spain, 2006), edited by M. Sanz-Solé et al., Eur. Math. Soc., Zürich, 2006.

[Kimura 2005] S.-I. Kimura, "Chow groups are finite dimensional, in some sense", *Math. Ann.* **331**:1 (2005), 173–201.

[Kontsevich 2005] M. Kontsevich, "Non-commutative motives", video of talk at the Institute for Advanced Study, 2005, https://video.ias.edu/Geometry-and-Arithmetic-Kontsevich.

[Kontsevich 2009] M. Kontsevich, "Notes on motives in finite characteristic", pp. 213–247 in *Algebra, arithmetic, and geometry: In honor of Yu. I. Manin*, vol. II, edited by Y. Tschinkel and Y. G. Zarhin, Progr. Math. **270**, Birkhäuser, Boston, 2009.

[Kontsevich 2010] M. Kontsevich, "Mixed noncommutative motives", Lecture notes, University of Miami, 2010. Talk at the Workshop on Homological Mirror Symmetry.

[Kuznetsov and Polishchuk 2011] A. Kuznetsov and A. Polishchuk, "Exceptional collections on isotropic grassmannians", preprint, 2011. arXiv 1110.5607

[Lunts and Orlov 2010] V. A. Lunts and D. O. Orlov, "Uniqueness of enhancement for triangulated categories", *J. Amer. Math. Soc.* **23**:3 (2010), 853–908.

[Manin 1968] J. I. Manin, "Correspondences, motifs and monoidal transformations", *Mat. Sb. (N.S.)* **77**:119 (1968), 475–507. In Russian; translated in *Math. USSR, Sb.* **6**:1968 (1969), 439–470.

[Manin and Smirnov 2013] Y. I. Manin and M. N. Smirnov, "On the derived category of $\overline{M}_{0,n}$", *Izv. Ross. Akad. Nauk Ser. Mat.* **77**:3 (2013), 93–108. In Russian; translated in *Izv. Math.* **77**:3 (2013), 525–540.

[Marcolli and Tabuada 2011] M. Marcolli and G. Tabuada, "Noncommutative numerical motives, Tannakian structures, and motivic Galois groups", preprint, 2011. To appear in *J. Euro. Math. Soc.* arXiv 1110.2438

[Marcolli and Tabuada 2012] M. Marcolli and G. Tabuada, "Kontsevich's noncommutative numerical motives", *Compos. Math.* **148**:6 (2012), 1811–1820.

[Marcolli and Tabuada 2014a] M. Marcolli and G. Tabuada, "Jacobians of noncommutative motives", *Mosc. Math. J.* **14**:3 (2014), 577–594.

[Marcolli and Tabuada 2014b] M. Marcolli and G. Tabuada, "Noncommutative Artin motives", *Selecta Math. (N.S.)* **20**:1 (2014), 315–358.

[Marcolli and Tabuada 2014c] M. Marcolli and G. Tabuada, "Noncommutative motives, numerical equivalence, and semi-simplicity", *Amer. J. Math.* **136**:1 (2014), 59–75.

[Marcolli and Tabuada 2014d] M. Marcolli and G. Tabuada, "Unconditional motivic Galois groups and Voevodsky's nilpotence conjecture in the noncommutative world", preprint, 2014. arXiv 1112.5422v1

[Marcolli and Tabuada 2015] M. Marcolli and G. Tabuada, "From exceptional collections to motivic decompositions via noncommutative motives", *J. Reine Angew. Math* **701** (2015), 153–167.

[Milne 2007] J. S. Milne, "Quotients of Tannakian categories", *Theory Appl. Categ.* **18**:21 (2007), 654–664.

[O'Sullivan 2005] P. O'Sullivan, "The structure of certain rigid tensor categories", *C. R. Math. Acad. Sci. Paris* **340**:8 (2005), 557–562.

[Paranjape 1994] K. H. Paranjape, "Cohomological and cycle-theoretic connectivity", *Ann. of Math.* (2) **139**:3 (1994), 641–660.

[Quillen 1973] D. Quillen, "Higher algebraic K-theory, I", pp. 85–147 in *Algebraic K-theory, I: Higher K-theories* (Seattle, WA, 1972), edited by H. Bass, Lecture Notes in Math. **341**, Springer, Berlin, 1973.

[Rost 1990] M. Rost, "Some new results on the Chow group of quadrics", unpublished manuscript, 1990, http://www.math.uni-bielefeld.de/~rost/chowqudr.html.

[Schlichting 2006] M. Schlichting, "Negative K-theory of derived categories", *Math. Z.* **253**:1 (2006), 97–134.

[Tabuada 2005] G. Tabuada, "Invariants additifs de DG-catégories", *Int. Math. Res. Not.* **2005**:53 (2005), 3309–3339.

[Tabuada 2010a] G. Tabuada, "Generalized spectral categories, topological Hochschild homology and trace maps", *Algebr. Geom. Topol.* **10**:1 (2010), 137–213.

[Tabuada 2010b] G. Tabuada, "On Drinfeld's dg quotient", *J. Algebra* **323**:5 (2010), 1226–1240.

[Tabuada 2012] G. Tabuada, "The fundamental theorem via derived Morita invariance, localization, and $\mathbb{A}^1$-homotopy invariance", *J. K-Theory* **9**:3 (2012), 407–420.

[Tabuada 2013] G. Tabuada, "Chow motives versus noncommutative motives", *J. Noncommut. Geom.* **7**:3 (2013), 767–786.

[Tabuada 2014] G. Tabuada, "Additive invariants of toric and twisted projective homogeneous varieties via noncommutative motives", *J. Algebra* **417** (2014), 15–38.

[Tabuada and Van den Bergh 2015] G. Tabuada and M. Van den Bergh, "Noncommutative motives of Azumaya algebras", *J. Inst. Math. Jussieu* **14**:2 (2015), 379–403.

[Thomason and Trobaugh 1990] R. W. Thomason and T. Trobaugh, "Higher algebraic K-theory of schemes and of derived categories", pp. 247–435 in *The Grothendieck Festschrift*, vol. III, edited by P. Cartier et al., Progr. Math. **88**, Birkhäuser, Boston, 1990.

[Ueda 2005] K. Ueda, "Stokes matrices for the quantum cohomologies of Grassmannians", *Int. Math. Res. Not.* **2005**:34 (2005), 2075–2086.

[Voevodsky 1995] V. Voevodsky, "A nilpotence theorem for cycles algebraically equivalent to zero", *Int. Math. Res. Not.* **1995**:4 (1995), 187–198.

[Waldhausen 1985] F. Waldhausen, "Algebraic K-theory of spaces", pp. 318–419 in *Algebraic and geometric topology* (New Brunswick, NJ, 1983), edited by A. Ranicki et al., Lecture Notes in Math. **1126**, Springer, Berlin, 1985.

[Weibel 1989] C. A. Weibel, "Homotopy algebraic K-theory", pp. 461–488 in *Algebraic K-theory and algebraic number theory* (Honolulu, HI, 1987), edited by M. R. Stein and R. K. Dennis, Contemp. Math. **83**, Amer. Math. Soc., Providence, RI, 1989.

matilde@caltech.edu
Mathematics Department, Caltech, 1200 E. California Blvd, Mail Code 253-37, Pasadena, CA 91125, United States

tabuada@math.mit.edu
Department of Mathematics, Massachusetts Institute of Technology, 77 Massachusetts Avenue, Cambridge, MA 02139, United States

Departamento de Matemática, Faculdade de Ciências e Tecnologia, Universidade Nova de Lisboa, Quinta da Torre, 2829-516 Caparica, Portugal

Centro de Matemática e Aplicações, Faculdade de Ciências e Tecnologia, Universidade Nova de Lisboa, Quinta da Torre, 2829-516 Caparica, Portugal

Commutative Algebra and Noncommutative Algebraic Geometry, I
MSRI Publications
Volume **67**, 2015

Infinite graded free resolutions

JASON MCCULLOUGH AND IRENA PEEVA

This paper is a survey on infinite graded free resolutions. We discuss their numerical invariants: Betti numbers, regularity, slope (rate), and rationality of Poincaré series. We also cover resolutions over complete intersections, Golod rings and Koszul rings.

1. Introduction

This paper is an expanded version of three talks given by I. Peeva during the Introductory Workshop in Commutative Algebra at MSRI in August 2013. It is a survey on infinite graded free resolutions, and includes many open problems and conjectures.

The idea of associating a free resolution to a finitely generated module was introduced in two famous papers by Hilbert [1890; 1893]. He proved Hilbert's Syzygy Theorem (Theorem 4.9), which says that the minimal free resolution of every finitely generated graded module over a polynomial ring is finite. Since then, there has been a lot of progress on the structure and properties of finite free resolutions. Much less is known about the properties of infinite free resolutions. Such resolutions occur abundantly since most minimal free resolutions over a graded nonlinear quotient ring of a polynomial ring are infinite. The challenges in studying them come from:

- The structure of infinite minimal free resolutions can be quite intricate.
- The methods and techniques for studying finite free resolutions usually do not work for infinite free resolutions. As noted by Avramov [1992]: "there seems to be a need for a whole new arsenal of tools."
- Computing examples with computer algebra systems is usually useless since we can only compute the beginning of a resolution and this is nonindicative for the structure of the entire resolution.

Peeva is partially supported by NSF grant DMS-1100046 and by a Simons Fellowship. McCullough is partially supported by a Rider University Summer Research Fellowship for Summer 2013. Both authors were partially supported by NSF grant 0932078000 while in residence at MSRI.
MSC2010: 13D02.
Keywords: syzygies, free resolutions.

Most importantly, there is a need for new insights and guiding conjectures. Coming up with reasonable conjectures is a major challenge on its own.

Because of space limitation, we have covered topics and results selectively. We survey open problems and results on Betti numbers (Section 4), resolutions over complete intersections (Section 5), rationality of Poincaré series and Golod rings (Section 6), regularity (Section 7), Koszul rings (Section 8), and slope (Section 9).

Some lectures and expository papers on infinite free resolutions are given in [Avramov 1992; 1998; Conca et al. 2013; Dao 2013; Fröberg 1999; Gulliksen and Levin 1969; Polishchuk and Positselski 2005; Peeva 2011; 2007].

2. Notation

Throughout, we use the following notation: $S = k[x_1, \dots, x_n]$ stands for a polynomial ring over a field k. We consider a quotient ring $R = S/I$, where I is an ideal in S. Furthermore, M stands for a finitely generated R-module. All modules are finitely generated unless otherwise stated. For simplicity, in the examples we may use x, y, z, etc. instead of $x_1, \dots, x_n$.

3. Free resolutions

Definition 3.1. *A left complex* $\boldsymbol{G}$ *of finitely generated free modules over* R *is a sequence of homomorphisms of finitely generated free* R*-modules*

$$\boldsymbol{G}: \quad \cdots \longrightarrow G_i \xrightarrow{d_i} G_{i-1} \longrightarrow \cdots \longrightarrow G_2 \xrightarrow{d_2} G_1 \xrightarrow{d_1} G_0,$$

such that $d_{i-1}d_i = 0$ *for all* i*. The collection of maps* $d = \{d_i\}$ *is called the differential of* $\boldsymbol{G}$*. The complex is sometimes denoted* $(\boldsymbol{G}, d)$*. The* i*-th Betti number of* $\boldsymbol{G}$ *is the rank of the module* G_i*. The homology of* $\boldsymbol{G}$ *is* $\mathrm{H}_i(\boldsymbol{G}) = \mathrm{Ker}(d_i)/\mathrm{Im}(d_{i+1})$*. The complex is exact at* G_i*, or at step* i*, if* $\mathrm{H}_i(\boldsymbol{G}) = 0$*. We say that* $\boldsymbol{G}$ *is acyclic if* $\mathrm{H}_i(\boldsymbol{G}) = 0$ *for all* $i > 0$*. A free resolution of a finitely generated* R*-module* M *is an acyclic left complex of finitely generated free* R*-modules*

$$\boldsymbol{F}: \quad \cdots \longrightarrow F_i \xrightarrow{d_i} F_{i-1} \longrightarrow \cdots \longrightarrow F_1 \xrightarrow{d_1} F_0,$$

such that $M \cong F_0/\mathrm{Im}(d_1)$.

The idea to associate a resolution to a finitely generated R-module M was introduced in Hilbert's famous papers [1890; 1893]. The key insight is that *a free resolution is a description of the structure of M* since it has the form

$$\cdots \to F_2 \xrightarrow{\begin{pmatrix}\text{a generating}\\ \text{system of the}\\ \text{relations on the}\\ \text{relations in } d_1\end{pmatrix}} F_1 \xrightarrow{\begin{pmatrix}\text{a generating}\\ \text{system of the}\\ \text{relations on the}\\ \text{generators of } M\end{pmatrix}} F_0 \xrightarrow{\begin{pmatrix}\text{a system of}\\ \text{generators}\\ \text{of } M\end{pmatrix}} M \to 0.$$

Therefore, the properties of M can be studied by understanding the properties and structure of a free resolution.

From a modern point of view, building a resolution amounts to repeatedly solving systems of polynomial equations. This is illustrated in Example 3.4 and implemented in Construction 3.3. It is based on the following observation.

Observation 3.2. If we are given a homomorphism $R^p \xrightarrow{A} R^q$, where A is the matrix of the map with respect to fixed bases, then describing the module $\operatorname{Ker}(A)$ is equivalent to solving the system of R-linear equations

$$A \begin{pmatrix} Y_1 \\ \vdots \\ Y_p \end{pmatrix} = 0$$

over R, where $Y_1, \ldots, Y_p$ are variables that take values in R.

Construction 3.3. We will show that every finitely generated R-module M has a free resolution. By induction on homological degree we will define the differential so that its image is the kernel of the previous differential.

Step 0: Set $M_0 = M$. We choose generators $m_1, \ldots, m_r$ of M_0 and set $F_0 = R^r$. Let $f_1, \ldots, f_r$ be a basis of F_0, and define

$$\begin{aligned} d_0 :\ & F_0 \to M, \\ & f_j \mapsto m_j \quad \text{for } 1 \le j \le r. \end{aligned}$$

Step i: By induction, F_{i-1} and d_{i-1} are defined. Set $M_i = \operatorname{Ker}(d_{i-1})$. We choose generators $w_1, \ldots, w_s$ of M_i and set $F_i = R^s$. Let $g_1, \ldots, g_s$ be a basis of F_i and define

$$\begin{aligned} d_i :\ & F_i \to M_i \subset F_{i-1}, \\ & g_j \mapsto w_j \quad \text{for } 1 \le j \le s. \end{aligned}$$

By construction $\operatorname{Ker}(d_{i-1}) = \operatorname{Im}(d_i)$.

The process described above may never terminate; in this case, we have an infinite free resolution.

Example 3.4. Let $S = k[x, y, z]$ and $J = (x^2z, xyz, yz^6)$. We will construct a free resolution of S/J over S.

Step 0: Set $F_0 = S$ and let $d_0 : S \to S/J$.

Step 1: The elements x^2z, xyz, yz^6 are generators of $\mathrm{Ker}(d_0)$. Denote by f_1, f_2, f_3 a basis of $F_1 = S^3$. Defining d_1 by

$$f_1 \mapsto x^2z, \quad f_2 \mapsto xyz, \quad f_3 \mapsto yz^6,$$

we obtain the beginning of the resolution:

$$F_1 = S^3 \xrightarrow{\begin{pmatrix} x^2z & xyz & yz^6 \end{pmatrix}} F_0 = S \to S/J \to 0.$$

Step 2: Next, we need to find generators of $\mathrm{Ker}(d_1)$. Equivalently, we have to solve the equation $d_1(X_1f_1 + X_2f_2 + X_3f_3) = 0$, or

$$X_1x^2z + X_2xyz + X_3yz^6 = 0,$$

where X_1, X_2, X_3 are indeterminates that take values in R. Computing with the computer algebra system Macaulay2, we find that $-yf_1 + xf_2$ and $-z^5f_2 + xf_3$ are homogeneous generators of $\mathrm{Ker}(d_1)$. Denote by g_1, g_2 a basis of $F_2 = S^2$. Defining d_2 by

$$g_1 \mapsto -yf_1 + xf_2, \quad g_2 \mapsto -z^5f_2 + xf_3,$$

we obtain the next step in the resolution:

$$F_2 = S^2 \xrightarrow{\begin{pmatrix} -y & 0 \\ x & -z^5 \\ 0 & x \end{pmatrix}} F_1 = S^3 \xrightarrow{\begin{pmatrix} x^2z & xyz & yz^6 \end{pmatrix}} F_0 = S.$$

Step 3: Next, we need to find generators of $\mathrm{Ker}(d_2)$. Equivalently, we have to solve $d(Y_1g_1 + Y_2g_2) = 0$; so we get the system of equations

$$\begin{aligned} -Y_1y &= 0, \\ Y_1x - Y_2z^5 &= 0, \\ Y_2x &= 0, \end{aligned}$$

where Y_1, Y_2 are indeterminates that take values in R. Clearly, the only solution is $Y_1 = Y_2 = 0$, so $\mathrm{Ker}(d_2) = 0$.

Thus, we obtain the free resolution

$$0 \to S^2 \xrightarrow{\begin{pmatrix} -y & 0 \\ x & -z^5 \\ 0 & x \end{pmatrix}} S^3 \xrightarrow{\begin{pmatrix} x^2z & xyz & yz^6 \end{pmatrix}} S.$$

The next theorem shows that any two free resolutions of a finitely generated R-module are homotopy equivalent.

Theorem 3.5 (see [Peeva 2011, Theorem 6.8]). *For every two free resolutions $\boldsymbol{F}$ and $\boldsymbol{F}'$ of a finitely generated R-module M there exist homomorphisms of complexes $\varphi : \boldsymbol{F} \to \boldsymbol{F}'$ and $\psi : \boldsymbol{F}' \to \boldsymbol{F}$ inducing* $\mathrm{id} : M \to M$*, such that $\varphi\psi$ is homotopic to* $\mathrm{id}_{\boldsymbol{F}'}$ *and $\psi\varphi$ is homotopic to* $\mathrm{id}_{\boldsymbol{F}}$.

4. Minimality and Betti numbers

In the rest of the paper, we will use the *standard grading* of the polynomial ring $S = k[x_1, \ldots, x_n]$. Set $\deg(x_i) = 1$ for each i. A monomial $x_1^{a_1} \ldots x_n^{a_n}$ has *degree* $a_1 + \cdots + a_n$. For $i \in \mathbb{N}$, we denote by S_i the k-vector space spanned by all monomials of degree i. In particular, $S_0 = k$. A polynomial f is called *homogeneous* if $f \in S_i$ for some i, and in this case we say that f has *degree* i or that f is a *form* of degree i and write $\deg(f) = i$. By convention, 0 is a homogeneous element with arbitrary degree. Every polynomial $f \in S$ can be written uniquely as a finite sum of nonzero homogeneous elements, called the *homogeneous components of* f. This provides a direct sum decomposition $S = \bigoplus_{i \in \mathbb{N}} S_i$ of S as a k-vector space with $S_i S_j \subseteq S_{i+j}$. A proper ideal J in S is called *graded* if it has a system of homogeneous generators, or equivalently, $J = \bigoplus_{i \in \mathbb{N}} (S_i \cap J)$; the k-vector spaces $J_i = S_i \cap J$ are called the *graded components* of the ideal J.

If I is a graded ideal, then the quotient ring $R = S/I$ inherits the grading from S, so $R_i \cong S_i / I_i$ for all i. Furthermore, an R-module M is called *graded* if it has a direct sum decomposition $M = \bigoplus_{i \in \mathbb{Z}} M_i$ as a k-vector space and $R_i M_j \subseteq M_{i+j}$ for all $i, j \in \mathbb{Z}$. The k-vector spaces M_i are called the *graded components* of M. An element $m \in M$ is called *homogeneous* if $m \in M_i$ for some i, and in this case we say that m has degree i and write $\deg(m) = i$. A homomorphism between graded R-modules $\varphi : M \to N$ is called *graded of degree* 0 if $\varphi(M_i) \subseteq N_i$ for all $i \in \mathbb{Z}$. It is easy to see that the kernel, cokernel, and image of a graded homomorphism are graded modules.

We use the following convention: For $p \in \mathbb{Z}$, the module M *shifted p degrees* is denoted by $M(-p)$ and is the graded R-module such that $M(-p)_i = M_{i-p}$ for all i. In particular, the generator $1 \in R(-p)$ has degree p since $R(-p)_p = R_0$.

Notation 4.1. In the rest of the paper, we assume that S is standard graded, I is a graded ideal in S, the quotient ring $R = S/I$ is graded, and M is a finitely generated graded R-module.

A complex $\boldsymbol{F}$ of finitely generated graded free modules

$$\boldsymbol{F} : \cdots \longrightarrow F_{i+1} \xrightarrow{d_{i+1}} F_i \xrightarrow{d_i} F_{i-1} \longrightarrow \cdots$$

is called *graded* if the modules F_i are graded and each d_i is a graded homomorphism of degree 0. In this case, the module $\boldsymbol{F}$ is actually bigraded since we have a homological degree and an internal degree, so we may write

$$F_i = \bigoplus_{j \in \mathbb{Z}} F_{i,j} \quad \text{for each } i.$$

An element in $F_{i,j}$ is said to have *homological degree* i and *internal degree* j.

Example 4.2. The graded version of the resolution of $S/(x^2z, xyz, yz^6)$ in Example 3.4 is

$$0 \longrightarrow S(-4) \oplus S(-8) \xrightarrow{\begin{pmatrix} -y & 0 \\ x & -z^5 \\ 0 & x \end{pmatrix}} S(-3)^2 \oplus S(-7) \xrightarrow{\begin{pmatrix} x^2z & xyz & yz^6 \end{pmatrix}} S.$$

A graded free resolution can be constructed following Construction 3.3 by choosing a homogeneous set of generators of the kernel of the differential at each step.

The graded component in a fixed internal degree j of a graded complex is a subcomplex which consists of k-vector spaces; see [Peeva 2011, 3.7]. Thus, the grading yields the following useful criterion for exactness: a graded complex is exact if and only if each of its graded components is an exact sequence of k-vector spaces.

Another important advantage of having a grading is that Nakayama's Lemma holds, leading to the foundational Theorem 4.4. The proof of Nakayama's Lemma for local rings is longer (see [Matsumura 1989, Section 2]); for graded rings, the lemma follows immediately from the observation that a finitely generated R-module has a minimal generator of minimal degree, and we include the short proof.

Nakayama's Lemma 4.3. *If J is a proper graded ideal in R and M is a finitely generated graded R-module such that $M = JM$, then $M = 0$.*

Proof. Suppose that $M \neq 0$. We choose a finite minimal system of homogeneous generators of M. Let m be an element of minimal degree in that system. It follows that $M_j = 0$ for $j < \deg(m)$. Since J is a proper ideal, we conclude that every homogeneous element in JM has degree strictly greater than $\deg(m)$. This contradicts to $m \in M = JM$. □

Theorem 4.4 (see [Peeva 2011, Theorem 2.12]). *Let M be a finitely generated graded R-module. Consider the graded k-vector space $\overline{M} = M/(x_1, \dots, x_n)M$. Homogeneous elements $m_1, \dots, m_r \in M$ form a minimal system of homogeneous* *generators of M if and only if their images in $\overline{M}$ form a basis. Every minimal system of homogeneous generators of M has* $\dim_k(\overline{M})$ *elements.*

In particular, Theorem 4.4 shows that every minimal system of generators of M has the same number of elements.

Definition 4.5. *A graded free resolution of a finitely generated graded R-module M is minimal if*

$$d_{i+1}(F_{i+1}) \subseteq (x_1, \ldots, x_n)F_i \quad \text{for all } i \geq 0.$$

This means that no invertible elements (nonzero constants) appear in the differential matrices.

The word "minimal" refers to the properties in the next two results. On the one hand, Theorem 4.6 shows that minimality means that at each step in Construction 3.3 we make an optimal choice, that is, we choose a minimal system of generators of the kernel in order to construct the next differential. On the other hand, Theorem 4.7 shows that minimality means that we have a smallest resolution which lies (as a direct summand) inside any other resolution of the module.

Theorem 4.6 (see [Peeva 2011, Theorem 3.4]). *The graded free resolution constructed in Construction* 3.3 *is minimal if and only if at each step we choose a minimal homogeneous system of generators of the kernel of the differential. In particular, every finitely generated graded R-module has a minimal graded free resolution.*

Theorem 4.7 (see [Peeva 2011, Theorem 3.5]). *Let M be a finitely generated graded R-module, and $\boldsymbol{F}$ be a minimal graded free resolution of M. If $\boldsymbol{G}$ is any graded free resolution of M, we have a direct sum of complexes $\boldsymbol{G} \cong \boldsymbol{F} \oplus \boldsymbol{P}$ for some complex $\boldsymbol{P}$, which is a direct sum of short trivial complexes*

$$0 \longrightarrow R(-p) \xrightarrow{\ 1\ } R(-p) \longrightarrow 0$$

possibly placed in different homological degrees.

The minimal graded free resolution of M is unique up to an isomorphism and has the form

$$\cdots \to F_2 \xrightarrow{\begin{pmatrix}\text{a } \textit{minimal}\\ \text{generating}\\ \text{system of the}\\ \text{relations on the}\\ \text{relations in } d_1\end{pmatrix}} F_1 \xrightarrow{\begin{pmatrix}\text{a } \textit{minimal}\\ \text{generating}\\ \text{system of the}\\ \text{relations on the}\\ \text{generators of } M\end{pmatrix}} F_0 \xrightarrow{\begin{pmatrix}\text{a } \textit{minimal}\\ \text{system of}\\ \text{generators}\\ \text{of } M\end{pmatrix}} M \to 0.$$

The properties of that resolution are closely related to the properties of M. A core area in commutative algebra is devoted to describing the properties of minimal free resolutions and relating them to the structure of the resolved modules. This area has many relations with and applications in other mathematical fields, especially algebraic geometry.

Free (or projective) resolutions exist over many rings (we can also consider noncommutative rings). However, the concept of a *minimal* free resolution needs in particular that each minimal system of generators of the module has the same number of elements, and that property follows from Nakayama's Lemma 4.3. For this reason, the theory of minimal free resolutions is developed in the local and in the graded cases where Nakayama's Lemma holds. This paper is focused on the graded case.

Definition 4.8. *Let $(\mathbf{F}, d)$ be a minimal graded free resolution of a finitely generated graded R-module M. Set $\mathrm{Syz}_0^R(M) = M$. For $i \geq 1$ the submodule*

$$\mathrm{Im}(d_i) = \mathrm{Ker}(d_{i-1}) \cong \mathrm{Coker}(d_{i+1})$$

of F_{i-1} is called the i-th syzygy module of M and is denoted $\mathrm{Syz}_i^R(M)$. Its elements are called i-th syzygies. Note that if $f_1, \dots, f_p$ is a basis of F_i, then the elements $d_i(f_1), \dots, d_i(f_p)$ form a minimal system of homogeneous generators of $\mathrm{Syz}_i^R(M)$.

Theorem 4.7 shows that the minimal graded free resolution is the smallest graded free resolution in the sense that the ranks of its free modules are less than or equal to the ranks of the corresponding free modules in an arbitrary graded free resolution of the resolved module. The i-th Betti number of M over R is

$$b_i^R(M) = \mathrm{rank}(F_i).$$

Observe that the differentials in the complexes $\mathbf{F} \otimes_R k$ and $\mathrm{Hom}_R(\mathbf{F}, k)$ are zero, and therefore

$$b_i^R(M) = \dim_k\big(\mathrm{Tor}_i^R(M, k)\big) = \dim_k\big(\mathrm{Ext}_R^i(M, k)\big) \quad \textit{for every } i.$$

Often it is very difficult to obtain a description of the differential. In such cases, we try to get some information about the numerical invariants of the resolution — the Betti numbers.

The length of the minimal graded free resolution is measured by the projective dimension, defined by

$$\mathrm{pd}_R(M) = \max\{i \mid b_i^R(M) \neq 0\}.$$

Hilbert introduced free resolutions motivated by invariant theory and proved the following important result.

Hilbert's Syzygy Theorem 4.9 (see [Peeva 2011, Theorem 15.2]). *The minimal graded free resolution of a finitely generated graded S-module is finite and its length is at most n.*

A more precise version of this is the Auslander–Buchsbaum formula, which states that

$$\mathrm{pd}_S(M) = n - \mathrm{depth}(M),$$

for any finitely generated graded S-module M (see [Peeva 2011, 15.3]).

It turns out that the main source of graded *finite* free resolutions are polynomial rings:

Auslander–Buchsbaum–Serre Regularity Criterion 4.10 (see [Eisenbud 1995, Theorem 19.12]). *The following are equivalent*:

(1) *Every finitely generated graded R-module has finite projective dimension.*

(2) $\mathrm{pd}_R(k) < \infty$.

(3) $R = S/I$ *is a polynomial ring, that is, I is generated by linear forms.*

This is a homological criterion for a ring to be regular. In the introduction to his book *Commutative Ring Theory*, Matsumura [1989] states that he considers Auslander–Buchsbaum–Serre's Criterion to be one of the top three results in commutative algebra.

Infinite minimal free resolutions appear abundantly over quotient rings. The simplest example of a minimal infinite free resolution is perhaps resolving $R/(x)$ over the quotient ring $R = k[x]/(x^2)$, which yields

$$\cdots \longrightarrow R \xrightarrow{x} R \xrightarrow{x} R \xrightarrow{x} R.$$

A homological criterion for complete intersections was obtained by Gulliksen. We say that a Betti sequence $\{b_i^R(M)\}$ is *polynomially bounded* if there exists a polynomial $g \in \mathbb{N}[x]$ such that $b_i^R(M) \le g(i)$ for $i \gg 0$.

Gulliksen's CI Criterion 4.11 [Gulliksen 1971; 1974; 1980]. *The following are equivalent*:

(1) *The Betti numbers of k are polynomially bounded.*

(2) *The Betti numbers of every finitely generated graded R-module are polynomially bounded.*

(3) *R is a complete intersection.*

One might hope to get similar homological criteria for Gorenstein rings and other interesting classes of rings. However, the type of growth of the Betti numbers of k cannot distinguish such rings: we will see in 6.12 that Serre's Inequality (6.11) implies that for every finitely generated graded R-module M, there exists a real number $\beta > 1$ such that $b_i^R(M) \le \beta^i$ for $i \ge 1$ (see [Avramov 1998, Corollary 4.1.5.]); thus, the Betti numbers grow at most exponentially. We say that $\{b_i^R(M)\}$ *grow exponentially* if there exists a real number $\alpha > 1$ such that $\alpha^i \le b_i^R(M)$ for $i \gg 0$. Avramov [1998] proved that the Betti numbers of k grow exponentially if R is not a complete intersection.

A general question on the Betti numbers is:

Open-Ended Problem 4.12. *How do the properties of the Betti sequence $\{b_i^R(M)\}$ relate to the structure of the minimal free resolution of M, the structure of M, and the structure of R?*

Auslander–Buchsbaum–Serre's Criterion 4.10 and Gulliksen's CI Criterion 4.11 are important results of this type.

The condition that an infinite minimal free resolution has bounded Betti numbers is very strong. Such resolutions do not occur over every quotient ring R, so one might ask which quotient rings admit such resolutions [Avramov 1992, Problem 4]. It is very interesting to explore what are the implications on the structure of the resolution.

Open-Ended Problem 4.13 [Eisenbud 1980]. *What causes bounded Betti numbers in an infinite minimal free resolution?*

Eisenbud [1980] conjectured that every such resolution is eventually periodic and its period is 2; we say that M is *periodic of period p* if $\mathrm{Syz}_p^R(M) \cong M$. The following counterexample was constructed:

Example 4.14 [Gasharov and Peeva 1990]. Let $0 \neq \alpha \in k$. Set

$$R = k[x_1, x_2, x_3, x_4]/(\alpha x_1 x_3 + x_2 x_3,\ x_1 x_4 + x_2 x_4,\ x_3 x_4,\ x_1^2,\ x_2^2,\ x_3^2,\ x_4^2),$$

and consider

$$\boldsymbol{T} : \cdots \to R(-3)^2 \xrightarrow{d_3} R(-2)^2 \xrightarrow{d_2} R(-1)^2 \xrightarrow{d_1} R^2 \to 0,$$

where

$$d_i = \begin{pmatrix} x_1 & \alpha^i x_3 + x_4 \\ 0 & x_2 \end{pmatrix}.$$

The complex $\boldsymbol{T}$ is acyclic and minimally resolves the module $M := \mathrm{Coker}(d_1)$. Moreover, M is shown to have period equal to the order of α in k^*, thus yielding resolutions with arbitrary period and with no period. Other counterexamples over Gorenstein rings are given in [Gasharov and Peeva 1990] as well. All of the examples have constant Betti numbers, supporting the following question which remains a mystery.

Problem 4.15 [Ramras 1980]. *Is it true that if the Betti numbers of a finitely generated graded R-module are bounded, then they are eventually constant?*

The following subproblem could be explored with the aid of computer computations.

Problem 4.16. *Does there exist a periodic module with nonconstant Betti numbers?*

Eisenbud [1980] proved that if the period is $p = 2$, then the Betti numbers are constant. The case $p = 3$ is open.

Problem 4.15 was extended by Avramov as follows:

Problem 4.17 [Avramov 1992, Problem 9]. *Is it true that the Betti numbers of every finitely generated graded R-module are eventually nondecreasing?*

In particular, Ramras [1980] asked whether $\{b_i^R(M)\}$ being unbounded implies that $\lim_{i\to\infty} b_i^R(M) = \infty$.

A positive answer to Problem 4.17 is known in some special cases: for example, for $M = k$ by a result of Gulliksen [1980], over complete intersections by a result of Avramov, Gasharov and Peeva [1997], when R is Golod by a result of Lescot [1990], for $R = S/I$ such that the integral closure of I is strictly smaller than the integral closure of $(I : (x_1, \ldots, x_n))$ by a result of Choi [1990], and for rings with $(x_1, \ldots, x_n)^3 = 0$ by a result of Lescot [1985].

For an infinite sequence of nonzero Betti numbers, one can ask how they change and how they behave asymptotically. Several such questions have been raised in [Avramov 1992; 1998].

5. Complete intersections

Throughout this section we assume that R is a graded complete intersection, that is, $R = S/(f_1, \ldots, f_c)$ and $f_1, \ldots, f_c$ is a homogeneous regular sequence.

The numerical properties of minimal free resolutions over complete intersections are well-understood:

Theorem 5.1 [Gulliksen 1974; Avramov 1989; Avramov, Gasharov and Peeva 1997]. *Let M be a finitely generated graded R-module. The Poincarè series $\mathrm{P}_M^R(t) = \sum_{i\geq 0} b_i^R(M)t^i$ is rational and has the form*

$$\mathrm{P}_M^R(t) = \frac{g(t)}{(1-t^2)^c},$$

for some polynomial $g(t) \in \mathbb{Z}[t]$. The Betti numbers $\{b_i^R(M)\}$ are eventually nondecreasing and are eventually given by two polynomials (one for the odd Betti numbers and one for the even Betti numbers) of the same degree and the same leading coefficient.

Example 5.2. This is an example where the Betti numbers cannot be given by a single polynomial. Consider the complete intersection $R = k[x, y]/(x^3, y^3)$ and the module $M = R/(x, y)^2$. By [Avramov 1994, Section 2.1] we get

$$b_i^R(M) = \tfrac{3}{2}i + 1 \quad \text{for even } i \geq 0,$$
$$b_i^R(M) = \tfrac{3}{2}i + \tfrac{3}{2} \quad \text{for odd } i \geq 1.$$

The minimal free resolution of k has an elegant structure discovered by Tate. His construction provides the minimal free resolution of k over any R, but if R is not a complete intersection, then the construction is an algorithm building the resolution inductively on homological degree.

Tate's Resolution 5.3 [Tate 1957]. *We will describe Tate's resolution of k over a complete intersection. Write the homogeneous regular sequence*

$$f_j = a_{j1}x_1 + \cdots + a_{jn}x_n, \quad 1 \leq j \leq c,$$

with coefficients $a_{ij} \in S$. Let $\boldsymbol{F}' = R \otimes_S \mathbf{K}$, where $\mathbf{K}$ is the Koszul complex resolving k over S. We may think of $\mathbf{K}$ as being the exterior algebra on variables $e_1, \ldots, e_n$, such that the differential maps e_i to x_i. In F_1' we have cycles

$$a_{j1}e_1 + \cdots + a_{jn}e_n, \quad 1 \leq j \leq c.$$

For simplicity, we assume $\operatorname{char}(k) = 0$. *Set $\boldsymbol{F} = \boldsymbol{F}'[y_1, \ldots, y_c]$ and*

$$d(y_j) = a_{j1}e_1 + \cdots + a_{jn}e_n.$$

The minimal free resolution of k is

$$\boldsymbol{F} = R\langle e_1, \ldots, e_n\rangle[y_1, \ldots, y_c] = (R \otimes_S \mathbf{K})[y_1, \ldots, y_c],$$

with differential defined by

$$d(e_{i_1} \cdots e_{i_j} y_1^{s_1} \cdots y_c^{s_c}) = d(e_{i_1} \cdots e_{i_j}) y_1^{s_1} \cdots y_c^{s_c} \\ + (-1)^j \sum_{\substack{1 \leq p \leq c \\ s_p \geq 1}} d(y_p) e_{i_1} \cdots e_{i_j} y_1^{s_1} \cdots y_p^{s_p - 1} \cdots y_c^{s_c}.$$

where $d(e_{i_1} \cdots e_{i_j})$ is the Koszul differential. In particular, the Poincarè series of k over the complete intersection is

$$\mathrm{P}_k^R(t) = \frac{(1+t)^n}{(1-t^2)^c}.$$

If $\operatorname{char}(k) \neq 0$, *then in the construction above instead of the polynomial algebra $R[y_1, \ldots, y_c]$ we have to take a divided power algebra.*

The study of infinite minimal free resolutions over complete intersections is focused on the asymptotic properties of the resolutions because for every $p > 0$, there exist examples where the first p steps do not agree with the asymptotic behavior:

Example 5.4 [Eisenbud 1980]. Consider the complete intersection

$$R = S/(x_1^2, \dots, x_n^2).$$

By Tate's Resolution 5.3, we have Tate's minimal free resolution $\boldsymbol{F}$ of k. It shows that the Betti numbers of k are strictly increasing. The dual $\boldsymbol{F}^* = \mathrm{Hom}(\mathbf{F}, R)$ is a minimal injective resolution of $\mathrm{Hom}(\mathbf{F}, R) \cong \mathrm{socle}(R) = (x_1 \cdots x_n) \cong k$. Gluing $\boldsymbol{F}$ and $\mathbf{F}^*$ we get a doubly infinite exact sequence of free R-modules

$$\cdots \longrightarrow F_2 \longrightarrow F_1 \longrightarrow R \xrightarrow{x_1 \cdots x_n} R \longrightarrow F_1^* \longrightarrow F_2^* \longrightarrow \cdots.$$

Thus, for any p,

$$\cdots \longrightarrow F_2 \longrightarrow F_1 \longrightarrow R \longrightarrow R \longrightarrow F_1^* \longrightarrow F_2^* \longrightarrow \cdots \longrightarrow F_p^*$$

is a minimal free resolution over R in which the first p Betti numbers are strictly decreasing, but after the $(p+1)$-st step the Betti numbers are strictly increasing.

Further examples exhibiting complex behavior of the Betti numbers at the beginning of a minimal free resolution are given in [Avramov, Gasharov and Peeva 1997]. Even though the beginning of a minimal free resolution can be unstructured and very complicated, the known results show that stable patterns occur eventually. Thus, instead of studying the entire resolution $\boldsymbol{F}$ we consider the *truncation*

$$\boldsymbol{F}_{\geq p}: \quad \cdots \longrightarrow F_i \xrightarrow{d_i} F_{i-1} \longrightarrow \cdots \longrightarrow F_{p+1} \xrightarrow{d_{p+1}} F_p,$$

for sufficiently large p. From that point of view, Hilbert's Syzygy Theorem (Theorem 4.9) says that over S every minimal free resolution is eventually the zero-complex.

Eisenbud [1980] described the asymptotic structure of minimal free resolutions over a hypersurface. He introduced the concept of a *matrix factorization* for a homogeneous $f \in S$: it is a pair of square matrices (d, h) with entries in S such that

$$dh = hd = f\,\mathrm{id}.$$

The module $\mathrm{Coker}(d)$ is called the *matrix factorization module* of (d, h).

Theorem 5.5 [Eisenbud 1980]. *With the notation above, the minimal $S/(f)$-resolution of the matrix factorization module is*

$$\cdots \to R^a \xrightarrow{d} R^a \xrightarrow{h} R^a \xrightarrow{d} R^a \xrightarrow{h} R^a \xrightarrow{d} R^a,$$

where $R = S/(f)$ and a is the size of the square matrices d, h.

After ignoring finitely many steps at the beginning, every minimal free resolution over the hypersurface ring $R = S/(f)$ is of this type. More precisely, if $\boldsymbol{F}$ is a minimal graded free resolution, then for every $p \gg 0$ the truncation $\boldsymbol{F}_{\geq p}$ minimally resolves some matrix factorization module and so it is described by a matrix factorization for the element f.

Matrix factorizations have amazing applications in many areas. Kapustin and Li [2003] started the use of matrix factorizations in string theory following an idea of Kontsevich; see [Aspinwall 2013] for a survey. A major discovery was made by Orlov [2004], who showed that matrix factorizations can be used to study Kontsevich's homological mirror symmetry by giving a new description of singularity categories. Matrix factorizations also have applications in the study of Cohen–Macaulay modules and singularity theory, cluster algebras and cluster tilting, Hodge theory, Khovanov–Rozansky homology, moduli of curves, quiver and group representations, and other topics.

A new conjecture on the size of matrix factorizations is recently introduced in [Eisenbud and Peeva $\geq$ 2015b]. Part of the motivation is that it implies a version of the Buchsbaum–Eisenbud–Horroks conjecture and also the Betti Degree Conjecture 5.9.

Minimal free resolutions of high syzygies over a codimension two complete intersection $S/(f_1, f_2)$ were constructed by Avramov and Buchweitz [2000a, 5.5] as quotient complexes. Eisenbud and Peeva [$\geq$ 2015a] provide a construction without using a quotient and give an explicit formula for the differential.

Recently, Eisenbud and Peeva [2015, Definition 1.1] introduced the concept of matrix factorization (d, h) for a regular sequence $f_1, \dots, f_c$ of any length. They constructed the minimal free resolutions of the matrix factorization module $\text{Coker}(R \otimes_S d)$ over S and over the complete intersection $R := S/(f_1, \dots, f_c)$. The infinite minimal free resolution over R is more complicated than the one over a hypersurface ring, but it is still nicely structured and exhibits two patterns — one pattern for odd homological degrees and another pattern for even homological degrees. They proved that asymptotically, every minimal free resolution over R is of this type. More precisely:

Theorem 5.6 [Eisenbud and Peeva 2015]. *If $\boldsymbol{F}$ is a minimal free resolution over the graded complete intersection $R = S/(f_1, \dots, f_c)$, then for every $p \gg 0$ the truncation $\boldsymbol{F}_{\geq p}$ resolves a matrix factorization module and so it is described by a matrix factorization.*

This structure explains and reproves the numerical results above.

One of the main tools in the study of free resolutions over a complete intersection are the CI operators. Let $(\boldsymbol{V}, \partial)$ be a complex of free modules over R.

Consider a lifting $\widetilde{V}$ of V to S, that is, a sequence of free modules $\widetilde{V}_i$ and maps $\tilde{\partial}_{i+1} : \widetilde{V}_{i+1} \to \widetilde{V}_i$ such that $\partial = R \otimes_S \tilde{\partial}$. Since $\partial^2 = 0$ we can choose maps $\tilde{t}_j : \widetilde{V}_{i+1} \to \widetilde{V}_{i-1}$, where $1 \leq j \leq c$, such that

$$\tilde{\partial}^2 = \sum_{j=1}^{c} f_j \tilde{t}_j.$$

The *CI operators*, sometimes called Eisenbud operators, are

$$t_j := R \otimes_S \tilde{t}_j.$$

This construction was introduced by Eisenbud in [1980]. Different constructions of the CI operators are discussed in [Avramov and Sun 1998]. Since

$$\sum_{j=1}^{c} f_j \tilde{t}_j \tilde{\partial} = \tilde{\partial}^3 = \sum_{j=1}^{c} f_j \tilde{\partial} \tilde{t}_j,$$

and the f_i form a regular sequence, it follows that each t_j commutes with the differential ∂, and thus each t_j defines a map of complexes $V[-2] \to V$; see [Eisenbud 1980, 1.1]. It was shown by Eisenbud [1980, 1.2 and 1.5] that the operators t_j are, up to homotopy, independent of the choice of liftings, and also that they commute up to homotopy.

Conjecture 5.7 [Eisenbud 1980]. *Let M be a finitely generated graded R-module, and let F be its graded minimal R-free resolution. There exists a choice of CI operators on a sufficiently high truncation $F_{\geq p}$ that commute.*

The original conjecture was for the resolution F (not for a truncation), and a counterexample to that was provided in [Avramov, Gasharov and Peeva 1997].

If V is an R-free resolution of a finitely generated graded R-module M, then the CI operators t_j induce well-defined, commutative maps χ_j on $\mathrm{Ext}_R(M, k)$ and thus make $\mathrm{Ext}_R(M, k)$ into a module over the polynomial ring $\mathcal{R} := k[\chi_1, \ldots, \chi_c]$, where the variables χ_j have degree 2. The χ_j are also called CI operators. By [Eisenbud 1980, Proposition 1.2], the action of χ_j can be defined using any CI operators on any R-free resolution of M. Since the χ_j have degree 2, we may split the Ext module into even degree and odd degree parts:

$$\mathrm{Ext}_R(N, k) = \mathrm{Ext}_R^{\mathrm{even}}(M, k) \oplus \mathrm{Ext}_R^{\mathrm{odd}}(M, k).$$

A version of the following result was first proved by Gulliksen [1974], who used a different construction of CI operators on Ext. A short proof using the above construction of CI operators is provided in [Eisenbud and Peeva 2015, Theorem 4.5].

Theorem 5.8 [Gulliksen 1974; Eisenbud 1980; Avramov and Sun 1998; Eisenbud and Peeva 2015, Theorem 4.5]. *Let M be a finitely generated graded R-module. The action of the CI operators makes* $\mathrm{Ext}_R(M, k)$ *into a finitely generated graded* $k[\chi_1, \dots, \chi_c]$*-module.*

The structure of the Ext module is studied in [Avramov and Buchweitz 2000b]. Avramov and Iyengar [2007] proved that the support variety (defined by the annihilator) of $\mathrm{Ext}_R(M, k)$ can be anything. Every sufficiently high truncation of $\mathrm{Ext}_R(M, k)$ is linearly presented, and its defining equations are described in terms of homotopies by Eisenbud and Peeva [$\geq$ 2015a].

By multiplicity $\mathrm{mult}(\mathrm{Ext}_R^*(N, k))$ we mean

$$\mathrm{mult}(\mathrm{Ext}_R^{\mathrm{even}}(M, k)) = \mathrm{mult}(\mathrm{Ext}_R^{\mathrm{odd}}(M, k)),$$

computed with respect to the standard grading of $k[\chi_1, \dots, \chi_c]$ with $\deg(\chi_i) = 1$ for each i. The *Betti degree* of M is the multiplicity $\mathrm{mult}(\mathrm{Ext}_R^*(M, k))$.

The Betti Degree Conjecture 5.9 [Avramov and Buchweitz 2000a, Conjecture 7.5]. *Let M be a graded finitely generated R-module. The Betti degree of M satisfies the inequality*

$$\mathrm{mult}(\mathrm{Ext}_R^*(M, k)) \geq 2^{\dim(\mathrm{Ext}_R(M,k))-1}.$$

We close this section by bringing up that the Eisenbud–Huneke Question 9.10 has a positive answer for k over R, and also for any finitely generated graded module if the forms in the regular sequence $f_1, \dots, f_c$ are of the same degree, but is open otherwise (including in the codimension two case):

Question 5.10 [Eisenbud and Huneke 2005, Question A]. *Let M be a finitely generated graded R-module and suppose that the forms in the regular sequence $f_1, \dots, f_c$ do not have the same degree. Does there exists a number u and bases of the free modules in the minimal graded R-free resolution $\mathbf{F}$ of M, such that for all $i \geq 0$ the entries in the matrix of the differential d_i have degrees $\leq u$?*

6. Rationality and Golod rings

We now focus on resolving the simplest possible module, namely k. The next construction provides a free resolution.

The Bar Resolution 6.1 (see [Mac Lane 1963]). *The bar resolution is an explicit construction which resolves k over any ring R, but usually provides a highly nonminimal free resolution. Let $\widetilde{R}$ be the cokernel of the canonical inclusion of vector spaces $k \to R$. For $i \geq 0$ set $B_i = R \otimes_k \widetilde{R} \otimes_k \cdots \otimes_k \widetilde{R}$, where we have i factors $\widetilde{R}$. The left factor R gives B_i a structure of a free R-module. Fix a basis $\mathfrak{R}$ of R over k such that $1 \in \mathfrak{R}$. Let $r \in R$ and $r_1, \dots, r_i \in \mathfrak{R}$. We* *denote by* $r[r_1 \mid \dots \mid r_i]$ *the element* $r \otimes_k r_1 \otimes_k \cdots \otimes_k r_i$ *in* B_i, *replacing* $\otimes_k$ *by a vertical bar; in particular,* $B_0 = R$ *with* $r[\ ] \in B_0$ *identified with* $r \in R$. *Note that* $r[r_1 \mid \dots \mid r_i] = 0$ *if some* $r_j = 1$ *or* $r = 0$. *Consider the sequence*

$$\boldsymbol{B}: \quad \cdots \to B_i \to B_{i-1} \to \cdots \to B_0 = R \to k \to 0,$$

with differential d defined by

$$d_i(r[r_1 \mid \dots \mid r_i]) = rr_1[r_2 \mid \dots \mid r_i] + \sum_{1 \le j \le i-1} (-1)^j r[r_1 \mid \dots \mid r_j r_{j+1} \mid \dots \mid r_i].$$

The differential is well-defined since if $r_j = 1$ *for some* $j > 1$, *then the terms*

$$(-1)^j r[r_1 \mid \dots \mid r_j r_{j+1} \mid \dots \mid r_i],$$
$$(-1)^{j-1} r[r_1 \mid \dots \mid r_{j-1} r_j \mid \dots \mid r_i]$$

cancel and all other terms vanish; similarly for $j = 1$. *Exactness may be proved by constructing an explicit homotopy; see* [Mac Lane 1963].

The minimal free resolution of k over S is the Koszul complex. It has an elegant and simple structure. In contrast, the situation over quotient rings is complicated; the structure of the *minimal* free resolution of k is known in some cases, but has remained mysterious in general. We start the discussion of the properties of that resolution by focusing on its Betti numbers. When we have infinitely many Betti numbers of a module M, we may study their properties via the *Poincaré series*

$$\mathrm{P}_M^R(t) = \sum_{i \ge 0} b_i^R(M) t^i.$$

The first natural question to consider is:

Open-Ended Problem 6.2. *Are the structure and invariants of an infinite minimal graded free resolution encoded in finite data?*

The main peak in this direction was:

The Serre–Kaplansky Problem 6.3. Is the Poincaré series of the residue field k over R rational? The question was originally asked for finitely generated commutative local Noetherian rings.

A Poincaré series $\mathrm{P}_M^R(t)$ is a rational function of the complex variable t if $\mathrm{P}_M^R(t) = f(t)/g(t)$ for two complex polynomials $f(t)$, $g(t)$ with $g(0) \neq 0$. By Fatou's Theorem, we have that the polynomials can be chosen with integer coefficients.

The Serre–Kaplansky Problem 6.3 was a central question in commutative algebra for many years. The high enthusiasm for research on the problem was motivated on the one hand by the expectation that the answer is positive (the problem was often considered a conjecture) and on the other hand by a result of Gulliksen [1972] who proved that a positive answer for all such rings implies the rationality of the Poincaré series of any finitely generated module. Additional interest was generated by a result of Anick and Gulliksen [1985], who reduced the rationality question to rings with the cube of the maximal ideal being zero.

Note that Yoneda multiplication makes $\mathrm{Ext}_R(k, k)$ a graded (by homological degree) k-algebra, and the Hilbert series of that algebra is the Poincaré series $\mathrm{P}_k^R(t)$. Problems of rationality of Poincaré and Hilbert series were stated by several mathematicians: by Serre and Kaplansky for local Noetherian rings, by Kostrikin and Shafarevich for the Hochschild homology of a finite-dimensional nilpotent k-algebra, by Govorov for finitely presented associative graded algebras, by Serre and Moore for simply connected complexes; see the survey by Babenko [1986].

An example of an irrational Poincaré series was first constructed by Anick [1980].

Example 6.4 [Anick 1982]. The Poincaré series $\mathrm{P}_k^R(t)$ is irrational for

$$R=k[x_1,\ldots,x_5]/\big(x_1^2, x_2^2, x_4^2, x_5^2, x_1x_2, x_4x_5, x_1x_3+x_3x_4+x_2x_5, (x_1,\ldots,x_n)^3\big)$$

if $\mathrm{char}(k) \neq 2$; in $\mathrm{char}(k) = 2$ we add x_3^2 to the defining ideal.

Since then several other such examples have been found and they exist even over Gorenstein rings; see, for example, [Bøgvad 1983]. Surveys on nonrationality are given by Anick [1988] and Roos [1981]. At present, it is not clear how wide spread such examples are. We do not have a feel for which of the following cases holds:

(1) Most Poincaré series are rational, and irrational Poincaré series occur rarely in specially crafted examples.

(2) Most Poincaré series are irrational, and there are some nice classes of rings (for example, Golod rings, complete intersections) where we have rationality.

(3) Both rational and irrational Poincaré series occur widely.

One would like to have results showing whether the Poincaré series are rational generically, or are irrational generically. A difficulty in even posing meaningful problems and conjectures is that currently we do not know a good concept of "generic".

The situation is clear for generic Artinian Gorenstein rings by [Rossi and Şega 2014], and also when we have a lot of combinatorial structure: the cases of monomial and toric quotients. Backelin [1982] proved that the Poincaré series of k over $R = S/I$ is rational if I is generated by monomials. His result was extended to all modules:

Theorem 6.5 [Lescot 1988]. *The Poincaré series of every finitely generated graded module over $R = S/I$ is rational if I is generated by monomials.*

An ideal I is called *toric* if it is the kernel of a map

$$S \longrightarrow k[m_1, \ldots, m_n] \subset k[t_1, \ldots, t_r]$$

that maps each variable x_i to a monomial m_i; in that case, $R = S/I$ is called a *toric ring*. Gasharov, Peeva and Welker [2000] proved that the Poincaré series of k is rational for generic toric rings; see Theorem 6.16(4). However, in contrast to the monomial case, toric ideals with irrational Poincaré series were found:

Example 6.6 [Roos and Sturmfels 1998]. Set $S = k[x_0, \ldots, x_9]$ and let I be the kernel of the homomorphism

$$k[x_1, \ldots, x_9] \to k[t^{36}, t^{33}s^3, t^{30}s^6, t^{28}s^8, t^{26}s^{10}, t^{25}s^{11}, t^{24}s^{12}, t^{18}s^{18}, s^{36}],$$

that sends the variables x_i to the listed monomials in t and s. Computer computation shows that I is generated by 12 quadrics. It defines a projective monomial curve. Roos–Sturmfels showed that the Poincaré series of k is irrational over S/I.

Example 6.7 [Fröberg and Roos 2000; Löfwall, Lundqvist and Roos 2015]. Set $S = k[x_1, \ldots, x_7]$ and let I be the kernel of the homomorphism

$$k[x_1, \ldots, x_7] \to k[t^{18}, t^{24}, t^{25}, t^{26}, t^{28}, t^{30}, t^{33}],$$

that sends the variables x_i to the listed monomials in t. Computer computation shows that I is generated by 7 quadrics and 4 cubics. It defines an affine monomial curve. The Poincaré series of k is irrational over S/I.

Open-Ended Problem 6.8. *The above results motivate the question of whether there are classes of rings (other than toric rings, monomial quotients, and Artinian Gorenstein rings) whose generic objects are Golod.*

We now go back to the discussion of rationality over a graded ring $R = S/I$. Jacobsson, Stoltenberg and Hensen proved [1985] that the sequence of Betti numbers of any finitely generated graded R-module is primitive recursive. The class of primitive recursive functions is countable.

Theorem 6.16 provides interesting classes of rings for which $\mathrm{P}_k^R(t)$ is rational. It is also known that it is rational if R is a complete intersection by a result of Tate [1957], if R is one link from a complete intersection by a result of Avramov [1978], and in other special cases.

Inspired by Open-Ended Problem 6.2 one can consider the following problem:

Open-Ended Problem 6.9. *Relate the properties of the infinite graded minimal free resolution of k over S/I to the properties of the finite minimal graded free resolution of S/I over the ring S.*

One option is to explore in the following general direction:

Open-ended strategy 6.10. One can take conjectures or results on finite minimal graded free resolutions and try to prove analogues for infinite minimal graded free resolutions.

Another option is to study the relations between the infinite minimal free resolution of k over S/I and the finite minimal free resolution of S/I over the ring S. There is a classical Cartan–Eilenberg spectral sequence relating the two resolutions [Cartan and Eilenberg 1956]:

$$\mathrm{Tor}_p^R(M, \mathrm{Tor}_q^S(R,k)) \Longrightarrow \mathrm{Tor}_{p+q}^S(M,k);$$

see [Avramov 1998, Section 3] for a detailed treatment and other spectral sequences.

Using that spectral sequence, Serre derived the inequality

$$\mathrm{P}_k^{S/I}(t) \preccurlyeq \frac{(1+t)^n}{1-t^2\,\mathrm{P}_I^S(t)}, \tag{6.11}$$

where $\preccurlyeq$ denotes coefficient-wise comparison of power series; see [Avramov 1998, Proposition 3.3.2]. Eagon constructed a free resolution of k over R whose generating function is the right-hand side of Serre's Inequality; see [Gulliksen and Levin 1969, Section 4.1]. We will see later in this section that the resolution is minimal over Golod rings.

Rationality and Growth of Betti Numbers 6.12. *Suppose that $\{b_i\}$ is a sequence of integer numbers and $\sum_i b_i t^i = f(t)/g(t)$ for some polynomials $f(t), g(t) \in \mathbb{Q}[t]$. Set $a = \deg(g)$. Let $h(t) = g(t^{-1})t^{\deg(g)}$; we may assume that the leading coefficient of h is 1 by scaling f if necessary, and write*

$$h(t) = t^a - h_1 t^{a-1} - h_2 t^{a-2} - \cdots - h_a.$$

Then the numbers b_i satisfy the recurrence relation

$$b_i = h_1 b_{i-1} + \cdots + h_a b_{i-a} \quad \textit{for } i \gg 0.$$

Thus, we have a recursive sequence. Let $r_1, \ldots, r_s$ be the roots of $h(t)$ with multiplicities $m_1, \ldots, m_s$ respectively. We have a formula for the numbers b_i in terms of the roots (see [Eisen 1969, Chapter III, Section 4] *and* [Markushevich

1975]), *namely,*

$$b_i = \sum_{1\le j\le s} r_j^i(c_{j,1} + c_{j,2}i + \cdots + c_{j,m_j}i^{m_j-1}) = \sum_{\substack{1\le j\le s\\ 0\le q\le m_j-1}} r_j^i c_{j,q+1} i^q, \quad (6.13)$$

where the coefficients c_{jq} *are determined in order to fit the initial conditions of the recurrence. It follows that the sequence* $\{b_i\}$ *is exponentially bounded, that is, there exists a real number* $\beta > 1$ *such that* $b_i \le \beta^i$ *for* $i \ge 1$*. Hence, Serre's Inequality* (6.11) *implies that for every finitely generated graded R-module M the sequence of Betti numbers* $\{b_i^R(M)\}$ *is exponentially bounded.*

Now, suppose that M is a module with a rational Poincaré series, and set $b_i := b_i^R(M)$*. By* (6.13) *it follows that one of the following two cases holds*:

- *If* $|r_j| \le 1$ *for all roots, then the Betti sequence* $\{b_i\}$ *is polynomially bounded, that is, there exists a polynomial* $e(t)$ *such that* $b_i \le e(i)$ *for* $i \gg 0$.
- *If there exists a root with* $|r_j| > 1$*, then the sequence* $\left\{\sum_{q\le i} b_q\right\}$ *grows exponentially* (*we say that* $\{b_q\}$ *grows weakly exponentially*)*, that is, there exists a real number* $\alpha > 1$ *so that* $\alpha^i \le \sum_{q\le i} b_q$ *for* $i \gg 0$*, by* [Avramov 1978].

We may wonder how the Betti numbers grow if the Poincaré series is not rational. Such questions have been raised in [Avramov 1992; 1998].

In the rest of the section, we discuss Golod rings, which provide many classes of rings over which Poincaré series are rational.

Definition 6.14. *A ring is called Golod if equality holds in Serre's Inequality* (6.11). *In particular, k has a rational Poincaré series in that case. Sometimes, we say that I is Golod if* $R = S/I$ *is.*

Golodness is encoded in the finite data given by the Koszul homology

$$H(\boldsymbol{K} \otimes_S S/I),$$

where $\mathbf{K} := \mathbf{K}(\mathrm{x}_1, \ldots, \mathrm{x}_n; \mathrm{S})$ is the Koszul complex resolving k over S, as follows. We define Massey operations on $\boldsymbol{K} \otimes_S S/I$ in the following way: Let $\mathcal{M}$ be a set of homogeneous (with respect to both the homological and the internal degree) elements in $\boldsymbol{K} \otimes S/I$ that form a basis of $H(\boldsymbol{K} \otimes_S S/I)$. For every $z_i, z_j \in \mathcal{M}$ we define $\mu_2(z_i, z_j)$ to be the homology class of $z_i z_j$ in $H(\boldsymbol{K} \otimes_S S/I)$ and call μ_2 the 2-*fold Massey operation*. This is just the multiplication in $H(\boldsymbol{K} \otimes_S S/I) \cong \mathrm{Tor}^S(S/I, k)$ and is sometimes called the *Koszul product*. Fix an $r > 2$. If all q-fold Massey operations vanish for all $q < r$, then we define the *r-fold Massey operation* as follows: for every $z_1, \ldots, z_j \in \mathcal{M}$

with $j < r$ choose a homogeneous $y_{z_1,\dots,z_j} \in \boldsymbol{K} \otimes S/I$ such that

$$d(y_{z_1,\dots,z_j}) = \mu_j(z_1, \dots, z_j),$$

and note that the homological degree of $y_{z_1,\dots,z_j}$ is $-1 + \sum_{v=1}^{j}(\deg(z_v) + 1)$; then set

$$\begin{aligned}\mu_r(z_1, \dots, z_r) := z_1 y_{z_2,\dots,z_r} \\ + \sum_{2\le s\le r-2} (-1)^{\sum_{v=1}^{s}(\deg(z_v)+1)} y_{z_1,\dots,z_s} y_{z_{s+1},\dots,z_r} \\ +(-1)^{\sum_{v=1}^{r-1}(\deg(z_v)+1)} y_{z_1,\dots,z_{r-1}} z_r.\end{aligned}$$

Note that Massey operations respect homological and internal degree. Massey operations are sometimes called *Massey products*. The r-fold products exist if and only if all lower products vanish. It was shown by Golod [1962] that all Massey products vanish (exist) exactly when the ring S/I is Golod; see [Gulliksen and Levin 1969].

Frank Moore pointed out that there are no known examples of rings which are not Golod and for which the second Massey product vanishes. Berglund and Jöllenbeck [2007] showed that the vanishing of the Koszul product (second Massey operation) is equivalent to Golodness if I is generated by monomials.

Golod's Resolution 6.15. If R is Golod, then Eagon's free resolution of k is minimal. This was proved by Golod [1962] (see [Gulliksen and Levin 1969]) who also provided an explicit formula for the differential using Massey products; see [Avramov 1998, Theorem 5.2.2]. It is known that the Ext-algebra over a Golod ring R is finitely presented by a result of Sjödin [1985].

The following is a list of some Golod rings.

Theorem 6.16.

(1) *Since Massey operations respect internal degree, it is easy to see that if an ideal I is generated in one degree and its minimal free resolution over S is linear (that is, the entries in the differential matrices are linear forms), then S/I is a Golod ring. In particular, $S/(x_1, \dots, x_n)^p$ is a Golod ring for all $p > 1$.*

(2) *Craig Huneke (personal communication) observed that using degree-reasons, one can show that if I is an ideal such that $I_i = 0$ for $i < r$ and* $\operatorname{reg}(I) \le 2r - 3$, *then S/I is a Golod ring.*

(3) [Herzog, Reiner and Welker 1999] *If I is a componentwise linear ideal (that is, I_p has a linear minimal free resolution for every p), then S/I is Golod. This includes the class of Gotzmann ideals.*

(4) [Gasharov, Peeva and Welker 2000] *If I is a generic toric ideal, then S/I is a Golod ring.*

(5) [Herzog and Steurich 1979] *If for two proper graded ideals we have $JJ' = J \cap J'$, then JJ' is Golod.*

(6) [Aramova and Herzog 1996; Peeva 1996] *An ideal L generated by monomials in S is called* 0-*Borel fixed (also referred to as strongly stable) if whenever m is a monomial in L and x_i divides m, then $x_j(m/x_i) \in L$ for all $1 \le j < i$. The interest in such ideals comes from the fact that generic initial ideals in characteristic zero are* 0-*Borel fixed. If L is a* 0-*Borel fixed ideal contained in $(x_1, \dots, x_n)^2$, then S/L is a Golod ring.*

(7) [Berglund and Jöllenbeck 2007] *If I is an ideal generated by monomials, then S/I is Golod if and only if the product on $H(\boldsymbol{K} \otimes S/I) = \mathrm{Tor}^S(S/I, k)$ is trivial.*

(8) [Fakhary and Welker 2012] *For any two proper monomial ideals I and J in S, the ring S/IJ is Golod.*

(9) [Herzog and Huneke 2013] *Let I be a graded ideal. For every $q \ge 2$ the rings S/I^q, $S/I^{(q)}$ and $S/\widetilde{I^q}$ are Golod, where $I^{(q)}$ and $\widetilde{I^q}$ denote the q-th symbolic and saturated powers of I, respectively. The proof hinges on a new definition, whereby the authors call an ideal I strongly Golod if $\partial(I)^2 \subset I$, where $\partial(I)$ denotes the ideal which is generated by all partial derivatives of the generators of I, and proceed to show that strongly Golod ideals are Golod. For large powers of ideals the result was previously proved by Herzog, Welker and Yassemi* [2011].

Open-Ended Problem 6.17. *It is of interest to find other nice classes of rings which are Golod.*

Theorem 6.16(5), (8) and (9) suggests the following open problem, which was first raised by Welker (personal communication).

Problem 6.18. *Is the product of any two proper graded ideals Golod?*

Recent work on the topic leads to the following question:

Problem 6.19 (Craig Huneke, personal communication). *If I is (strongly) Golod, then is the integral closure of I Golod as well?*

Over a Golod ring, we have rationality not only for the Poincaré series of k but for all Poincaré series:

Theorem 6.20 (Lescot 1990; see [Avramov 1998, Theorem 5.3.2]). *Let $R = S/I$ be a Golod ring. If M is a graded finitely generated R-module, then its Poincaré series is*

$$\mathrm{P}_M^R(t) = h(t) + \frac{\mathrm{P}_{M'}^S(t)}{1 - t^2\,\mathrm{P}_I^S(t)},$$

where $h(t)$ is a polynomial in $\mathbb{N}[t] \cup 0$ of degree $\leq n$, and the polynomial $\mathrm{P}^S_{M'}(t)$ of degree $\leq n$ is the Poincaré series over S of a syzygy M' of M over R. In particular, the Poincaré series of all graded finitely generated modules over R have common denominator

$$1 - t^2\, \mathrm{P}^S_I(t).$$

The property about the common denominator in the above theorem does not hold in general:

Example 6.21 [Roos 2005]. There exists a ring R defined by quadrics, such that the rational Poincaré series over R do not have a common denominator. Such examples are provided in [Roos 2005, Theorem 2.4]. For example,

$$\begin{aligned} R &= k[x, y, z, u]/(x^2, y^2, z^2, u^2, xy, zu), \\ R &= k[x, y, z, u]/(x^2, z^2, u^2, xy, zu), \\ R &= k[x, y, z, u]/(x^2, u^2, xy, zu). \end{aligned}$$

7. Regularity

Definition 7.1. *We will define the graded Betti numbers, which are a refined version of the Betti numbers. Let $\boldsymbol{F}$ be the minimal graded free resolution of a finitely generated graded R-module M. We may write*

$$F_i = \bigoplus_{p \in \mathbb{Z}} R(-p)^{b_{i,p}},$$

for each i. Therefore, the resolution is

$$\boldsymbol{F} : \quad \cdots \longrightarrow \bigoplus_{p \in \mathbb{Z}} R(-p)^{b_{i,p}} \xrightarrow{d_i} \bigoplus_{p \in \mathbb{Z}} R(-p)^{b_{i-1,p}} \longrightarrow \cdots .$$

The numbers $b_{i,p}$ are called the graded Betti numbers of M and denoted $b^R_{i,p}(M)$. We say that $b^R_{i,p}(M)$ is the Betti number in homological degree i and internal degree p. We have that

$$b^R_{i,p}(M) = \dim_k(\mathrm{Tor}^R_i(M, k)_p) = \dim_k(\mathrm{Ext}^i_R(M, k)_p).$$

The graded Poincaré series of M over R is

$$\mathrm{P}^R_M(t, z) = \sum_{i \geq 0,\, p \in \mathbb{Z}} b^R_{i,p}(M) t^i z^p.$$

There is a graded version of Serre's Inequality (6.11):

$$\mathrm{P}^R_M(t, z) \preccurlyeq \frac{\mathrm{P}^S_M(t, z)}{1 - t^2\, \mathrm{P}^S_I(t, z)}. \tag{7.2}$$

Often we consider the *Betti table*, defined as follows: The columns are indexed from left to right by homological degree starting with homological degree zero. The rows are indexed increasingly from top to bottom starting with the minimal degree of an element in a minimal system of homogeneous generators of M. The entry in position i, j is $b^R_{i,i+j}(M)$. Note that the Betti numbers $b^R_{i,i}(M)$ appear in the top row if M is generated in degree 0. This format of the table is meaningful since the minimality of the resolution implies that $b^R_{i,p}(M) = 0$ for $p < i + c$ if c is the minimal degree of an element in a minimal system of homogeneous generators of M. For example, a module M generated in degrees ≥ 0 has Betti table of the form

	0	1	2	...
0:	$b_{0,0}$	$b_{1,1}$	$b_{2,2}$	...
1:	$b_{0,1}$	$b_{1,2}$	$b_{2,3}$	...
2:	$b_{0,2}$	$b_{1,3}$	$b_{2,4}$	...
3:	$b_{0,3}$	$b_{1,4}$	$b_{2,5}$	...
⋮	⋮	⋮	⋮	

In Example 3.4, the Betti table of S/J is

	0	1	2
0:	1	-	-
1:	-	-	-
2:	-	2	1
3:	-	-	-
4:	-	-	-
5:	-	-	-
6:	-	1	1

where we put - instead of zero.

We may ignore the zeros in a Betti table and consider the shape in which the nonzero entries lie. In Example 3.4 the shape of the Betti table is determined by

	0	1	2
0:	*		
1:			
2:		*	*
3:			
4:			
5:			
6:		*	*

Open-Ended Problem 7.3. *What are the possible shapes of Betti tables either over a fixed ring, or of a fixed class of modules?*

Two basic invariants measuring the shape of a Betti table are the projective dimension and the regularity: The projective dimension $\mathrm{pd}_R(M)$ is the index of the last nonzero column of the Betti table, and thus it measures the length of the table. The width of the table is measured by the index of the last nonzero row of the Betti table, and it is another well-studied numerical invariant: the *Castelnuovo–Mumford regularity* of M, which is

$$\mathrm{reg}_R(M) = \sup\{j \mid b^R_{i,i+j}(M) \neq 0\}.$$

In Example 3.4 we have

	0	1	2
0:	1	-	-
1:	-	-	-
2:	-	2	1
3:	-	-	-
4:	-	-	-
5:	-	-	-
6:	-	1	1

$\mathrm{reg}(S/J) = 6$

$\mathrm{pd}(S/J) = 2$

Hilbert's Syzygy Theorem 4.9 implies that every finitely generated graded module over the polynomial ring S has finite regularity. If the module M has finite length, then

$$\mathrm{reg}_R(M) = \sup\{j \mid M_j \neq 0\};$$

see [Eisenbud 2005, Section 4B]. Regularity is among the most interesting and important numerical invariants of M, and it has attracted a lot of attention and work both in commutative algebra and algebraic geometry.

It is natural to ask for an analogue of Auslander–Buchsbaum–Serre's Criterion 4.10 to characterize the rings over which all modules have finite regularity. It is given by the following two results.

Theorem 7.4 [Avramov and Eisenbud 1992]. *If R is a Koszul algebra, that is, $\mathrm{reg}_R(k) = 0$, then for every graded R-module M we have*

$$\mathrm{reg}_R(M) \leq \mathrm{reg}_S(M).$$

Theorem 7.5 [Avramov and Peeva 2001]. *The following are equivalent*:

(1) *Every finitely generated graded R-module has finite regularity.*

(2) *The residue field k has finite regularity.*

(3) *R is a Koszul algebra.*

As noted above in Theorem 7.4, Koszul algebras are defined by the vanishing of the regularity of k. They are the topic discussed in the next section.

Open-Ended Problem 7.6. *It would be interesting to find analogues over Koszul rings of conjectures/results on regularity over a polynomial ring.*

We give an example: In a recent paper, Ananyan and Hochster [2012] showed that the projective dimension of an ideal generated by a fixed number r of quadrics in a polynomial ring is bounded by a number independent of the number of variables. This solved a problem of Stillman [Peeva and Stillman 2009, Problem 3.14] in the case of quadrics. Then by a result of Caviglia (see [Peeva 2011, Theorem 29.5]), it follows that the regularity of an ideal generated by r quadrics in a polynomial ring is bounded by a number independent of the number of variables. One can ask for an analogue to Stillman's conjecture (which is for polynomial rings) for infinite free resolutions over a Koszul ring. Caviglia (personal communication) observed that if we fix integer numbers r and q, and consider an ideal J generated by r quadrics in a Koszul algebra $R = S/I$ defined by q quadrics, then by Theorem 7.4 we have $\text{reg}_R(J) \le \text{reg}_S(I + J)$, which is bounded by the result of Ananyan–Hochster since $I + J$ is generated by $r + q$ quadrics. Thus, the regularity of an ideal generated by r quadrics in a Koszul algebra with a fixed number of defining equations is bounded by a formula independent of the number of variables. If the number of defining equations of the Koszul algebra is not fixed, then the property fails to hold by an example constructed by McCullough [2013].

8. Koszul rings

Definition 8.1. *Following* [Priddy 1970], *we say that R is Koszul if* $\text{reg}_R(k) = 0$. *Equivalently, R is Koszul if the minimal graded free resolution of k over R is linear, that is, the entries in the matrices of the differential are linear forms. These rings have played an important role in several mathematical fields. It is easy to see that if* $R = S/I$ *is Koszul, then the ideal I is generated by quadrics and linear forms; we ignore the linear forms by assuming* $I \subseteq (x_1, \dots, x_n)^2$. *We say that I is Koszul when* S/I *is Koszul.*

Example 8.2. If $f_1, \dots, f_c$ is a regular sequence of quadrics, then by Tate's Resolution 5.3 it follows that $S/(f_1, \dots, f_c)$ is a Koszul ring.

Example 8.3. Suppose I is generated by quadrics and has a linear minimal free resolution over S. Then R is Golod by Theorem 6.16(1). It follows that R is Koszul.

Theorem 7.4 implies:

Corollary 8.4 [Avramov and Eisenbud 1992]. *A sufficiently high truncation $M_{\geq p}$ of a graded finitely generated module M over a Koszul algebra R has a linear minimal R-free resolution, that is, the entries in the matrices of the differential are linear forms.*

Note that R_i is a k-vector space since $R_0 = k$ and $R_0 R_i \subseteq R_i$. The generating function

$$i \mapsto \dim_k(R_i)$$

is called the *Hilbert function* of R and is studied via the *Hilbert series*

$$\mathrm{Hilb}_R(t) = \sum_{i \geq 0} \dim_k(R_i) t^i .$$

The Hilbert function encodes important information about R, for example, its dimension and multiplicity. Hilbert introduced resolutions in order to compute Hilbert functions; see [Peeva 2011, Section 16]. The same kind of computation works over a Koszul ring and yields the following result.

Theorem 8.5 [see Polishchuk and Positselski 2005, Chapter 2, Section 2; Fröberg 1999]. *If the ring R is Koszul, then the Poincaré series of k is related to the Hilbert series of R as follows*

$$\mathrm{P}_k^R(t) = \frac{1}{\mathrm{Hilb}_R(-t)} .$$

Example 8.6. Not all ideals generated by quadrics define Koszul rings. The relation in Theorem 8.5 can be used to show that particular rings are not Koszul. Consider

$$R = k[x, y, z, w]/(x^2, y^2, z^2, w^2, xy + xz + xw).$$

This is an Artinian ring with Hilbert series $\mathrm{Hilb}_R(t) = 1 + 4t + 5t^2 + t^3$. One computes

$$\frac{1}{\mathrm{Hilb}_R(-t)} = 1 + 4t + 11t^2 + 25t^3 + 49t^4 + 82t^5 + 108t^6 + 71t^7 - 174t^8 \cdots .$$

Hence, R cannot be Koszul; if it were, the previous expression would be its Poincaré series, which cannot have any negative coefficients.

Next we will see that Theorem 8.5 is an expression of duality. Suppose that I is generated by quadrics; in that case we say that the algebra R *is quadratic*. Let $y_1, \dots, y_n$ be indeterminates (recall that n is the number of variables in S), and denote by V the vector space spanned by them. Write $R = k\langle V\rangle/(W)$, where $k\langle V\rangle = k \oplus V \oplus (V \otimes_k V) \oplus \cdots$ is the tensor algebra on V and $W \subset V \otimes_k V$ is the vector space spanned by the quadrics generating I and the commutator relations $y_i \otimes y_j - y_j \otimes y_i$ for $i \neq j$. The *dual algebra* of R is the quadratic algebra $R^! = k\langle V^*\rangle/(W^\perp)$, where V^* is the dual vector space of V and $W^\perp \subset (V \otimes_k V)^*$ is the two-sided ideal of forms that vanish on W; see [Polishchuk and Positselski 2005, Chapter I, Section 2]. For example, the dual algebra of the polynomial ring S is an exterior algebra. We denote $z_1, \dots, z_n$ the basis of V^* dual to the basis $y_1, \dots, y_n$ of V. Computing the generators of $W^\perp$ amounts to linear algebra computations: Set $[z_i, z_i] = z_i^2$, and $[z_i, z_j] = z_i z_j + z_j z_i$ for $i \neq j$. If $R = k[x_1, \dots, x_n]/(f_1, \dots, f_r)$, where

$$f_p = \sum_{i \le j} a_{pij} x_i x_j,$$

then choose a basis (c_{qij}) of the solutions to the linear system of equations

$$\sum_{i \le j} a_{pij} X_{ij} = 0, \quad p = 1, \dots, r,$$

and then $R^! = k\langle z_1, \dots, z_n\rangle/(g_1, \dots, g_s)$, where

$$g_q = \sum_{i \le j} c_{qij} [z_i, z_j], \quad q = 1, \dots, s.$$

Example 8.7. Let $R = S/I$, where I is generated by quadratic monomials. Then $R^! = k\langle z_1, \dots, z_n\rangle/T$, where T is generated by all z_i^2 such that $x_i^2 \notin I$ and $z_i z_j + z_j z_i$ such that $i \neq j$ and $x_i x_j \notin I$.

Example 8.8. Let $R = k[x, y, z]/I$ with

$$I = (x^2, y^2, xy + xz, xy + yz).$$

Then $R^! = k\langle X, Y, Z\rangle/T$ with

$$T = (Z^2, XY + YX - XZ - ZX - YZ - ZY).$$

While it is not obvious that the ideal I has a quadratic Gröbner basis, it is apparent that T is generated by a noncommutative quadratic Gröbner basis. It follows that both R and $R^!$ are Koszul. The role of Gröbner bases is explained after Example 8.13.

The dual algebra can be defined for any (not necessarily commutative) graded k-algebra generated by finitely many generators of degree 1 and with relations generated in degree 2; it is easy to see that $(R^!)^! \cong R$. The notions of grading, resolution and Koszulness extend to the noncommutative setting. It then follows that a quadratic algebra is Koszul if and only if its dual algebra is Koszul [Priddy 1970]. Furthermore, Löfwall [1986] proved that $R^!$ is isomorphic to the *diagonal subalgebra* $\sum_i \operatorname{Ext}^i_R(k,k)_i$ of the Yoneda algebra $\operatorname{Ext}_R(k,k)$. If R is Koszul, then $\operatorname{Ext}^i_R(k,k)_j = 0$ for $i \neq j$, so $R^!$ is the entire Yoneda algebra $\operatorname{Ext}_R(k,k)$ and hence $\mathrm{P}^R_k(t) = \operatorname{Hilb}_{R^!}(t)$. Therefore, the formula in Theorem 8.5 can be written

$$\operatorname{Hilb}_R(t)\operatorname{Hilb}_{R^!}(-t) = 1. \tag{8.9}$$

Examples of non-Koszul quadratic algebras for which the above formula holds were constructed by Roos [1995] (see [Positselksi 1995] and [Piontkovskiĭ 2001] for noncommutative examples).

Example 8.10 [Roos 1995, Case B]. Consider the ring

$$R = k[x,y,z,u,v,w]/(x^2+xy,\ x^2+yz,\ xz,\ z^2,\ zu+yv,\ zv,\ uw+v^2).$$

Roos proved that (8.9) holds for R, but R is not Koszul.

It would have been very helpful if one could recognize whether a ring is Koszul or not by just looking at the beginning of the infinite minimal free resolution of k (for example, by computing the beginning of the resolution by computer). Unfortunately, this does not work out. Roos constructed for each integer $q \geq 3$ a quotient $Q(q)$ of a polynomial ring in 6 variables subject to 11 quadratic relations, so that the minimal free resolution of k over $Q(q)$ is linear for the first q steps and has a nonlinear q-th Betti number:

Example 8.11 [Roos 1993]. Choose a number $2 \leq q \in \mathbb{N}$. Let the ring R be

$$\frac{\mathbb{Q}[x,y,z,u,v,w]}{\left(x^2, xy, yz, z^2, zu, u^2, uv, vw, w^2, xz+qzw-uw, zw+xu+(q-2)uw\right)}.$$

Roos proved that

$$\begin{aligned} b^R_{i,j}(k) &= 0 \quad \text{for } j \neq i \text{ and } i \leq q, \\ b^R_{q+1,q+2}(k) &\neq 0. \end{aligned}$$

Generalized Koszul Resolution 8.12 ([Priddy 1970]; also see [Beilinson et al. 1996, 2.8.1] and [Manin 1987]). *If R is Koszul, then the minimal free resolution of k over R can be described by the generalized Koszul complex (also called the Priddy complex) constructed by Priddy* [1970]. *The j-th term of the complex is*

$$\boldsymbol{K}_j(R,k) := R \otimes_k (R^!_j)^*,$$

where $-^*$ *stands for taking a vector space dual. The differential is defined by*

$$\boldsymbol{K}_{j+1}(R,k) = R \otimes_k (R^!_{j+1})^* \to \boldsymbol{K}_j(R,k) = R \otimes_k (R^!_j)^*,$$
$$r \otimes \varphi \mapsto \sum_{1 \le i \le n} r x_i \otimes \varphi \tilde{z}_i,$$

where $\varphi\tilde{z}_i \in (R^!_j)^*$ *is defined by* $\varphi\tilde{z}_i(e) = \varphi(e z_i)$ *for* $e \in R^!_j$ *(thus,* $\varphi\tilde{z}_i$ *is the composition of* φ *after multiplication by* z_i *on the right).*

In Example 8.7, we considered the case when $R = S/I$ and I is generated by quadratic monomials. In that case, the generalized Koszul complex is described in [Fröberg 1975].

Example 8.13. Let $R = k[x_1, x_2]/(x_1^2, x_1x_2)$. Then R is Koszul and

$$R^! = k\langle z_1, z_2\rangle/(z_2^2).$$

Hence we can resolve k minimally over R by the generalized Koszul complex. Here we compute the first few terms in the resolution. We fix k-bases for $(R^!_i)^*$:

$$R^!_0 = \operatorname{span}\langle 1^*\rangle, \quad R^!_1 = \operatorname{span}\langle z_1^*, z_2^*\rangle, \quad R^!_2 = \operatorname{span}\big\langle (z_1^2)^*, (z_2z_1)^*, (z_1z_2)^*\big\rangle,$$

which we identify with R-bases of $\boldsymbol{K}_i(R,k)$. We then compute the beginning of $\boldsymbol{K}(R,k)$ as

$$\cdots \longrightarrow R^3 \xrightarrow{\begin{pmatrix} x_1 & 0 & x_2 \\ 0 & x_1 & 0 \end{pmatrix}} R^2 \xrightarrow{(\,x_1 \ x_2\,)} R.$$

Koszul rings were introduced by Priddy and he also introduced an approach very similar to using quadratic Gröbner bases. The following result is well-known and often used in proofs that a ring is Koszul:

Theorem 8.14. *An ideal with a quadratic Gröbner basis is Koszul.*

This follows from the fact that the Betti numbers of k over R are less or equal than the Betti numbers of k over $S/\operatorname{in}(I)$ for any initial ideal I (see [Peeva 2011, Theorem 22.9]) and the following result.

Theorem 8.15 [Fröberg 1975]. *Every ideal generated by quadratic monomials is Koszul.*

Two important examples using a quadratic Gröbner basis are described below. The r*-th Veronese ring* is

$$V_{c,r} = \bigoplus_{i=0}^{\infty} T_{ir} = k[\text{all monomials of degree } r \text{ in } c \text{ variables}],$$

and it defines the r-th *Veronese embedding of* $\boldsymbol{P}^{c-1}$. Bărcănescu and Manolache [1981] showed that the defining toric ideal of every Veronese ring has a quadratic Gröbner basis. Thus, the Veronese rings are Koszul.

The toric ideal of the Segre embedding of $\boldsymbol{P}^p \times \boldsymbol{P}^q$ in $\boldsymbol{P}^{pq+p+q}$ is generated by the (2×2)-minors of a $\big((p+1) \times (q+1)\big)$-matrix of indeterminates

$$\{x_{i,j} \mid 1 \leq i \leq p+1,\ 1 \leq j \leq q+1\}.$$

Bărcănescu and Manolache [1981] showed that there exists a quadratic Gröbner basis. Thus, the Segre rings are also Koszul.

There exist examples of Koszul rings for which there is no quadratic Gröbner basis:

Example 8.16 (Conca, personal communication). The ring

$$k[x, y, z, w]/(xz,\ x^2 - xw,\ yw,\ yz + xw,\ y^2)$$

is Koszul. However, it has no quadratic Gröbner basis even after change of coordinates because there is no quadratic monomial ideal with the same Hilbert function as the ideal $(xz,\ x^2 - xw,\ yw,\ yz + xw,\ y^2)$, which can be easily verified by computer.

Example 8.17 [Caviglia 2009]. The ideal

$$\begin{aligned} I = \big(&x_8^2 - x_3x_9, x_5x_8 - x_6x_9, x_1x_8 - x_9^2, x_5x_7 - x_2x_8, x_4x_7 - x_6x_9,\\ &x_1x_7 - x_4x_8, x_6^2 - x_2x_7, x_4x_6 - x_2x_9, x_3x_6 - x_7^2, x_1x_6 - x_5x_9, x_3x_5 - x_6x_8,\\ &x_3x_4 - x_7x_9, x_4^2 - x_1x_5, x_2x_4 - x_5^2, x_2x_3 - x_6x_7, x_1x_3 - x_8x_9, x_1x_2 - x_4x_5\big) \end{aligned}$$

is the toric ideal, that is the kernel of the homomorphism

$$k[x_1, \ldots, x_9] \longrightarrow k\big[\text{all monomials of degree 3 except } abc \text{ in } k[a, b, c]\big]$$

sending the variables $x_1, \ldots, x_9$ to the cubic monomials

$$a^3, b^3, c^3, a^2b, ab^2, b^2c, c^2b, c^2a, a^2c,$$

respectively. This example was introduced by Sturmfels and is very similar to the cubic Veronese. It is called the pinched Veronese. But in contrast to the cubic Veronese, the ideal I has no quadratic Gröbner basis in these variables (it is not known if a quadratic Gröbner basis exists after a change of variables), which Sturmfels verified by computer computation. Caviglia proved that I is Koszul. The example was revisited in two papers: Caviglia and Conca [2013] classify the projections of the Veronese cubic surface to $\mathbb{P}^8$ whose coordinate rings are Koszul, and Vu [2013] proved Koszulness for a more general class of ideals.

Proving cases with no known quadratic Gröbner basis can be challenging. Recently, Nguyen and Vu [2015] introduced a method using Fröbenius-like epimorphisms. Another possibility might be to use filtrations. There are various versions in which this method can be used. The method was formally introduced by Conca, Trung and Valla [2001] with the name "Koszul filtration", although it had been used by other authors previously. If there exists a Koszul filtration of R, then R is Koszul. We define a version of a Koszul filtration: Fix a graded ideal I in S. Let $\mathcal{H}$ be a set of tuples $(L; l)$, where L is a linear ideal (that is, L is generated by linear forms) in $R = S/I$ and l is a linear form in L. Denote by $\overline{\mathcal{H}}$ the set of linear ideals appearing in the tuples in $\mathcal{H}$. A *Koszul filtration* of R is a set $\mathcal{H}$ such that the following two conditions are satisfied:

(1) $(x_1, \ldots, x_n) \in \overline{\mathcal{H}}$.

(2) If $(L; l) \in \mathcal{H}$ and $L \neq 0$, then there exists a proper subideal $N \subset L$ such that $L = (N, l)$, $(N : l) \in \overline{\mathcal{H}}$ and $N \in \overline{\mathcal{H}}$.

Note that we do not assume that the ideal I is generated by quadrics.

Open-Ended Problem 8.18. *It is an ever tantalizing problem to find more classes of Koszul rings and to develop new approaches that can be used to show that a ring is Koszul in the absence of a quadratic Gröbner basis.*

Here is a sample conjecture:

Conjecture 8.19 [Bøgvad 1994]. *The toric ring of a smooth projectively normal toric variety is Koszul.*

The idea to consider linear minimal free resolutions of k naturally leads to the consideration of linear minimal free resolutions of other modules. We say that M has a *linear* (or a p-linear) minimal free resolution if $b_{ij}^R(M) = 0$ for all i and $j \neq i + p$; in particular, M is generated in degree p in this case. Equivalently, M is generated in one degree and has linear entries in the matrices of the differentials (in any basis) of its minimal free resolution. As in Theorem 8.5, a straightforward computation shows that

$$t^p \, \mathrm{P}_M^R(t) = (-1)^p \frac{\mathrm{Hilb}_M(-t)}{\mathrm{Hilb}_R(-t)},$$

for such modules.

We close this section by outlining a problem on Koszul rings coming from the theory of hyperplane arrangements. A set $\mathcal{A} = \bigcup_{i=1}^s H_i \subseteq \mathbb{C}^r$ is a *central hyperplane arrangement* if each H_i is a hyperplane containing the origin. Arnold considered the case when $\mathcal{A}$ is a braid arrangement and constructed the cohomology algebra of the complement. For any central hyperplane arrangement, Orlik and Solomon [1980] provided a description of the cohomology algebra

$A := \mathbf{H}^*(\mathbb{C}^r \setminus \mathscr{A}; \mathbb{C})$ of the complement of $\mathscr{A}$; it is a quotient of an exterior algebra by a combinatorially determined ideal. Namely, if E is the exterior algebra on n variables $e_1, \ldots, e_n$ over $\mathbb{C}$, then the *Orlik-Solomon algebra* is $A = E/J$, where J is generated by the elements

$$\partial(e_{i_1 \ldots i_p}) = \sum_{\substack{1 \le q \le p \\ \operatorname{codim}(H_{i_1} \cap \cdots \cap H_{i_p}) < p}} (-1)^{q-1} e_{i_1} \wedge \cdots \wedge \widehat{e_{i_q}} \wedge \cdots \wedge e_{i_p},$$

and $\widehat{e_{i_q}}$ means that e_{i_q} is omitted. In the introduction to [Hirzebruch 1983], Hirzebruch wrote: "The topology of the complement of an arrangement of lines in the projective plane is very interesting, the investigation of the fundamental group of the complement very difficult." The fundamental group $\pi_1(X)$ of the complement $X = \mathbb{C}^r \setminus \mathscr{A}$ is interesting, complicated, and few results are known about it. Let

$$Z_1 = \pi_1(X), \ldots, Z_{i+1} = [Z_i, \pi_1(X)], \ldots$$

be the *lower central series* and set $\varphi_i = \operatorname{rank}(Z_i/Z_{i+1})$. For supersolvable arrangements, Falk and Randell [1985] have shown that these numbers are determined by the Orlik–Solomon algebra A through the LCS (Lower Central Series) Formula

$$\prod_{j=1}^{\infty} (1 - t^j)^{\varphi_j} = \sum_{i \ge 0} (-t)^i \dim A_i.$$

It was first noted by Shelton and Yuzvinsky [1997] that the formula holds precisely when the algebra A is Koszul, that is, when $b^A_{i,i+j}(\mathbb{C})$ vanish for $j \neq 0$. As described in [Falk and Randell 2000], much progress has been made on the investigation of the fundamental group $\pi_1(X)$ of the complement, but the following challenging problem [Falk and Randell, Problem 2.2] remains open: *Does there exist a nonsupersolvable central hyperplane arrangement for which the LCS Formula holds*? Peeva showed [2003] that a central hyperplane arrangement is supersolvable if and only if J has a quadratic Gröbner basis with respect to some monomial order. Thus, the above problem is equivalent to:

Problem 8.20. *Does there exist a central hyperplane arrangement for which A is Koszul but J does not have any quadratic Gröbner basis*?

9. Slope and shifts

The following simple example shows that infinite regularity can occur if R is not a polynomial ring: resolving $R/(x^2)$ over $R = k[x]/(x^4)$ we get

$$\cdots \longrightarrow R(-6) \xrightarrow{x^2} R(-4) \xrightarrow{x^2} R(-2) \xrightarrow{x^2} R.$$

The infinite Betti table is

	0	1	2	3	...
0:	1	-	-	-	...
1:	-	1	-	-	...
2:	-	-	1	-	...
3:	-	-	-	1	...
⋮	⋮	⋮	⋮	⋮	⋱

(9.1)

If the regularity is infinite, we study another numerical invariant, called slope. The concept was introduced by Backelin [1988], who defined and studied a different version called rate. It is easier to visualize the slope. Consider the *maximal shift at step i*

$$t_i(M) = \max\{j \mid \operatorname{Tor}_{i,j}^{S/I}(M,k) \neq 0\}$$

and the *adjusted maximal shift*

$$r_i(M) = \max\{j \mid \operatorname{Tor}_{i,i+j}^{S/I}(M,k) \neq 0\},$$

so $r_i(M) = t_i(M) - i$. Note that $r_0(M)$ is then the maximal degree of an element in a minimal system of generators of M. Following Eisenbud (personal communication), we consider the *slope*

$$\operatorname{slope}_R(M) = \sup\left\{\frac{r_i(M) - r_0(M)}{i} \,\middle|\, i \geq 1\right\}, \tag{9.2}$$

which is the minimal absolute value of the slope of a line in the Betti table through position $(0, r_0(M))$ and such that there are only zeros below it. For example, in the Betti table (9.1) above we consider the following line with slope -1:

	0	1	2	3	...
0:	1	-	-	-	...
1:	-	1	-	-	...
2:	-	-	1	-	...
3:	-	-	-	1	...
⋮	⋮	⋮	⋮	⋮	⋱

In (9.2) we start measuring the slope at homological degree 1 because if we start in homological degree 0 then we can make a dramatic change of the invariant by simply increasing by a large number q the degrees of the elements in a minimal system of generators of M, while the structure of the minimal free resolution will remain the same (the graded Betti numbers will get shifted by q). Note that our definition of slope is slightly different than the one introduced by

Avramov, Conca and Iyengar [2010] which is measuring the slope in a different Betti table with entries $b_{i,j}$ instead of our entries $b_{i,i+j}$.

Straightforward computation using Serre's Inequality (7.2) implies the following result.

Theorem 9.3. *Every finitely generated graded R-module has finite slope over R.*

In some situations it might be helpful to consider the slope of a syzygy module

$$\operatorname{slope}_R(\operatorname{Syz}_s(M)) = \sup\left\{\frac{r_{i+s}(M) - r_s(M)}{i} \,\middle|\, i \geq 1\right\} \quad \text{for a fixed } s,$$

or measure the slope starting at a later place v by

$$\operatorname{slope}_R(M, v) = \sup\left\{\frac{r_i(M) - r_0(M)}{i} \,\middle|\, i \geq v\right\}.$$

Both concepts lead to the following open problem, recently raised by Conca:

Open-Ended Problem 9.4 (Conca, personal communication, 2012). *Describe the asymptotic properties of slope for particular classes of rings.*

The first version of the concept slope was introduced by Backelin [1988] and it is the *rate* of a module defined by

$$\operatorname{rate}_R(M) = \sup\left\{\frac{t_i(M) - 1}{i - 1} \,\middle|\, i \geq 2\right\}.$$

Clearly,

$$\operatorname{rate}_R(k) = \operatorname{slope}_R\big((x_1, \dots, x_n)\big) + 1.$$

He also considered

$$\operatorname{slant}_S(R) = \sup\left\{\frac{t_i(R)}{i} \,\middle|\, i \geq 1\right\},$$

which he denoted by $\operatorname{rate}(\varphi)$ for $\varphi : S \to R$. Backelin proved some inequalities, which are usually not sharp.

Theorem 9.5 [Backelin 1988, Theorem 1; Avramov, Conca and Iyengar 2010, Proposition 1.2].

$$\operatorname{slope}_R(M) \leq \max\big\{\operatorname{slope}_S(M), \operatorname{slope}_S(R)\big\},$$
$$\operatorname{slant}_S(R) \leq \operatorname{rate}_R(k) + 1.$$

Note that $\text{rate}_R(k) = 1$ is equivalent to R being Koszul. Thus, if R is Koszul then Theorem 9.5 shows that

$$t_i^S(S/I) \le 2i,$$

for every $i \ge 1$. The following inequality is conjectured:

Conjecture 9.6 [Avramov, Conca and Iyengar 2015, Introduction]. *Suppose that $R = S/I$ is Koszul. Then*

$$t_{i+j}^S(S/I) \le t_i^S(S/I) + t_j^S(S/I),$$

for all $i \ge 1$, $j \ge 1$.

See [Herzog and Srinivasan 2013] and [McCullough 2012] for related results.

Open-Ended Problem 9.7. *It would be interesting to study the properties of the shifts over R.*

The rate is known in very few cases, for example:

Theorem 9.8. *Let I be a graded ideal generated in degrees $\le r$ and such that it has a minimal generator in degree r.*

(1) [Eisenbud, Reeves and Totaro 1994] *If I is generated by monomials then* $\text{rate}_{S/I}(k) = r - 1$.

(2) [Gasharov, Peeva and Welker 2000] *If I is a generic toric ideal, then we have* $\text{rate}_{S/I}(k) = r - 1$.

In both cases, the rate is achieved at the beginning of the free resolution.

Open-Ended Problem 9.9. *Determine the slope (rate) for other nice classes of quotient rings, or obtain upper bounds on it.*

We close this section in a related direction with the following interesting problem.

Eisenbud–Huneke Question 9.10 [Eisenbud and Huneke 2005, Question A]. Let M be a finitely generated graded R-module. Does there exists a number u and bases of the free modules in the minimal graded R-free resolution $\boldsymbol{F}$ of M, such that for all $i \ge 0$ the entries in the matrix of the differential d_i have degrees $\le u$?

Acknowledgements

We thank Luchezar Avramov, David Eisenbud, and Craig Huneke for helpful discussions.

References

[Ananyan and Hochster 2012] T. Ananyan and M. Hochster, "Ideals generated by quadratic polynomials", *Math. Res. Lett.* **19**:1 (2012), 233–244.

[Anick 1980] D. Anick, "Construction d'espaces de lacets et d'anneaux locaux à séries de Poincaré–Betti non rationnelles", *C. R. Acad. Sci. Paris Sér. A-B* **290**:16 (1980), A729–A732.

[Anick 1982] D. J. Anick, "A counterexample to a conjecture of Serre", *Ann. of Math.* (2) **115**:1 (1982), 1–33.

[Anick 1988] D. J. Anick, "Recent progress in Hilbert and Poincaré series", pp. 1–25 in *Algebraic topology–rational homotopy* (Louvain-la-Neuve, 1986), edited by Y. Felix, Lecture Notes in Math. **1318**, Springer, Berlin, 1988.

[Anick and Gulliksen 1985] D. J. Anick and T. H. Gulliksen, "Rational dependence among Hilbert and Poincaré series", *J. Pure Appl. Algebra* **38**:2-3 (1985), 135–157.

[Aramova and Herzog 1996] A. Aramova and J. Herzog, "Koszul cycles and Eliahou–Kervaire type resolutions", *J. Algebra* **181**:2 (1996), 347–370.

[Aspinwall 2013] P. S. Aspinwall, "Some applications of commutative algebra to string theory", pp. 25–56 in *Commutative algebra*, edited by I. Peeva, Springer, New York, 2013.

[Avramov 1978] L. L. Avramov, "Small homomorphisms of local rings", *J. Algebra* **50**:2 (1978), 400–453.

[Avramov 1989] L. L. Avramov, "Modules of finite virtual projective dimension", *Invent. Math.* **96**:1 (1989), 71–101.

[Avramov 1992] L. L. Avramov, "Problems on infinite free resolutions", pp. 3–23 in *Free resolutions in commutative algebra and algebraic geometry* (Sundance, UT, 1990), edited by D. Eisenbud and C. Huneke, Res. Notes Math. **2**, Jones and Bartlett, Boston, 1992.

[Avramov 1994] L. L. Avramov, "Local rings over which all modules have rational Poincaré series", *J. Pure Appl. Algebra* **91**:1-3 (1994), 29–48.

[Avramov 1998] L. L. Avramov, "Infinite free resolutions", pp. 1–118 in *Six lectures on commutative algebra* (Bellaterra, 1996), edited by J. Elias et al., Progr. Math. **166**, Birkhäuser, Basel, 1998.

[Avramov and Buchweitz 2000a] L. L. Avramov and R.-O. Buchweitz, "Homological algebra modulo a regular sequence with special attention to codimension two", *J. Algebra* **230**:1 (2000), 24–67.

[Avramov and Buchweitz 2000b] L. L. Avramov and R.-O. Buchweitz, "Support varieties and cohomology over complete intersections", *Invent. Math.* **142**:2 (2000), 285–318.

[Avramov and Eisenbud 1992] L. L. Avramov and D. Eisenbud, "Regularity of modules over a Koszul algebra", *J. Algebra* **153**:1 (1992), 85–90.

[Avramov and Iyengar 2007] L. L. Avramov and S. B. Iyengar, "Constructing modules with prescribed cohomological support", *Illinois J. Math.* **51**:1 (2007), 1–20.

[Avramov and Peeva 2001] L. L. Avramov and I. Peeva, "Finite regularity and Koszul algebras", *Amer. J. Math.* **123**:2 (2001), 275–281.

[Avramov and Sun 1998] L. L. Avramov and L.-C. Sun, "Cohomology operators defined by a deformation", *J. Algebra* **204**:2 (1998), 684–710.

[Avramov, Conca and Iyengar 2010] L. L. Avramov, A. Conca, and S. B. Iyengar, "Free resolutions over commutative Koszul algebras", *Math. Res. Lett.* **17**:2 (2010), 197–210.

[Avramov, Conca and Iyengar 2015] L. L. Avramov, A. Conca, and S. B. Iyengar, "Subadditivity of syzygies of Koszul algebras", *Math. Ann.* **361**:1-2 (2015), 511–534.

[Avramov, Gasharov and Peeva 1997] L. L. Avramov, V. N. Gasharov, and I. V. Peeva, "Complete intersection dimension", *Inst. Hautes Études Sci. Publ. Math.* 86 (1997), 67–114.

[Babenko 1986] I. K. Babenko, "Problems of growth and rationality in algebra and topology", *Uspekhi Mat. Nauk* **41**:2(248) (1986), 95–142. In Russian; translated in *Russ. Math. Surv.* **41**:2 (1986), 117–175.

[Backelin 1982] J. Backelin, "Les anneaux locaux à relations monomiales ont des séries de Poincaré–Betti rationnelles", *C. R. Acad. Sci. Paris Sér. I Math.* **295**:11 (1982), 607–610.

[Backelin 1988] J. Backelin, "Relations between rates of growth of homologies", Reports of the Department of Mathematics **25**, Univ. of Stockholm, 1988.

[Bărcănescu and Manolache 1981] Ş. Bărcănescu and N. Manolache, "Betti numbers of Segre–Veronese singularities", *Rev. Roumaine Math. Pures Appl.* **26**:4 (1981), 549–565.

[Beilinson et al. 1996] A. Beilinson, V. Ginzburg, and W. Soergel, "Koszul duality patterns in representation theory", *J. Amer. Math. Soc.* **9**:2 (1996), 473–527.

[Berglund and Jöllenbeck 2007] A. Berglund and M. Jöllenbeck, "On the Golod property of Stanley–Reisner rings", *J. Algebra* **315**:1 (2007), 249–273.

[Bøgvad 1983] R. Bøgvad, "Gorenstein rings with transcendental Poincaré-series", *Math. Scand.* **53**:1 (1983), 5–15.

[Bøgvad 1994] R. Bøgvad, "Some homogeneous coordinate rings that are Koszul algebras", Reports of the Department of Mathematics **6**, Univ. of Stockholm, 1994.

[Cartan and Eilenberg 1956] H. Cartan and S. Eilenberg, *Homological algebra*, Princeton University Press, 1956.

[Caviglia 2009] G. Caviglia, "The pinched Veronese is Koszul", *J. Algebraic Combin.* **30**:4 (2009), 539–548.

[Caviglia and Conca 2013] G. Caviglia and A. Conca, "Koszul property of projections of the Veronese cubic surface", *Adv. Math.* **234** (2013), 404–413.

[Choi 1990] S. Choi, "Betti numbers and the integral closure of ideals", *Math. Scand.* **66**:2 (1990), 173–184.

[Conca et al. 2013] A. Conca, E. De Negri, and M. E. Rossi, "Koszul algebras and regularity", pp. 285–315 in *Commutative algebra*, edited by I. Peeva, Springer, New York, 2013.

[Conca, Trung and Valla 2001] A. Conca, N. V. Trung, and G. Valla, "Koszul property for points in projective spaces", *Math. Scand.* **89**:2 (2001), 201–216.

[Dao 2013] H. Dao, "Some homological properties of modules over a complete intersection, with applications", pp. 335–371 in *Commutative algebra*, edited by I. Peeva, Springer, New York, 2013.

[Eisen 1969] M. Eisen, *Elementary combinatorial analysis*, Gordon and Breach, New York, 1969.

[Eisenbud 1980] D. Eisenbud, "Homological algebra on a complete intersection, with an application to group representations", *Trans. Amer. Math. Soc.* **260**:1 (1980), 35–64.

[Eisenbud 1995] D. Eisenbud, *Commutative algebra: With a view toward algebraic geometry*, Graduate Texts in Mathematics **150**, Springer, New York, 1995.

[Eisenbud 2005] D. Eisenbud, *The geometry of syzygies: A second course in commutative algebra and algebraic geometry*, Graduate Texts in Mathematics **229**, Springer, New York, 2005.

[Eisenbud and Huneke 2005] D. Eisenbud and C. Huneke, "A finiteness property of infinite resolutions", *J. Pure Appl. Algebra* **201**:1-3 (2005), 284–294.

[Eisenbud and Peeva 2015] D. Eisenbud and I. Peeva, "Matrix factorizations for complete intersections and minimal free resolutions", preprint, 2015. arXiv 1306.2615

[Eisenbud and Peeva $\geq$ 2015a] D. Eisenbud and I. Peeva, "Formulas for minimal free resolutions over complete intersections". In preparation.

[Eisenbud and Peeva $\geq$ 2015b] D. Eisenbud and I. Peeva, "Quadratic complete intersections". In preparation.

[Eisenbud, Reeves and Totaro 1994] D. Eisenbud, A. Reeves, and B. Totaro, "Initial ideals, Veronese subrings, and rates of algebras", *Adv. Math.* **109**:2 (1994), 168–187.

[Fakhary and Welker 2012] S. A. S. Fakhary and V. Welker, "The Golod property for products and high symbolic powers of monomial ideals", preprint, 2012. arXiv 1209.2577

[Falk and Randell 1985] M. Falk and R. Randell, "The lower central series of a fiber-type arrangement", *Invent. Math.* **82**:1 (1985), 77–88.

[Falk and Randell 2000] M. Falk and R. Randell, "On the homotopy theory of arrangements, II", pp. 93–125 in *Arrangements—Tokyo 1998*, edited by M. Falk and H. Terao, Adv. Stud. Pure Math. **27**, Kinokuniya, Tokyo, 2000.

[Fröberg 1975] R. Fröberg, "Determination of a class of Poincaré series", *Math. Scand.* **37**:1 (1975), 29–39.

[Fröberg 1999] R. Fröberg, "Koszul algebras", pp. 337–350 in *Advances in commutative ring theory* (Fez, 1997), edited by D. E. Dobbs et al., Lecture Notes in Pure and Appl. Math. **205**, Dekker, New York, 1999.

[Fröberg and Roos 2000] R. Fröberg and J.-E. Roos, "An affine monomial curve with irrational Poincaré–Betti series", *J. Pure Appl. Algebra* **152**:1-3 (2000), 89–92.

[Gasharov and Peeva 1990] V. N. Gasharov and I. V. Peeva, "Boundedness versus periodicity over commutative local rings", *Trans. Amer. Math. Soc.* **320**:2 (1990), 569–580.

[Gasharov, Peeva and Welker 2000] V. Gasharov, I. Peeva, and V. Welker, "Rationality for generic toric rings", *Math. Z.* **233**:1 (2000), 93–102.

[Golod 1962] E. S. Golod, "Homologies of some local rings", *Dokl. Akad. Nauk SSSR* **144** (1962), 479–482.

[Gulliksen 1971] T. H. Gulliksen, "A homological characterization of local complete intersections", *Compositio Math.* **23** (1971), 251–255.

[Gulliksen 1972] T. H. Gulliksen, "Massey operations and the Poincaré series of certain local rings", *J. Algebra* **22** (1972), 223–232.

[Gulliksen 1974] T. H. Gulliksen, "A change of ring theorem with applications to Poincaré series and intersection multiplicity", *Math. Scand.* **34** (1974), 167–183.

[Gulliksen 1980] T. H. Gulliksen, "On the deviations of a local ring", *Math. Scand.* **47**:1 (1980), 5–20.

[Gulliksen and Levin 1969] T. H. Gulliksen and G. Levin, *Homology of local rings*, Queen's Paper in Pure and Applied Mathematics **20**, Queen's University, Kingston, Ont., 1969.

[Herzog and Huneke 2013] J. Herzog and C. Huneke, "Ordinary and symbolic powers are Golod", *Adv. Math.* **246** (2013), 89–99.

[Herzog and Srinivasan 2013] J. Herzog and H. Srinivasan, "A note on the subadditivity problem for maximal shifts in free resolutions", preprint, 2013. arXiv 1303.6214

[Herzog and Steurich 1979] J. Herzog and M. Steurich, "Golodideale der Gestalt $\mathfrak{a}\cap\mathfrak{b}$", *J. Algebra* **58**:1 (1979), 31–36.

[Herzog et al. 2011] J. Herzog, V. Welker, and S. Yassemi, "Homology of powers of ideals: Artin–Rees numbers of syzygies and the Golod property", preprint, 2011. arXiv 1108.5862v1

[Herzog, Reiner and Welker 1999] J. Herzog, V. Reiner, and V. Welker, "Componentwise linear ideals and Golod rings", *Michigan Math. J.* **46**:2 (1999), 211–223.

[Hilbert 1890] D. Hilbert, "Ueber die Theorie der algebraischen Formen", *Math. Ann.* **36**:4 (1890), 473–534.

[Hilbert 1893] D. Hilbert, "Ueber die vollen Invariantensysteme", *Math. Ann.* **42**:3 (1893), 313–373.

[Hirzebruch 1983] F. Hirzebruch, "Arrangements of lines and algebraic surfaces", pp. 113–140 in *Arithmetic and geometry, II*, edited by J. Coates and S. Helgason, Progr. Math. **36**, Birkhäuser, Boston, 1983.

[Jacobsson and Stoltenberg-Hansen 1985] C. Jacobsson and V. Stoltenberg-Hansen, "Poincaré–Betti series are primitive recursive", *J. London Math. Soc.* (2) **31**:1 (1985), 1–9.

[Kapustin and Li 2003] A. Kapustin and Y. Li, "D-branes in Landau–Ginzburg models and algebraic geometry", *J. High Energy Phys.* **2003**:12 (2003), 1–43.

[Lescot 1985] J. Lescot, "Asymptotic properties of Betti numbers of modules over certain rings", *J. Pure Appl. Algebra* **38**:2-3 (1985), 287–298.

[Lescot 1988] J. Lescot, "Séries de Poincaré des modules multi-gradués sur les anneaux monomiaux", pp. 155–161 in *Algebraic topology—rational homotopy* (Louvain-la-Neuve, 1986), edited by Y. Felix, Lecture Notes in Math. **1318**, Springer, Berlin, 1988.

[Lescot 1990] J. Lescot, "Séries de Poincaré et modules inertes", *J. Algebra* **132**:1 (1990), 22–49.

[Löfwall 1986] C. Löfwall, "On the subalgebra generated by the one-dimensional elements in the Yoneda Ext-algebra", pp. 291–338 in *Algebra, algebraic topology and their interactions* (Stockholm, 1983), edited by J.-E. Roos, Lecture Notes in Math. **1183**, Springer, Berlin, 1986.

[Löfwall, Lundqvist and Roos 2015] C. Löfwall, S. Lundqvist, and J.-E. Roos, "A Gorenstein numerical semi-group ring having a transcendental series of Betti numbers", *J. Pure Appl. Algebra* **219**:3 (2015), 591–621.

[Mac Lane 1963] S. Mac Lane, *Homology*, Die Grundlehren der mathematischen Wissenschaften **114**, Academic Press, New York, 1963.

[Manin 1987] Y. I. Manin, "Some remarks on Koszul algebras and quantum groups", *Ann. Inst. Fourier (Grenoble)* **37**:4 (1987), 191–205.

[Markushevich 1975] A. I. Markushevich, *Recursion sequences*, Little Mathematics Library **1**, Mir Publishers, Moscow, 1975.

[Matsumura 1989] H. Matsumura, *Commutative ring theory*, 2nd ed., Cambridge Studies in Advanced Mathematics **8**, Cambridge University Press, 1989.

[McCullough 2012] J. McCullough, "A polynomial bound on the regularity of an ideal in terms of half of the syzygies", *Math. Res. Lett.* **19**:3 (2012), 555–565.

[McCullough 2013] J. McCullough, "Stillman's question for exterior algebras", preprint, 2013. arXiv 1307.8162

[Nguyen and Vu 2015] H. Nguyen and T. Vu, "Koszul algebras and the Frobenius endomorphism", preprint, 2015. arXiv 1303.5160

[Orlik and Solomon 1980] P. Orlik and L. Solomon, "Combinatorics and topology of complements of hyperplanes", *Invent. Math.* **56**:2 (1980), 167–189.

[Orlov 2004] D. O. Orlov, "Triangulated categories of singularities and D-branes in Landau–Ginzburg models", *Tr. Mat. Inst. Steklova* **246**:Algebr. Geom. Metody, Svyazi i Prilozh. (2004), 240–262.

[Peeva 1996] I. Peeva, "0-Borel fixed ideals", *J. Algebra* **184**:3 (1996), 945–984.

[Peeva 2003] I. Peeva, "Hyperplane arrangements and linear strands in resolutions", *Trans. Amer. Math. Soc.* **355**:2 (2003), 609–618.

[Peeva 2007] I. Peeva, "Infinite free resolutions over toric rings", pp. 233–247 in *Syzygies and Hilbert functions*, edited by I. Peeva, Lect. Notes Pure Appl. Math. **254**, Chapman & Hall/CRC, Boca Raton, FL, 2007.

[Peeva 2011] I. Peeva, *Graded syzygies*, Algebra and Applications **14**, Springer, London, 2011.

[Peeva and Stillman 2009] I. Peeva and M. Stillman, "Open problems on syzygies and Hilbert functions", *J. Commut. Algebra* **1**:1 (2009), 159–195.

[Piontkovskiĭ 2001] D. I. Piontkovskiĭ, "On Hilbert series of Koszul algebras", *Funktsional. Anal. i Prilozhen.* **35**:2 (2001), 64–69. In Russian; translated in *Funct. Anal. Appl.* **35**:2 (2001), 133–137.

[Polishchuk and Positselski 2005] A. Polishchuk and L. Positselski, *Quadratic algebras*, University Lecture Series **37**, Amer. Math. Soc., Providence, RI, 2005.

[Positselksi 1995] L. E. Positsel′skiĭ, "The correspondence between Hilbert series of quadratically dual algebras does not imply their having the Koszul property", *Funktsional. Anal. i Prilozhen.* **29**:3 (1995), 83–87. In Russian; translated in *Funct. Anal. Appl.* **29**:3 (1995), 213–217.

[Priddy 1970] S. B. Priddy, "Koszul resolutions", *Trans. Amer. Math. Soc.* **152** (1970), 39–60.

[Ramras 1980] M. Ramras, "Sequences of Betti numbers", *J. Algebra* **66**:1 (1980), 193–204.

[Roos 1981] J.-E. Roos, "Homology of loop spaces and of local rings", pp. 441–468 in *18th Scandinavian Congress of Mathematicians* (Aarhus, 1980), edited by E. Balslev, Progr. Math. **11**, Birkhäuser, Boston, 1981.

[Roos 1993] J.-E. Roos, "Commutative non–Koszul algebras having a linear resolution of arbitrarily high order: Applications to torsion in loop space homology", *C. R. Acad. Sci. Paris Sér. I Math.* **316**:11 (1993), 1123–1128.

[Roos 1995] J.-E. Roos, "On the characterisation of Koszul algebras: Four counterexamples", *C. R. Acad. Sci. Paris Sér. I Math.* **321**:1 (1995), 15–20.

[Roos 2005] J.-E. Roos, "Good and bad Koszul algebras and their Hochschild homology", *J. Pure Appl. Algebra* **201**:1-3 (2005), 295–327.

[Roos and Sturmfels 1998] J.-E. Roos and B. Sturmfels, "A toric ring with irrational Poincaré–Betti series", *C. R. Acad. Sci. Paris Sér. I Math.* **326**:2 (1998), 141–146.

[Rossi and Şega 2014] M. E. Rossi and L. M. Şega, "Poincaré series of modules over compressed Gorenstein local rings", *Adv. Math.* **259** (2014), 421–447.

[Shelton and Yuzvinsky 1997] B. Shelton and S. Yuzvinsky, "Koszul algebras from graphs and hyperplane arrangements", *J. London Math. Soc.* (2) **56**:3 (1997), 477–490.

[Sjödin 1985] G. Sjödin, "The Ext-algebra of a Golod ring", *J. Pure Appl. Algebra* **38**:2-3 (1985), 337–351.

[Tate 1957] J. Tate, "Homology of Noetherian rings and local rings", *Illinois J. Math.* **1** (1957), 14–27.

[Vu 2013] T. Vu, "The Koszul property of pinched Veronese varieties", preprint, 2013. arXiv 1309.3033

jmccullough@rider.edu

Department of Mathematics, Rider University, 2083 Lawrenceville Road, Lawrenceville, NJ 08648, United States

Department of Mathematics, Malott Hall, Cornell University, Ithaca, NY 14853, United States

Commutative Algebra and Noncommutative Algebraic Geometry, I
MSRI Publications
Volume **67**, 2015

Poincaré–Birkhoff–Witt theorems

ANNE V. SHEPLER AND SARAH WITHERSPOON

We sample some Poincaré–Birkhoff–Witt theorems appearing in mathematics. Along the way, we compare modern techniques used to establish such results, for example, the composition-diamond lemma, Gröbner basis theory, and the homological approaches of Braverman and Gaitsgory and of Polishchuk and Positselski. We discuss several contexts for PBW theorems and their applications, such as Drinfeld–Jimbo quantum groups, graded Hecke algebras, and symplectic reflection and related algebras.

1. Introduction

Poincaré [1900] published a fundamental result on Lie algebras that would prove a powerful tool in representation theory: A Lie algebra embeds into an associative algebra that behaves in many ways like a polynomial ring. Capelli [1890] had proven a special case of this theorem, for the general linear Lie algebra, ten years earlier. Birkhoff [1937] and Witt [1937] independently formulated and proved versions of the theorem that we use today, although neither author cited this earlier work. The result was called the Birkhoff–Witt theorem for years and then later the Poincaré–Witt theorem (see [Cartan and Eilenberg 1956]) before Bourbaki [1960] prompted use of its current name, the *Poincaré–Birkhoff–Witt theorem*.

The original theorem on Lie algebras was greatly expanded over time by a number of authors to describe various algebras, especially those defined by quadratic-type relations (including Koszul rings over semisimple algebras). Poincaré–Birkhoff–Witt theorems are often used as a springboard for investigating the representation theory of algebras. These theorems are used to

- reveal an algebra as a deformation of another, well-behaved algebra,
- posit a convenient basis (of "monomials") for an algebra, and
- endow an algebra with a canonical homogeneous (or graded) version.

This material is based upon work supported by the National Science Foundation under grant no. 0932078000, while Witherspoon was in residence at the Mathematical Sciences Research Institute (MSRI) in Berkeley, California, during the spring semester of 2013. Shepler was partially supported by NSF grant #DMS-1101177. Witherspoon was partially supported by NSF grant #DMS-1101399.

In this survey, we sample some of the various Poincaré–Birkhoff–Witt theorems, applications, and techniques used to date for proving these results. Our survey is not intended to be all-inclusive; we instead seek to highlight a few of the more recent contributions and provide a helpful resource for users of Poincaré–Birkhoff–Witt theorems, which we henceforth refer to as *PBW theorems.*

We begin with a quick review in Section 2 of the original PBW theorem for enveloping algebras of Lie algebras. We next discuss PBW properties for quadratic algebras in Section 3, and for Koszul algebras in particular, before turning to arbitrary finitely generated algebras in Section 4. We recall needed facts on Hochschild cohomology and algebraic deformation theory in Section 5, and more background on Koszul algebras is given in Section 6. Sections 7–8 outline techniques for proving PBW results recently used in more general settings, some by way of homological methods and others via the composition-diamond lemma (and Gröbner basis theory). One inevitably is led to similar computations when applying any of these techniques to specific algebras, but with different points of view. Homological approaches can help to organize computations and may contain additional information, while approaches using Gröbner basis theory are particularly well-suited for computer computation. We focus on some classes of algebras in Sections 9 and 10 of recent interest: Drinfeld–Jimbo quantum groups, Nichols algebras of diagonal type, symplectic reflection algebras, rational Cherednik algebras, and graded (Drinfeld) Hecke algebras. In Section 11, we mention applications in positive characteristic (including algebras built on group actions in the modular case) and other generalizations that mathematicians have only just begun to explore.

We take all tensor products over an underlying field k unless otherwise indicated and assume all algebras are associative k-algebras with unity. Note that although we limit discussions to finitely generated algebras over k for simplicity, many remarks extend to more general settings.

2. Lie algebras and the classical PBW theorem

All PBW theorems harken back to a classical theorem for universal enveloping algebras of Lie algebras established independently by Poincaré [1900], Birkhoff [1937] and Witt [1937]. In this section, we recall this original PBW theorem in order to set the stage for other PBW theorems and properties; for comprehensive historical treatments, see [Grivel 2004; Ton-That and Tran 1999].

A finite dimensional *Lie algebra* is a finite dimensional vector space $\mathfrak{g}$ over a field k together with a binary operation $[\ ,\ ] : \mathfrak{g} \times \mathfrak{g} \to \mathfrak{g}$ satisfying

(i) (antisymmetry) $[v, v] = 0$, and

(ii) (Jacobi identity) $[u, [v, w]] + [v, [w, u]] + [w, [u, v]] = 0$, for all $u, v, w \in \mathfrak{g}$. Condition (i) implies $[v, w] = -[w, v]$ for all v, w in $\mathfrak{g}$ (and is equivalent to this condition in all characteristics other than 2).

The *universal enveloping algebra* $U(\mathfrak{g})$ of $\mathfrak{g}$ is the associative algebra generated by the vectors in $\mathfrak{g}$ with relations $vw - wv = [v, w]$ for all v, w in $\mathfrak{g}$, that is,

$$U(\mathfrak{g}) = T(\mathfrak{g})\big/(v \otimes w - w \otimes v - [v, w] : v, w \in \mathfrak{g}),$$

where $T(\mathfrak{g})$ is the tensor algebra of the vector space $\mathfrak{g}$ over k. It can be defined by a universal property: $U(\mathfrak{g})$ is the (unique up to isomorphism) associative algebra such that any linear map ϕ from $\mathfrak{g}$ to an associative algebra A satisfying $[\phi(v), \phi(w)] = \phi([v, w])$ for all $v, w \in \mathfrak{g}$ factors through $U(\mathfrak{g})$. (The bracket operation on an associative algebra A is given by $[a, b] := ab - ba$ for all $a, b \in A$.) As an algebra, $U(\mathfrak{g})$ is filtered, under the assignment of degree 1 to each vector in $\mathfrak{g}$.

Original PBW theorem. *A Lie algebra $\mathfrak{g}$ embeds into its universal enveloping algebra $U(\mathfrak{g})$, and the associated graded algebra of $U(\mathfrak{g})$ is isomorphic to $S(\mathfrak{g})$, the symmetric algebra on the vector space $\mathfrak{g}$.*

Thus the original PBW theorem compares a universal enveloping algebra $U(\mathfrak{g})$ to an algebra of (commutative) polynomials. Since monomials form a k-basis for a polynomial algebra, the original PBW theorem is often rephrased in terms of a *PBW basis* (with tensor signs between vectors dropped):

PBW basis theorem. *Let $v_1, \ldots, v_n$ be an ordered k-vector space basis of the Lie algebra $\mathfrak{g}$. Then $\{v_1^{a_1} \cdots v_n^{a_n} : a_i \in \mathbb{N}\}$ is a k-basis of the universal enveloping algebra $U(\mathfrak{g})$.*

Example 2.1. The Lie algebra $\mathfrak{sl}_2(\mathbb{C})$ consists of 2×2 matrices of trace 0 with entries in $\mathbb{C}$ under the bracket operation on the associative algebra of all 2×2 matrices. The standard basis of $\mathfrak{sl}_2(\mathbb{C})$ is

$$e = \begin{pmatrix} 0 & 1 \\ 0 & 0 \end{pmatrix}, \quad f = \begin{pmatrix} 0 & 0 \\ 1 & 0 \end{pmatrix}, \quad h = \begin{pmatrix} 1 & 0 \\ 0 & -1 \end{pmatrix},$$

for which $[e, f] = h$, $[h, e] = 2e$, $[h, f] = -2f$. Thus $U(\mathfrak{sl}_2(\mathbb{C}))$ is the associative $\mathbb{C}$-algebra generated by three symbols that we also denote by e, f, h (abusing notation) subject to the relations $ef - fe = h$, $he - eh = 2e$, $hf - fh = -2f$. It has $\mathbb{C}$-basis $\{e^a h^b f^c : a, b, c \in \mathbb{N}\}$.

Proofs of the original PBW theorem vary (and by how much is open to interpretation). The interested reader may wish to consult, for example, [Cartan and Eilenberg 1956; Dixmier 1977; Humphreys 1972; Jacobson 1962; Varadarajan 1984]. Jacobson [1941] proved a PBW theorem for restricted enveloping algebras in positive characteristic. Higgins [1969] gives references and a comprehensive

PBW theorem over more general ground rings. A PBW theorem for Lie superalgebras goes back to Milnor and Moore [1965] (see also [Kac 1977]). Grivel's historical article [2004] includes further references on generalizations to other ground rings, to Leibniz algebras, and to Weyl algebras. In Sections 7 and 8 below, we discuss two proof techniques particularly well suited to generalization: a combinatorial approach through the composition-diamond lemma and a homological approach through algebraic deformation theory. First we lay some groundwork on quadratic algebras.

3. Homogeneous quadratic algebras

Many authors have defined the notions of PBW algebra, PBW basis, PBW deformation, or PBW property in order to establish theorems like the original PBW theorem in more general settings. Let us compare a few of these concepts, beginning in this section with those defined for *homogeneous* quadratic algebras.

Quadratic algebras. Consider a finite dimensional vector space V over k with basis $v_1, \ldots, v_n$. Let T be its tensor algebra over k, that is, the free k-algebra $k\langle v_1, \ldots, v_n\rangle$ generated by the v_i. Then T is an $\mathbb{N}$-graded k-algebra with

$$T^0 = k, \quad T^1 = V, \quad T^2 = V \otimes V, \quad T^3 = V \otimes V \otimes V, \text{ etc.}$$

We often omit tensor signs in writing elements of T as is customary in noncommutative algebra, for example, writing x^3 for $x \otimes x \otimes x$ and xy for $x \otimes y$.

Suppose P is a set of filtered (nonhomogeneous) relations in degree 2,

$$P \subseteq T^0 \oplus T^1 \oplus T^2,$$

and let $I = (P)$ be the 2-sided ideal in T generated by P. The quotient $A = T/I$ is a *nonhomogeneous quadratic algebra*. If P consists of elements of homogeneous degree 2, that is, $P \subseteq T^2$, then A is a *homogeneous quadratic algebra*. Thus a quadratic algebra is just an algebra whose relations are generated by (homogeneous or nonhomogeneous) quadratic expressions.

We usually write each element of a finitely presented algebra $A = T/I$ as a coset representative in T, suppressing mention of the ideal I. Then a *k-basis* for A is a subset of T representing cosets modulo I which form a basis for A as a k-vector space. Some authors say a quadratic algebra has a *PBW basis* if it has the same k-basis as a universal enveloping algebra, that is, if $\{v_1^{a_1} \cdots v_n^{a_n} : a_i \in \mathbb{N}\}$ is a basis for A as a k-vector space. Such algebras include Weyl algebras, quantum/skew polynomial rings, some iterated Ore extensions, some quantum groups, etc.

Priddy's PBW algebras. Priddy [1970] gave a broader definition of PBW basis for homogeneous quadratic algebras in terms of any ordered basis of V (say, $v_1 < v_2 < \cdots < v_n$) in establishing the notion of Koszul algebras. (A quadratic algebra is *Koszul* if the boundary maps in its minimal free resolution have matrix entries that are linear forms; see Section 6.) Priddy first extended the ordering degree-lexicographically to a monomial ordering on the tensor algebra T, where we regard pure tensors in $v_1, \ldots, v_n$ as monomials. He then called a k-vector space basis for $A = T/I$ a *PBW basis* (and the algebra A a *PBW algebra*) if the product of any two basis elements either lay again in the basis or could be expressed modulo I as a sum of larger elements in the basis. In doing so, Priddy [1970, Theorem 5.3] gave a class of Koszul algebras which is easy to study:

Theorem 3.1. *If a homogeneous quadratic algebra has a PBW basis, then it is Koszul.*

Polishchuk and Positselski [2005, Chapter 4, Section 1] reframed Priddy's idea; we summarize their approach using the notion of leading monomial LM of any element of T written in terms of the basis $v_1, \ldots, v_n$ of V. Suppose the set of generating relations P is a subspace of T^2. Consider those monomials that are not divisible by the leading monomial of any generating quadratic relation:

$$\mathcal{B}_P = \{\text{monomials } m \in T : \mathrm{LM}(a) \nmid m \text{ for all } a \in P\}.$$

Polishchuk and Positselski call $\mathcal{B}_P$ a *PBW basis* of the quadratic algebra A (and A a *PBW algebra*) whenever $\mathcal{B}_P$ is a k-basis of A.

Gröbner bases. Priddy's definition and the reformulation of Polishchuk and Positselski immediately call to mind the theory of Gröbner bases. Recall that a set $\mathcal{G}$ of nonzero elements generating an ideal I is called a (noncommutative) *Gröbner basis* if the leading monomial of each nonzero element of I is divisible by the leading monomial of some element of $\mathcal{G}$ with respect to a fixed monomial (i.e., term) ordering (see [Mora 1986] or [Li 2012]). (Gröbner bases and Gröbner–Shirshov bases were developed independently in various contexts by Shirshov [1962a], Hironaka [1964a; 1964b], Buchberger [1965], Bokut [1976] and Bergman [1978].) A Gröbner basis $\mathcal{G}$ is *quadratic* if it consists of homogeneous elements of degree 2 (i.e., lies in T^2) and it is *minimal* if no proper subset is also a Gröbner basis. A version of the composition-diamond lemma for associative algebras (see Section 8) implies that if $\mathcal{G}$ is a Gröbner basis for I, then

$$\mathcal{B}_{\mathcal{G}} = \{\text{monomials } m \in T : \mathrm{LM}(a) \nmid m \text{ for all } a \in \mathcal{G}\}$$

is a k-basis for $A = T(V)/I$.

Example 3.2. Let A be the $\mathbb{C}$-algebra generated by symbols x, y with a single generating relation $xy = y^2$. Set $V = \mathbb{C}\text{-span}\{x, y\}$ and $P = \{xy - y^2\}$ so that $A = T(V)/(P)$. A Gröbner basis $\mathcal{G}$ for the ideal $I = (P)$ with respect to the degree-lexicographical monomial ordering with $x < y$ is infinite:

$$\mathcal{G} = \{yx^n y - x^{n+1} y : n \in \mathbb{N}\},$$

$$\mathcal{B}_P = \{\text{monomials } m \in T \text{ that are not divisible by } y^2\},$$

$$\mathcal{B}_{\mathcal{G}} = \{\text{monomials } m \in T \text{ that are not divisible by } yx^n y \text{ for any } n \in \mathbb{N}\}.$$

Hence, A is not a PBW algebra using the ordering $x < y$ since $\mathcal{B}_{\mathcal{G}}$ is a $\mathbb{C}$-basis for A but $\mathcal{B}_P$ is not.

If we instead take some monomial ordering with $x > y$, then $\mathcal{G} = P$ is a Gröbner basis for the ideal $I = (P)$ and $\mathcal{B}_{\mathcal{G}} = \mathcal{B}_P$ is a $\mathbb{C}$-basis of A:

$$\begin{aligned}\mathcal{B}_P = \mathcal{B}_{\mathcal{G}} &= \{\text{monomials } m \in T \text{ that are not divisible by } xy\} \\ &= \{y^a x^b : a, b \in \mathbb{N}\}.\end{aligned}$$

Hence A is a PBW algebra using the ordering $y < x$.

Quadratic Gröbner bases. How do the sets of monomials $\mathcal{B}_P$ and $\mathcal{B}_{\mathcal{G}}$ compare after fixing an appropriate monomial ordering? Suppose $\mathcal{G}$ is a minimal Gröbner basis for $I = (P)$ (which implies that no element of $\mathcal{G}$ has leading monomial dividing that of another). Then $\mathcal{B}_{\mathcal{G}} \subset \mathcal{B}_P$, and the reverse inclusion holds whenever $\mathcal{G}$ is quadratic (since then $\mathcal{G}$ must be a subset of the subspace P). Since each graded piece of A is finite dimensional over k, a PBW basis thus corresponds to a quadratic Gröbner basis:

$$\mathcal{B}_P \text{ is a PBW basis of } A \iff \mathcal{B}_{\mathcal{G}} = \mathcal{B}_P \iff \mathcal{G} \text{ is quadratic.}$$

Thus authors sometimes call any algebra defined by an ideal of relations with a quadratic Gröbner basis a PBW algebra. In any case, such an algebra is Koszul [Anick 1986; Bruns et al. 1994; Fröberg 1999; Loday and Vallette 2012]:

Theorem 3.3. *Any quadratic algebra whose ideal of relations has a* (*noncommutative*) *quadratic Gröbner basis is Koszul.*

Backelin (see [Polishchuk and Positselski 2005, Chapter 4, Section 3]) gave an example of a Koszul algebra defined by an ideal of relations with no quadratic Gröbner basis. Eisenbud, Reeves and Totaro [Eisenbud et al. 1994, p. 187] gave an example of a commutative Koszul algebra whose ideal of relations does not have a quadratic Gröbner basis with respect to *any* ordering, even after a change of basis (see also [Fröberg 1999]).

We relate Gröbner bases and PBW theorems for *nonhomogeneous* algebras in Section 8.

4. Nonhomogeneous algebras: PBW deformations

Algebras defined by generators and relations are not naturally graded, but merely filtered, and often one wants to pass to some graded or homogeneous version of the algebra for quick information. There is more than one way to do this in general. The original PBW theorem shows that the universal enveloping algebra of a Lie algebra has one natural homogeneous version. Authors apply this idea to other algebras, saying that an algebra satisfies a *PBW property* when graded versions are isomorphic and call the original algebra a *PBW deformation* of this graded version. We make these notions precise in this section and relate them to the work of Braverman and Gaitsgory and of Polishchuk and Positselski on Koszul algebras in the next section.

Filtered algebras. Again, consider an algebra A generated by a finite dimensional vector space V over a field k with some defining set of relations P. (More generally, one might consider a module over a group algebra or some other k-algebra.) Let $T = \bigoplus_{i\geq 0} T^i$ be the tensor algebra over V and let $I = (P)$ be the two-sided ideal of relations so that

$$A = T/I.$$

If I is homogeneous, then the quotient algebra A is graded. In general, I is nonhomogeneous and the quotient algebra is only filtered, with i-th filtered component $F^i(A) = F^i(T/I) = (F^i(T) + I)/I$ induced from the filtration on T obtained by assigning degree one to each vector in V (i.e., $F^i(T) = T^0 \oplus T^1 \oplus \ldots \oplus T^i$).

Homogeneous versions. One associates to the filtered algebra A two possibly different graded versions. On one hand, we cross out lower order terms in the *generating* set P of relations to obtain a homogeneous version of the original algebra. On the other hand, we cross out lower order terms in each element of the *entire* ideal of relations. Then *PBW conditions* are precisely those under which these two graded versions of the original algebra coincide, as we recall next.

The *associated graded algebra* of A,

$$\mathrm{gr}(A) = \bigoplus_{i\geq 0} F^i(A) \big/ F^{i-1}(A),$$

is a graded version of A which does not depend on the choice of generators P of the ideal of relations I. (We set $F^{-1} = \{0\}$.) The associated graded algebra may be realized concretely by projecting each element in the ideal I onto its leading homogeneous part (see [Li 2012, Theorem 3.2]):

$$\operatorname{gr}(T/I) \cong T/(\mathrm{LH}(I)),$$

where $\mathrm{LH}(S) = \{\mathrm{LH}(f) : f \in S\}$ for any $S \subseteq T$ and $\mathrm{LH}(f)$ picks off the leading (or highest) homogeneous part of f in the graded algebra T. (Formally, $\mathrm{LH}(f) = f_d$ for $f = \sum_{i=1}^{d} f_i$ with each f_i in T^i and f_d nonzero.) Those looking for a shortcut may be tempted instead simply to project elements of the generating set P onto their leading homogeneous parts. A natural surjection (of graded algebras) always arises from this homogeneous version of A determined by P to the associated graded algebra of A:

$$T/(\mathrm{LH}(P)) \twoheadrightarrow \operatorname{gr}(T/I).$$

PBW deformations. We say the algebra T/I is a *PBW deformation* of its homogeneous version $T/(\mathrm{LH}(P))$ (or satisfies the *PBW property* with respect to P) when the above surjection is also injective, that is, when the associated graded algebra and the homogeneous algebra determined by P coincide (see [Braverman and Gaitsgory 1996]):

$$T/(\mathrm{LH}(I)) \cong \operatorname{gr}(T/I) \cong T/(\mathrm{LH}(P)).$$

In the next section, we explain the connections among PBW deformations, graded (and formal) deformations, and Hochschild cohomology.

In this language, the original PBW theorem for universal enveloping algebras asserts that the set

$$P = \{v \otimes w - w \otimes v - [v, w] : v, w \in V\}$$

gives rise to a quotient algebra $T/(P)$ that is a PBW deformation of the commutative polynomial ring $S(V)$, for V the underlying vector space of a Lie algebra. Here, each element of V has degree 1 so that the relations are nonhomogeneous of degree 2 and $T/(P)$ is a nonhomogeneous quadratic algebra.

We include an example next to show how the PBW property depends on choice of generating relations P defining the algebra T/I. (But note that if A satisfies the PBW property with respect to some generating set P of relations, then the subspace P generates is unique; see [Shepler and Witherspoon 2014, Proposition 2.1].)

Example 4.1. We mention a filtered algebra that exhibits the PBW property with respect to one generating set of relations but not another. Consider the (noncommutative) algebra A generated by symbols x and y with defining relations $xy = x$ and $yx = y$:

$$A = k\langle x, y\rangle/(xy - x, yx - y),$$

where $k\langle x, y\rangle$ is the free k-algebra generated by x and y. The algebra A *does not* satisfy the PBW property with respect to the generating relations $xy - x$ and $yx - y$. Indeed, the relations imply that $x^2 = x$ and $y^2 = y$ in A and thus the associated graded algebra $\mathrm{gr}(A)$ is trivial in degree two while the homogeneous version of A is not (as x^2 and y^2 represent nonzero classes). The algebra A *does* exhibit the PBW property with respect to the larger generating set $\{xy - x, yx - y, x^2 - x, y^2 - y\}$ since

$$\mathrm{gr}\, A \cong k\langle x, y\rangle \big/ (xy, yx, x^2, y^2).$$

Examples 8.1 and 8.2 explain this recovery of the PBW property in terms of Gröbner bases and the composition-diamond lemma.

5. Deformation theory and Hochschild cohomology

In the last section, we saw that an algebra defined by nonhomogeneous relations is called a *PBW deformation* when the homogeneous version determined by generating relations coincides with its associated graded algebra. How may one view formally the original nonhomogeneous algebra as a *deformation* of its homogeneous version? In this section, we begin to fit PBW deformations into the theory of algebraic deformations. We recall the theory of deformations of algebras and Hochschild cohomology, a homological tool used to predict deformations and prove PBW properties.

Graded deformations. Let t be a formal parameter. A *graded deformation* of a graded k-algebra A is a graded associative $k[t]$-algebra A_t (for t in degree 1) which is isomorphic to $A[t] = A \otimes_k k[t]$ as a $k[t]$-module with

$$A_t|_{t=0} \cong A$$

as an algebra. If we specialize t to an element of k in the algebra A_t, then we may no longer have a graded algebra, but a filtered algebra instead.

PBW deformations may be viewed as graded deformations: Each PBW deformation is a graded deformation of its homogeneous version with parameter t specialized to some element of k. Indeed, given a finitely generated algebra $A = T/(P)$, we may insert a formal parameter t of degree 1 throughout the defining relations P to make each relation homogeneous and extend scalars to $k[t]$; the result yields a graded algebra B_t over $k[t]$ with $A = B_t|_{t=1}$ and $B = B_t|_{t=0}$, the homogeneous version of A. One may verify that if A satisfies the PBW property, then this interpolating algebra B_t also satisfies a PBW condition over $k[t]$ and that B_t and $B[t]$ are isomorphic as $k[t]$-modules. Thus as B_t is an associative graded algebra, it defines a graded deformation of B.

Suppose A_t is a graded deformation of a graded k-algebra A. Then up to isomorphism, A_t is just the vector space $A[t]$ together with some associative multiplication given by

$$a * b = ab + \mu_1(a \otimes b)t + \mu_2(a \otimes b)t^2 + \cdots , \tag{5.1}$$

where ab is the product of a and b in A, and for each i, μ_i is a linear map from $A \otimes A$ to A of degree $-i$, extended to be $k[t]$-linear. The degree condition on the maps μ_i are forced by the fact that A_t is graded for t in degree 1. (One sometimes considers a *formal* deformation, defined over formal power series $k[[t]]$ instead of polynomials $k[t]$.)

The condition that the multiplication $*$ in $A[t]$ be associative imposes conditions on the functions μ_i which are often expressed using Hochschild cohomology. For example, comparing coefficients of t in the equation $(a * b) * c = a * (b * c)$, we see that μ_1 must satisfy

$$a\mu_1(b \otimes c) + \mu_1(a \otimes bc) = \mu_1(ab \otimes c) + \mu_1(a \otimes b)c \tag{5.2}$$

for all $a, b, c \in A$. We see below that this condition implies that μ_1 is a Hochschild 2-cocycle. Comparing coefficients of t^2 yields a condition on μ_1, μ_2 called the *first obstruction*, comparing coefficients of t^3 yields a condition on μ_1, μ_2, μ_3 called the *second obstruction*, and so on. (See [Gerstenhaber 1964].)

Hochschild cohomology. Hochschild cohomology is a generalization of group cohomology well suited to noncommutative algebras. It gives information about an algebra A viewed as a bimodule over itself, thus capturing right and left multiplication, and predicts possible multiplication maps μ_i that could be used to define a deformation of A. One may define the Hochschild cohomology of a k-algebra concretely as Hochschild cocycles modulo Hochschild coboundaries by setting

$$\text{Hochschild } i\text{-cochains} = \{\text{linear functions } \phi : \underbrace{A \otimes \cdots \otimes A}_{i \text{ times}} \to A\}$$

(i.e., multilinear maps $A \times \cdots \times A \to A$) with linear boundary operator

$$\delta^*_{i+1} : i\text{-cochains} \to (i+1)\text{-cochains}$$

given by

$$\begin{aligned}(\delta^*_{i+1}\phi)(a_0 \otimes \cdots \otimes a_i) &= a_0\phi(a_1 \otimes \cdots \otimes a_i)\\ &+ \sum_{0 \le j \le i-1} (-1)^{j+1}\phi(a_0 \otimes \cdots \otimes a_{j-1} \otimes a_j a_{j+1} \otimes a_{j+2} \otimes \cdots \otimes a_i)\\ &\qquad + (-1)^{i+1}\phi(a_0 \otimes \cdots \otimes a_{i-1})a_i .\end{aligned}$$

We identify A with the set of 0-cochains. Then

$$\mathrm{HH}^i(A) := \mathrm{Ker}\,\delta_{i+1}^* / \mathrm{Im}\,\delta_i^*.$$

We are interested in other concrete realizations of Hochschild cohomology giving isomorphic cohomology groups. Formally, we view any k-algebra A as a bimodule over itself, that is, a right A^e-module where A^e is its enveloping algebra, $A \otimes A^{\mathrm{op}}$, for A^{op} the opposite algebra of A. The Hochschild cohomology of A is then just

$$\mathrm{HH}^\bullet(A) = \mathrm{Ext}^\bullet_{A^e}(A, A).$$

This cohomology is often computed using the A-bimodule bar resolution of A:

$$\cdots \longrightarrow A^{\otimes 4} \xrightarrow{\delta_2} A^{\otimes 3} \xrightarrow{\delta_1} A^{\otimes 2} \xrightarrow{\delta_0} A \longrightarrow 0, \tag{5.3}$$

where δ_0 is the multiplication in A, and, for each $i \geq 1$,

$$\delta_i(a_0 \otimes \cdots \otimes a_{i+1}) = \sum_{j=0}^{i} (-1)^j a_0 \otimes \cdots \otimes a_j a_{j+1} \otimes \cdots \otimes a_{i+1}$$

for $a_0, \ldots, a_{i+1}$ in A. We take the homology of this complex after dropping the initial term A and applying $\mathrm{Hom}_{A \otimes A^{\mathrm{op}}}(-, A)$ to obtain the above description of Hochschild cohomology in terms of Hochschild cocycles and coboundaries, using the identification

$$\mathrm{Hom}_{A \otimes A^{\mathrm{op}}}(A \otimes A^{\otimes i} \otimes A, A) \cong \mathrm{Hom}_k(A^{\otimes i}, A).$$

6. Koszul algebras

We wish to extend the original PBW theorem for universal enveloping algebras to other nonhomogeneous quadratic algebras. When is a given algebra a PBW deformation of another well-understood and well-behaved algebra? Can we replace the polynomial algebra in the original PBW theorem by any homogeneous quadratic algebra, provided it is well-behaved in some way? We turn to Koszul algebras as a wide class of quadratic algebras generalizing the class of polynomial algebras. In this section, we briefly recall the definition of a Koszul algebra.

6.1. ***Koszul complex.*** An algebra S is a *Koszul algebra* if the underlying field k admits a linear S-free resolution, that is, one with boundary maps given by matrices whose entries are linear forms. Equivalently, S is a Koszul algebra if S is quadratic with generating vector space V and generating relations $R \subseteq V \otimes V$ and the following complex of left S-modules is acyclic:

$$\cdots \longrightarrow K_3(S) \longrightarrow K_2(S) \longrightarrow K_1(S) \longrightarrow K_0(S) \longrightarrow k \longrightarrow 0, \tag{6.1}$$

where $K_0(S) = S$, $K_1(S) = S \otimes V$, $K_2(S) = S \otimes R$, and for $i \geq 3$,

$$K_i(S) = S \otimes \left(\bigcap_{j=0}^{i-2} V^{\otimes j} \otimes R \otimes V^{\otimes(i-2-j)} \right).$$

The differential is that inherited from the bar resolution of k as an S-bimodule,

$$\cdots \xrightarrow{\partial_4} S^{\otimes 4} \xrightarrow{\partial_3} S^{\otimes 3} \xrightarrow{\partial_2} S^{\otimes 2} \xrightarrow{\partial_1} S \xrightarrow{\varepsilon} k \longrightarrow 0, \tag{6.2}$$

where ε is the augmentation map ($\varepsilon(v) = 0$ for all v in V) and for each $i \geq 1$,

$$\partial_i(s_0 \otimes \cdots \otimes s_i) = (-1)^i \varepsilon(s_i) s_0 \otimes \cdots \otimes s_{i-1} + \sum_{j=0}^{i-1} (-1)^j s_0 \otimes \cdots \otimes s_j s_{j+1} \otimes \cdots \otimes s_i.$$

(Note that for each i, $K_i(S)$ is an S-submodule of $S^{\otimes(i+1)}$.)

Bimodule Koszul complex. Braverman and Gaitsgory gave an equivalent definition of Koszul algebra via the bimodule Koszul complex: Let

$$\widetilde{K}_i(S) = K_i(S) \otimes S, \tag{6.3}$$

an S^e-module (equivalently S-bimodule) where $S^e = S \otimes S^{op}$. Then $\widetilde{K}_\bullet(S)$ embeds into the bimodule bar resolution (5.3) whose i-th term is $S^{\otimes(i+2)}$, and S is Koszul if and only if $\widetilde{K}_\bullet(S)$ is a bimodule resolution of S. Thus we may obtain the Hochschild cohomology $\mathrm{HH}^\bullet(S)$ of S (which contains information about its deformations) by applying $\mathrm{Hom}_{S^e}(-, S)$ either to the Koszul resolution $\widetilde{K}_\bullet(S)$ or to the bar resolution (5.3) of S as an S^e-module (after dropping the initial nonzero terms of each) and taking homology. We see in the next section how these resolutions and the resulting cohomology are used in homological proofs of a generalization of the PBW theorem from [Braverman and Gaitsgory 1996; Polishchuk and Positselski 2005; Positselski 1993].

7. Homological methods and deformations of Koszul algebras

Polishchuk and Positselski [2005; Positselski 1993] and Braverman and Gaitsgory [1996] extended the idea of the original PBW theorem for universal enveloping algebras to other nonhomogeneous quadratic algebras by replacing the polynomial algebra in the theorem by an arbitrary Koszul algebra. They stated conditions for a version of the original PBW theorem to hold in this greater generality and gave homological proofs. (Polishchuk and Positselski [2005] in fact gave two proofs, one homological that goes back to [Positselski 1993] and another using distributive lattices.) We briefly summarize these two homological approaches in this section and discuss generalizations.

Theorem of Polishchuk and Positselski, Braverman and Gaitsgory. As in the last sections, let V be a finite dimensional vector space over a field k and let T be its tensor algebra over k with i-th filtered component $F^i(T)$. Consider a subspace P of $F^2(T)$ defining a nonhomogeneous quadratic algebra

$$A = {}^{T}\big/_{(P)}.$$

Let $R = \mathrm{LH}(P) \cap T^2$ be the projection of P onto the homogeneous component of degree 2, and set

$$S = {}^{T}\big/_{(R)},$$

a homogeneous quadratic algebra (the homogeneous version of A as in Section 4). Then A is a PBW deformation of S when $\mathrm{gr}\, A$ and S are isomorphic as graded algebras.

Braverman and Gaitsgory [1996] and also Polishchuk and Positselski [2005; Positselski 1993] gave a generalization of the PBW theorem as:

Theorem 7.1. *Let A be a nonhomogeneous quadratic algebra, $A = T/(P)$, and $S = T/(R)$ its corresponding homogeneous quadratic algebra. Suppose S is a Koszul algebra. Then A is a PBW deformation of S if, and only if, the following two conditions hold*:

(I) $P \cap F^1(T) = \{0\}$, *and*

(J) $(F^1(T) \cdot P \cdot F^1(T)) \cap F^2(T) = P$.

We have chosen the notation of Braverman and Gaitsgory. The necessity of conditions (I) and (J) can be seen by direct algebraic manipulations. Similarly, direct computation shows that if (I) holds, then (J) is equivalent to (i), (ii), and (iii) of Theorem 7.2 below. Braverman and Gaitsgory used algebraic deformation theory to show that these conditions are also sufficient. Polishchuk and Positselski used properties of an explicit complex defined using the Koszul dual of S. The conditions (i), (ii), (iii) facilitate these connections to homological algebra, and they are easier in practice to verify than checking (J) directly. But in order to state these conditions, we require a canonical decomposition for elements of P: Condition (I) of Theorem 7.1 implies that every element of P can be written as the sum of a nonzero element of R (of degree 2), a linear term, and a constant term, that is, there exist linear functions $\alpha : R \to V$, $\beta : R \to k$ for which

$$P = \{r - \alpha(r) - \beta(r) \mid r \in R\}.$$

One may then rewrite Condition (J) and reformulate Theorem 7.1 as follows.

Theorem 7.2. *Let A be a nonhomogeneous quadratic algebra, $A = T/(P)$, and $S = T/(R)$ its corresponding homogeneous quadratic algebra. Suppose S is a*

Koszul algebra. Then A is a PBW deformation of S if, and only if, the following conditions hold:

(I) $P \cap F^1(T) = \{0\}$,

(i) $\mathrm{Im}(\alpha \otimes \mathrm{id} - \mathrm{id} \otimes \alpha) \subseteq R$,

(ii) $\alpha \circ (\alpha \otimes \mathrm{id} - \mathrm{id} \otimes \alpha) = -(\beta \otimes \mathrm{id} - \mathrm{id} \otimes \beta)$,

(iii) $\beta \circ (\alpha \otimes \mathrm{id} - \mathrm{id} \otimes \alpha) = 0$,

where the maps $\alpha \otimes \mathrm{id} - \mathrm{id} \otimes \alpha$ *and* $\beta \otimes \mathrm{id} - \mathrm{id} \otimes \beta$ *are defined on the subspace* $(R \otimes V) \cap (V \otimes R)$ *of* T.

We explain next how the original PBW theorem is a consequence of Theorem 7.2. Indeed, Polishchuk and Positselski [2005, Chapter 5, Sections 1 and 2] described the "self-consistency conditions" (i), (ii), and (iii) of the theorem as generalizing the Jacobi identity for Lie brackets.

Example 7.3. Let $\mathfrak{g}$ be a finite dimensional complex Lie algebra, $A = U(\mathfrak{g})$ its universal enveloping algebra, and $S = S(\mathfrak{g})$. Then R has $\mathbb{C}$-basis all $v \otimes w - w \otimes v$ for v, w in V, and $\alpha(v \otimes w - w \otimes v) = [v, w]$, $\beta \equiv 0$. Condition (I) is equivalent to antisymmetry of the bracket. Condition (J) is equivalent to the Jacobi identity, with (i), (ii) expressing the condition separately in each degree in the tensor algebra ($\beta \equiv 0$ in this case). More generally, there are examples with $\beta \not\equiv 0$, for instance, the Sridharan enveloping algebras [Sridharan 1961].

Homological proofs. We now explain how Braverman and Gaitsgory and Polishchuk and Positselski used algebraic deformation theory and Hochschild cohomology to prove that the conditions of Theorem 7.2 are sufficient. Braverman and Gaitsgory constructed a graded deformation S_t interpolating between S and A (i.e., with $S = S_t|_{t=0}$ and $A = S_t|_{t=1}$), implying that $\mathrm{gr}(A) \cong S$ as graded algebras. They constructed the deformation S_t as follows.

- They identified α with a choice of first multiplication map μ_1 and β with a choice of second multiplication map μ_2, via the canonical embedding of the bimodule Koszul resolution (6.3) into the bar resolution (5.3) of S. (In order to do this, one must extend α, β (respectively, μ_1, μ_2) to be maps on a larger space via an isomorphism $\mathrm{Hom}_k(R, S) \cong \mathrm{Hom}_{S^e}(S \otimes R \otimes S, S)$ (respectively, $\mathrm{Hom}_k(S \otimes S, S) \cong \mathrm{Hom}_{S^e}(S^{\otimes 4}, S)$.)
- Condition (i) is then seen to be equivalent to μ_1 being a Hochschild 2-cocycle (i.e., satisfies Equation (5.2)).
- Condition (ii) is equivalent to the vanishing of the first obstruction.
- Condition (iii) is equivalent to the vanishing of the second obstruction.

- All other obstructions vanish automatically for a Koszul algebra due to the structure of its Hochschild cohomology.
- Thus there exist maps μ_i for $i > 2$ defining an associative multiplication $*$ (as in Equation (5.1)) on $S[t]$.

Positselski [1993, Theorem 3.3] (see also [Polishchuk and Positselski 2005, Proposition 5.7.2]) gave a different homological proof of Theorem 7.2. Let B be the Koszul dual $S^! := \mathrm{Ext}^*_S(k,k)$ of S. Then $S \cong B^! := \mathrm{Ext}^*_B(k,k)$. Polishchuk defined a complex whose terms are the same as those in the bar resolution of B but with boundary maps modified using the functions $\alpha : R \to V$, $\beta : R \to k$ by first identifying β with an element h of B^2 and α with a dual to a derivation d on B. The conditions (i), (ii), and (iii) on α, β correspond to conditions on d, h, under which Positselski called B a CDG-algebra. The idea is that CDG-algebra structures on B are dual to PBW deformations of S. Positselski's proof relies on the Koszul property of S (equivalently of B) to imply collapsing of a spectral sequence with $E^2_{p,q} = \mathrm{Ext}_B^{-q,p}(k,k)$. The sequence converges to the homology of the original complex for B. Koszulness implies the only nonzero terms occur when $p+q=0$, and we are left with the homology of the total complex in degree 0. By its definition this is simply the algebra A, and it follows that $\mathrm{gr}\, A \cong B^! \cong S$.

Generalizations and extensions. Theorem 7.2 describes nonhomogeneous quadratic algebras whose quadratic versions are Koszul. What if one replaces the underlying field by an arbitrary ring? Etingof and Ginzburg [2002] noted that Braverman and Gaitsgory's proof of Theorem 7.2 is in fact valid more generally for Koszul rings over semisimple subrings as defined by Beilinson, Ginzburg and Soergel [Beilinson et al. 1996]. They chose their semisimple subring to be the complex group algebra $\mathbb{C}G$ of a finite group G acting symplectically and their Koszul ring to be a polynomial algebra $S(V)$. They were interested in the case $\alpha \equiv 0$ for their applications to symplectic reflection algebras (outlined in Section 10 below). Halbout, Oudom and Tang [Halbout et al. 2011] state a generalization of Theorem 7.2 in this setting that allows nonzero α (i.e., allows relations defining the algebra A to set commutators of vectors in V to a combination of group algebra elements and vectors). A proof using the Koszul ring theory of Beilinson, Ginzburg and Soergel and the results of Braverman and Gaitsgory is outlined in [Shepler and Witherspoon 2012a] for arbitrary group algebras over the complex numbers. We also included a second proof there for group algebras over arbitrary fields (of characteristic not 2) using the composition-diamond lemma (described in the next section), which has the advantage that it is characteristic free. We adapted the program of Braverman and Gaitsgory to arbitrary nonhomogeneous quadratic algebras and Koszul rings defined over nonsemisimple rings in [Shepler and Witherspoon 2014], including group rings kG where the characteristic of k divides the order of the group G.

The theory of Braverman and Gaitsgory was further generalized to algebras that are N-Koszul (all relations homogeneous of degree N plus a homological condition) over semisimple or von Neumann regular rings by a number of authors (see [Berger and Ginzburg 2006; Fløystad and Vatne 2006; Herscovich et al. 2014]). Cassidy and Shelton [2007] generalized the theory of Braverman and Gaitsgory in a different direction, to graded algebras over a field satisfying a particular homological finiteness condition (not necessarily having all relations in a single fixed degree).

8. The composition-diamond lemma and Gröbner basis theory

PBW theorems are often proven using diamond or composition lemmas and the theory of (noncommutative) Gröbner bases. Diamond lemmas predict existence of a canonical normal form in a mathematical system. Often one is presented with various ways of simplifying an element to obtain a normal form. If two different ways of rewriting the original element result in the same desired reduced expression, one is reminded of diverging paths meeting like the sides of the shape of a diamond. Diamond lemmas often originate from Newman's lemma [1942] for graph theory. Shirshov [1962a; 1962b] gave a general version for Lie polynomials which Bokut (see [Bokut 1976] and [Bokut and Chen 2014]) extended to associative algebras in 1976, using the term "composition lemma". Around the same time,[1] Bergman [1978] developed a similar result which he instead called the diamond lemma.

Both the diamond lemma and composition lemma are easy to explain but difficult to state precisely without the formalism absorbed by Gröbner basis theory. In fact, the level of rigor necessary to carefully state and prove these results can be the subject of debate. Bergman himself writes that the lemma "has been considered obvious and used freely by some ring-theorists... but others seem unaware of it and write out tortuous verifications." (Some authors are reminded of life in a lunatic asylum when making the basic idea rigorous; see [Hellström and Silvestrov 2000].) We leave careful definitions to any one of numerous texts (for example, see [Beidar et al. 1996], [Bueso et al. 2003], or [Loday and Vallette 2012]) and instead present the intuitive idea behind the result developed by Shirshov, Bokut, and Bergman.

The result of Bokut (and Shirshov). We first give the original result of Bokut (see [1976, Proposition 1 and Corollary 1]), who used a degree-lexicographical monomial ordering (also see [Bokut and Kukin 1994]).

[1] Bokut cites a preprint by Bergman.

Original composition lemma. *Suppose a set of relations P defining a k-algebra A is "closed under composition". Then the set of monomials that do not contain the leading monomial of any element of P as a subword is a k-basis of A.*

Before explaining the notion of "closed under composition," we rephrase the results of Bokut in modern language using Gröbner bases to give a PBW-like basis as in Section 3 (see, for example, [Green 1994] or [Mora 1986]). Fix a monomial ordering on a free k-algebra T and again write $\mathrm{LM}(p)$ for the leading monomial of any p in T. We include the converse of the lemma which can be deduced from the work of Shirshov and Bokut and was given explicitly by Bergman, who used arbitrary monomial orderings.

Gröbner basis version of composition lemma. *The set P is a (noncommutative) Gröbner basis of the ideal I it generates if and only if*

$$\mathcal{B}_P = \{\textit{monomials } m \textit{ in } T : \ m \textit{ not divisible by any } \mathrm{LM}(p),\ p \in P\}$$

is a k-basis for the algebra $A = T/I$.

Example 8.1. Let A be the k-algebra generated by symbols x and y and relations $xy = x$ and $yx = y$ (Example 4.1):

$$A = k\langle x, y\rangle \big/ (xy - x,\ yx - y).$$

Let P be the set of defining relations, $P = \{xy - x,\ yx - y\}$, and consider the degree-lexicographical monomial ordering with $x > y$. Then P is *not* a Gröbner basis of the ideal it generates since $x^2 - x = x(yx - y) - (xy - x)(x - 1)$ lies in the ideal (P) and has leading monomial x^2, which does not lie in the ideal generated by the leading monomials of the elements of P. Indeed, $\mathcal{B}_P$ contains both x^2 and x and hence can not be a basis for A. We set $P' = \{xy - x,\ yx - y,\ x^2 - x,\ y^2 - y\}$ to obtain a Gröbner basis of (P). Then $\mathcal{B}_{P'} = \{\text{monomials } m : m \text{ not divisible by } xy,\ yx,\ x^2,\ y^2\}$ is a k-basis for the algebra A.

Resolving ambiguities. Bergman focused on the problem of resolving ambiguities that arise when trying to rewrite elements of an algebra using different defining relations. Consider a k-algebra A defined by a finite set of generators and a finite set of relations

$$m_1 = f_1,\ m_2 = f_2,\ \ldots,\ m_k = f_k,$$

where the m_i are monomials (in the set of generators of A) and the f_i are linear combinations of monomials. Suppose we prefer the right side of our relations and try to eradicate the m_i whenever possible in writing the elements of A in terms of its generators. Can we define the notion of a canonical form for every element of A by deciding to replace each m_i by f_i whenever possible? We say an expression for an element of A is *reduced* if no m_i appears (as a subword anywhere), that is, when no further replacements using the defining relations of A are possible. The idea of a *canonical form* for A then makes sense if the set of reduced expressions is a k-basis for A, that is, if every element can be written uniquely in reduced form.

A natural ambiguity arises: If a monomial m contains both m_1 and m_2 as (overlapping) subwords, do we "reduce" first m_1 to f_1 or rather first m_2 to f_2 by replacing? (In the last example, the word xyx contains overlapping subwords xy and yx.) If the order of application of the two relations does not matter and we end up with the same reduced expression, then we say the *(overlap) ambiguity was resolvable*. The composition-diamond lemma states that *knowing certain ambiguities resolve is enough to conclude that a canonical normal form exists for all elements in the algebra.*

Example 8.2. Again, let A be the k-algebra generated by symbols x and y and relations $xy = x$ and $yx = y$ (Example 4.1). We decide to eradicate xy and yx whenever possible in expressions for the elements of A using just the defining relations. On one hand, we may reduce xyx to x^2 (using the first relation); on the other hand, we may reduce xyx to xy (using the second relation) then to x (using the first relation). The words x and x^2 can not be reduced further using just the defining relations, so we consider them both "reduced". Yet they represent the same element xyx of A. Thus, a canonical "reduced" form does not make sense given this choice of defining relations for the algebra A.

The result of Bergman. One makes the above notions precise by introducing a monomial ordering and giving formal definitions for ambiguities, reduction, rewriting procedures, resolving, etc. We consider the quotient algebra $A = T/(P)$ where T (a tensor algebra) is the free k-algebra on the generators of A and P is a (say) finite set of generating relations. We single out a monomial m_i in each generating relation, writing

$$P = \{m_i - f_i : 1 \le i \le k\},$$

and choose a monomial ordering so that m_i is the leading monomial of each $m_i - f_i$ (assuming such an ordering exists). Then the reduced elements are exactly those spanned by $\mathcal{B}_P$. If all the ambiguities among elements of P are resolvable, we obtain a PBW-like basis, but Bokut and Bergman give a condition that is easier to check. Instead of choosing to replace monomial m_i by f_i or monomial m_j by f_j when they both appear as subwords of a monomial m, we make *both* replacements separately and take the difference. If we can express this difference as a linear combination of elements p in the ideal (P) with $\mathrm{LM}(p) < m$, then we say the ambiguity was resolvable *relative to the ordering*. (Bokut used "closed under composition" to describe this condition along with minimality of P.) See [Bergman 1978, Theorem 1.2].

Diamond lemma idea. *The following are equivalent*:

- *The set of reduced words is a* k*-basis of* $T/(P)$.
- *All ambiguities among elements of* P *are resolvable.*
- *All ambiguities among elements of* P *are resolvable relative to the ordering.*
- *Every element in* (P) *can be reduced to zero in* $T/(P)$ *by just using the relations in* P.

In essence, the lemma says that if the generating set of relations P is well-behaved with respect to some monomial ordering, then one can define a canonical form just by checking that nothing goes wrong with the set P instead of checking for problems implied by the whole ideal (P). Thus, resolving ambiguities is just another way of testing for a Gröbner basis (see [Green 1994]): The set P is a Gröbner basis for the ideal (P) if and only if all ambiguities among elements of P are resolvable.

Applications. Although the idea of the composition-diamond lemma can be phrased in many ways, the hypothesis to be checked in the various versions of the lemma requires very similar computations in application. One finds precursors of the ideas underlying the composition-diamond lemma in the original proofs given by Poincaré, Birkhoff, and Witt of the PBW theorem for universal enveloping algebras of Lie algebras. These techniques and computations have been used in a number of other settings. For example, explicit PBW conditions are given for Drinfeld Hecke algebras (which include symplectic reflection algebras) by Ram and Shepler [2003]; see Section 10. In [Shepler and Witherspoon 2012a], we studied the general case of algebras defined by relations which set commutators to lower order terms using both a homological approach and the composition-diamond lemma (as it holds in arbitrary characteristic). These algebras, called *Drinfeld orbifold algebras*, include Sridharan enveloping algebras, Drinfeld Hecke algebras, enveloping algebras of Lie algebras, Weyl algebras, and twists of these algebras with group actions. Gröbner bases were used explicitly by Levandovskyy and Shepler [2014] in replacing a commutative polynomial algebra by a skew (or quantum) polynomial algebra in the theory of Drinfeld Hecke algebras. Bergman's diamond lemma was used by Khare [2007] to generalize the Drinfeld Hecke algebras of Section 10 from the setting of group actions to that of algebra actions.

Of course the composition-diamond lemma and Gröbner–Shirshov bases have been used to explore many different kinds of algebras (and in particular to find

PBW-like bases) that we will not discuss here. See [Bokut and Kukin 1994; Bokut and Chen 2014] for many such examples.

Note that some authors prove PBW theorems by creating a space upon which the algebra in question acts (see, e.g., [Humphreys 1972] or [Griffeth 2010, first version]). Showing that the given space is actually a module for the algebra requires checking certain relations that are similar to the conditions that one must check before invoking the composition-diamond lemma.

9. Drinfeld–Jimbo quantum groups and related Hopf algebras

Quantized enveloping algebras (that is, Drinfeld–Jimbo quantum groups [Drinfeld 1987; Jimbo 1985]) are deformations of universal enveloping algebras of Lie algebras. (Technically, they are bialgebra deformations rather than algebra deformations.) PBW bases for these algebras were discovered by many, including Lusztig [1990b; 1990a; 1990c] in a very general setting, and De Concini and Kac [1990], who defined a corresponding algebra filtration. Although there are many incarnations of these algebras, we restrict ourselves to the simply laced case and to algebras over the complex numbers for ease of notation. We state a PBW theorem in this context and refer the reader to the literature for more general statements (see, e.g., [Lusztig 1990c]).

Quantum groups. Let $\mathfrak{g}$ be a finite dimensional semisimple complex Lie algebra of rank n with symmetric Cartan matrix (a_{ij}). Let q be a nonzero complex number, $q \neq \pm 1$. (Often q is taken to be an indeterminate instead.) The *quantum group* $U_q(\mathfrak{g})$ is the associative $\mathbb{C}$-algebra defined by generators

$$E_1, \dots, E_n, F_1, \dots, F_n, K_1^{\pm 1}, \dots, K_n^{\pm 1}$$

and relations

$$\begin{gathered}
K_i^{\pm 1} K_j^{\pm 1} = K_j^{\pm 1} K_i^{\pm 1}, \quad K_i K_i^{-1} = 1 = K_i^{-1} K_i, \\
K_i E_j = q^{a_{ij}} E_j K_i, \quad K_i F_j = q^{-a_{ij}} F_j K_i, \\
E_i F_j - F_j E_i = \delta_{ij} \frac{K_i - K_i^{-1}}{q - q^{-1}},
\end{gathered}$$

$$\begin{aligned}
E_i^2 E_j - (q+q^{-1}) E_i E_j E_i + E_j E_i^2 &= 0 && \text{if } a_{ij} = -1, \\
E_i E_j &= E_j E_i && \text{if } a_{ij} = 0, \\
F_i^2 F_j - (q+q^{-1}) F_i F_j F_i + F_j F_i^2 &= 0 && \text{if } a_{ij} = -1, \\
F_i F_j &= F_j F_i && \text{if } a_{ij} = 0.
\end{aligned}$$

The last two sets of relations are called the quantum Serre relations.

Let W be the Weyl group of $\mathfrak{g}$. Fix a reduced expression of the longest element w_0 of W. This choice yields a total order on the set Φ^+ of positive roots, $\beta_1, \ldots, \beta_m$. To each $\alpha \in \Phi^+$, Lusztig [1990b; 1990a; 1990c] assigned an element E_α (respectively, F_α) in $U_q(\mathfrak{g})$ determined by this ordering that is an iterated braided commutator of the generators $E_1, \ldots, E_n$ (respectively, $F_1, \ldots, F_n$). These "root vectors" then appear in a PBW basis:

PBW theorem for quantum groups. *There is a basis of $U_q(\mathfrak{g})$ given by*

$$\{E_{\beta_1}^{a_1} \cdots E_{\beta_m}^{a_m} K_1^{b_1} \cdots K_n^{b_n} F_{\beta_1}^{c_1} \cdots F_{\beta_m}^{c_m} : a_i, c_i \geq 0, b_i \in \mathbb{Z}\}.$$

Moreover, there is a filtration on the subalgebra $U_q^{>0}(\mathfrak{g})$ (respectively, $U_q^{<0}(\mathfrak{g})$) generated by $E_1, \ldots, E_n$ (respectively, $F_1, \ldots, F_n$) for which the associated graded algebra is isomorphic to a skew polynomial ring.

The skew polynomial ring to which the theorem refers is generated by the images of the E_α (respectively, F_α), with relations $E_\alpha E_\beta = q_{\alpha\beta} E_\beta E_\alpha$ (respectively, $F_\alpha F_\beta = q_{\alpha\beta} F_\beta F_\alpha$) where each $q_{\alpha\beta}$ is a scalar determined by q and by α, β in Φ^+.

Example 9.1. The algebra $U_q^{>0}(\mathfrak{sl}_3)$ is generated by E_1, E_2. Let

$$E_{12} := E_1 E_2 - q E_2 E_1.$$

Then, as a consequence of the quantum Serre relations, $E_1 E_{12} = q^{-1} E_{12} E_1$ and $E_{12} E_2 = q^{-1} E_2 E_{12}$, and, by definition of E_{12}, we also have $E_1 E_2 = q E_2 E_1 + E_{12}$. In the associated graded algebra, this last relation becomes $E_1 E_2 = q E_2 E_1$. The algebras $U_q^{>0}(\mathfrak{sl}_n)$ are similar, however in general the filtration on $U_q^{>0}(\mathfrak{g})$ stated in the theorem is more complicated.

Proofs and related results. There are several proofs in the literature of the first statement of the above theorem and related results, beginning with [Khoroshkin and Tolstoy 1991; Lusztig 1990b, 1990a; 1990c; Takeuchi 1990; Yamane 1989]. These generally involve explicit computations facilitated by representation theory. Specifically, one obtains representations of $U_q(\mathfrak{g})$ from those of the corresponding Lie algebra $\mathfrak{g}$ by deformation, and one then uses what is known in the classical setting to obtain information about $U_q(\mathfrak{g})$. Ringel [1996] gave a different approach via Hall algebras. The filtration and structure of the associated graded algebra of $U^{>0}(\mathfrak{g})$ was first given by De Concini and Kac [1990]. For a general "quantum PBW theorem" that applies to some of these algebras, see [Berger 1992].

In case q is a root of unity (of order ℓ), there are finite dimensional versions of Drinfeld–Jimbo quantum groups. The *small quantum group* $u_q(\mathfrak{g})$ may be defined as the quotient of $U_q(\mathfrak{g})$ by the ideal generated by all E_α^ℓ, F_α^ℓ, $K_\alpha^\ell - 1$.

This finite dimensional algebra has k-basis given by elements in the PBW basis of the above theorem for which $0 \leq a_i, b_i, c_i < \ell$.

The existence of PBW bases for $U_q(\mathfrak{g})$ and $u_q(\mathfrak{g})$ plays a crucial role in their representation theory, just as it does in the classical setting of Lie algebras. Bases of finite dimensional simple modules and other modules are defined from weight vectors and PBW bases [Lusztig 1990a]. R-matrices may be expressed in terms of PBW basis elements [Drinfeld 1987; Jimbo 1985; Rosso 1989]. Computations of cohomology take advantage of the structure provided by the PBW basis and filtration (see, e.g., [Ginzburg and Kumar 1993], based on techniques developed for restricted Lie algebras [Friedlander and Parshall 1983]).

More generally, PBW bases and some Lie-theoretic structure appear in a much larger class of Hopf algebras. Efforts to understand finite dimensional Hopf algebras of various types led in particular to a study of those arising from underlying Nichols algebras. Consequently, a classification of some types of pointed Hopf algebras was completed by Andruskiewitsch and Schneider [2010], Heckenberger [2006] and Rosso [1998]. A Nichols algebra is a "braided" graded Hopf algebra that is connected, generated by its degree 1 elements, and whose subspace of primitive elements is precisely its degree 1 component. The simplest Nichols algebras are those of "diagonal type", and these underlie the Drinfeld–Jimbo quantum groups and the Hopf algebras in the above-mentioned classification. These algebras have PBW bases just as does $U_q^{>0}(\mathfrak{g})$ or $u_q^{>0}(\mathfrak{g})$; a proof given by Kharchenko [1999] uses a combinatorial approach such as that in Section 8.

10. Symplectic reflection algebras, rational Cherednik algebras, and graded (Drinfeld) Hecke algebras

Drinfeld [1986] and Lusztig [1988; 1989] originally defined the algebras now variously called symplectic reflection algebras, rational Cherednik algebras, and graded (Drinfeld) Hecke algebras, depending on context. These are PBW deformations of group extensions of polynomial rings (skew group algebras) defined by relations that set commutators of vectors to elements of a group algebra. Lusztig explored the representation theory of these algebras when the acting group is a Weyl group. Crawley-Boevey and Holland [1998] considered subgroups of $\mathrm{SL}_2(\mathbb{C})$ and studied subalgebras of these algebras in relation to corresponding orbifolds. Initial work on these types of PBW deformations for arbitrary groups began with [Etingof and Ginzburg 2002] and [Ram and Shepler 2003]. Gordon [2003] used the rational Cherednik algebra to prove a version of the $n!$-conjecture for Weyl groups and the representation theory of these algebras remains an active area. (See [Brown 2003; Gordon 2010; 2008; Rouquier 2005].)

We briefly recall and compare these algebras. (See also [Chlouveraki 2015] for a survey of symplectic reflection algebras and rational Cherednik algebras in the context of Hecke algebras and representation theory.)

Let G be a group acting by automorphisms on a k-algebra S. The *skew group algebra* $S\#G$ (also written as a semidirect product $S \rtimes G$) is the k-vector space $S \otimes kG$ together with multiplication given by $(r \otimes g)(s \otimes h) = r({}^g s) \otimes gh$ for all r, s in S and g, h in G, where ${}^g s$ is the image of s under the automorphism g.

Drinfeld's "Hecke algebra". Suppose G is a finite group acting linearly on a finite dimensional vector space V over $k = \mathbb{C}$ with symmetric algebra $S(V)$. Consider the quotient algebra

$$\mathcal{H}_\kappa = T(V)\#G \big/ (v_1 \otimes v_2 - v_2 \otimes v_1 - \kappa(v_1, v_2) : v_1, v_2 \in V)$$

defined by a bilinear parameter function $\kappa : V \times V \to \mathbb{C}G$. We view $\mathcal{H}_\kappa$ as a filtered algebra by assigning degree one to vectors in V and degree zero to group elements in G. Then the algebra $\mathcal{H}_\kappa$ is a PBW deformation of $S(V)\#G$ if its associated graded algebra is isomorphic to $S(V)\#G$. Drinfeld [1986] originally defined these algebras for arbitrary groups, and he also counted the dimension of the parameter space of such PBW deformations for Coxeter groups. For more information and a complete characterization of parameters κ yielding the PBW property for arbitrary groups, see [Etingof and Ginzburg 2002; Ram and Shepler 2003; Shepler and Witherspoon 2008; 2012a; 2012b].

Example 10.1. Let V be a vector space of dimension 3 with basis v_1, v_2, v_3, and let G be the symmetric group S_3 acting on V by permuting the chosen basis elements. The following is a PBW deformation of $S(V)\#G$, where (ijk) denotes a 3-cycle in S_3:

$$\mathcal{H}_\kappa = T(V)\#S_3 \big/ \big(v_i \otimes v_j - v_j \otimes v_i - (ijk) + (jik) : \{i, j, k\} = \{1, 2, 3\}\big).$$

Lusztig's graded affine Hecke algebra. While exploring the representation theory of groups of Lie type, Lusztig [1988; 1989] defined a variant of the affine Hecke algebra for Weyl groups which he called "graded" (as it was obtained from a particular filtration of the affine Hecke algebra). He gave a presentation for this algebra $\mathbb{H}_\lambda$ using the same generators as those for Drinfeld's Hecke algebra $\mathcal{H}_\kappa$, but he gave relations preserving the structure of the polynomial ring and altering that of the skew group algebra. (Drinfeld's relations do the reverse.) The *graded affine Hecke algebra* $\mathbb{H}_\lambda$ (or simply the *graded Hecke algebra*) for a finite Coxeter group G acting on a finite dimensional complex vector space V (in its natural reflection representation) is the $\mathbb{C}$-algebra generated by the polynomial algebra $S(V)$ together with the group algebra $\mathbb{C}G$ with relations

$$gv = {}^g v g + \lambda_g(v) g$$

for all v in V and g in a set $\mathcal{S}$ of simple reflections (generating G) where λ_g in V^* defines the reflecting hyperplane ($\ker \lambda_g \subseteq V$) of g and $\lambda_g = \lambda_{hgh^{-1}}$ for all h in G. (Recall that a *reflection* on a finite dimensional vector space is just a nonidentity linear transformation that fixes a hyperplane pointwise.)

Note that for g representing a fixed conjugacy class of reflections, the linear form λ_g is only well-defined up to a nonzero scalar. Thus one often fixes once and for all a choice of linear forms $\lambda = \{\lambda_g\}$ defining the orbits of reflecting hyperplanes (usually expressed using Demazure/BGG operators) and then introduces a formal parameter by which to rescale. This highlights the degree of freedom arising from each orbit; for example, one might replace

$$\lambda_g(v) \quad \text{by } c_g\langle v, \alpha_g^\vee\rangle = c_g\left(\frac{v - {}^g v}{\alpha_g}\right)$$

for some conjugation invariant formal parameter c_g after fixing a G-invariant inner product and root system $\{\alpha_g : g \in \mathcal{S}\} \subset V$ with coroot vectors $\alpha_g^\vee$. (Note that for any reflection g, the vector $(v - {}^g v)$ is a nonzero scalar multiple of α_g and so the quotient of $v - {}^g v$ by α_g is a scalar.) Each graded affine Hecke algebra $\mathbb{H}_\lambda$ is filtered with vectors in degree one and group elements in degree zero and defines a PBW deformation of $S(V)\#G$.

finite group	any $G \le \mathrm{GL}(V)$	Coxeter $G \le \mathrm{GL}(V)$
algebra	$\mathcal{H}_\kappa$ (Drinfeld)	$\mathbb{H}_\lambda$ (Lusztig)
generated by	V and $\mathbb{C}G$	V and $\mathbb{C}G$
with relations	$gv = {}^g v g$, $vw = wv + \kappa(v, w)$ $(\forall v, w \in V, \forall g \in G)$	$gv = {}^g v g + \lambda_g(v) g$, $vw = wv$ $(\forall v, w \in V, \forall g \in \mathcal{S})$

Comparing algebras. Ram and Shepler [2003] showed that Lusztig's graded affine Hecke algebras are a special case of Drinfeld's construction: For each parameter λ, there is a parameter κ so that the filtered algebras $\mathbb{H}_\lambda$ and $\mathcal{H}_\kappa$ are isomorphic (see [Shepler and Witherspoon 2015]). Etingof and Ginzburg [2002] rediscovered Drinfeld's algebras with focus on groups G acting symplectically (in the context of orbifold theory). They called algebras $\mathcal{H}_\kappa$ satisfying the PBW property *symplectic reflection algebras*, giving necessary and sufficient conditions on κ. They used the theory of Beilinson, Ginzburg and Soergel [Beilinson et al. 1996] of Koszul rings to generalize Braverman and Gaitsgory's conditions to the setting where the ground field is replaced by the semisimple group ring $\mathbb{C}G$. (The skew group algebra $S(V)\#G$ is Koszul as a ring over the semisimple subring $\mathbb{C}G$.) Ram and Shepler [2003] independently gave necessary and sufficient PBW conditions on κ (for arbitrary groups acting linearly over $\mathbb{C}$) and classified all such quotient algebras for complex reflection groups. Their proof relies on the composition-diamond lemma. (See Sections 7 and 8 for a comparison of these two techniques for showing PBW properties.) Both approaches depend on the fact that the underlying field $k = \mathbb{C}$ has characteristic zero (or, more generally, has characteristic that does not divide the order of the group G). See Section 11 for a discussion of PBW theorems in the modular setting when $\mathbb{C}$ is replaced by a field whose characteristic divides $|G|$.

Rational Cherednik algebras. The rational Cherednik algebra is a special case of a quotient algebra $\mathcal{H}_\kappa$ satisfying the PBW property (in fact, a special case of a symplectic reflection algebra) for a reflection group acting diagonally on the reflection representation and its dual ("doubled up"). These algebras are regarded as "doubly degenerate" versions of the double affine Hecke algebra introduced by Cherednik [1995] to solve the Macdonald (constant term) conjectures in combinatorics. We simply recall the definition here in terms of reflections and hyperplane arrangements.

Suppose G is a finite group generated by reflections on a finite dimensional complex vector space V. (If G is a Coxeter group, then extend the action to one over the complex numbers.) Then the induced diagonal action of G on $V \oplus V^*$ is generated by *bireflections* (linear transformations that each fix a subspace of codimension 2 pointwise), that is, by *symplectic reflections* with respect to a natural symplectic form on $V \oplus V^*$.

Let $\mathcal{R}$ be the set of all reflections in G acting on V. For each reflection s in $\mathcal{R}$, let α_s in V and α_s^* in V^* be eigenvectors ("root vectors") each with nonidentity eigenvalue. We define an algebra generated by $\mathbb{C}G$, V, and V^* in which vectors in V commute with each other and vectors in V^* commute with each other, but passing a vector from V over one from V^* gives a linear combination of reflections (and the identity). As parameters, we take a scalar t and a set of scalars $\boldsymbol{c} = \{c_s : s \in \mathcal{R}\}$ with $c_s = c_{hsh^{-1}}$ for all h in G. The *rational Cherednik algebra* $\boldsymbol{H}_{t,\boldsymbol{c}}$ with parameters t, $\boldsymbol{c}$ is then the $\mathbb{C}$-algebra generated by the vectors in V and V^* together with the group algebra $\mathbb{C}G$ satisfying the relations

$$gu = {}^g u g, \quad uu' = u'u,$$
$$vv^* = v^*v + tv^*(v) - \sum_{s \in \mathcal{R}} c_s \alpha_s^*(v) v^*(\alpha_s) s,$$

for any g in G, v in V, v^* in V^*, and any u, u' both in V or both in V^*. Note that α_s and α_s^* are only well-defined up to a nonzero scalar, and we make some conjugation invariant choice of normalization in this definition, say, by assuming that $\alpha_s^*(\alpha_s) = 1$. One often replaces $\mathbb{C}$ by $\mathbb{C}[t, \boldsymbol{c}]$ to work in a formal parameter space.

The relations defining the rational Cherednik algebra are often given in terms of the arrangement of reflecting hyperplanes $\mathcal{A}$ for G acting on V. For each hyperplane H in $\mathcal{A}$, choose a linear form α_H^* in V^* defining H (so $H = \ker \alpha_H^*$) and let α_H be a nonzero vector in V perpendicular to H with respect to some fixed G-invariant inner product. Then the third defining relation of $\boldsymbol{H}_{t,c}$ can be rewritten (without a choice of normalization) as

$$vv^* - v^*v = tv^*(v) - \sum_{H \in \mathcal{A}} \frac{\alpha_H^*(v)v^*(\alpha_H)}{\alpha_H^*(\alpha_H)} \big(c_{s_H} s_H + c_{s_H^2} s_H^2 + \cdots + c_{s_H^{a_H}} s_H^{a_H}\big),$$

where s_H is the reflection in G about the hyperplane H of maximal order $a_H + 1$. Again, this is usually expressed geometrically in terms of the inner product on V and induced product on V^*:

$$\frac{\alpha_H^*(v)v^*(\alpha_H)}{\alpha_H^*(\alpha_H)} = \frac{\langle v, \alpha_H^\vee\rangle\langle \alpha_H, v^*\rangle}{\langle \alpha_H, \alpha^\vee\rangle}.$$

The PBW theorem then holds for the algebra $\boldsymbol{H}_{t,c}$ (see [Etingof and Ginzburg 2002]):

PBW theorem for rational Cherednik algebras. *The rational Cherednik algebra $\boldsymbol{H}_{t,c}$ is isomorphic to $S(V) \otimes S(V^*) \otimes \mathbb{C}G$ as a complex vector space for any choices of parameters t and $\mathbf{c}$, and its associated graded algebra is isomorphic to $(S(V) \otimes S(V^*))\#G$.*

Connections between rational Cherednik algebras and other fields of mathematics are growing stronger. For example, Gordon and Griffeth [2012] link the Fuss–Catalan numbers in combinatorics to the representation theory of rational Cherednik algebras. These investigations also bring insight to the classical theory of complex reflection groups, especially to the perplexing question of why some reflection groups acting on n-dimensional space can be generated by n reflections (called "well-generated" or "duality" groups) and others not. (See [Berkesch et al. 2013; Gorsky et al. 2014; Shan et al. 2014] for other recent applications.)

11. Positive characteristic and nonsemisimple ground rings

Algebras displaying PBW properties are quite common over ground fields of positive characteristic and nonsemisimple ground rings, but techniques for establishing PBW theorems are not all equally suited for work over arbitrary fields and rings. We briefly mention a few results of ongoing efforts to establish and apply PBW theorems in these settings.

The algebras of Section 10 make sense in the *modular setting*, that is, when the characteristic of k is a prime dividing the order of the finite group G. In this case, however, the group algebra kG is not semisimple, and one must take more care in proofs. PBW conditions on κ were examined by Griffeth [2010] by construction of an explicit $\mathcal{H}_\kappa$-module, as is done in one standard proof of the PBW theorem for universal enveloping algebras. (See also [Bazlov and Berenstein 2009] for a generalization.) The composition-diamond lemma, being characteristic free, applies in the modular setting; see [Shepler and Witherspoon 2012a] for a proof of the PBW property using this lemma that applies to graded (Drinfeld) Hecke algebras over fields of arbitrary characteristic. (Gröbner bases are explicitly used in [Levandovskyy and Shepler 2014].) Several authors consider representations of rational Cherednik algebras in the modular setting: for example, Balagovic and Chen [2013], Griffeth [2010], and Norton [2013].

The theory of Beilinson, Ginzburg, and Soergel of Koszul rings over semisimple subrings, used in Braverman–Gaitsgory style proofs of PBW theorems, does not apply directly to the modular setting. However it may be adapted using a larger complex replacing the Koszul complex: In [Shepler and Witherspoon 2014], we used this approach to generalize the Braverman–Gaitsgory argument to arbitrary Koszul algebras with finite group actions. This replacement complex has an advantage over the composition-diamond lemma or Gröbner basis theory arguments in that it contains information about potentially new types of deformations that do not occur in the nonmodular setting.

Other constructions generalize the algebras of Section 10 to algebras over ground rings that are not necessarily semisimple. Etingof, Gan, and Ginzburg [2005] considered deformations of algebras that are extensions of polynomial rings by acting algebraic groups or Lie algebras. They used a Braverman–Gaitsgory approach to obtain a Jacobi condition by realizing the acting algebras as inverse limits of finite dimensional semisimple algebras. Gan and Khare [2007] investigated actions of $U_q(\mathfrak{sl}_2)$ on the quantum plane (a skew polynomial algebra), and Khare [2007] looked at actions of arbitrary cocommutative algebras on polynomial rings. In both cases PBW theorems were proven using the composition-diamond lemma. A general result for actions of (not necessarily semisimple) Hopf algebras on Koszul algebras is contained in [Walton and Witherspoon 2014] with a Braverman–Gaitsgory style proof. See also [He et al. 2015] for a PBW theorem using a somewhat different complex in a general setting of Koszul rings over not necessarily semisimple ground rings. One expects yet further generalizations and applications of the ubiquitous and potent PBW theorem.

References

[Andruskiewitsch and Schneider 2010] N. Andruskiewitsch and H.-J. Schneider, "On the classification of finite-dimensional pointed Hopf algebras", *Ann. of Math. (2)* **171**:1 (2010), 375–417.

[Anick 1986] D. J. Anick, "On the homology of associative algebras", *Trans. Amer. Math. Soc.* **296**:2 (1986), 641–659.

[Balagović and Chen 2013] M. Balagović and H. Chen, "Representations of rational Cherednik algebras in positive characteristic", *J. Pure Appl. Algebra* **217**:4 (2013), 716–740.

[Bazlov and Berenstein 2009] Y. Bazlov and A. Berenstein, "Braided doubles and rational Cherednik algebras", *Adv. Math.* **220**:5 (2009), 1466–1530.

[Beidar et al. 1996] K. I. Beidar, W. S. Martindale, III, and A. V. Mikhalev, *Rings with generalized identities*, Monographs and Textbooks in Pure and Applied Mathematics **196**, Marcel Dekker, New York, 1996.

[Beilinson et al. 1996] A. Beilinson, V. Ginzburg, and W. Soergel, "Koszul duality patterns in representation theory", *J. Amer. Math. Soc.* **9**:2 (1996), 473–527.

[Berger 1992] R. Berger, "The quantum Poincaré–Birkhoff–Witt theorem", *Comm. Math. Phys.* **143**:2 (1992), 215–234.

[Berger and Ginzburg 2006] R. Berger and V. Ginzburg, "Higher symplectic reflection algebras and non-homogeneous N-Koszul property", *J. Algebra* **304**:1 (2006), 577–601.

[Bergman 1978] G. M. Bergman, "The diamond lemma for ring theory", *Adv. in Math.* **29**:2 (1978), 178–218.

[Berkesch et al. 2013] C. Berkesch, S. Griffeth, and S. Sam, "Jack polynomials as fractional quantum Hall states and the Betti numbers of the $(k+1)$-equals ideals", preprint, 2013. arXiv 1303.4126

[Birkhoff 1937] G. Birkhoff, "Representability of Lie algebras and Lie groups by matrices", *Ann. of Math. (2)* **38**:2 (1937), 526–532.

[Bokut 1976] L. A. Bokut, "Imbeddings into simple associative algebras", *Algebra i Logika* **15**:2 (1976), 117–142, 245. In Russian; translated in *Algebra and Logic* **15**:2 (1976), 73–90.

[Bokut and Chen 2014] L. A. Bokut and Y. Chen, "Gröbner–Shirshov bases and their calculation", *Bull. Math. Sci.* **4**:3 (2014), 325–395.

[Bokut and Kukin 1994] L. A. Bokut and G. P. Kukin, *Algorithmic and combinatorial algebra*, vol. 255, Mathematics and its Applications, Kluwer, Dordrecht, 1994.

[Bourbaki 1960] N. Bourbaki, *Groupes et algèbres de Lie, Chapitre 1: Algèbres de Lie*, Actualités Sci. Ind. **1285**, Hermann, Paris, 1960.

[Braverman and Gaitsgory 1996] A. Braverman and D. Gaitsgory, "Poincaré–Birkhoff–Witt theorem for quadratic algebras of Koszul type", *J. Algebra* **181**:2 (1996), 315–328.

[Brown 2003] K. A. Brown, "Symplectic reflection algebras", *Irish Math. Soc. Bull.* **50** (2003), 27–49.

[Bruns et al. 1994] W. Bruns, J. Herzog, and U. Vetter, "Syzygies and walks", pp. 36–57 in *Commutative algebra* (Trieste, 1992), World Sci. Publ., River Edge, NJ, 1994.

[Buchberger 1965] B. Buchberger, *Ein Algorithmus zum Auffinden der Basiselemente des Restklassenringes nach einem nulldimensionalen Polynomideal*, Ph.D. thesis, University of Innsbruck, 1965.

[Bueso et al. 2003] J. Bueso, J. Gómez-Torrecillas, and A. Verschoren, *Algorithmic methods in non-commutative algebra: applications to quantum groups*, Mathematical Modelling **17**, Kluwer, Dordrecht, 2003.

[Capelli 1890] A. Capelli, "Sur les Opérations dans la théorie des formes algébriques", *Math. Ann.* **37**:1 (1890), 1–37.

[Cartan and Eilenberg 1956] H. Cartan and S. Eilenberg, *Homological algebra*, Princeton University Press, Princeton, NJ, 1956.

[Cassidy and Shelton 2007] T. Cassidy and B. Shelton, "PBW-deformation theory and regular central extensions", *J. Reine Angew. Math.* **610** (2007), 1–12.

[Cherednik 1995] I. Cherednik, "Double affine Hecke algebras and Macdonald's conjectures", *Ann. of Math. (2)* **141**:1 (1995), 191–216.

[Chlouveraki 2015] M. Chlouveraki, "Hecke algebras and symplectic reflection algebras", pp. 95–139 in *Commutative Algebra and Noncommutative Algebraic Geometry*, vol. 1, edited by D. Eisenbud et al., Mathematical Sciences Research Institute Publications **67**, 2015.

[Crawley-Boevey and Holland 1998] W. Crawley-Boevey and M. P. Holland, "Noncommutative deformations of Kleinian singularities", *Duke Math. J.* **92**:3 (1998), 605–635.

[De Concini and Kac 1990] C. De Concini and V. G. Kac, "Representations of quantum groups at roots of 1", pp. 471–506 in *Operator algebras, unitary representations, enveloping algebras, and invariant theory* (Paris, 1989), Progr. Math. **92**, Birkhäuser, Boston, 1990.

[Dixmier 1977] J. Dixmier, *Enveloping algebras*, North-Holland Mathematical Library **14**, North-Holland, Amsterdam, 1977.

[Drinfeld 1986] V. G. Drinfeld, "Degenerate affine Hecke algebras and Yangians", *Funktsional. Anal. i Prilozhen.* **20**:1 (1986), 69–70. In Russian; translated in *Functional Anal. Appl.* **20**:1 (1986), 62–64.

[Drinfeld 1987] V. G. Drinfeld, "Quantum groups", pp. 798–820 in *Proceedings of the International Congress of Mathematicians* (Berkeley, 1986), vol. 1, Amer. Math. Soc., Providence, RI, 1987.

[Eisenbud et al. 1994] D. Eisenbud, A. Reeves, and B. Totaro, "Initial ideals, Veronese subrings, and rates of algebras", *Adv. Math.* **109**:2 (1994), 168–187.

[Etingof and Ginzburg 2002] P. Etingof and V. Ginzburg, "Symplectic reflection algebras, Calogero–Moser space, and deformed Harish-Chandra homomorphism", *Invent. Math.* **147**:2 (2002), 243–348.

[Etingof et al. 2005] P. Etingof, W. L. Gan, and V. Ginzburg, "Continuous Hecke algebras", *Transform. Groups* **10**:3-4 (2005), 423–447.

[Fløystad and Vatne 2006] G. Fløystad and J. E. Vatne, "PBW-deformations of N-Koszul algebras", *J. Algebra* **302**:1 (2006), 116–155.

[Friedlander and Parshall 1983] E. M. Friedlander and B. J. Parshall, "On the cohomology of algebraic and related finite groups", *Invent. Math.* **74**:1 (1983), 85–117.

[Fröberg 1999] R. Fröberg, "Koszul algebras", pp. 337–350 in *Advances in commutative ring theory* (Fez, 1997), Lecture Notes in Pure and Appl. Math. **205**, Dekker, New York, 1999.

[Gan and Khare 2007] W. L. Gan and A. Khare, "Quantized symplectic oscillator algebras of rank one", *J. Algebra* **310**:2 (2007), 671–707.

[Gerstenhaber 1964] M. Gerstenhaber, "On the deformation of rings and algebras", *Ann. of Math. (2)* **79** (1964), 59–103.

[Ginzburg and Kumar 1993] V. Ginzburg and S. Kumar, "Cohomology of quantum groups at roots of unity", *Duke Math. J.* **69**:1 (1993), 179–198.

[Gordon 2003] I. Gordon, "On the quotient ring by diagonal invariants", *Invent. Math.* **153**:3 (2003), 503–518.

[Gordon 2008] I. G. Gordon, "Symplectic reflection alegebras", pp. 285–347 in *Trends in representation theory of algebras and related topics*, EMS Ser. Congr. Rep., Eur. Math. Soc., Zürich, 2008.

[Gordon 2010] I. G. Gordon, "Rational Cherednik algebras", pp. 1209–1225 in *Proceedings of the International Congress of Mathematicians, III*, Hindustan Book Agency, New Delhi, 2010.

[Gordon and Griffeth 2012] I. G. Gordon and S. Griffeth, "Catalan numbers for complex reflection groups", *Amer. J. Math.* **134**:6 (2012), 1491–1502.

[Gorsky et al. 2014] E. Gorsky, A. Oblomkov, J. Rasmussen, and V. Shende, "Torus knots and the rational DAHA", *Duke Math. J.* **163**:14 (2014), 2709–2794.

[Green 1994] E. L. Green, "An introduction to noncommutative Gröbner bases", pp. 167–190 in *Computational algebra* (Fairfax, VA, 1993), Lecture Notes in Pure and Appl. Math. **151**, Dekker, New York, 1994.

[Griffeth 2010] S. Griffeth, "Towards a combinatorial representation theory for the rational Cherednik algebra of type $G(r, p, n)$", *Proc. Edinb. Math. Soc. (2)* **53**:2 (2010), 419–445.

[Grivel 2004] P.-P. Grivel, "Une histoire du théorème de Poincaré–Birkhoff–Witt", *Expo. Math.* **22**:2 (2004), 145–184.

[Halbout et al. 2011] G. Halbout, J.-M. Oudom, and X. Tang, "Deformations of orbifolds with noncommutative linear Poisson structures", *Int. Math. Res. Not. IMRN* **2011**:1 (2011), 1–39.

[He et al. 2015] J. He, F. V. Oystaeyen, and Y. Zhang, "PBW deformations of Koszul algebras over a nonsemisimple ring", *Mathematische Zeitschrift* **279**:1-2 (2015), 185–210.

[Heckenberger 2006] I. Heckenberger, "The Weyl groupoid of a Nichols algebra of diagonal type", *Invent. Math.* **164**:1 (2006), 175–188.

[Hellström and Silvestrov 2000] L. Hellström and S. D. Silvestrov, *Commuting elements in q-deformed Heisenberg algebras*, World Sci. Publ., River Edge, NJ, 2000.

[Herscovich et al. 2014] E. Herscovich, A. Solotar, and M. Suárez-Álvarez, "PBW-deformations and deformations à la Gerstenhaber of N-Koszul algebras", *J. Noncommut. Geom.* **8**:2 (2014), 505–539.

[Higgins 1969] P. J. Higgins, "Baer invariants and the Birkhoff–Witt theorem", *J. Algebra* **11** (1969), 469–482.

[Hironaka 1964a] H. Hironaka, "Resolution of singularities of an algebraic variety over a field of characteristic zero, I", *Ann. of Math. (2)* **79** (1964), 109–203.

[Hironaka 1964b] H. Hironaka, "Resolution of singularities of an algebraic variety over a field of characteristic zero, II", *Ann. of Math. (2)* **79** (1964), 205–326.

[Humphreys 1972] J. E. Humphreys, *Introduction to Lie algebras and representation theory*, Graduate Texts in Mathematics **9**, Springer, New York, 1972.

[Jacobson 1941] N. Jacobson, "Restricted Lie algebras of characteristic p", *Trans. Amer. Math. Soc.* **50** (1941), 15–25.

[Jacobson 1962] N. Jacobson, *Lie algebras*, Interscience Tracts in Pure and Applied Mathematics **10**, Interscience Publishers, New York, 1962.

[Jimbo 1985] M. Jimbo, "A q-difference analogue of $U(\mathfrak{g})$ and the Yang–Baxter equation", *Lett. Math. Phys.* **10**:1 (1985), 63–69.

[Kac 1977] V. G. Kac, "Lie superalgebras", *Advances in Math.* **26**:1 (1977), 8–96.

[Kharchenko 1999] V. K. Kharchenko, "A quantum analogue of the Poincaré–Birkhoff–Witt theorem", *Algebra Log.* **38**:4 (1999), 476–507, 509. In Russian; translated in *Algebra and Logic* **38**:4 (1999), 259–276.

[Khare 2007] A. Khare, "Drinfeld–Hecke algebras over cocommutative algebras", preprint, 2007. arXiv 0705.2067

[Khoroshkin and Tolstoy 1991] S. M. Khoroshkin and V. N. Tolstoy, "Universal R-matrix for quantized (super)algebras", *Comm. Math. Phys.* **141**:3 (1991), 599–617.

[Levandovskyy and Shepler 2014] V. Levandovskyy and A. V. Shepler, "Quantum Drinfeld Hecke algebras", *Canad. J. Math.* **66**:4 (2014), 874–901.

[Li 2012] H. Li, *Gröbner bases in ring theory*, World Sci. Publ., Hackensack, NJ, 2012.

[Loday and Vallette 2012] J.-L. Loday and B. Vallette, *Algebraic operads*, Grundlehren der Math. Wissenschaften **346**, Springer, Heidelberg, 2012.

[Lusztig 1988] G. Lusztig, "Cuspidal local systems and graded Hecke algebras. I", *Inst. Hautes Études Sci. Publ. Math.* **67** (1988), 145–202.

[Lusztig 1989] G. Lusztig, "Affine Hecke algebras and their graded version", *J. Amer. Math. Soc.* **2**:3 (1989), 599–635.

[Lusztig 1990a] G. Lusztig, "Canonical bases arising from quantized enveloping algebras", *J. Amer. Math. Soc.* **3**:2 (1990), 447–498.

[Lusztig 1990b] G. Lusztig, "Finite-dimensional Hopf algebras arising from quantized universal enveloping algebra", *J. Amer. Math. Soc.* **3**:1 (1990), 257–296.

[Lusztig 1990c] G. Lusztig, "Quantum groups at roots of 1", *Geom. Dedicata* **35**:1-3 (1990), 89–113.

[Milnor and Moore 1965] J. W. Milnor and J. C. Moore, "On the structure of Hopf algebras", *Ann. of Math. (2)* **81** (1965), 211–264.

[Mora 1986] F. Mora, "Groebner bases for noncommutative polynomial rings", pp. 353–362 in *Algebraic algorithms and error correcting codes (Grenoble, 1985)*, Lecture Notes in Comput. Sci. **229**, Springer, Berlin, 1986.

[Newman 1942] M. H. A. Newman, "On theories with a combinatorial definition of "equivalence."", *Ann. of Math. (2)* **43** (1942), 223–243.

[Norton 2013] E. Norton, "Symplectic reflection algebras in positive charactersitic as Ore extensions", preprint, 2013. arXiv 1302.5411

[Poincaré 1900] H. Poincaré, "Sur les groupes continus", *Trans. Cambr. Philos. Soc.* **18** (1900), 220–255.

[Polishchuk and Positselski 2005] A. Polishchuk and L. Positselski, *Quadratic algebras*, vol. 37, University Lecture Series, Amer. Math. Soc., Providence, RI, 2005.

[Positselski 1993] L. E. Positselski, "Nonhomogeneous quadratic duality and curvature", *Funktsional. Anal. i Prilozhen.* **27**:3 (1993), 57–66, 96. In Russian; translated in *Funct. Anal. Appl.* **27**:3 (1993), 197–204.

[Priddy 1970] S. B. Priddy, "Koszul resolutions", *Trans. Amer. Math. Soc.* **152** (1970), 39–60.

[Ram and Shepler 2003] A. Ram and A. V. Shepler, "Classification of graded Hecke algebras for complex reflection groups", *Comment. Math. Helv.* **78**:2 (2003), 308–334.

[Ringel 1996] C. M. Ringel, "PBW-bases of quantum groups", *J. Reine Angew. Math.* **470** (1996), 51–88.

[Rosso 1989] M. Rosso, "An analogue of P.B.W. theorem and the universal R-matrix for $U_h\mathrm{sl}(N+1)$", *Comm. Math. Phys.* **124**:2 (1989), 307–318.

[Rosso 1998] M. Rosso, "Quantum groups and quantum shuffles", *Invent. Math.* **133**:2 (1998), 399–416.

[Rouquier 2005] R. Rouquier, "Representations of rational Cherednik algebras", pp. 103–131 in *Infinite-dimensional aspects of representation theory and applications*, Contemp. Math. **392**, Amer. Math. Soc., Providence, RI, 2005.

[Shan et al. 2014] P. Shan, M. Varagnolo, and E. Vasserot, "Koszul duality of affine Kac–Moody algebras and cyclotomic rational double affine Hecke algebras", *Adv. Math.* **262** (2014), 370–435.

[Shepler and Witherspoon 2008] A. V. Shepler and S. Witherspoon, "Hochschild cohomology and graded Hecke algebras", *Trans. Amer. Math. Soc.* **360**:8 (2008), 3975–4005.

[Shepler and Witherspoon 2012a] A. V. Shepler and S. Witherspoon, "Drinfeld orbifold algebras", *Pacific J. Math.* **259**:1 (2012), 161–193.

[Shepler and Witherspoon 2012b] A. V. Shepler and S. Witherspoon, "Group actions on algebras and the graded Lie structure of Hochschild cohomology", *J. Algebra* **351** (2012), 350–381.

[Shepler and Witherspoon 2014] A. V. Shepler and S. Witherspoon, "A Poincaré–Birkhoff–Witt theorem for quadratic algebras with group actions", *Trans. Amer. Math. Soc.* **366**:12 (2014), 6483–6506.

[Shepler and Witherspoon 2015] A. V. Shepler and S. Witherspoon, "PBW Deformations of Skew Group Algebras in Positive Characteristic", *Algebras and Representation Theory* **18**:1 (2015), 257–280.

[Shirshov 1962a] A. I. Shirshov, "On the bases of a free Lie algebra", *Algebra i Logika Sem.* **1**:1 (1962), 14–19.

[Shirshov 1962b] A. I. Shirshov, "Some algorithmic problem for Lie algebras", *Sibirsk Mat. Zh.* **3**:2 (1962), 292–296. In Russian; translated in *SIGSAM Bull.* **33** (1999), 3–6.

[Sridharan 1961] R. Sridharan, "Filtered algebras and representations of Lie algebras", *Trans. Amer. Math. Soc.* **100** (1961), 530–550.

[Takeuchi 1990] M. Takeuchi, "The q-bracket product and quantum enveloping algebras of classical types", *J. Math. Soc. Japan* **42**:4 (1990), 605–629.

[Ton-That and Tran 1999] T. Ton-That and T.-D. Tran, "Poincaré's proof of the so-called Birkhoff–Witt theorem", *Rev. Histoire Math.* **5**:2 (1999), 249–284.

[Varadarajan 1984] V. S. Varadarajan, *Lie groups, Lie algebras, and their representations*, Graduate Texts in Mathematics **102**, Springer, New York, 1984. Reprint of the 1974 edition.

[Walton and Witherspoon 2014] C. Walton and S. Witherspoon, "Poincaré–Birkhoff–Witt deformations of smash product algebras from Hopf actions on Koszul algebras", *Algebra Number Theory* **8**:7 (2014), 1701–1731.

[Witt 1937] E. Witt, "Treue Darstellung Liescher Ringe", *J. Reine Angew. Math.* **177** (1937), 152–160.

[Yamane 1989] H. Yamane, "A Poincaré–Birkhoff–Witt theorem for quantized universal enveloping algebras of type A_N", *Publ. Res. Inst. Math. Sci.* **25**:3 (1989), 503–520.

ashepler@unt.edu — *1155 Union Circle #311430, Denton, TX 76203-5017 United States*

sjw@math.tamu.edu — *Department of Mathematics, Texas A&M University, Mailstop 3368, College Station, TX 77843-3368, United States*

Commutative Algebra and Noncommutative Algebraic Geometry, I
MSRI Publications
Volume **67**, 2015

Frobenius splitting in commutative algebra

KAREN E. SMITH AND WENLIANG ZHANG

Frobenius splitting has inspired a number of techniques in commutative algebra, algebraic geometry, and representation theory. This is an introduction to the subject for beginners. We discuss the local theory (Frobenius map for rings) and the global theory (extension to schemes), test ideals, and explore connections with the Cohen–Macaulay property.

The purpose of these lectures is to give a gentle introduction to Frobenius splitting, or more broadly "Frobenius techniques", for beginners. Frobenius splitting has inspired a vast arsenal of techniques in commutative algebra, algebraic geometry, and representation theory. Many related techniques have been developed by different camps of researchers, often using different language and notation. Although there are great number of technical papers and books written over the past forty years, many of the most elegant ideas, and the connections between them, have coalesced only in the past decade. We wish to bring this emerging simplicity to the uninitiated.

Our story of Frobenius splitting begins in the 1970s, with the proof of the celebrated Hochster–Roberts theorem on the Cohen–Macaulayness of rings of invariants [Hochster and Roberts 1974]. This proof, in turn, was inspired by Peskine and Szpiro's ingenious use [1973] of the iterated Frobenius — or p-th power — map to prove a constellation of "intersection conjectures" due to Serre. Mehta and Ramanathan [1985] coined the term "Frobenius splitting" a decade later in a beautiful paper which moved beyond the affine case to prove theorems about Schubert varieties and other important topics in the representation theory of algebraic groups. Although these "characteristic p techniques" are powerful also for proving theorems for algebras and varieties over fields of characteristic zero, we focus on the prime characteristic case, since the technique of reduction to characteristic p has now become fairly standard.

Section 1 treats Frobenius splitting in the local algebraic spirit of Hochster and Roberts, where it provides a tool for controlling the singularities of a local ring. We prove the Hochster–Roberts theorem, giving what is essentially the tight closure proof of [Hochster and Huneke 1990] without explicitly mentioning tight closure. Roughly the point is that a ring of invariants (of a linearly reductive group) is a direct summand of a polynomial ring, and as such inherits a strong form of Frobenius splitting called *F-regularity*, which in turn, implies Cohen–Macaulayness.

Section 2 considers Frobenius splitting for schemes in the spirit of Mehta and Ramanathan, emphasizing how the Frobenius map can be used to study *global* properties of a smooth projective scheme. For example, we show how Frobenius splitting leads to powerful vanishing theorems for cohomology of line bundles, and prove structure theorems for Frobenius split and globally F-regular projective varieties. We explain how these local and global Frobenius tools are equivalent, despite developing independently during the last decades of the twentieth century. (This separate development is evidenced by the surprising disjointness of the two monographs "Tight closure and its applications" [Huneke 1996] and "Frobenius splitting methods in geometry and representation theory" [Brion and Kumar 2005]). The two schools independently discovered the same ideas, though often in language almost impenetrable to the other: Ramanathan's [1991] notion of *Frobenius splitting along a divisor* is closely related to Hochster and Huneke's [1989] *strong F-regularity*; compatibly split subschemes turn out to be essentially dual to modules with Frobenius action in commutative algebra; and criteria for Frobenius splitting of projective varieties (in terms of pluri-canonical sections) amount to dual criteria in terms of Frobenius actions on the injective hull of the residue field in a local ring. Both points of view are enhanced by understanding the connections between them.

Sections 3 and 4 treat *test ideals,* special ideals (subschemes) carrying information about the action of Frobenius. Although test ideals first arose on the commutative algebra side as a technical component of local tight closure theory, more recent work of Karl Schwede [2010a] has led to deeper understanding of the test ideal in the broader context of Frobenius splitting for schemes. Test ideals can be viewed as a prime characteristic analog of multiplier ideals [Smith 2000b; Hara 2001] and also of log-canonical centers for complex varieties [Schwede 2010a], depending on the context. Indeed in recent years, this connection is of increasing interest in the minimal model program in characteristic p. Section 3 develops the "absolute test ideal" as one ideal in a distinguished lattice of ideals well-behaved with respect to the Frobenius map. Section 4 develops test ideals for *pairs* in the important special case where the ambient ring is regular. We present simple proofs of all the basic properties in the setting of an *regular* ambient regular ring (much more technical proofs are scattered throughout the literature as special cases of more general results due to Hara, Watanabe, Takagi and Yoshida). We include a self-contained development of an asymptotic theory of test ideals analogous to the story of asymptotic multiplier ideals developed in [Ein et al. 2001], including an application to the behavior of symbolic powers of ideals in a regular ring.

This article is not in any way intended to compete with the excellent surveys [Schwede and Tucker 2012] or [Blickle and Schwede 2013], both of which contain a more technical and extensive survey of recent ideas. There are also older surveys such as [Smith 1997b], which explain more about reduction to characteristic p in this context and the connections between singularities in the minimal model program and characteristic p techniques. Other possible surveys of interest include Huneke's lectures [1996] on tight closure, Brion and Kumar's text [2005] on Mehta–Ramanathan Frobenius splitting, Swanson's notes [2002] on tight closure, Huneke's survey [2013] on F-signature and Hilbert–Kunz multiplicities, the surveys [Benito et al. 2013] (more basic) or [Mustaţă 2012] (more advanced) on log canonical and F-threshold, or Holger Brenner's survey [2008] on geometric methods in tight closure theory.

1. The Frobenius map for rings: the local theory

Let R be any commutative ring of prime characteristic p. The Frobenius map is the p-th power map:

$$F : R \to R, \quad r \mapsto r^p.$$

Because $(r+s)^p = r^p + s^p$ in characteristic p, the Frobenius map is a *ring homomorphism.* Its image is the subring R^p of all elements of R that are p-th powers. We thus have an inclusion of rings $R^p \hookrightarrow R$.

Our goal is to use the Frobenius map — or more precisely the *R^p-module structure of R* — to understand the singularities of the ring R. Typically, thankfully, this module is finitely generated:

Definition 1.1. A ring R of positive characteristic p is said to be *F-finite* if R is a finitely generated module over its subring R^p.

F-finiteness is a mild assumption, usually satisfied in "geometric" settings. For example, a perfect field k is F-finite of course: by definition $k^p = k$. Every ring finitely generated over an F-finite ring is F-finite. In particular, finitely generated algebras over perfect fields are F-finite. Moreover, it is easy to see that the class of F-finite rings is closed under localization, surjective image, completion at a maximal ideal, and finite extensions.

Already we can observe that one of the most basic singularities of the ring R can be detected by Frobenius. Namely R is *reduced* (meaning that 0 is the only nilpotent) if and only if the Frobenius map is injective.

A less trivial observation, and indeed the starting point for using Frobenius to classify singularities, is the famous 1969 theorem of Ernst Kunz:

Theorem 1.2 [Kunz 1969]. *An F-finite ring R is regular if and only if R is locally free as an R^p-module.*[1]

Example 1.3. Let R be the ring $\mathbb{F}_p[x, y]$. Considered as a module over the subring $R^p = \mathbb{F}_p[x^p, y^p]$, it is easy to see that R is a free R^p-module. Indeed, the monomials

$$\{x^a y^b \mid 0 \leq a, b < p\}$$

form a free basis. Similarly, a polynomial ring R in d variables over $\mathbb{F}_p$ is a free-module over R^p of rank p^d.

As we will see, it is possible to classify the singularities of R according to how far the R^p-module R is from free. This is one of the ways in which the local theory of "Frobenius techniques" takes shape.

Notation. For simplicity, let us assume that R is a domain. In this case, the inclusion of R^p-modules $R^p \hookrightarrow R$ is entirely equivalent to the inclusion of R-modules $R \hookrightarrow R^{1/p}$, where $R^{1/p}$ is the subring of p-th roots of elements of R in an algebraic closure of the fraction field of R (note that each $r \in R$ has a *unique* p-th root). Thus to understand the R^p module structure of R is to understand the R-module structure of $R^{1/p}$. Both are equivalent to viewing R as an R-module via restriction of scalars by the Frobenius map $F : R \to R$. When using this last point of view, it can be useful to notationally distinguish the source and target copies of R; our favorite way (because it is consistent with the standard notation used for maps of schemes as in [Hartshorne 1977]) is to use F_*R for the target copy of R, so that the Frobenius map is denoted $F : R \to F_*R$. It is worth being open to any of these notations, since depending on the situation, one may be more illuminating than another.[2] On the other hand, sometimes it is convenient

[1] More generally, even if R is not F-finite, Kunz shows that a ring of characteristic $p > 0$ is regular if and only if its Frobenius map is flat. F-finite rings are always excellent [Kunz 1976].

[2] In the tight closure literature, the notation F_*R is often replaced by ${}^1R$, so the Frobenius map is written $R \to {}^1R$, with the second copy of R denoting R as an abelian group with R-module structure $r \cdot x = r^p x$ where $r \in R$ and $x \in {}^1R$. The notation $R^{1/p}$ can also be used to denote the target copy of R in general; if R is reduced, each element has a unique p-th root inside the total ring of quotients so that the Frobenius map becomes the inclusion $R \hookrightarrow R^{1/p}$, but this notation can be misleading if R is not reduced because then the Frobenius map $R \to R^{1/p}$ is not injective.

not to notationally distinguish the source and target of Frobenius; this point of view is at the heart of Blickle's theory of Cartier modules. See Remark 3.15.

1A. *Splitting.* Let $R \to S$ be any homomorphism of rings. Considering S as an R-module via restriction of scalars, we can ask whether or not this map *splits* in the category of R-modules.

Definition 1.4. We say that $R \to S$ *splits* if there is an R-module map $S \xrightarrow{\phi} R$ such that the composition

$$R \to S \xrightarrow{\phi} R$$

is the identity map on R. Equivalently, $R \to S$ splits if there exists ϕ in $\mathrm{Hom}_R(S, R)$ such that $\phi(1) = 1$.

If $R \to S$ splits, then it is obviously injective, so we often restrict attention to inclusions of rings. Given an inclusion $R \hookrightarrow S$, we also say that *R is a direct summand of S* to mean that the map splits. This concept is important because many nice properties of rings pass to direct summands.

Example 1.5. Let G be a finite group acting on a ring S. Let S^G denote the ring of invariants, that is, the subring of elements of S that are fixed by the action of G. The reader will easily show that the map

$$\varphi : S \to S^G, \quad \varphi(s) = \frac{1}{|G|} \sum_{g \in G} g \cdot s,$$

gives a splitting of $S^G \hookrightarrow S$, provided that $|G|$ is invertible in S.

Definition 1.6. A ring R of characteristic p is *Frobenius split* (or *F-split*) if the Frobenius map splits. Explicitly, a reduced ring R is Frobenius split if the ring inclusion $R^p \hookrightarrow R$ splits as a map of R^p-modules. Equivalently, a reduced ring R is Frobenius split if there exists $\pi \in \mathrm{Hom}_R(R^{1/p}, R)$ such that $\pi(1) = 1$.

Example 1.7. The ring $R = \mathbb{F}_p[x, y]$ is Frobenius split. Indeed, we have seen that R is free over the subring $R^p = \mathbb{F}_p[x^p, y^p]$, with basis $\{x^a y^b\}$ where $0 \le a, b < p$. Any projection onto the summand generated by the basis element $1 = x^0 y^0$ gives a splitting.

Frobenius splitting is a local condition on a ring:[3]

Lemma 1.8. *Let R be any F-finite ring of prime characteristic. The locus of points P in Spec R such that R_P is Frobenius split is an open set. In particular, R is Frobenius split if and only if for all maximal (equivalently, all prime) ideals P in Spec R, the local ring R_P is Frobenius split.*

[3]But use caution: on a nonaffine scheme the Frobenius map can split locally at each point but not globally! For example, a smooth projective curve is locally Frobenius split, but not globally Frobenius split if the genus is greater than one; see Example 2.16.

The proof of Lemma 1.8 is easy, following from the fact that $R^{1/p}$ is a finitely generated R module, so we leave it to the reader. Using Lemma 1.8, it is not hard to prove the following generalization of Example 1.7:

Proposition 1.9. *Every F-finite regular ring is Frobenius split.*

Indeed, for a local ring (R, m), we can think of regularity as the condition that the R-module $R^{1/p}$ decomposes completely into a direct sum of copies of R, whereas Frobenius splitting is the condition that $R^{1/p}$ contains *at least one direct sum copy* of R.

The property of Frobenius splitting is passed on to direct summands:

Proposition 1.10. *Let $R \subset S$ be any inclusion of rings of characteristic p which splits in the category of R-modules. If S is Frobenius split, then so is R.*

Proof. We have a commutative diagram:

$$\begin{array}{ccc} R & \hookrightarrow & S \\ \uparrow & & \uparrow \\ R^p & \hookrightarrow & S^p \end{array}$$

If we denote the splitting of $R \hookrightarrow S$ by ϕ, then the map $R^p \hookrightarrow S^p$ is also split, by the map ϕ^p defined by taking the p-th powers of everything. Our assumption that S is Frobenius split amounts to the existence of an S^p-linear map $\pi : S \to S^p$ sending 1 to 1. The composition $\phi^p \circ \pi$, when restricted to R, gives an R^p-linear map from R to R^p sending 1 to 1. Thus R is also Frobenius split. □

With Property 1.10 in hand, it is easy to construct examples of Frobenius split rings which are not regular: a direct summand of a regular ring is always Frobenius split but not usually regular. We give an example.

Example 1.11. For any graded ring $R = \bigoplus_{n \in \mathbb{N}} R_n$, the inclusion of any Veronese subring

$$R^{(d)} = \bigoplus_{n \in \mathbb{N}} R_{dn} \hookrightarrow R$$

splits. So a Veronese subring of a polynomial ring is Frobenius split in any characteristic, although such a subring is rarely regular. For instance, the ring $k[x, y, z]/(xz - y^2) \cong k[u^2, uv, v^2] \subset k[u, v]$, being the second Veronese subring of a polynomial ring, is Frobenius split in every characteristic (but never regular).

Remark 1.12. Frobenius splitting was first systematically studied by Hochster and Roberts [1974; 1976]. The term *Frobenius split*, however, was introduced in the beautiful paper [Mehta and Ramanathan 1985], which interpreted many of these ideas in a projective setting. Hochster and Roberts actually introduced a slightly more technical notion called *F-purity*, which (as they show) is equivalent to Frobenius splitting under the F-finite hypothesis. For non-F-finite rings, the notion of F-purity is *a priori* weaker than Frobenius splitting; however, we do not know a single (excellent) example of an F-pure ring which is not Frobenius split.

1B. ***Iterations of Frobenius and F-regularity.*** The real power of Frobenius emerges when we iterate it. The composition of the Frobenius map with itself is obviously a ring homomorphism sending each element r to $(r^p)^p = r^{p^2}$. More generally, for each natural number e, the iteration of Frobenius e times is the ring homomorphism

$$F^e : R \to R, \quad r \mapsto r^{p^e}.$$

The images of each of these iterates produces an infinite descending chain of subrings

$$R \supset R^p \supset R^{p^2} \supset R^{p^3} \supset \dots.$$

The original ring R can be viewed as a module over each of these subrings R^{p^e}. Indeed, assuming that R is F-finite, then also R is finitely generated as an R^{p^e}-module for each e. Again, understanding the R^{p^e}-module structure of the R^{p^e}-module R is essentially the same as understanding the R-module structure of the R-module R^{1/p^e} (or $F^e_* R$).

If R is F-finite, Kunz's theorem implies that R is regular if and only if the R-modules R^{1/p^e} are all locally free. Classes of "F-singularities" can be defined depending on the extent to which the R^{1/p^e} fail to be locally free. The first of these are the *F-regular* rings, which have many direct sum copies of R in R^{1/p^e} as e gets larger:

Definition 1.13 [Hochster and Huneke 1989]. An F-finite domain R is *strongly F-regular*, or simply *F-regular*,[4] if for every nonzero element $f \in R$ there exists $e \in \mathbb{N}$ such that the R-module inclusion $Rf^{1/p^e} \hookrightarrow R^{1/p^e}$ splits. Put differently, this means that for all nonzero f, there exists $e \in \mathbb{N}$ and $\phi \in \operatorname{Hom}_R(R^{1/p^e}, R)$ such that $\phi(f^{1/p^e}) = 1$.

An F-regular ring therefore, may not be free when considered as a module over R^{p^e}, but it will have *many* summands isomorphic to R^{p^e}. Indeed, every nonzero element of R will generate an R^{p^e}-module direct summand of R for sufficiently large e.

[4]For brevity, we often drop the qualifier "strongly" in the text. Hochster and Huneke introduced three flavors of F-regularity — weak F-regularity, F-regularity and strong F-regularity — and conjectured all to be equivalent. This is still not known in general; however it is known for Gorenstein rings [Hochster and Huneke 1989], graded rings [Lyubeznik and Smith 1999], and even more generally; see [Aberbach 2002; Maccrimmon 1996]. In any case, because our bias is that strong F-regularity is the central notion and we treat only that notion, we drop the clumsy modifier and frequently write "F-regular" instead of strongly F-regular, begging forgiveness of Hochster and Huneke.

Proposition 1.14. *Every F-regular ring is Frobenius split.*

Proof. Taking f to be 1, we know that there exists an e such that $R \hookrightarrow R^{1/p^e}$ splits. Restricting the splitting $\phi : R^{1/p^e} \to R$ to the subring $R^{1/p}$, we have a splitting of the inclusion $R^{1/p} \hookrightarrow R$. Thus R is Frobenius split. □

The proof of the following lemma is a straight-forward exercise, using the fact that R^{1/p^e} is a finitely generated R-module.

Lemma 1.15. [Hochster and Huneke 1989, 3.1,3.2] Let R be an F-finite ring of characteristic p.

(1) R is F-regular if and only if R_m is F-regular for every maximal (equivalently, prime) ideal m of R.

(2) A local ring (R, m) is F-regular if and only if the completion $\hat{R}$ of R at its maximal ideal is F-regular.

Proposition 1.16. *An F-finite regular domain*[5] *is strongly F-regular.*

Proof. Since F-regularity can be checked locally at each prime (Lemma 1.15), there is no loss of generality in assuming that (R, m) is local.

The proof is a simple application of Nakayama's lemma [Atiyah and Macdonald 1969, Proposition 2.8]. What does Nakayama's lemma say about the finitely generated R-module $M = R^{1/p^e}$? It says that an element f^{1/p^e} is part of a minimal generating set of R^{1/p^e} as an R-module if and only if it is not in mR^{1/p^e}, which in turn happens if and only if f is not in the ideal $m^{[p^e]}$ in R, where $m^{[p^e]}$ denotes the ideal of R generated by the p^e-th powers of the elements of m. Since $\bigcap_e m^{[p^e]} \subset \bigcap_e m^e = 0$, this means that for each fixed nonzero $f \in R$, we can always find an e such that f^{1/p^e} is a part of a set of minimal generators for R^{1/p^e} over R. This observation holds quite generally, whether or not R is regular.

Now if R is regular, then R^{1/p^e} is a free R-module, so that such a minimal generator f^{1/p^e} for R^{1/p^e} over R will necessary be part of a free basis for R^{1/p^e} over R. This means that f^{1/p^e} spans a free R-module summand of R^{1/p^e}. Since this holds for every nonzero f (with possibly larger e), we conclude that R is F-regular. □

The analog of Proposition 1.10 holds for F-regular rings, with essentially the same proof:

Proposition 1.17. *Let $R \subset S$ be any inclusion of rings of characteristic p which splits in the category of R-modules. If S is F-regular, then R is F-regular.*

[5] An arbitrary F-finite ring (not a domain) can be defined as strongly F-regular if for every f not in any minimal prime, there exists an e and an $\phi \in Hom_R(F^e_* R, R)$ such that $\phi(f) = 1$. However, this is not an essential generalization of the theory: it is easy to check that an F-regular ring is a product of F-regular domains; see [Hochster and Huneke 1989].

In particular, Veronese subrings of a polynomial ring are F-regular, as are rings of invariants of finite groups whose order is coprime to the characteristic. See Example 1.11.

The power of Proposition 1.17 stems from the nice properties of F-regular rings:

Theorem 1.18 [Hochster and Huneke 1989]. *All F-regular rings are Cohen–Macaulay and normal.*

For those not already enamored by Cohen–Macaulay singularities, we have included an Appendix discussing this crucially important if slightly technical condition.

Proof. Without loss of generality, we can assume that (R, m) is an F-regular local domain. First we prove that R is Cohen–Macaulay, i.e., each system of parameters of R is a regular sequence. By Lemma 1.15, we can assume that R is complete, and therefore has a coefficient field, K. By the Cohen Structure Theorem, R is module finite over the subring A of formal power series over K in any system of parameters $x_1, \dots, x_d$.

Suppose we have a relation on our system of parameters, $x_i z \in (x_1, \dots, x_{i-1})R$. Let B be the intermediate ring generated by z over A. Note that $B \cong A[t]/(g(t))$, where g is the minimal polynomial of $z \in B$ over A, so that B is a hypersurface ring and in particular, Cohen–Macaulay. To summarize, we have module finite extensions

$$A = K[[x_1, \dots, x_d]] \hookrightarrow B = A[z] \hookrightarrow R,$$

where B is Cohen–Macaulay with regular sequence $x_1, \dots, x_d$.

Since $B \hookrightarrow R$ is finite, there is a B-linear map $\psi : R \to B$ sending 1 to, say, $b \neq 0$. Raising our relation $zx_i \in (x_1, \dots, x_{i-1})R$ to the p^e-th power, we have $z^{p^e} x_i^{p^e} \in (x_1^{p^e}, \dots, x_{i-1}^{p^e})R$. Applying ψ to this relation, we have $bz^{p^e} x_i^{p^e} \in (x_1^{p^e}, \dots, x_{i-1}^{p^e})B$ in B. But because B is Cohen–Macaulay, we can divide out the $x_i^{p^e}$ to get $bz^{p^e} \in (x_1^{p^e}, \dots, x_{i-1}^{p^e})B$. Expanding to R, we have equivalently

$$b^{1/p^e} z \in (x_1, \dots, x_{i-1})R^{1/p^e}. \tag{1.18.1}$$

Now using the F-regularity of R, there is an e and $\phi \in \operatorname{Hom}_R(R^{1/p^e}, R)$ such that $\phi(b^{1/p^e}) = 1$. Applying ϕ to (1.18.1) we have $z \in (x_1, \dots, x_{i-1})R$. Thus R is Cohen–Macaulay.[6]

Next, we tackle normality. Fix an element x/y in the fraction field of R integral over R. We must show that y divides x in R. Since x/y is integral over

[6]For readers familiar with local cohomology, we leave as an exercise to find a slick local cohomology proof that F-regular rings are Cohen–Macaulay.

R, there is an equation

$$(x/y)^m + r_1(x/y)^{m-1} + \cdots + r_m = 0,$$

where each r_j is in R. Raising both sides of this equation to the p^e-th power, we can see that $(x/y)^{p^e}$ is also integral over R for all $e \geq 1$. Since the integral closure $\overline{R}$ of R is module-finite over R, there is a $c \in R$ such that $c\overline{R} \subseteq R$; in particular, $c(x/y)^{p^e} \in R$ for all $e \geq 1$, i.e., $cx^{p^e} \in (y^{p^e})$ for all $e \geq 1$. Hence $cx^{p^e} = ry^{p^e}$ for some $r \in R$. Therefore we have

$$c^{1/p^e} x = r^{1/p^e} y \tag{1.18.2}$$

Since R is F-regular, there is an e and $\phi_e \in \mathrm{Hom}_R(R^{1/p^e}, R)$ such that $\phi_e(c^{1/p^e}) = 1$. Applying ϕ_e to (1.18.2) we have $x \in (y)$. This shows that $x/y \in R$ and finishes the proof. □

To summarize, we now have proved the following implications among classes of singularities:

{Regular} $\Longrightarrow$ {F-regular} $\Longrightarrow$ {Frobenius split, Cohen–Macaulay, Normal}.

In addition, we have shown that both Frobenius splitting and F-regularity descend to direct summands. This is all that is needed to prove the Hochster–Roberts theorem, at least in characteristic p.

1C. *The Hochster–Roberts theorem.*

Theorem 1.19 [Hochster and Roberts 1974]. *Fix any ground field k. Let G be a linearly reductive algebraic group over k acting on a regular Noetherian k-algebra S. Then the ring of invariants*

$$S^G := \{f \in S \mid f \circ g = f \textit{ for all } g \in G\};$$

is Cohen–Macaulay.

For example, let V be a finite dimensional representation of a linearly reductive group G. The Hochster–Roberts theorem guarantees that the ring of invariants for the induced action of G on the symmetric algebra of V is Cohen–Macaulay. From a practical point of view, this means that the invariants form a graded finitely generated *free* module over some polynomial subring.

Geometrically, the point of the Hochster–Roberts theorem is that when a reasonable group acts on a smooth variety, the "quotient variety" will have reasonably nice singularities. Indeed, let X be a smooth (affine) variety on which the group G acts by regular maps. Then there is an induced action on the coordinate ring S, and more or less by definition, the "quotient variety" is the unique variety X/G whose coordinate ring is S^G. The Cohen–Macaulayness of S^G is a niceness condition on the singularities of the quotient.[7]

Linear reductive groups. By definition, a *linearly reductive group* is an algebraic group with the property that every finite dimensional representation is completely reducible, that is, it decomposes as a direct sum of irreducible representations. In characteristic zero, linearly reductive is the same as reductive, so includes all semisimple algebraic groups. In particular, all finite groups, all tori, and all matrix groups such as GL_n and SL_n are linearly reductive over a field of characteristic zero. Over a field of positive characteristic p, linearly reductive groups are less abundant: tori and finite groups whose order is not divisible by p, as well as extensions of these. See [Nagata 1961] for more information.

The point for us is this: if G is a linearly reductive group acting on a regular k-algebra S, then the inclusion of the ring of invariants S^G in S splits.

Using this, it is easy to prove the Hochster–Roberts theorem in the prime characteristic case. We refer to the original paper [Hochster and Roberts 1974] for the reduction to prime characteristic.

Proof of the Hochster–Roberts theorem in prime characteristic. Because S^G is a *direct summand* of the ring S, the Hochster–Roberts theorem follows immediately from the following:

Theorem 1.20. *Let $R \subset S$ be a split inclusion of rings of positive characteristic. If S is regular, then R is F-regular, hence Cohen–Macaulay.*

This theorem in turn is simply a stringing together of Proposition 1.16, which tells us S is F-regular, Proposition 1.17, which guarantees that the property of F-regularity is passed on to the direct summand S^G, and finally Theorem 1.18, which tells us that S^G is therefore Cohen–Macaulay and normal. □

While the Hochster–Roberts theorem is most interesting in characteristic zero (since that is where the most interesting groups are found), its original proof fundamentally uses characteristic p. Later, Boutot gave a different proof of the characteristic zero case, which does not use reduction to prime characteristic, although it still exploits the philosophy of "splitting" we discuss here [Boutot 1987]. This philosophy is further expounded in [Kovács 2000].

---

[7]Some caution is in order here: one can not usually put the structure of a variety on the set of G-orbits of X, although in a sense that can be made precise, Spec S^G is the algebraic variety "closest to being a quotient" — it behaves as a quotient in a categorical sense. If G is finite (and its order is not divisible by p), the topological space Spec S^G is the topological quotient of Spec S by G. In the nonaffine case, the situation is even more complicated, and there are several "quotients," which depend on a choice of *linearization* of the action. This is the huge and beautiful theory of geometric invariant theory, or GIT [Mumford et al. 1994].

Example 1.21. Let G be the two-element group $\{\pm 1\}$ under multiplication. Let G act on k^2, where k is a field whose characteristic is not two, in the obvious way by multiplication: $-1 \cdot (x, y) = (-x, -y)$. The induced action on the coordinate ring $k[x, y]$ is the same: -1 acts by multiplying both x and y by -1, so that a monomial $x^a y^b$ is sent to $(-1)^{a+b} x^a y^b$. In particular, the invariant ring is the subring generated by polynomials of even degree, or

$$k[x, y]^G = k[x^2, xy, y^2] \cong k[u, v, w]/(v^2 - uw).$$

According to the Hochster–Roberts theorem and its proof, this ring is F-regular in all finite characteristics, and hence Cohen–Macaulay. Standard "reduction to characteristic p techniques" guarantee the ring is Cohen–Macaulay also when k has characteristic zero.

Example 1.22. Let X be the variety of all $2 \times n$ complex matrices, so $X \cong \mathbb{C}^{2n}$. Let $G = \mathrm{SL}_2$ act on X by left multiplication. The ring of invariants for the induced action of SL_2 on $\mathbb{C}[x_{11}, \ldots, x_{1n}, x_{21}, \ldots, x_{2n}]$ is generated by all the 2×2 subdeterminants of the matrix of indeterminates:

$$\begin{pmatrix} x_{11} & x_{12} & \cdots & x_{1n} \\ x_{21} & x_{22} & \cdots & x_{2n} \end{pmatrix}.$$

This is the homogeneous coordinate ring for the Plücker embedding of the Grassmannian of 2-dimensional subspaces of $\mathbb{C}^n$. The Hochster–Roberts Theorem guarantees that this ring is Cohen–Macaulay. More generally, the Plücker ring of the Grassmannian of d-dimensional subspaces of $\mathbb{C}^n$ is Cohen–Macaulay, since it is the ring of invariants for SL_d acting on $\mathbb{C}^{d \times n}$ by left multiplication. In characteristic p, the group SL is not linearly reductive and the ring of invariants does not split. Nonetheless, this ring is F-regular in all prime characteristics [Hochster and Huneke 1994], hence Cohen–Macaulay.

1D. *F-signature.* A numerical refinement of F-regularity called the *F-signature* sharpens the classification of F-singularities by measuring the *growth rate* of the rank of a maximal free summand of the R-module R^{1/p^e} as e goes to infinity. This was first studied in [Smith and Van den Bergh 1997].

Fix a local F-finite domain R. For each natural number e, we can decompose the R module R^{1/p^e} as a direct sum of indecomposable modules, and count the number a_e of summands of R^{1/p^e} that are isomorphic to R. For a Frobenius split ring R, there is always at least one. If R is regular, all summands are isomorphic to R, so a_e is equal to the rank of R^{1/p^e} over R. For arbitrary R, the (generic) rank of R^{1/p^e} over R obviously bounds the number a_e. For an F-regular ring, we expect many summands of R^{1/p^e} isomorphic to R, so we expect a_e to be relatively large, and to grow with e. In fact, as it turns out, we can define the

F-signature of R to be

$$s(R) = \lim_{e\to\infty} \frac{a_e}{\delta^e},$$

where δ is the generic rank of R over R^p, that is $\delta = [K : K^p]$ where K is the fraction field of R. This limit exists [Tucker 2012], and is at most one.

The F-signature can be used to classify F-regular rings. Indeed, Huneke and Leuschke proved that the F-signature is one if and only if R is regular in the paper [Huneke and Leuschke 2002] that coined the term "F-signature". Furthermore, the F-signature is positive if and only if R is F-regular [Aberbach and Leuschke 2003]. Thus each F-regular ring has an F-signature strictly between zero and one; the closer the F-signature is to one, the "less singular" the ring is. For example, for rational double points such $xy = z^{n+1}$, the F-signature is $1/(n+1)$ [Huneke and Leuschke 2002], reflecting the fact that the singularity is "worse" for larger n. Many more computations of this type can be found in [Huneke and Leuschke 2002] and later [Yao 2006]. Formulas for the F-signature of toric varieties are worked out in Von Korff's PhD thesis [Von Korff 2012]; see also [Watanabe and Yoshida 2004] and [Singh 2005]. Tucker vastly generalizes and simplifies much of the literature on F-signature in [Tucker 2012].

There are many interesting open questions about the F-signature. No known examples of nonrational F-signatures are known (though some expect that they exist). Also, Florian Enescu (personal communication) has suggested that there may be an upper bound on the F-signature of a nonregular ring depending only on the dimension: for $d \geq 2$, the singularity defined by $x_0^2 + x_1^2 + \cdots + x_d^2$ (characteristic $\neq 2$) has F-signature $1 - \frac{1}{d}$, and no d-dimensional singularities of larger F-signature are known. The F-signature is closely related to the Hilbert–Kunz multiplicity, a subject pioneered by Paul Monsky [1983]; see [Huneke 2013] or [Brenner 2013]. Further developments, including generalizations of F-signature to pairs, are covered by Blickle, Schwede and Tucker in [Blickle et al. 2012; 2013].

1E. ***Frobenius splitting in characteristic zero and connections with singularities in birational geometry.*** We briefly recall the standard technique for extending these ideas to algebras over fields of characteristic zero. Let $\mathbb{C}$ denote any field of characteristic 0.

Let R be a finitely generated $\mathbb{C}$-algebra. Fix a presentation

$$R \cong \mathbb{C}[x_1, \ldots, x_n]/(f_1, \ldots, f_r).$$

Let A be the $\mathbb{Z}$-subalgebra of $\mathbb{C}$ generated by all coefficients of the polynomials $f_1, \ldots, f_r$, and set

$$R_A = A[x_1, \ldots, x_n]/(f_1, \ldots, f_r).$$

Since A is a finitely generated $\mathbb{Z}$-algebra, the residue field of A at each of its maximal ideals is finite. The map Spec $R_A \to$ Spec A can be viewed as a "family of models" of the original algebra R. The closed fibers of this map are characteristic p schemes (of varying p) whereas the generic fiber is a flat base change from the original R. Roughly speaking, R is F-regular or Frobenius split if most (or at least a dense set) of the closed models have this property. More precisely:

Definition 1.23. Let R, A, R_A be as above. The ring R is said to have *Frobenius split type* (or *F-regular type*) if there is a Zariski dense set of maximal ideals μ in Spec A such that $A/\mu \otimes_A R_A$ is Frobenius split (or F-regular).[8]

Although it is not completely obvious, Definition 1.23 does not depend on the presentation of R, nor on the choice of A. See [Hochster and Huneke 2006].

Example 1.24. (1) The ring $\mathbb{C}[x, y, z]/(y^2 - xz)$ has F-regular type. In fact, taking $A = \mathbb{Z}$, the closed fibers of the family are the rings $\mathbb{F}_p[x, y, z]/(y^2 - xz)$, which are F-regular for *every* prime number p. See also Example 1.11 and Proposition 1.17.

(2) The ring $\mathbb{C}[x, y, z]/(x^3 + y^3 + z^3)$ has Frobenius split type, but *not* F-regular type. Indeed, $\mathbb{F}_p[x, y, z]/(x^3 + y^3 + z^3)$ can be checked to be Frobenius split after reduction mod p whenever $p \equiv 1$ mod 3, so that there is an infinite set of prime numbers p (hence dense set of Spec $\mathbb{Z}$) for which the "reduction mod p" is Frobenius split.[9] On the other hand, for every $p \geq 5$ and every e, one can show that there is *no map* sending x^{1/p^e} to 1. So this ring is not F-regular type.

Connections with the singularities in the minimal model program. Amazingly, the properties of Frobenius splitting and F-regularity in characteristic zero turn out to be closely related to a number of important issues studied independently in algebraic geometry, including log canonical and log terminal singularities, and ultimately positivity and multiplier ideals as well. For example:

Theorem 1.25 [Smith 1997a; Hara 1998; Mehta and Srinivas 1997; Elkik 1981]. *Let R be a Gorenstein domain finitely generated over a field of characteristic zero.*

(1) *R has F-regular type if and only if R has log terminal singularities.*

(2) *If R has Frobenius split type, then R has log canonical singularities.*

[8] In the literature, this is usually called *dense* F-split (F-regular) type. The related condition that $A/\mu \otimes_A R_A$ is F-split (F-regular) for a Zariski *open* set of maximal ideal μ in Spec A is called *open* F-split (F-regular) type.

[9] If $p \equiv 2$ mod 3, the ring is not Frobenius split, so R has *dense* F-split type, but not *open* F-split type. Compare Example 2.6. On the other hand, it is expected that dense and open F-regular type are equivalent; this is known in Gorenstein rings and related settings; see [Smith 1997a] and [Hara and Watanabe 2002].

We do not digress to discuss all the relevant definitions here, but refer instead to the literature. For Gorenstein varieties, log terminal is equivalent to rational singularities [Elkik 1981], which may be more familiar.

Theorem 1.25 is proven more generally in [Hara and Watanabe 2002]. Statement (1) is closely related to the equivalence of rational singularities with F-rational type, proved in [Smith 1997a; Hara 1998; Mehta and Srinivas 1997]. Statement (2) is closely related to the fact F-injective type implies Du Bois singularities, proved in [Schwede 2009b].

The converse of Statement (2) is conjectured to hold in general as well. This long-standing open question related to an important conjecture linking the *F-pure threshold* and the *log canonical threshold*, a very rich area of research with a huge literature. While still wide open, it is worth pointing out that Mustaţă and Srinivas [2011] have reduced the question to an interesting conjecture with roots in Example 2.6. Although we cannot go into the F-pure threshold (or it generalization to the "F-jumping numbers") here, fortunately there are already extensive recent surveys, including the survey for beginners [Benito et al. 2013] and the more advanced research survey [Mustaţă 2012]. The F-pure threshold is very difficult to compute, often with complicated fractal-like behavior; see, for example, [Hernández 2014; 2015; Hernández and Texiera 2015] for concrete computations of F-thresholds. There are few general results, but some can be found in [Bhatt and Singh 2015] and [Hernández et al. 2014].

2. Frobenius for schemes: the global theory

Let X denote the affine scheme Spec R, where R is a ring of prime characteristic p. Like any map between rings, the Frobenius map induces a map of schemes, which we also denote F. As a map of the underlying topological space, $F : X \to X$ is the *identity* map, but the associated map of sheaves $\mathcal{O}_X \to F_*\mathcal{O}_X$ is induced by the p-th power map of R.

Of course, the p-th power map is compatible with localization, so that the Frobenius map on affine charts can be patched together to get a Frobenius map for *any scheme X of characteristic p*. This Frobenius map is the identity map on the underlying topological space of X while the corresponding map of sheaves of rings $\mathcal{O}_X \to F_*\mathcal{O}_X$ is the p-th power map locally on sections.

The sheaf $F_*\mathcal{O}_X$ is quasicoherent on X. Consistent with the terminology for rings, we say that a scheme X is *F-finite* if the sheaf $F_*\mathcal{O}_X$ is *coherent.* Our main interest is when X is a variety over a perfect field k of characteristic p; such a variety is always F-finite.[10]

[10]The Frobenius map is always a *scheme map,* but not usually a morphism of varieties over k, because it is not linear over k (unless, for example, $k = \mathbb{F}_p$). If we insist on working with maps

As in the affine case, the sheaf $F_*\mathcal{O}_X$ carries a remarkable amount of information about the scheme X. For example, Theorem 1.2 implies that an F-finite scheme X is regular if and only if the coherent $\mathcal{O}_X$-module $F_*\mathcal{O}_X$ is locally free. That is, a variety over a perfect field is smooth if and only if the coherent sheaf $F_*\mathcal{O}_X$ is a *vector bundle* over X.

Similarly, we can define a scheme X to be *locally Frobenius split* if the map $\mathcal{O}_X \to F_*\mathcal{O}_X$ splits locally in a neighborhood of each point, or equivalently, if the corresponding map on stalks splits for each $p \in X$. Likewise, we can define X to be *locally F-regular* if the stalks are all F-regular. Since Frobenius splitting and F-regularity are *local* properties for affine schemes (by Lemmas 1.8 and 1.15), all the results from the previous section give corresponding local results for an arbitrary F-finite scheme of prime characteristic. For example, a locally F-regular scheme is normal and Cohen–Macaulay by Theorem 1.18.

It is much stronger, of course, to require a *global* splitting of the Frobenius map $\mathcal{O}_X \to F_*\mathcal{O}_X$. Not surprisingly, a global splitting of Frobenius has strong consequences for the global geometry of X. This is the topic of our Section 2.

Definition 2.1. The scheme X is *Frobenius split*[11] if the Frobenius map

$$\mathcal{O}_X \to F_*\mathcal{O}_X$$

splits as a map of $\mathcal{O}_X$-modules. This means that there exists a map $F_*\mathcal{O}_X \to \mathcal{O}_X$ of sheaves of $\mathcal{O}_X$-modules, such that the composition $\mathcal{O}_X \to F_*\mathcal{O}_X \to \mathcal{O}_X$ is the identity map.

The global consequences of splitting Frobenius, and indeed the term *Frobenius split,* were first treated systematically by Mehta and Ramanathan in [Mehta and Ramanathan 1985]; see also [Haboush 1980]. While inspired by Hochster and Roberts' paper ten years prior, which focused on the local case, Mehta and Ramanathan were motivated by the possibility of understanding the *global* geometry of Schubert varieties and related objects in algebrogeometric representation theory; see, e.g., [Mehta and Ramanathan 1985] or [Ramanan and Ramanathan 1985; Mehta and Ramanathan 1988]. This idea was very fruitful, leading the Indian school of algebrogeometric representation theory to many important results now chronicled in the book [Brion and Kumar 2005]. In Section 2B, we formally show how the local and global points of view converge by translating global splittings of a projective variety X into local splittings "at the vertex of the cone" over X.

of varieties, we can force the Frobenius map to be defined over k by changing base to make this so; this is called the relative Frobenius map. See, e.g., [Mehta and Ramanathan 1985; Brion and Kumar 2005].

[11]We will say "globally Frobenius split" if there is any possibility of confusion.

Example 2.2. Projective space $\mathbb{P}^n_k$ is Frobenius split in every positive characteristic. Indeed, any (homogeneous) splitting of Frobenius for the polynomial ring $k[x_0, \dots, x_n]$ induces a splitting of the corresponding Frobenius map of sheaves $\mathcal{O}_{\mathbb{P}^n} \to F_* \mathcal{O}_{\mathbb{P}^n}$.

Frobenius split varieties satisfy strong vanishing theorems:

Theorem 2.3. *Let X be a Frobenius split scheme. If $\mathcal{L}$ is an invertible sheaf on X such that $H^i(X, \mathcal{L}^n) = 0$ for $n \gg 0$, then $H^i(X, \mathcal{L}) = 0$.*

Corollary 2.4. *Let $\mathcal{L}$ be a ample invertible sheaf on a Frobenius split projective variety X. Then $H^i(X, \mathcal{L})$ vanishes for all $i > 0$, and if X is Cohen–Macaulay (e.g., smooth), then also $H^i(X, \omega_X \otimes \mathcal{L})$ vanishes for all $i > 0$.*

Proof of Theorem 2.3 and its corollaries. By definition, the map $\mathcal{O}_X \to F^e_* \mathcal{O}_X$ splits. So tensoring with $\mathcal{L}$, we also have a splitting of

$$\mathcal{L} \to F^e_* \mathcal{O}_X \otimes \mathcal{L} = F^e_* F^{e*} \mathcal{L} = F^e_* \mathcal{L}^{p^e}.$$

Here, the first equality follows from the projection formula [Hartshorne 1977, Exercise II 5.2(d)]; the second equality $F^{e*}\mathcal{L} = \mathcal{L}^{p^e}$ holds because pulling back under Frobenius raises transition functions to the p-th power. Since $\mathcal{L}$ is a direct summand of $F^e_* \mathcal{L}^{p^e}$, it follows that the cohomology $H^i(X, \mathcal{L})$ is a direct summand of $H^i(X, F^e_* \mathcal{L}^{p^e}) = H^i(X, \mathcal{L}^{p^e})$ for all e. But $H^i(X, \mathcal{L}^{p^e}) = 0$ for large e by our hypothesis, so that $H^i(X, \mathcal{L}) = 0$.

The corollary follows by Serre vanishing [ibid., Proposition III 5.3]: an ample invertible sheaf $\mathcal{L}$ on a projective variety X satisfies $H^i(X, \mathcal{L}^n) = 0$ for large n and positive i. The second statement follows by Serre duality [ibid., III §7]: Serre vanishing ensures $H^i(X, \omega_X \otimes \mathcal{L}^n)$ vanishes for large n, hence its dual $H^{\dim X - i}(X, \mathcal{L}^{-n})$ vanishes; by the Theorem also $H^{\dim X - i}(X, \mathcal{L}^{-1})$ vanishes, and hence so does its dual $H^i(X, \omega_X \otimes \mathcal{L})$. □

Proving a variety is Frobenius split is therefore a worthwhile endeavor. One useful criterion is essentially due to Hochster and Roberts:

Proposition 2.5. *A projective variety X is Frobenius split if and only if the induced map $H^{\dim X}(X, \omega_X) \to H^{\dim X}(X, F^* \omega_X)$ is injective.*

Proof. Hochster and Roberts actually stated a local version of this proposition: to check purity of any finite map of algebras $R \to S$, where R is local, it is enough to show that the map $R \otimes E \to S \otimes E$ remains injective after tensoring with the injective hull E of the residue field of R. This statement reduces to the statement of Proposition 2.5 by taking R to be the localization of a homogeneous coordinate ring of R at the unique homogeneous maximal ideal. See [Smith 1997b, 4.10.2]. □

Example 2.6. An elliptic curve over a perfect field of prime characteristic is Frobenius split if and only if it is *ordinary.* Indeed, ordinary means that the natural map induced by Frobenius $H^1(X, \mathcal{O}_X) \to H^1(X, F^*\mathcal{O}_X)$ is injective, so the statement follows from Proposition 2.5 since the canonical bundle of an elliptic curve is trivial. Thus in genus one, there are infinitely many Frobenius split curves, as well as infinitely many non-Frobenius split curves, depending on the Hasse-invariant of the curve. See, for example, [Hartshorne 1977, IV §4 Exercise 4.14], and compare Example 1.24(2).

2A. ***Global F-regularity.*** We define a global analog of F-regularity for arbitrary integral F-finite schemes of prime characteristic p. For any effective Weil divisor D on a normal variety, there is an obvious inclusion $\mathcal{O}_X \hookrightarrow \mathcal{O}_X(D)$. Thus for any e we have an inclusion $F^e_*\mathcal{O}_X \hookrightarrow F^e_*\mathcal{O}_X(D)$, which we can precompose with the iterated Frobenius map to get a map

$$\mathcal{O}_X \to F^e_*\mathcal{O}_X \hookrightarrow F^e_*\mathcal{O}_X(D).$$

Definition 2.7 [Smith 2000a]. An F-finite normal[12] scheme X is called *globally F-regular* if, for all effective Weil divisors D, there is an e such that the composition

$$\mathcal{O}_X \to F^e_*\mathcal{O}_X \hookrightarrow F^e_*\mathcal{O}_X(D)$$

splits as a map of $\mathcal{O}_X$-modules.

Globally F-regular varieties are strongly Frobenius split, in the sense that not only are they Frobenius split but there are typically *many* splittings of Frobenius. Indeed, suppose that X is globally F-regular. Then for *any* effective divisor D, there is an e and a map $F^e_*\mathcal{O}_X(D) \xrightarrow{\phi} \mathcal{O}_X$ such that the composition

$$\mathcal{O}_X \to F^e_*\mathcal{O}_X \overset{\iota}{\hookrightarrow} F^e_*\mathcal{O}_X(D) \xrightarrow{\phi} \mathcal{O}_X$$

is the identity map. Thus we can view the composition $\phi \circ \iota$ as a splitting of the (iterated) Frobenius $\mathcal{O}_X \to F^e_*\mathcal{O}_X$, which just happens to factor through $F^e_*\mathcal{O}_X(D)$. Thus there are actually *many* splittings of the (iterated) Frobenius, as these are typically different maps for different D. Of course, the Frobenius itself splits as well (not just its iterates): we have a factorization

$$\mathcal{O}_X \hookrightarrow F_*\mathcal{O}_X \hookrightarrow F^e_*\mathcal{O}_X,$$

so the splitting can be restricted to $F_*\mathcal{O}_X$.

[12] If X is quasiprojective, we can drop the normal from the definition and assume instead that D is an effective Cartier divisor. This produces the same definition, because splitting along all Cartier divisors will imply normality (by Theorem 1.18) as well as splitting along all Weil divisors (since given an effective Weil divisor D, we can always find an effective Cartier divisor D' such that $\mathcal{O}_X(D) \subset \mathcal{O}_X(D')$).

Thus we have proved the following analog of Proposition 1.14:

Proposition 2.8. *A globally F-regular scheme X is always Frobenius split.*

Thus a globally F-regular variety can be viewed as belonging to a restricted class of Frobenius split varieties, in which there are many different splittings of (the iterated) Frobenius — indeed so many that, for *every* Weil divisor on X, we can find a splitting factoring through $F_*^e\mathcal{O}_X(D)$. Such a Frobenius splitting is said to be a *Frobenius splitting along the divisor D* [Ramanathan 1991; Ramanan and Ramanathan 1985].

Remark 2.9. The reader will easily verify that for affine schemes, the local and global definitions of F-regularity are equivalent. Let us point out only this much: given an effective Cartier divisor D, we can chose a sufficiently small affine chart so that D has local defining equation f on Spec R. This means that $\mathcal{O}_X(D)$ is the (sheaf corresponding to the) invertible R-module $R \cdot (1/f)$. Then the map $\mathcal{O}_X \to F_*^e\mathcal{O}_X(D)$ of Definition 2.7 corresponds to the R-module map

$$R \to [R \cdot (1/f)]^{1/p^e}, \quad 1 \mapsto 1,$$

which splits if and only if the map $R \to R^{1/p^e}$ sending 1 to f^{1/p^e} splits. This is Definition 2.7.

For a nonaffine scheme X, the global splitting of the map $\mathcal{O}_X \to F_*^e\mathcal{O}_X(D)$ is a *strong condition, which can not be checked locally at stalks.* Thus X can fail to be globally F-regular, and usually does, even when it is locally F-regular at each point: globally F-regular varieties are rare even among smooth projective varieties. For example, all smooth projective curves are locally F-regular (because all local rings are regular), but the only smooth projective curve which is globally F-regular is $\mathbb{P}^1$. This follows immediately from the vanishing theorem:

Corollary 2.10. *If $\mathcal{L}$ is a nef*[13] *invertible sheaf on a globally F-regular projective variety, then $H^i(X, \mathcal{L})$ vanishes for all $\mathcal{L}$ and all $i > 0$.*

Now to see that $\mathbb{P}^1$ is the only globally F-regular projective curve, note that the degree of the canonical bundle on a curve of genus g is $2g-2$ so the canonical bundle is nef when the genus is positive. But since $H^1(X, \omega_X) = 1$ for all connected curves, Corollary 2.10 prohibits a curve of positive genus from being globally F-regular.

[13] By definition, an invertible sheaf on a curve is nef if it has nonnegative degree; an invertible sheaf on a higher dimensional variety is nef if its restriction to every algebraic curve in the variety is nef. Ample line bundles are always nef. Nef line bundles play an important role in higher dimensional birational geometry, being the "limits of ample divisors". See Section 1.4 of [Lazarsfeld 2004].

Proof. The proof is similar to the proof of Theorem 2.3, so we only sketch it, referring to [Smith 2000a, Theorem 4.2] for details. For any effective D, we can use the splitting of a map $\mathcal{O}_X \to F^e_* \mathcal{O}_X(D)$ to show that if $\mathcal{L}$ is any invertible sheaf such that $H^i(X, \mathcal{L}^n \otimes \mathcal{O}_X(D))$ vanishes for all large n and some effective D, then also $H^i(X, \mathcal{L})$ vanishes. Now the corollary follows because if $\mathcal{L}$ is nef, there is an effective D such that all $\mathcal{L}^n \otimes \mathcal{O}_X(D)$ are ample by [Lazarsfeld 2004, Corollary 1.4.10]. □

In practice, we do not have to check splitting for *all* divisors D to establish global F-regularity:

Theorem 2.11 [Smith 2000a]. *A projective variety X is globally F-regular if for some ample divisor D' containing the singular locus of X and all divisors D of the form mD' for $m \gg 0$, there is an e such that the map $\mathcal{O}_X \to F^e_* \mathcal{O}_X(D)$ splits as a map of $\mathcal{O}_X$-modules.*

Proof. This is a global adaptation of the following local theorem of Hochster and Huneke: If c is an element of an F-finite domain R such that R_c is regular, then c has some power f such that R is F-regular if and only if there exists an e and an R-linear map $R^{1/p^e} \to R$ sending f^{1/p^e} to 1. More general and more effective versions of this theorem are proved in [Smith 2000a] and [Schwede and Smith 2010, Theorem 3.9]. □

2B. *Local versus global splitting.* The precise relationship between local and global Frobenius splitting is clarified by the following theorem, which states roughly that a projective variety is Frobenius split or globally F-regular if and only if "its affine cone" has that property:

Theorem 2.12. *Let $X \subset \mathbb{P}^n$ be a normally embedded projective variety over a field of characteristic p. Then X is Frobenius split (respectively, globally F-regular) if and only if the corresponding homogenous coordinate ring is Frobenius split (respectively, globally F-regular).*

Example 2.13. Grassmannian varieties of any dimensions and characteristic are globally F-regular. Indeed, the homogeneous coordinate ring for the Plücker embedding of any Grassmannian is F-regular [Hochster and Huneke 1994]. More generally, all Schubert varieties are globally F-regular [Lauritzen et al. 2006]. Compare Example 1.22.

Example 2.14. A normal projective toric variety (of any characteristic) is globally F-regular. The point is that there is a torus-invariant ample divisor, so some multiple of it will give a normal embedding into projective space. The corresponding homogenous coordinate ring is a normal domain generated by monomials. All finitely generated normal rings generated by monomials are strongly F-regular since they are direct summands of the corresponding polynomial ring [Bruns and Herzog 1993, Exercise 6.1.10]. Since the homogenous coordinate ring is F-regular, the projective toric variety is globally F-regular by Theorem 2.15. See also Example 2.20.

The proof of Theorem 2.12 is clearest in the following context, which is only slightly more general.

Theorem 2.15. *Let X be any projective scheme over a perfect field. The following are equivalent*:

(1) *X is Frobenius split.*

(2) *The ring $S_{\mathcal{L}} = \bigoplus_{n\in\mathbb{N}} H^0(X, \mathcal{L}^n)$ is Frobenius split for all invertible sheaves $\mathcal{L}$.*

(3) *The section ring $S_{\mathcal{L}} = \bigoplus_{n\in\mathbb{N}} H^0(X, \mathcal{L}^n)$ is Frobenius split for some ample invertible sheaf $\mathcal{L}$.*

Likewise, a projective variety X is globally F-regular if and only if some (equivalently, every) section ring $S_{\mathcal{L}}$ with respect to an ample invertible sheaf $\mathcal{L}$ is F-regular.

Proof of Theorem 2.15. If $\mathcal{O}_X \to F_*\mathcal{O}_X$ splits, then the same is true after tensoring with any invertible sheaf $\mathcal{L}$. So as in the proof of Theorem 2.3,

$$\mathcal{L} \to F_*\mathcal{O}_X \otimes \mathcal{L} = F_*F^*\mathcal{L} = F_*\mathcal{L}^p,$$

splits. Likewise, we have a splitting after tensoring with the sheaf of algebras $\bigoplus_{n\in\mathbb{N}} \mathcal{L}^n$. Taking global sections produces a Frobenius splitting for the ring $S_{\mathcal{L}}$. So (1) implies (2). Also (2) obviously implies (3).

To see that (3) implies (1), we fix an ample invertible sheaf $\mathcal{L}$ on X. In particular, this means that X is the scheme Proj $S_{\mathcal{L}}$, and coherent sheaves on X correspond to finitely generated $\mathbb{Z}$-graded $S_{\mathcal{L}}$-graded modules (up to agreement in large degree). Now, if $S_{\mathcal{L}}$ is Frobenius split, then we can find a homogeneous $S_{\mathcal{L}}$-linear splitting $S_{\mathcal{L}}^{1/p} \xrightarrow{\pi} S_{\mathcal{L}}$ such that the composition

$$S \hookrightarrow S_{\mathcal{L}}^{1/p} \xrightarrow{\pi} S_{\mathcal{L}} \tag{2.15.1}$$

is the identity map. Note that $S_{\mathcal{L}}^{1/p}$ can be viewed as naturally $\frac{1}{p}\mathbb{N}$-graded, by defining the degree of $s^{1/p}$ to be $\frac{1}{p}\deg s$. Consider the graded S-submodule $[S^{1/p}]_{\mathbb{N}}$ of $S^{1/p}$ of elements of integer degree: this includes all the elements of S, but also elements of the form $(s)^{1/p}$, where s is *not* a p-th power in S but its degree is a multiple of p. The graded map of S-modules

$$S \hookrightarrow [S^{1/p}]_{\mathbb{N}}$$

corresponds to the Frobenius map of coherent sheaves $\mathcal{O}_X \to F_*\mathcal{O}_X$ on X. The point now is that restricting the map $S_{\mathcal{L}}^{1/p} \xrightarrow{\pi} S_{\mathcal{L}}$ to the subgroup $[S^{1/p}]_{\mathbb{N}}$, the composition of maps of graded S-modules

$$S \hookrightarrow [S^{1/p}]_{\mathbb{N}} \xrightarrow{\pi} S$$

gives a graded splitting of S-modules, whose corresponding map of coherent sheaves on X gives a splitting of Frobenius for X. The proof for global F-regularity is similar. See [Smith 2000a] or [Schwede and Smith 2010] for details. □

2C. *Frobenius splittings and anticanonical divisors.* Summarizing the situation for curves, we see that the existence of Frobenius splittings appears to be related to positivity of the anticanonical divisor:

Example 2.16. Consider smooth projective curves over a perfect field of prime characteristic.

(1) A genus zero curve is *always* globally F-regular.

(2) A genus one curve is never globally F-regular, and it is Frobenius split if and only if it is an ordinary elliptic curve.

(3) Higher genus curves are *never* Frobenius split (hence nor globally F-regular).

Indeed, there is a natural sense in which the sheaf of "potential Frobenius splittings" is a sheaf of pluri-anticanonical forms. The following crucial fact, first appearing in this guise in [Mehta and Ramanathan 1985], is at the heart of many ideas in both the local and global theories:

Lemma 2.17. *Let X be a normal*[14] *projective variety over a perfect field. Then we have*

$$\mathcal{H}om_{\mathcal{O}_X}(F^e_*\mathcal{O}_X, \mathcal{O}_X) \cong F^e_*\omega_X^{1-p^e}.$$

Proof. First assume that X is smooth. We have a natural isomorphism

$$\mathcal{H}om_{\mathcal{O}_X}(F^e_*\mathcal{O}_X, \mathcal{O}_X) \cong \mathcal{H}om_{\mathcal{O}_X}(F^e_*\mathcal{O}_X, \omega_X) \otimes \omega_X^{-1}.$$

By Grothendieck duality for the finite map $F^e : X \to X$, we also have a natural isomorphism

$$\mathcal{H}om_{\mathcal{O}_X}(F^e_*\mathcal{O}_X, \omega_X) \cong F^e_*\mathcal{H}om_{\mathcal{O}_X}(\mathcal{O}_X, \omega_X) \cong F^e_*\omega_X,$$

[14] If X is not smooth, the notation ω_X^n denotes the unique reflexive sheaf which agrees with the n-th tensor power of ω_X on the smooth locus; equivalently, it is the double dual of the n-th tensor power of ω_X, or equivalently, $\mathcal{O}_X(nK_X)$ where K_X is the Weil divisor agreeing with a canonical divisor on the smooth locus.

so that

$$\mathcal{H}om_{\mathcal{O}_X}(F^e_*\mathcal{O}_X, \mathcal{O}_X) \cong F^e_*\omega_X \otimes \omega_X^{-1} \cong F^e_*(\omega_X \otimes F^{e*}\omega_X^{-1}),$$

with the last isomorphism coming from the projection formula. Finally we have

$$\mathcal{H}om_{\mathcal{O}_X}(F^e_*\mathcal{O}_X, \mathcal{O}_X) \cong F^e_*(\omega_X^{1-p^e}),$$

since pulling back an invertible sheaf under Frobenius amounts to raising it to the p-th power.

Now, even if X is not smooth, this proof is essentially valid. Indeed, we can carry out the same argument on the smooth locus of X, to produce the desired natural isomorphism of sheaves there. Since both sheaves $\mathcal{H}om_{\mathcal{O}_X}(F^e_*\mathcal{O}_X, \mathcal{O}_X)$ and $F^e_*(\omega_X^{1-p^e})$ are reflexive sheaves on the normal variety X, this isomorphism extends uniquely to an isomorphism of $\mathcal{O}_X$-modules over all X. □

Any Frobenius splitting is a map $F_*\mathcal{O}_X \to \mathcal{O}_X$, and hence a nonzero global section of $\mathcal{H}om_{\mathcal{O}_X}(F_*\mathcal{O}_X, \mathcal{O}_X) \cong F_*(\omega_X^{1-p})$, which in turn, is a nonzero section of $H^0(X, \omega_X^{1-p})$. Thus, if X is Frobenius split, we expect nonzero sections of ω_X^{1-p}.

If ω_X is ample, then the sheaves ω_X^{1-p} are dual to ample, and can have no global sections:

Corollary 2.18. *A smooth projective variety X with ample canonical bundle is never Frobenius split.*

Even if ω_X^{1-p} has global sections, it is not so obvious which of these might correspond to a splitting of Frobenius. Mehta and Ramanathan [1985, Proposition 6] found a nice criterion:

Proposition 2.19. *Let X be a normal projective variety. A section*

$$s \in H^0(X, \omega_X^{1-p})$$

corresponds to a Frobenius splitting of X if and only if there exists a smooth point $x \in X$ at which s has a nonzero residue. Explicitly, for a global section $s \in H^0(X, \omega_X^{1-p})$, we can write the germ of s at the point x as $s = f(dx_1 \wedge \cdots \wedge dx_n)^{1-p}$ where $x_1, \ldots, x_n$ are a regular sequence of parameters at x and $f \in \mathcal{O}_{X,x}$. Now s is a splitting of Frobenius if and only if the power series expansion of f in the coordinates x_i has a nonzero $(x_1x_2 \ldots x_n)^{p-1}$ term.

Summarizing this in divisor language: the nonzero mappings $F_*\mathcal{O}_X \to \mathcal{O}_X$ correspond to effective divisors in the linear system $|(1-p)K_X|$. Given a particular divisor D in $|(1-p)K_X|$, it is a splitting of Frobenius if and only if in local analytic coordinates at some smooth point $x \in X$, the divisor D is $(p-1)$ times a simple normal crossing divisor whose components intersect exactly in $\{x\}$.

Example 2.20. One easy case in which Frobenius splitting can be established using Proposition 2.19 is when a smooth projective variety X of dimension n admits n effective divisors $D_1, \dots, D_n$ meeting transversely at a point of X and whose sum is an anti canonical divisor. Projective space obviously has this property (taking $-K_X$ to be the sum of the coordinate hyperplanes). Similarly, a projective toric variety is Frobenius split as well, since $-K_X$ is the sum of all the torus invariant divisors [Fulton 1993, page 85]. Again, we recover that fact that smooth projective toric varieties are Frobenius split.

Similarly, if X is globally F-regular, we expect many global sections of $\omega_X^{1-p^e}$: for each D, the splitting $F_*^e\mathcal{O}_X(D) \xrightarrow{t} \mathcal{O}_X$ induces a different splitting $F_*^e\mathcal{O}_X \to \mathcal{O}_X$, so gives rise to a nonzero element global section of $\omega_X^{1-p^e}$. Varying D, we get many sections of $\omega_X^{1-p^e}$—so many that they grow polynomially in p^e:

Corollary 2.21. *If a smooth projective variety is globally F-regular, then its anticanonical bundle is big.*[15]

Proof. For any Weil divisor D on a normal variety X, the proof of Lemma 2.17 immediately generalizes to give a natural isomorphism

$$\mathcal{H}om_{\mathcal{O}_X}(F_*^e\mathcal{O}_X(D), \mathcal{O}_X) \cong F_*^e(\omega_X^{1-p^e}(-D)).$$

Now, let X be globally F-regular, and let A be any ample effective Cartier divisor. By definition, there exists an e and a global nonzero map $t : F_*^e\mathcal{O}_X(A) \xrightarrow{t} \mathcal{O}_X$ splitting the composition $\mathcal{O}_X \to F_*^e\mathcal{O}_X(A)$. The map t can be viewed thus be viewed as a nonzero global section of $\mathcal{H}om_{\mathcal{O}_X}(F_*^e\mathcal{O}_X(A), \mathcal{O}_X)$, hence a nonzero element of

$$H^0(X, F_*^e(\omega_X^{1-p^e}(-A))) = H^0(X, \omega_X^{1-p^e}(-A)) \subset H^0(X, \omega_X^{1-p^e}).$$

Let E be the effective divisor of the section t, so that $E \in |(1-p^e)K_X - A|$, whereby

$$E + A \in |(1-p^e)K_X|.$$

Finally, we see that $-K_X$ is $\mathbb{Q}$-linearly equivalent to $\frac{1}{p^e-1}(A+E)$, so that $-K_X$ is $\mathbb{Q}$-linearly equivalent to "ample plus effective". That is, $-K_X$ is big by Corollary 2.2.7 in [Lazarsfeld 2004]. □

Unfortunately, bigness of $\omega_X^{1-p^e}$ is not sufficient for global F-regularity. For example, a ruled surface over an elliptic curve is never strongly F-regular, but its anticanonical divisor can be big. See [Schwede and Smith 2010, Example 6.7].

---

[15] A line bundle $\mathcal{L}$ on a projective variety X is *big* if the space of global sections $H^0(X, \mathcal{L}^n)$ grows as a polynomial of degree $\dim X$ in n; See [Lazarsfeld 2004, Section 2.2].

On the other hand, there is strengthened form of "almost amplitude of $-K_X$" which guarantees enough good sections of $\omega_X^{1-p^e}$ to find lots of splittings of Frobenius:

Definition 2.22. A normal projective variety X is *log Fano* if there exists an effective $\mathbb{Q}$-divisor Δ on X such that

(1) $-K_X - \Delta$ is ample; and

(2) the pair (X, Δ) has (at worst) Kawamata log terminal singularities.[16]

If X is smooth and ω_X^{-1} ample (that is, if X is Fano), then X is log Fano: we can take $\Delta = 0$. For log Fano varieties in general, $-K_X$ is not ample, but it is close to ample in the sense that it is big and even more: it is "close" to the ample cone in the sense that we can obtain it from the ample divisor $(-K_X - \Delta)$ by adding only the very small effective $\mathbb{Q}$-divisor Δ whose singularities are highly controlled.

We can now state a pair of theorems which can be viewed as a sort of geometric characterization of globally F-regular varieties:

Theorem 2.23 [Smith 2000a; Schwede and Smith 2010]. *If X is a globally F-regular projective variety of characteristic p, then X is log Fano.*

The converse isn't quite true because of irregularities in small characteristic. For example, the cubic hypersurface defined by $x^3 + y^3 + z^3 + z^3$ in $\mathbb{P}^3$ is obviously a smooth Fano variety (hence log Fano) in every characteristic $p \neq 3$. But it is not globally F-regular or even Frobenius split in characteristic two. However, it *is* globally F-regular for all characteristics $p \geq 5$. In general we have

Theorem 2.24 [Smith 2000a; Schwede and Smith 2010]. *If X is a log Fano variety of characteristic zero, then X has globally F-regular type.*

Remarkably, the converse to Theorem 2.24 is open: we do not know whether a globally F-regular type variety must be log Fano. This may seem surprising at first glance. If X has globally F-regular type, then in each characteristic p model, the proof of Theorem 2.23 constructs a "witness" divisor Δ_p establishing that the pair (X_p, Δ_p) is log Fano. But Δ_p depends on p and there is no *a priori* reason that the Δ_p all come from some divisor Δ on the characteristic zero variety X.

Conjecture 2.25. *A projective globally F-regular type variety* (*of characteristic zero*) *is log Fano.*

[16]Kawamata log terminal singularity is usually defined in characteristic 0, but it can be defined in any characteristic by considering *all* birational proper maps as follows. Let X be a normal variety and Δ be an effective $\mathbb{Q}$-divisor on X. Then (X, Δ) is called *Kawamata log terminal* if (a) $K_X + \Delta$ is $\mathbb{Q}$-Cartier, and (b) for *all* birational proper maps $\pi : Y \to X$, choosing K_Y so that $\pi_* K_Y = K_X$, each coefficient of $\pi^*(K_X + \Delta) - K_Y$ is strictly less than 1.

Gongyo, Okawa, Sannai and Takagi prove Conjecture 2.25 under the additional hypothesis that the variety is a $\mathbb{Q}$-factorial Mori dream space, by applying the minimal model program [Gongyo et al. 2015]. This gives urgency to another interesting question: are globally F-regular type varieties (of characteristic zero) Mori dream spaces? Moreover, since log Fano spaces (of characteristic zero) are Mori dream spaces by [Birkar et al. 2010, Corollary 1.3.2], the answer is necessarily *yes* if Conjecture 2.25 is true. What about in characteristic p?

Question 2.26. Assume that X is globally F-regular. Is it true that the Picard group of X is finitely generated? Is it true that the Cox ring of X is always finitely generated?

Similarly, there are related questions and results about the geometry of Frobenius split projective varieties. For example:

Theorem 2.27 [Schwede and Smith 2010]. *If X is a normal Frobenius split projective variety of characteristic p, then X is log Calabi-Yau. This means that X admits an effective $\mathbb{Q}$-divisor such that (X, Δ) is log canonical* [17] *and $K_X + \Delta$ is $\mathbb{Q}$-linearly equivalent to the trivial divisor.*

Again, the converse fails because of irregularities in small characteristic. The same cubic hypersurface defined by $x^3 + y^3 + z^3 + z^3$ in $\mathbb{P}^3$ is not Frobenius split in characteristic two, but it is a smooth log Calabi Yau variety. However, we do expect an analog of Theorem 2.24 to hold. We conjecture:

Conjecture 2.28 [Schwede and Smith 2010]. *If X is a log Calabi Yau variety of characteristic zero, then X has Frobenius split type.*

2D. *Pairs.* We have discussed Frobenius splitting and F-regularity in an absolute setting: these were defined as properties of a scheme X. However, in the decade since the last MSRI special year in commutative algebra, a theory of "F-singularities of pairs" has flourished, inspired by the rich theory of pairs developed in birational geometry [Kollár 1997]. The idea to extend Frobenius splitting and F-regularity to pairs was a major breakthrough, pioneered by Nobuo Hara and Kei-ichi Watanabe in [Hara and Watanabe 2002]. Although we do not have space to include a careful treatment of this generalization here, we briefly outline the definitions and main ideas.

By *pair* in this context, we have in mind a normal irreducible scheme X of finite type over a perfect field, together with a $\mathbb{Q}$-divisor Δ on X. (Another variant considers pairs $(X, \mathfrak{a}^t)$ consisting of an ambient scheme X together with

[17] Log canonical is usually defined in characteristic 0, but it can be defined in any characteristic similarly to how we defined Kawamata log terminal singularities. We require instead that each coefficient of $\pi^*(K_X + \Delta) - K_Y$ is at most 1.

a sheaf of ideals $\mathfrak{a}$ and a rational exponent t.[18]) In the geometric setting, an additional assumption — namely that $K_X + \Delta$ is $\mathbb{Q}$-Cartier — is usually imposed, because one of the main techniques in birational geometry involves pulling back divisors to different birational models (and only $\mathbb{Q}$-Cartier divisors can be sensible pulled back). One possible advantage of the algebraic notion of pairs is that it is not necessary to assume that $K_X + \Delta$ is $\mathbb{Q}$-Cartier, although alternatives have also been proposed directly in the world of birational geometry as well; see [de Fernex and Hacon 2009].

What do pairs have to do with Frobenius splittings? Given an $\mathcal{O}_X$-linear map $\phi : F^e_* \mathcal{O}_X \to \mathcal{O}_X$, we have already seen how it gives rise to a global section of $F^e_* \omega_X^{1-p^e}$, hence an effective divisor $\tilde{D}$-linearly equivalent to $(1 - p^e)K_X$. If we set $\Delta = \frac{1}{p^e-1}\tilde{D}$, the pair (X, Δ) can be interpreted as more or less equivalent to the data of the map ϕ. If D is an effective Cartier divisor through which our map ϕ factors, then (as in the proof of Corollary 2.21), we can view ϕ as a global section of $\mathcal{O}_X((1 - p^e)K_X + D)$, hence an effective divisor $\tilde{D}$-linearly equivalent to $(1 - p^e)K_X + D$. Setting $\Delta = \frac{1}{p^e-1}\tilde{D}$, we have that Δ is a $\mathbb{Q}$-divisor satisfying $K_X + \Delta$ is $\mathbb{Q}$-Cartier, since it is $\mathbb{Q}$-linearly equivalent to the Cartier $\mathbb{Q}$-divisor $\frac{1}{p^e-1}D$. We refer to the exceptionally clear exposition of this idea, with deep applications to understanding the behavior of test ideals under finite morphisms, in the paper [Schwede and Tucker 2014a].

The definition of (local or global) F-regularity and Frobenius splitting can be generalized to pairs as follows:

Definition 2.29. Let X be a normal F-finite variety, and Δ an effective $\mathbb{Q}$-divisor on X.

(1) The pair (X, Δ) is sharply Frobenius split (respectively locally sharply Frobenius split) if there exists an $e \in \mathbb{N}$ such that the natural map

$$\mathcal{O}_X \to F^{p^e}_* \mathcal{O}_X(\lceil (p^e - 1)\Delta \rceil)$$

splits as an map of sheaves of $\mathcal{O}_X$-modules (respectively, splits locally at each stalk).

(2) The pair (X, Δ) is globally (respectively, locally) F-regular if for all effective divisors D, there exists an $e \in \mathbb{N}$ such that the natural map

$$\mathcal{O}_X \to F^{p^e}_* \mathcal{O}_X(\lceil (p^e - 1)\Delta \rceil + D)$$

splits as an map of sheaves of $\mathcal{O}_X$-modules (respectively, splits locally at each stalk).

[18] There are even triples $(X, \Delta, \mathfrak{a}^t)$ incorporating aspects of both variants.

Remark 2.30. A slightly different definition of Frobenius splitting for a pair (X, Δ) was first given in [Hara and Watanabe 2002]. The variant here, which fits better into our context, was introduced in [Schwede 2010b].

The local properties of F-regularity and F-purity for pairs turn out to be closely related to the properties of log terminality and log canonicity that arose independently, in the minimal model program, in the 1980's. The absolute versions of the following theorems were already mentioned at the end of the first section.

Theorem 2.31 [Hara and Watanabe 2002]. *Let (X, Δ) be a pair where X is a normal variety of prime characteristic and Δ is a $\mathbb{Q}$-divisor such that $K_X + \Delta$ is $\mathbb{Q}$-Cartier.*

(1) *If (X, Δ) is a locally F-regular pair, then (X, Δ) is Kawamata log terminal.*

(2) *If (X, Δ) is a locally sharply Frobenius split pair, then (X, Δ) is log canonical.*

Similarly, there are global versions: Theorems 2.23 and 2.27 also hold for "pairs". See [Schwede and Smith 2010].

In characteristic zero, the discussion of Section 1E generalizes easily to pairs. Given a pair (X, Δ), where now X is normal and essentially finite type over a field of characteristic *zero*, we can define the *pair* to be (locally or globally) *F-regular type* or (locally or globally) *sharply Frobenius split type* as in Definition 1.23. That is, we chose a ground ring A over which both X and Δ are defined, which gives rise to a pair (X_A, Δ_A) over X_A, and define the pair (X, Δ) to be of (locally or globally) F-regular-type if for a Zariski dense set of closed points in Spec A, the pair $(X_A \times Spec A/\mu, \Delta_A \bmod \mu)$ is (locally or globally) F-regular, where $\Delta_A \bmod \mu$ denotes the pullback of Δ_A to the closed fiber $X_A \times Spec A/\mu$. Similarly, we define sharp Frobenius splitting type. For details of this reduction to prime characteristic, see [Hara and Watanabe 2002], for example.

With the definitions in place, Theorem 2.31 implies characteristic zero versions. Let (X, Δ) be a pair where X is a normal variety of characteristic zero and Δ is a $\mathbb{Q}$-divisor satisfying $K_X + \Delta$ is $\mathbb{Q}$-Cartier. Then if (X, Δ) is of locally F-regular type, then (X, Δ) has klt singularities [Hara and Watanabe 2002]. See the cited papers for details.

As in the absolute case, there are converses (some conjectured) to all these results for pairs in characteristic zero. Hara and Watanabe [2002] show that klt pair (X, Δ) of characteristic zero has F-regular type; this leans heavily on some injectivity results for Frobenius acting on cohomology groups proved in [Hara 1998]; see also [Mehta and Srinivas 1997]. In [Schwede and Smith 2010], these results are used to prove a global version: a log Fano pair (X, Δ)

of characteristic zero is of globally F-regular type. The log canonical property has proved more elusive: it is conjectured that if a pair (X, Δ) of characteristic zero is log canonical, then it is of locally sharply Frobenius split type, but this questions has remained open since its inception in the nineties. See [Hara and Watanabe 2002]. If this local conjecture holds, then the global analog follows: a Calabi-Yau pair (X, Δ) of characteristic zero is of globally sharply Frobenius split type; see [Schwede and Smith 2010].

We can think of F-regularity as a "characteristic p analog" of klt singularities, and (at least conjecturally) F-splitting as a "characteristic p analog" of log canonical singularities. The analogy runs deep: F-pure thresholds become "characteristic p analogs" of log canonical thresholds, test ideals become "characteristic p analogs" of multiplier ideals, centers of sharp F-purity become "characteristic p analogs" of log canonicity, F-injectivity becomes a "characteristic p analog" of Du Bois singularities.

3. The test ideal

The *test ideal* is a distinguished ideal reflecting the Frobenius properties of a prime characteristic ring. For example, the test ideal defines the closed locus of Spec R consisting of points $\mathfrak{p}$ at which $R_{\mathfrak{p}}$ is not F-regular. Test ideals are "characteristic p analogs" of multiplier ideals in birational algebraic geometry [Smith 2000b; Hara 2001]; they define a distinguished "compatibly split subscheme" of a Frobenius split variety [Vassilev 1998; Schwede 2010a].

Test ideals can be defined very generally for pairs on more or less arbitrary Noetherian schemes of characteristic p. However, the theory becomes most transparent in two special cases, which are loosely the "classical commutative algebra case" and the "classical algebraic geometry case". In the classical commutative algebra case, the scheme is the spectrum of a local ring R and we are interested in the "absolute" test ideal. In this case, the test ideal $\tau(R)$ is essentially Hochster and Huneke's test ideal for tight closure.[19] In the classical algebraic geometry case, we are interested in the test ideal of a pair (X, Δ), where X is a smooth ambient scheme and Δ is an effective $\mathbb{Q}$-divisor on X (or $\Delta = \mathfrak{a}^t$ where $\mathfrak{a}$ is an ideal sheaf on X and t is a positive rational number). In this case, the test ideal $\tau(X, \Delta)$ turns out to be a "characteristic p analog" of the multiplier ideal in complex algebraic geometry, a subject described, for example, in Rob Lazarsfeld's MSRI introductory talks [2004].

[19]Our terminology differs slightly from the tight closure literature, where our test ideal would be called the "big test ideal" for "nonfinitistic tight closure"; also, it defines the non-strongly F-regular locus.

In this section, we explain the test ideal in the "classical commutative algebra" setting. We develop the test ideal as just one ideal in a lattice of ideals distinguished with respect to the Frobenius map. Our definition is not the traditional one due to Hochster and Huneke, but a newer twist (due essentially to Schwede [Schwede 2010a]) which is both illuminating and elegant, tying the ideas into Mehta and Ramanathan's theory of compatibly split ideals. Section 4 will treat test ideals in the "classical algebrogeometric" setting.

3A. *Compatible ideals.* Let R be an F-finite reduced ring of characteristic p.

Definition 3.1. Fix any R-linear map $\varphi : R^{1/p^e} \to R$. An ideal J of R is called *φ-compatible* if $\varphi(J^{1/p^e}) \subseteq J$.

Put differently, given an R-linear map $\varphi : R^{1/p^e} \to R$, consider the obvious diagram

$$\begin{array}{ccc} R^{1/p^e} & \xrightarrow{\ \varphi\ } & R \\ \downarrow & & \downarrow \\ (R/J)^{1/p^e} & \dashrightarrow & R/J, \end{array}$$

where the vertical arrows are the natural surjections. The bottom arrow can not be filled in to make a commutative diagram in general: it can be filled in if and only if J is φ-compatible. That is, an ideal J is φ-compatible if and only if the map $\varphi : R^{1/p^e} \to R$ descends to a map $(R/J)^{1/p^e} \to R/J$.

Example 3.2. Let $R = \mathbb{F}_p[x, y]$ and let $\phi : R^{1/p} \to R$ be the R-linear splitting defined by

$$\phi(x^{i/p}y^{j/p}) = \begin{cases} x^{i/p}y^{j/p} & \text{if } \frac{i}{p}, \frac{j}{p} \in \mathbb{N}, \\ 0 & \text{otherwise.} \end{cases}$$

As an exercise, the reader should check that the ideals (0), (x), (y), (xy), and (x, y) are all ϕ-compatible — in fact, they are the only ϕ-compatible ideals in R.

The following properties are straightforward:

Proposition 3.3. *Fix an R-linear map $\varphi : R^{1/p^e} \to R$.*

(1) *The set of φ-compatible ideals is closed under sum and intersection.*

(2) *The minimal primes of a φ-compatible ideal are φ-compatible.*

(3) *If φ is a Frobenius splitting, then all φ-compatible ideals are radical.*

Proof. We leave (1) as an easy exercise. For statement (2), let P be a minimal prime of a φ-compatible ideal J. Take any w not in P but in the intersection of all other primary components of J, that is, take $w \in (J : P) \setminus P$. For any $z \in P$, we need to show that $\varphi(z^{1/p^e}) \in P$. Now since $w^{p^e} z \in J$ and J is ϕ-compatible, we have

$$w\varphi(z^{1/p^e}) = \varphi(wz^{1/p^e}) \in \varphi(J^{1/p^e}) \subset J \subset P.$$

Since $w \notin P$, we conclude that $\varphi(z^{1/p^e}) \in P$. So P is φ-compatible.

Statement (3) is also easy: if J is ϕ compatible, the commutative diagram

$$\begin{array}{ccc} R^{1/p^e} & \xrightarrow{\quad\varphi\quad} & R \\ \downarrow & & \downarrow \\ (R/J)^{1/p^e} & \longrightarrow & R/J, \end{array}$$

shows that if φ is a Frobenius splitting, so is the induced map on R/J. Since Frobenius split rings are reduced, the ideal J is radical. □

Compatibly split subschemes. When φ is a Frobenius splitting, a φ-compatible ideal is often called a φ-compatibly split ideal, or (suppressing the dependence on φ) a compatibly split ideal. The subscheme it defines is called *a compatibly split subscheme.* The notion of *compatible Frobenius splitting* was first introduced in [Mehta and Ramanathan 1985]. In that language, a compatibly split ideal (sheaf) on a Frobenius split variety defines a compatibly split subscheme — a subscheme to which the given splitting of Frobenius restricts.

For a fixed splitting φ of Frobenius, the set of φ-compatibly split ideals is always a *finite* set of radical ideals. The finiteness is not obvious; see [Schwede 2009a] or [Kumar and Mehta 2009], or [Schwede and Tucker 2010]. Also [Enescu and Hochster 2008] and [Sharp 2007] contain a related dual fact.

3B. ***Uniformly compatible ideals.*** Of course, if R is Frobenius split, there may be many different splittings of Frobenius. Different splittings produce different compatibly split ideals. For example, by linearly changing coordinates in $\mathbb{F}_p[x, y]$, we can construct a different splitting of Frobenius, call it φ_{ab}, much like the one in Example 3.2 but centered instead on the point $(x - a, y - b)$. Its compatibly split ideals will be the ideals $(x - a, y - b)$, $(x - a)(x - b)$, $(x - a)$, $(y - b)$, and the zero ideal. These ideals are compatibly split with respect to φ_{ab} but not with the φ from Example 3.2.

The ideals which are compatible with respect to *every* R-linear map $R^{1/p^e} \to R$ play an essential role in our story:

Definition 3.4. An ideal J in an F-finite ring is *uniformly F compatible* if it is compatible with respect to *every* R-linear map $R^{1/p^e} \to R$, for all e.

The test ideal is a distinguished uniformly F compatible ideal:

Definition 3.5. The test ideal[20] of an F-finite Noetherian domain R is the *smallest* nonzero uniformly F compatible ideal. That is, the test ideal is the smallest nonzero ideal J that satisfies

$$\varphi(J^{1/p^e}) \subseteq J$$

for all $\varphi \in \mathrm{Hom}_R(R^{1/p^e}, R)$ and all $e \geq 1$. More generally, if R is not a domain, we define the test ideal as the smallest uniformly F compatible ideal not contained in any minimal prime.

Note that, in this definition, *it is not at all obvious that there exists a smallest such ideal:* why couldn't the set of all compatible ideals include an infinite descending chain of nonzero ideals? This is a deep result, essentially relying on an important lemma of Hochster and Huneke crucial to their proof of the existence of "completely stable test elements" (Compare the proof of Theorem 2.11.) For a summary of the proof, see the survey [Schwede and Tucker 2012].

Remark 3.6. The set of uniformly F compatibly ideals forms a lattice closed under sum and intersection, according to Proposition 3.3. This lattice has been studied before: in the local Gorenstein case, it is the precisely the lattice of F-ideals discussed in [Smith 1994]; more generally, it is the lattice of annihilators of $\mathcal{F}(E)$-modules in [Lyubeznik and Smith 2001]. However, those sources define the ideals as annihilators of certain Artinian R-modules with Frobenius action. Schwede's insight was that these ideals could be defined directly (dually to the original emphasis), thereby producing a more straightforward and global theory which neatly ties in with Mehta and Ramanathan's ideas on compatible Frobenius splitting.

Remark 3.7. Experts in tight closure can see easily how the definition of the test ideal here relates to the one in the literature, and why there is a unique smallest uniformly F compatible ideal, at least in the Gorenstein local case. Let (R, m) be a Gorenstein local domain of dimension d. As is well-known, the test ideal is the annihilator of the tight closure of zero in $H^d_m(R)$. In [Smith 1997a], the Frobenius stable submodules of $H^d_m(R)$ (including the tight closure of zero) are analyzed and their annihilators in R are dubbed "F-ideals"; there it is shown (also using test elements!) that there is a unique largest proper Frobenius stable submodule of $H^d_m(R)$, hence a unique smallest nonzero F-ideal, namely test ideal of R. The

[20]Again a reminder for experts in tight closure: this is equal to the "big" test ideal in the tight closure terminology. If R is complete local, for example, the test ideal we define here is the same as the annihilator of the nonfinitistic tight closure of zero in the injective hull of the residue field of R [Lyubeznik and Smith 2001]. Of course, all versions of test ideals in the tight closure theory are conjectured to be equal, and are known to be equal in many cases, including for Gorenstein R and graded R [Lyubeznik and Smith 1999].

uniformly F compatible ideals are precisely the F-ideals — that is, annihilators of submodules of the top local cohomology module $H^d_m(R)$ stable under Frobenius. This is not hard to check using Lemma 3.13; see [Schwede 2010a] or [Enescu and Hochster 2008, Theorem 4.1]. The non-Gorenstein case is treated in [Lyubeznik and Smith 2001]; the uniformly F compatible ideals are the annihilators of the $\mathcal{F}(E)$-modules there. Schwede includes a fairly comprehensive discussion of the connections between his uniformly F compatible ideals and existing ideas in the literature; see [Schwede 2010a].

Example 3.8. The test ideal of $\mathbb{F}_p[x, y]$ is the whole ring. Indeed, we have seen that every nonzero $c \in \mathbb{F}_p[x, y]$ can be taken to 1 by some R^{p^e}-linear map. So no nonzero proper ideal is uniformly F-compatible.

Theorem 3.9. *Let R be a reduced F-finite of characteristic $p > 0$.*

(1) *The test ideal behaves well under localization and completion: for any multiplicative set U, the ideals $\tau(RU^{-1})$ and $\tau(R)U^{-1}$ coincide in RU^{-1}, and for any prime ideal $\mathfrak{p}$, $\tau(\hat{R}_{\mathfrak{p}}) = \tau(R)\hat{R}_{\mathfrak{p}}$.*

(2) *R is F-regular if and only if its test ideal is trivial.*

(3) *The test ideal defines the closed locus of prime ideals $\mathfrak{p}$ in Spec R such that $R_{\mathfrak{p}}$ fails to be F-regular.*

The proof uses the following important lemma.

Lemma 3.10. Let c be an element of a reduced F-finite ring R. The ideal generated by all elements $\phi(c^{1/p^e})$ as we range over all e and all $\phi \in \mathrm{Hom}_R(R^{1/p^e}, R)$ is uniformly F compatible. In particular, fixing any $c \in \tau(R)$ but not in any minimal prime of R, the elements $\phi(c^{1/p^e})$ generate τ.

Proof of the lemma. Let J be the ideal generated by the $\phi(c^{1/p^e})$. We need to show that elements of the form $[r\phi(c^{1/p^e})]^{1/p^f}$ are taken into J by any $\psi \in \mathrm{Hom}_R(R^{1/p^f}, R)$. But

$$\psi[r^{1/p^f}[\phi(c^{1/p^e})]^{1/p^f}] = \psi[r^{1/p^f}[\phi^{1/p^f}(c^{1/p^{e+f}})] = (\psi \circ (\phi \circ r)^{1/p^f})(c^{1/p^{e+f}})$$

where $\psi \circ (\phi \circ r)^{1/p^f}$ is the R-linear map

$$R^{1/p^{e+f}} \xrightarrow{r^{1/p^f}} R^{1/p^{e+f}} \xrightarrow{\phi^{1/p^f}} R^{1/p^f} \xrightarrow{\psi} R.$$

The second equality above is satisfied because ϕ^{1/p^f} is R^{1/p^f}-linear. The second statement of the lemma follows by the minimality of the test ideal. The lemma is proved. ∎

Proof of Theorem 3.9. To prove (1), the point is that R^{1/p^e} is a finitely generated R-module, so that

$$\operatorname{Hom}_{RU^{-1}}((RU^{-1})^{1/p^e}, RU^{-1}) \cong \operatorname{Hom}_R(R^{1/p^e}, R) \otimes_R RU^{-1}.$$

Take any $c \in R$ (not in any minimal prime) such that $c \in \tau(R)$ and $\frac{c}{1} \in \tau(RU^{-1})$. By the lemma just proved, both $\tau(RU^{-1})$ and $\tau(R)U^{-1}$ are ideals of RU^{-1} generated by elements of the form

$$\frac{\phi(c^{1/p^e})}{1} = \frac{\phi}{1}\left(\left(\frac{c}{1}\right)^{1/p^e}\right)$$

as we range through all $\phi \in \operatorname{Hom}_R(R^{1/p^e}, R)$ and all $e \in \mathbb{N}$. That is, the ideals $\tau(RU^{-1}))$ and $\tau(R)U^{-1}$ coincide. The second statement follows similarly, since $\operatorname{Hom}_{\hat{R}_\mathfrak{p}}(\hat{R}_\mathfrak{p}^{1/p^e}, \hat{R}_\mathfrak{p}) \cong \operatorname{Hom}_{R_\mathfrak{p}}({R_\mathfrak{p}}^{1/p^e}, R_\mathfrak{p}) \otimes_{R_\mathfrak{p}} \hat{R}_\mathfrak{p}$.

For (2), assume for simplicity that R is a domain.[21] If R is F-regular, then for any nonzero c, there exists e and $\phi \in \operatorname{Hom}_R(R^{1/p^e}, R)$ such that $\phi(c^{1/p^e}) = 1$. This means that every uniformly F-compatible ideal contains 1. In particular, $\tau(R)$ is trivial. Conversely, assume $\tau(R)$ is trivial. By (1), also $\tau(R_m)$ is trivial for each maximal ideal m. For any nonzero element $c \in R_m$, the lemma implies that the elements $\phi(c^{1/p^e})$ can not be all contained in m as ϕ ranges over all $\operatorname{Hom}_{R_m}(R_m^{1/p^e}, R_m)$. Thus there exists $\phi \in \operatorname{Hom}_{R_m}(R_m^{1/p^e}, R_m)$ such that $\phi(c^{1/p^e})$ is a unit, and hence R_m is F-regular. Since this holds for each maximal ideal, we conclude that R is F-regular by Lemma 1.15.

Statement (3) follows from (1) and (2) together. □

The lattice of uniformly F compatible ideals is especially nice in a Frobenius split ring. The next result follows immediately from Proposition 3.3(3).

Corollary 3.11. *Every uniformly F compatible ideal in a Frobenius split ring is radical. In particular, the test ideal in a Frobenius split ring is radical.*

The converse is not true: the ring $R = \mathbb{F}_p[x, y, z]/(x^3 + y^3 + z^3)$ has test ideal (x, y, z) for all characteristics $p \neq 3$, but R is not Frobenius split if $p = 2 \bmod 3$. See [Smith 1995b, Example 6.3].

3C. *Splitting primes and centers of F-purity.* One might also wonder whether the *largest* proper ideal compatible with respect to all ϕ might also be of interest? We know a largest such exists by the Noetherian property of R, since the sum of uniformly F compatible ideals is uniformly F compatible. For Frobenius split R, this largest compatible ideal turns out to be the *splitting prime* of [Aberbach and Enescu 2005]. It is also the minimal center of F-purity in the language of [Schwede 2010a]. The prime uniformly compatible ideals are what Schwede

[21] Else replace "nonzero" by "not in any minimal prime" throughout.

calls *centers of F-purity*. He shows that they are "characteristic p analogs" of Kawamata's centers of log canonicity, and that they satisfy an analog Kawamata's subadjunction [1998]. See [Schwede 2009a].

3D. *The Frobenius filtration of a Frobenius split ring.* We have already observed that when R is a Frobenius split ring, the set of uniformly F compatible ideals forms a (finite) lattice of radical ideals closed under addition and intersection. An interesting observation of Janet Vassilev [1998] creates a distinguished chain in this lattice.

Lemma 3.12. If τ is a uniformly F compatible ideal of R, then the preimage in R of any compatibly split ideal of R/τ is compatibly split in R.

Proof. Let J be the preimage in R of a uniformly compatibly split ideal of R/τ. Let $\phi : R^{1/p^e} \to R$ be any R-module homomorphism. Because τ is uniformly F compatible in R, there is an induced map of R/τ-modules $\bar{\phi} : (R/\tau)^{1/p^e} \to R/\tau$, and because J/τ is uniformly F compatible in R/τ, $\bar{\phi}$ descends to a map $(R/J)^{1/p^e} \to R/J$. But this is exactly what it means that J is uniformly F compatible in R. □

To construct Vassilev's chain, start with a Frobenius split ring R, with test ideal τ_0. Because τ_0 is compatible with respect to some (indeed, every) Frobenius splitting, the ring R/τ_0 is also Frobenius split. Let τ_1 be the preimage of the test ideal $\tau(R/\tau_0)$ of R/τ_0 in R. By the lemma, τ_1 is also uniformly F compatible, so R/τ_1 is Frobenius split, and so its test ideal lifts to an ideal τ_2. Continuing in this way, we produce a chain $\tau_0 \subset \tau_1 \subset \cdots \subset \tau_t$ of radical ideals, all uniformly F compatible. Since the test ideal is never contained in a minimal prime, each ideal in the chain has strictly larger height than its predecessor, and since τ_0 defines the non-F-regular locus of R, we see that the length of Vassilev's chain is bounded by the dimension of the non-F-regular set of R.

3E. *Trace of Frobenius.* To check that an ideal is uniformly F compatible, we do not actually have to test compatibility with respect to *all* homomorphisms $\varphi : R^{1/p^e} \to R$. Since $\mathrm{Hom}_R(R^{1/p^e}, R)$ is a finitely generated R^{1/p^e}-module, it is enough to check compatibility with respect to a finite set of R^{1/p^e}-generators for each e. In the Gorenstein case, this takes an especially nice form:

Lemma 3.13. If (R, m) is an F-finite Gorenstein local ring, then $\mathrm{Hom}_R(R^{1/p}, R)$ is a cyclic $R^{1/p}$-module. An ideal J is uniformly compatible if and only if it is compatible with respect to an $R^{1/p}$-module generator for $\mathrm{Hom}_R(R^{1/p}, R)$.

Proof. The point is that if $R \to S$ is a finite map of rings with canonical module, there is an S-module isomorphism $\omega_S \cong \mathrm{Hom}_R(S, \omega_R)$ [Bruns and Herzog 1993, Theorem 3.7.7]. If R is a Gorenstein local ring, then R is a canonical module for R (and so of course $R^{1/p}$ is a canonical module for $R^{1/p}$), so the first statement follows.

By the same argument, each $\mathrm{Hom}_R(R^{1/p^e}, R)$ is a cyclic R^{1/p^e}-module. Moreover, if Ψ is a generator for $\mathrm{Hom}_R(R^{1/p}, R)$, the composition map

$$\Psi_e = \Psi \circ \Psi^{1/p} \circ \cdots \circ \Psi^{1/p^{e-1}}$$

is an R^{1/p^e}-generator for $\mathrm{Hom}_R(R^{1/p^e}, R)$. (For example, it is easy to check that Ψ_e is not in $m_{R^{1/p^e}} \mathrm{Hom}_R(R^{1/p^e}, R)$, so it must be a generator by Nakayama's lemma.) For example, Ψ_2 is the composition

$$R^{1/p^2} \xrightarrow{\Psi^{1/p}} R^{1/p} \xrightarrow{\Psi} R, \qquad r^{1/p^2} \mapsto [\Psi(r^{1/p})]^{1/p} \mapsto \Psi([\Psi(r^{1/p})]^{1/p}).$$

Now, consider an ideal J which is Ψ-compatible. Any $\varphi \in \mathrm{Hom}_R(R^{1/p^e}, R)$, can be written as $\Psi_e \circ r^{1/p^e}$. So

$$\varphi(J^{1/p^e}) = \Psi_e(r^{1/p^e} J^{1/p^e}) \subset \Psi_e(J^{1/p^e}),$$

which by definition of Ψ_e is the same as

$$\Psi_{e-1}(\Psi^{1/p^{e-1}}(J^{1/p^e})) = \Psi_{e-1}([\Psi(J^{1/p})]^{1/p^{e-1}}).$$

This is contained in $\Psi_{e-1}(J^{1/p^{e-1}})$, because $\Psi(J^{1/p}) \subset J$ by the Ψ-compatibility of J. Finally, this is contained in J by induction on e. Thus any Ψ-compatible ideal is uniformly F compatible. □

The generator in Lemma 3.13 is uniquely defined up to multiplication by a unit in R^{1/p^e}. It is sometimes abusively called the *trace of the Frobenius map*.

For any F-finite[22] ring R, we can dualize the Frobenius map $R \to R^{1/p}$ into ω_R:

$$\mathrm{Hom}_R(R^{1/p}, \omega_R) \longrightarrow \mathrm{Hom}_R(R, \omega_R),$$

which produces an R-module map

$$F_* \omega_{R^{1/p}} \to \omega_R,$$

the *trace* of Frobenius.[23] In notation more common in algebraic geometry: the dual of the Frobenius map $R \to F_* R$ is the trace map $F_* \omega_R \to \omega_R$. For smooth projective varieties, this is called the Cartier map. If R is local and Gorenstein, of course, this can be identified with a map $F_* R \to R$, which will be a generator for $\mathrm{Hom}_R(R^{1/p^e}, R)$.

[22] F-finite rings always admit a canonical module [Gabber 2004, 13.6].

[23] This map is only as canonical as the choice of ω_R, so the "the" is slightly misleading. Of course in geometric situations where the canonical module is defined by differential forms, there is a canonical choice.

Using ω_X has the advantage of globalizing; we have already encountered this idea in Section 2C. See [Schwede and Tucker 2014a] or [Brion and Kumar 2005] for more on the trace map.

Example 3.14. The trace of the Frobenius map is *not* usually a Frobenius splitting! For example, if $R = \mathbb{F}_p[x, y]$, we recall that the monomials $x^a y^b$ where $0 \leq a, b, \leq p-1$ form a basis for $F_* R$ over R. As the trace map, we can take the R-linear map $R^{1/p} \xrightarrow{\Psi} R$ sending $(xy)^{(p-1)/p}$ to 1 and all other monomials in the basis to zero. This is clearly *not* a Frobenius splitting. The Frobenius splitting of Example 1.7 can be obtained as $\Psi \circ (xy)^{(p-1)/p}$.

Remark 3.15. Blickle's *Cartier algebras* give another point of view on test ideals and uniformly F compatible ideals [Blickle 2013]. An R-module map $R^{1/p^e} \to R$ can be viewed as an additive map $R \xrightarrow{\phi} R$ satisfying $\phi(r^{p^e} x) = r\phi(x)$ for any $r, x \in R$. Blickle and Böckle [2011] dub this a p^{-e}-linear map. This point of view has the advantage that composition is slicker — the source and target are always R — so we can easily compose such maps. Indeed, the composition of p^{-e}- and p^{-f}-linear maps is easily checked to be p^{-e-f}-linear. The *Cartier algebra*[24] $\mathcal{C}(R)$ is the subalgebra of $\mathrm{Hom}_{\mathbb{Z}}(R, R)$ generated by all p^{-e}-linear maps (as we range over all e). Clearly R is a module over $\mathcal{C}(R)$, and clearly its $\mathcal{C}(R)$-submodules are precisely the uniformly F-compatible ideals. The trace map can also be easily interpreted in this language: in the Gorenstein local case, the trace Ψ_e of Lemma 3.13 is literally the composition of Ψ with itself e-times, so that Ψ generates $\mathcal{C}(R)$ as an R-algebra. Blickle [2013] develops the *ultimate generalization of test ideals* in by looking at submodules of R and other modules under various distinguished subalgebras of (variants of) the Cartier algebra.

Remark 3.16. The uniformly F compatible ideals have been studied for many years in the tight closure literature under many different names. They were first studied in the local Gorenstein case in [Smith 1997a], where they are called *F-ideals*, and for more general local rings in [Lyubeznik and Smith 2001], where they are descriptively called annihilators of $\mathcal{F}$-submodules of E. Both these papers have a dual point of view to our current perspective, which was first proposed in [Schwede 2010a], and it is not obvious that the definitions there produce precisely the uniformly F compatible ideals (see [Enescu and Hochster 2008, Theorem 4.1] for a proof). Schwede used the term *uniformly F-compatible* ideals, in a nod to the connection with Mehta and Ramanathan's notion of compatibly Frobenius split subschemes. In the same paper, Schwede

[24]Here we assume that R is reduced and of dimension greater than zero. In general, the definition of Cartier algebra is slightly more technical, but it reduces to this under very mild conditions. See [Blickle 2013].

shows how the prime compatible ideals (which he calls centers of sharp F-purity) can be viewed characteristic p analogs of log canonical centers. The term ϕ-compatible is lifted from the survey [Schwede and Tucker 2012]. Generalizations of uniformly F compatible ideals also come up in the work of Blickle (e.g., [Blickle 2013]) under the name of Cartier submodules and crystals; see the survey [Blickle and Schwede 2013].

4. Test ideals for pairs

As deeper connections between Frobenius splitting and singularities in birational geometry emerged, it was natural to look for generalizations of the characteristic p story to "pairs", the natural setting for much of the geometry. For example, with the realization that the multiplier ideal "reduces mod p to the test ideal" (when the former is defined; see [Smith 2000b; Hara 2001]), interest rose in defining test ideals for pairs, since this was the main setting for multiplier ideals. After Hara and Watanabe [2002] introduced Frobenius splitting for pairs, the theory of tight closure for pairs quickly developed in a series of technical papers by the Japanese school of tight closure, beginning about the time of the last decade's special year in commutative algebra at MSRI. In particular, a theory of test ideals for pairs was introduced in [Hara and Yoshida 2003] and [Hara and Takagi 2004a].

In this section, we introduce the theory of test ideals for pairs, focusing on the case where the ambient variety is smooth and affine — the "classical algebrogeometric setting". We do not use the traditional tight closure definition, but rather an equivalent definition first proposed in [Blickle et al. 2008]. By shunning the most general setting, and instead working in the simplest useful setting, we hope to highlight the elegance of test ideal arguments when the ambient ring is regular. In particular, we give elementary proofs of all the basic properties, several of which do not seem to have been noticed before. As an application, we include a self-contained proof of a well-known theorem on the behavior of symbolic powers of ideals in a regular ring following the analogous multiplier ideal proof in [Ein et al. 2001]. See also [Hara 2005].

Let R be an F-finite domain, and let $\mathfrak{a}$ be an ideal of R. For each nonnegative real number t, we associate an ideal[25]

$$\{t \in \mathbb{R}_{\geq 0}\} \rightsquigarrow \{\tau(R, \mathfrak{a}^t)\}_{\mathbb{R}_{\geq 0}}.$$

[25] If t is a natural number, the notation $\tau(R, \mathfrak{a}^t)$ could be interpreted to mean the test ideal of the ideal $\mathfrak{a}^t$ (with exponent 1) or to mean the test ideal of the ideal $\mathfrak{a}$ with exponent t. Fortunately, these ideals are the same (as will soon be revealed when we give the definition), so the danger of confusion is minimal.

In the classical commutative algebra case, $\mathfrak{a} = R$, and all $\tau(R, \mathfrak{a}^t)$ produce the same ideal, $\tau(R)$, the test ideal discussed in the previous section. For many years, this was the only test ideal in the literature. In the classical algebraic geometry setting, multiplier ideals are much easier to handle in the case where the ambient variety is *smooth*; indeed the emphasis has always been that case. So perhaps it should not be surprising that, returning the commutative algebra to the case where the ambient ring R is regular, arguments should simplify dramatically for test ideals as well.

4A. *Test ideals in ambient regular rings.* Let R be an F-finite regular domain, and $\mathfrak{a}$ any ideal of R. We first define the test ideal of a pair $\tau(R, \mathfrak{a}^t)$ in the special case where t is a positive rational number whose denominator is a power of p. The case of arbitrary t will be obtained by approximating t by a sequence of rational numbers whose denominators are powers of p. When R is clear from the context, we will often write $\tau(\mathfrak{a}^t)$.

For each R-linear map $\phi : R^{1/p^e} \to R$, we consider the image of $\mathfrak{a}$ under ϕ. That is, looking at the ideal $\mathfrak{a}^{1/p^e}$ as an R-submodule of R^{1/p^e}, we consider its image $\phi(\mathfrak{a}^{1/p^e}) \subset R$, which is of course an ideal of R. Ranging over all $\phi \in \mathrm{Hom}_R(R^{1/p^e}, R)$, we get the test ideal of $\mathfrak{a}$. Before stating this formally, we set up some notation:

$$\mathfrak{a}^{[1/p^e]} := \sum_{\phi \in \mathrm{Hom}_R(R^{1/p^e}, R)} \phi(\mathfrak{a}^{1/p^e}).$$

Lemma 4.1. For any ideal $\mathfrak{a}$ in a Frobenius split ring R, we have

$$\mathfrak{a}^{[1/p^e]} \subset (\mathfrak{a}^p)^{[1/p^{e+1}]},$$

with equality if $\mathfrak{a}$ is principal.

Proof. Fix a Frobenius splitting $\pi : R^{1/p} \to R$. There is a corresponding splitting

$$\pi^{1/p^e} : R^{1/p^{e+1}} \to R^{1/p^e}$$

defined by taking p^e-th roots of everything in sight. For any R-linear map $R^{1/p^e} \xrightarrow{\phi} R$, the composition

$$R^{1/p^{e+1}} \xrightarrow{\pi^{1/p^e}} R^{1/p^e} \xrightarrow{\phi} R$$

is an element of $\mathrm{Hom}_R(R^{1/p^{e+1}}, R)$. Now, the ideal $\mathfrak{a}^{[1/p^e]}$ is generated by elements of the form $\phi(x^{1/p^e})$, where $x \in \mathfrak{a}$. To see that all such elements are also in $(\mathfrak{a}^p)^{[1/p^{e+1}]}$, note simply that x^p is in $\mathfrak{a}^p$, and that the composition above sends $(x^p)^{1/p^{e+1}}$ to $\phi(x)$. This completes the proof. □

Now, given a rational number t whose denominator is a power of p, we can write

$$t = \frac{n}{p^e} = \frac{np}{p^{e+1}} = \frac{np^2}{p^{e+2}} = \cdots .$$

The lemma implies a corresponding increasing sequence of ideals:

$$(\mathfrak{a}^n)^{[1/p^e]} \subset (\mathfrak{a}^{np})^{[1/p^{e+1}]} \subset (\mathfrak{a}^{np^2})^{[1/p^{e+2}]} \subset \cdots \tag{4.1.1}$$

which must eventually stabilize by the Noetherian property of the ring. This sequence stabilizes to the test ideal:

Definition 4.2. Let R be an F-finite regular ring of characteristic p and let $\mathfrak{a}$ be an ideal of R. Fix any positive rational number whose denominator is a power of p, say n/p^e. The test ideal

$$\tau(\mathfrak{a}^{n/p^e}) = \bigcup_f (\mathfrak{a}^{np^f})^{[1/p^{e+f}]} = \sum_{\varphi \in \operatorname{Hom}_R(R^{1/p^{e+f}}, R)} \varphi((\mathfrak{a}^{np^f})^{1/p^{e+f}}).$$

That is, if we write the number $t = n/p^e$ with a sufficiently high power of p in the denominator, the test ideal $\tau(\mathfrak{a}^{n/p^e})$ is the ideal $(\mathfrak{a}^n)^{[1/p^e]}$ of R generated by the images of the ideal $(\mathfrak{a}^n)^{1/p^e} \subset R^{1/p^e}$ under all projections $R^{1/p^e} \to R$.

Remark 4.3. If R is Frobenius split and $\mathfrak{a}$ is principal, the sequence (4.1.1) above stabilizes immediately. That is, $\tau(\mathfrak{a}^{n/p^e}) := (\mathfrak{a}^n)^{[1/p^e]}$ for any representation of the fraction n/p^e.

Remark 4.4. Let $R = \mathbb{F}_p[[x, y]]$ and $\mathfrak{a} = (x, y)$. Show that $\tau(\mathfrak{a}^n) = (x, y)^{n-1}$ for all $n \in \mathbb{N}$.

The case of arbitrary t. Fix a positive real number t. Choose a *nonincreasing* sequence of rational numbers $\{t_n\}$ whose denominators are p-th powers.[26] This allows us to define the test ideal $\tau(\mathfrak{a}^t)$ because for t_n sufficiently close to t, the ideals $\tau(\mathfrak{a}^{t_n})$ will all coincide. Indeed, if $n/p^e > m/p^f$, then getting a common denominator, we have that $(\mathfrak{a}^{np^f})^{[1/p^{e+f}]} \subset (\mathfrak{a}^{mp^e})^{[1/p^{f+e}]}$. So if $n/p^e > m/p^f$, then clearly

$$\tau(\mathfrak{a}^{n/p^e}) \subset \tau(\mathfrak{a}^{m/p^f}). \tag{4.4.1}$$

Thus any decreasing sequence of positive rational numbers whose denominators are p-th powers must produce an ascending chain of ideals, which stabilizes by the Noetherian property of the ring. If two such descending sequences converge to the same real number t, it is clear again by property (4.4.1) that the corresponding chains of ideals must stabilize to the *same* ideal. Thus we can define:

[26] For example, we can take Hernandez's sequence of successive truncations of a nonterminating base p expansion for t; see [Hernández 2015].

Definition 4.5 [Blickle et al. 2008]. Let R be an F-finite regular ring of characteristic p and let $\mathfrak{a}$ be an ideal of R. For each $t \in \mathbb{R}_{\geq 0}$, we define

$$\tau(\mathfrak{a}^t) := \bigcup_{e \in \mathbb{N}} \tau(\mathfrak{a}^{\lceil tp^e \rceil / p^e}).$$

The sequence $\lceil tp^e \rceil / p^e$, as e runs through the natural numbers, is a decreasing sequence converging to the real number t. We have picked it for the sake of definitiveness; *any* such deceasing sequence can be used instead, they all produce the same ideal in light of the inclusion (4.4.1).

4B. *Properties of test ideals.* All the basic properties of test ideals for an ambient regular ring follow easily from the definition, using the flatness of Frobenius for regular rings.

Theorem 4.6. *Let R be an F-finite regular ring of characteristic p, with ideals $\mathfrak{a}$, $\mathfrak{b}$. The following properties of the test ideal hold*:

(1) $\mathfrak{a} \subseteq \mathfrak{b} \Rightarrow \tau(R, \mathfrak{a}^t) \subseteq \tau(R, \mathfrak{b}^t)$ *for all* $t \in \mathbb{R}_{>0}$.

(2) $t \geq t' \Rightarrow \tau(R, \mathfrak{a}^t) \subseteq \tau(R, \mathfrak{a}^{t'})$.

(3) $\tau((\mathfrak{a}^n)^t) = \tau(\mathfrak{a}^{nt})$ *for each positive integer n and each* $t \in \mathbb{R}_{>0}$.

(4) *Let W be a multiplicatively closed set in R, then*

$$\tau(R, \mathfrak{a}^t)RW^{-1} = \tau(RW^{-1}, (\mathfrak{a}RW^{-1})^t).$$

(5) *Let $\overline{\mathfrak{a}}$ denote the integral closure of $\mathfrak{a}$ in R. Then*

$$\tau(\overline{\mathfrak{a}}^t) = \tau(\mathfrak{a}^t) \text{ for all } t.$$

(6) *For each $t \in \mathbb{R}_{>0}$, there exists an $\varepsilon > 0$ such that $\tau(\mathfrak{a}^{t'}) = \tau(\mathfrak{a}^t)$ for all $t' \in [t, t+\varepsilon)$.*

(7) $\mathfrak{a} \subseteq \tau(\mathfrak{a})$.

(8) (Theorem of Briançon–Skoda[27]) *If $\mathfrak{a}$ can be generated by r elements, then for each integer $\ell \geq r$ we have*

$$\tau(\mathfrak{a}^{\ell}) = \mathfrak{a}\tau(\mathfrak{a}^{\ell-1}).$$

(9) (Restriction theorem) *Let $x \in R$ be a regular parameter and* $\mathfrak{a}$ mod x *denote the image of $\mathfrak{a}$ in $R/(x)$, then*

$$\tau((\mathfrak{a} \bmod x)^t) \subseteq \tau(\mathfrak{a}^t) \bmod x.$$

[27] In geometry circles, it is typical to refer to this statement as Skoda's theorem; we adopt the more generous tradition of commutative algebra. This type of statement has also been referred to as a "Briançon–Skoda theorem with coefficients". See, e.g., [Aberbach and Huneke 1996] or [Aberbach and Hosry 2011].

(10) (Subadditivity theorem) *If R is essentially of finite type over a perfect field, then $\tau(\mathfrak{a}^{tn}) \subseteq \tau(\mathfrak{a}^t)^n$ for all $t \in \mathbb{R}_{\geq 0}$ and all $n \in \mathbb{N}$.*[28]

Remark 4.7. In fact, the first six properties above hold more generally; this is basic for the first five, once the definitions have been made (see [Schwede and Tucker 2012]) and the sixth is proved in [Blickle et al. 2010]. There are various versions of the other properties as well in more general settings, but most require some sort of restriction on the singularities of R and the proofs tend to be very technical. See, e.g., [Hara and Yoshida 2003; Takagi and Yoshida 2008; Takagi 2006]. Our proofs mostly follow [Blickle et al. 2008]. The simple proof of (9) here is new (hence so is the proof of the corollary (10)), although the statements can be viewed as (very) special cases of much more technical results in [Takagi 2008] and [Hara and Yoshida 2003, Theorem 6.10(2)], respectively.

Proof. The first three properties follow immediately from the definitions, and the fourth is also straightforward. These are left to the reader.

The fifth follows easily from the following basic property of integral closure (see eg. [Huneke and Swanson 2006, Corollary 1.2.5]): there exists a natural number ℓ such that for all $n \in \mathbb{N}$, $(\overline{\mathfrak{a}})^{n+\ell} \subset \mathfrak{a}^n$. Fix this ℓ. We already know that $\tau(\mathfrak{a}^t) \subset \tau(\overline{\mathfrak{a}}^t)$ by property (1). For the reverse inclusion, note that since $(\lceil tp^e \rceil + \ell)/p^e$ is a decreasing sequence converging to t as e gets large, we have

$$\begin{aligned}\tau(\overline{\mathfrak{a}}^t) &= \bigcup_{e\in\mathbb{N}} \tau(\overline{\mathfrak{a}}^{(\lceil tp^e\rceil+\ell)/p^e}) = \bigcup_{e\in\mathbb{N}} \tau((\overline{\mathfrak{a}}^{\lceil tp^e\rceil+\ell})^{1/p^e}) \\ &\subset \bigcup_{e\in\mathbb{N}} \tau((\mathfrak{a}^{\lceil tp^e\rceil})^{1/p^e}) = \bigcup_{e\in\mathbb{N}} \tau(\mathfrak{a}^{\lceil tp^e\rceil/p^e}) = \tau(\mathfrak{a}^t),\end{aligned}$$

with the inclusion coming from property (1).

The sixth follows immediately from the Noetherian property of the ring. Since $(\lceil p^e t\rceil + 1)/p^e$ is a decreasing sequence converging to t, we can fix e large enough that $\tau(\mathfrak{a}^t)$ agrees with $\tau(\mathfrak{a}^{(\lceil p^e t\rceil+1)/p^e})$. By Property (2), for all t' in the interval $[t, (\lceil p^e t\rceil + 1)/p^e)$, we have $\tau(\mathfrak{a}^{(\lceil p^e t\rceil+1)/p^e}) \subset \tau(\mathfrak{a}^{t'}) \subset \tau(\mathfrak{a}^t)$. In other words, all three ideals are the same.

The seventh property is easy too: since $t = p^e/p^e$ for all e, we have $\tau(\mathfrak{a}) = (\mathfrak{a}^{p^e})^{[1/p^e]}$ for $e \gg 0$, which contains $\mathfrak{a}$ by Lemma 4.1.

The Briançon–Skoda property is also easy. Thinking of ℓ as $(\ell p^e)/p^e$ for large e, we have $\tau(\mathfrak{a}^\ell) = (\mathfrak{a}^{\ell p^e})^{[1/p^e]}$. But it is easy to see that $\mathfrak{a}^{\ell p^e} = \mathfrak{a}^{[p^e]}(\mathfrak{a}^{(\ell-1)p^e})$, where $\mathfrak{a}^{[p^e]}$ is the ideal generated by the p^e-th powers of the elements of $\mathfrak{a}$. (Indeed, if $\mathfrak{a}$ is generated by the elements $a_1, \dots, a_r$, then $\mathfrak{a}^{\ell p^e}$ is generated by

[28] More generally, our proof of the subadditivity property shows that for the *mixed test ideal* $\tau(\mathfrak{a}^t\mathfrak{b}^s)$ defined analogously as $\tau(\mathfrak{a}^{\lceil sp^e\rceil}\mathfrak{b}^{\lceil tp^e\rceil})^{[1/p^e]}$ for $e \gg 0$, we have $\tau(\mathfrak{a}^t\mathfrak{b}^s) \subseteq \tau(\mathfrak{a}^t)\tau(\mathfrak{b}^s)$ for all $t, s \in \mathbb{R}_{\geq 0}$.

the monomials $a_1^{i_1}\dots a_r^{i_r}$ of degree ℓp^e; if all exponents $i_j \le p^e-1$, then the $p^e\ell \le rp^e - r$, contradicting our assumption that $\ell \ge r$.) Now clearly

$$\tau(\mathfrak{a}^\ell) = (\mathfrak{a}^{\ell p^e})^{[1/p^e]} = (\mathfrak{a}^{[p^e]}\mathfrak{a}^{(\ell-1)p^e})^{[1/p^e]} = \mathfrak{a}(\mathfrak{a}^{(\ell-1)p^e})^{[1/p^e]} = \mathfrak{a}\tau(\mathfrak{a}^{\ell-1}).$$

The third equality here holds since by definition, for any ideal $\mathfrak{b}$, we have $\mathfrak{b}^{[1/p^e]}$ is the image of $\mathfrak{b}^{1/p^e}$ under the R-linear maps $R^{1/p^e}\to R$. In particular, $(\mathfrak{a}^{[p^e]}\mathfrak{b})^{[1/p^e]} = \mathfrak{a}\mathfrak{b}^{[1/p^e]}$.

Now the restriction property (9). Let us denote $R/(x)$ by $\overline{R}$; its elements are denoted $\overline{r}$ where r is any representative in R. Consider any $\overline{R}$-linear map $\overline{R}^{1/p^e} \longrightarrow \overline{R}$. We claim that this map lifts to a R-linear map $\phi : R^{1/p^e}\to R$. Indeed, consider the diagram of R-modules

$$\begin{array}{ccc} R^{1/p^e} & \twoheadrightarrow & \overline{R}^{1/p^e} \\ \vdots & & \downarrow \\ R & \twoheadrightarrow & \overline{R} \end{array}$$

where the horizontal arrows are the natural surjections, and the vertical arrow is the one we are given. Because the bottom arrow is surjective and R^{1/p^e} is a *projective* R module (by Kunz's Theorem 1.2), the composition map $R^{1/p^e}\to\overline{R}$ lifts to some $\phi : R^{1/p^e}\to R$ making the diagram commute. Thus it is reasonable to denote the given map $\overline{R}^{1/p^e}\longrightarrow\overline{R}$ by $\overline{\phi}$. For any $\overline{r}\in\overline{R}$, we have $\overline{\phi}(\overline{r}^{1/p^e}) = \overline{\phi(r^{1/p^e})}$.

With this observation in place, the restriction theorem is easy. Take any $\overline{y}\in\tau(\overline{\mathfrak{a}}^t)$. By definition, there is some $\overline{\phi} : \overline{R}^{1/p^e}\to\overline{R}$ such that $\overline{y} = \overline{\phi}(\overline{r}^{1/p^e})$, where $\overline{r}\in\overline{\mathfrak{a}}^t$. By the commutativity of the diagram, $\overline{y} = \overline{\phi(r^{1/p^e})}$, for some $r\in\mathfrak{a}^t$. That is, $\overline{y}\in\tau(\mathfrak{a}^t)$ *mod* (x). The restriction theorem is proved.

Finally, we observe that the subadditivity property follows formally from the restriction property in exactly the same way as for multiplier ideals; compare [Blickle and Lazarsfeld 2004]. Let $\mathfrak{a}$ and $\mathfrak{b}$ be ideals in a regular ring R essentially finitely generated over k. In $S = R\otimes_k R$, which is also regular, we have the ideal $\mathfrak{a}\otimes R + R\otimes\mathfrak{b}$. For any positive rational s,t, it is easy to check that

$$\tau(\mathfrak{a}^s\otimes R + R\otimes\mathfrak{b}^t) = \tau(\mathfrak{a}^s)\otimes R + R\otimes\tau(\mathfrak{b}^t).$$

Now, locally at each maximal ideal, the diagonal ideal $\Delta\subset R\otimes R$ is generated by a sequence of regular parameters. By the restriction property, at each maximal ideal we have

$$\tau((\mathfrak{a}^s\otimes R + R\otimes\mathfrak{b}^t)\, mod\ \Delta) \subset [\tau(\mathfrak{a}^s)\otimes R + R\otimes\tau(\mathfrak{b}^t)]\, mod\ \Delta.$$

Interpreting this in $R\otimes R/\Delta = R$ yields the inclusion $\tau(\mathfrak{a}^s\mathfrak{b}^t)\subset\tau(\mathfrak{a}^s)\tau(\mathfrak{b}^t)$. □

4C. *Asymptotic test ideals and an application to symbolic powers.* We now introduce an asymptotic version of the test ideal, analogous to the asymptotic multiplier ideal first defined in [Ein et al. 2001]. We will use this concept to give a simple proof of the following well-known theorem about the asymptotic behavior of symbolic powers.

Theorem 4.8 [Ein et al. 2001; Hochster and Huneke 2002]. *Let I be an unmixed (e.g., prime) ideal in $k[x_1, \dots, x_d]$. Then*

$$I^{(dn)} \subseteq I^n \text{ for all } n \in \mathbb{N}.$$

Our proof here is a straightforward and self-contained adaptation of the original multiplier ideal proof in [Ein et al. 2001] in characteristic zero. Hara had also adapted that proof to prime characteristic using test ideals in [Hara 2005], although the definitions and proofs are different and less self-contained than ours here. Hochster and Huneke gave a tight closure proof and generalized this result in the characteristic p case [Hochster and Huneke 2002]. See also [Takagi and Yoshida 2008].

For a prime ideal $\mathfrak{p}$ in a polynomial ring, the symbolic power $\mathfrak{p}^{(n)}$ is the ideal of all functions vanishing to order n on the variety defined by $\mathfrak{a}$. Put differently, the symbolic powers of a prime ideal $\mathfrak{p}$ in any ring R are defined by $\mathfrak{p}^{(n)} = \mathfrak{p}^n R_{\mathfrak{p}} \cap R$. For arbitrary $\mathfrak{a}$, we take a primary decomposition $\mathfrak{a}^N = \mathfrak{p}_1 \cap \cdots \cap \mathfrak{p}_n \cap Q_1 \cap \cdots \cap Q_m$ where P_i's are the minimal primary components and Q_j's are the embedded components, then define $\mathfrak{a}^{(N)} = \mathfrak{p}_1 \cap \cdots \cap \mathfrak{p}_n$.

Definition 4.9. A sequence of ideals $\{\mathfrak{a}_n\}_{n\in\mathbb{N}}$ is called a *graded sequence* of ideals if

$$\mathfrak{a}_n \mathfrak{a}_m \subseteq \mathfrak{a}_{n+m}$$

for all n, m.

It is easy to check that the symbolic powers $\{\mathfrak{a}^{(n)}\}_{n\in\mathbb{N}}$ of any ideal $\mathfrak{a}$ in any ring form a graded sequence. Graded sequences arise naturally in many contexts in algebraic geometry. For example, the sequence of base loci of the powers of a fixed line bundle form a graded sequence of ideals on a variety. See [Ein et al. 2001; 2003] or [Blickle and Lazarsfeld 2004] for many more examples.

Given any graded sequence of ideals $\{\mathfrak{a}_n\}$, it follows from the definition and Property 4.6(1) that for any positive λ,

$$\tau(\mathfrak{a}_n^{\lambda}) = \tau((\mathfrak{a}_n^{\lambda m})^{1/m}) \subseteq \tau(\mathfrak{a}_{mn}^{\lambda/m}).$$

In other words, the collection

$$\{\tau(\mathfrak{a}_m^{\lambda/m})\}_{m\in\mathbb{N}}$$

has the property that any two ideals are dominated by a third in the collection. Since R is noetherian, this collection must have a maximal element; this stable ideal is called the **asymptotic test ideal:**

Definition 4.10. The n-th asymptotic test ideal of the graded sequence $\{\mathfrak{a}_n\}_{n\in\mathbb{N}}$ is the ideal

$$\tau_\infty(R, \mathfrak{a}_n) := \sum_{\ell\in\mathbb{N}} \tau(R, \mathfrak{a}_{\ell n}^{1/\ell}),$$

which is equal to

$$\tau(R, \mathfrak{a}_{mn}^{1/m})$$

for sufficiently large and divisible m.

By definition, it is clear that $\tau_\infty(R, \mathfrak{a}_n)$ satisfies appropriate analogs of all the properties listed in Properties 4.6 — the asymptotic test ideal is a particular test ideal, after all. Especially we point out a consequence of the subadditivity theorem:

Corollary 4.11. *For any graded sequence in an F-finite regular ring R, we have* $\tau_\infty(R, \mathfrak{a}_{nm}) \subset (\tau_\infty(R, \mathfrak{a}_n))^m$ *for all* $n, m \in \mathbb{N}$.

Proof. Since $\tau_\infty(R, \mathfrak{a}_{nm}) := \tau(R, \mathfrak{a}_{nm\ell}^{1/\ell})$ for sufficiently divisible ℓ, we have

$$\tau_\infty(R, \mathfrak{a}_{nm}) = \tau(R, \mathfrak{a}_{nm\ell}^{1/\ell}) = \tau(R, \mathfrak{a}_{nm\ell}^{m/(m\ell)}) \subset \tau(R, \mathfrak{a}_{nm\ell}^{1/(m\ell)})^m,$$

with the inclusion following from the subadditivity property 4.6(10) for test ideals. Since ℓ here can be taken arbitrarily large and divisible, we have that $\tau(R, \mathfrak{a}_{nm\ell}^{1/(m\ell)}) = \tau_\infty(R, \mathfrak{a}_n)$. Thus

$$\tau_\infty(R, \mathfrak{a}_{nm}) \subset \tau_\infty(R, \mathfrak{a}_n)^m.$$

□

Proof of Theorem 4.8. We consider the graded sequence of ideals $\{I^{(n)}\}_{n\in\mathbb{N}}$. According to Properties 4.6(3), we have $I^{(dN)} \subseteq \tau_\infty(I^{(dN)})$. By Corollary 4.11, we have

$$\tau_\infty(I^{(dN)}) \subseteq \tau_\infty(I^{(d)})^N$$

for all N. Hence it is enough to check that $\tau_\infty(I^{(d)}) \subseteq I$. For this, we can check at each associated prime $\mathfrak{p}$ of I, which means essentially that we can assume that R is local and that I is primary to the maximal ideal; that is, we need to show that

$$\tau_\infty(R_\mathfrak{p}, (I^d R_\mathfrak{p})) \subset I R_\mathfrak{p}.$$

In $R_{\mathfrak{p}}$, there is a reduction of I that can be generated by $\dim(R_{\mathfrak{p}}) \leq d$ elements, and hence according to Properties 4.6(5) we may assume that I itself can be generated by d elements. Then the Briançon–Skoda property 4.6(8) tells us

$$\tau_{\infty}(R_{\mathfrak{p}}, (I^d R_{\mathfrak{p}})) \subseteq I.$$

This finishes the proof of our theorem. □

4D. ***The definition of the test ideal for a pair $(R, \mathfrak{a}^t)$ in general.*** The definition of the test ideal for a singular ambient ring can be adapted to the general case of pairs. We include the definition for completeness without getting into details; see [Schwede 2010a] or [Schwede and Tucker 2012] for more, including generalizations to "triples".

Definition 4.12. Let R be a reduced F-finite ring and let $\mathfrak{a}$ be an ideal of R. The test ideal $\tau(R, \mathfrak{a}^t)$ is defined to be the smallest ideal J not contained in any minimal prime that satisfies

$$\varphi((\mathfrak{a}^{\lceil t(p^e-1)\rceil} J)^{1/p^e}) \subseteq J$$

for all $\varphi \in \mathrm{Hom}_R(R^{1/p^e}, R)$ (ranging over all $e \geq 1$).

In particular, the test ideal $\tau(R, \mathfrak{a}^t)$ is defined to be the smallest nonzero ideal J not contained in any minimal prime that satisfies

$$\varphi(J^{1/p^e}) \subseteq J$$

for all $\varphi \in \mathrm{Hom}_R(R^{1/p^e}, R)$ which are in the submodule consisting of homomorphisms obtained by first precomposing with elements of in $(\mathfrak{a}^{\lceil t(p^e-1)\rceil})^{1/p^e}$, and all $e \geq 1$.

Again, the existence of such smallest nonzero ideal is a nontrivial statement; see [Schwede and Tucker 2012] and [Hara and Takagi 2004b] for the general proof.

Although it is not completely obvious, if R is regular, this gives the test ideal that we have already discussed in the previous subsection. Indeed, both definitions developed here are shown to agree with the tight closure definition of the test ideal in [Hara and Yoshida 2003] in their respective introductory papers, [Blickle et al. 2008] and [Schwede 2010a].

Further reading on test ideals. Much more is known about test ideals than we can discuss here, and the story of test ideals is very much still a work in progress. One notable paper is [Schwede and Tucker 2014b] which discusses a framework under which test ideals and multiplier ideals can be constructed in the same way, an idea begun in [Blickle et al. 2015]. On the other hand, the paper [Mustaţă and Yoshida 2009] includes a cautionary result: *every ideal in a regular ring is the test ideal of some ideal with some coefficient.* This indicates that test ideals are in some ways very different from multiplier ideals, since multiplier ideals are always integrally closed; see also [McDermott 2003].

A rich literature has evolved on the study of F-jumping numbers — analogs of the *jumping numbers for multiplier ideals* of [Ein et al. 2004]. As with multiplier ideals, as we increase the exponent t, the ideals $\tau(R, \mathfrak{a}^t)$ get deeper; the values of α such that $\tau(R, \mathfrak{a}^{\alpha-\epsilon})$ strictly contains $\tau(R, \mathfrak{a}^{\alpha})$ (for all positive ϵ) are called *F-jumping numbers.* The smallest F-jumping number is called the F-pure threshold. First introduced in [Hara and Yoshida 2003], one of the main questions has been whether or not the F-jumping numbers are always discrete and rational. The first major progress was the case of regular ambient rings [Blickle et al. 2008]; an exceptionally well-written account of the state of the art appears in [Schwede and Tucker 2014b]. See also [Blickle et al. 2009; 2010; Katzman et al. 2009; Schwede et al. 2012]. The F-jumping numbers are notoriously difficult to compute; see [Hernández 2015]. Just as jumping numbers for multiplier ideals (in characteristic zero) are roots of the Bernstein Sato polynomial [Ein et al. 2004], similar phenomena have been studied for F-jumping numbers; see, for example, [Mustaţă et al. 2005], [Blickle and Stäler 2015], or [Mustaţă 2009].

The connection between the test ideal and differential operators was first pointed out in [Smith 1995a], where it is shown that the test ideal is a D-module. There are deep connections between the lattice of uniformly F-compatible ideals and intersection homology D-module in characteristic p [Blickle 2004], and other works of Blickle and his collaborators. See also [Smith and Van den Bergh 1997].

Appendix: What does Cohen–Macaulay mean?

The Cohen–Macaulay property is so central to commutative algebra that the field has been jokingly called the "study of Cohen–Macaulayness". Cohen–Macaulayness is also important in algebraic geometry, representation theory and combinatorics, with many different characterizations. We briefly review three of these. See [Bruns and Herzog 1993] for a more in depth discussion.

First, Cohen–Macaulay is a local property — meaning that *we can define* a Noetherian ring R to be Cohen–Macaulay if all its local rings are Cohen–Macaulay. So we focus only on what it means for a *local* ring (R, m) to be Cohen–Macaulay.

Alternatively, if the reader prefers graded rings, one can take (R, m) to mean an $\mathbb{N}$-graded ring R, finitely generated over its zero-th graded piece R_0 (a field). In this case, m denotes the unique homogenous maximal (or irrelevant) ideal of R.

The standard textbook definition. A local ring (R, m) is Cohen–Macaulay if it admits a regular sequence[29] of length equal to the dimension of R. A sequence of elements $x_1, \ldots, x_d$ is regular if x_1 is not a zero divisor of R, and the image of x_i in $R/(x_1, \ldots, x_{i-1})$ is not a zero divisor for $i = 2, 3, \ldots, d$. (See [Bruns and Herzog 1993, Definitions 1.1.1 and 2.1.1].) Another point of view on regular sequences is this: the Koszul complex on a set of elements $\{x_1, \ldots, x_d\}$ is acyclic if and only if the elements form a regular sequence.

Regular sequences are useful for creating induction arguments using long exact sequences induced from the short exact sequences

$$0 \to R/(x_1, \ldots, x_{i-1}) \xrightarrow{\cdot x_i} R/(x_1, \ldots, x_{i-1}) \to R/(x_1, \ldots, x_i) \to 0.$$

In algebraic geometry, say when R is the homogeneous coordinate ring of a projective variety, this is the technique of "cutting down by hypersurface sections". This works best when the resulting intersections contain no embedded points — which is to say, the defining equations of the hypersurfaces form a regular sequence.

A possibly more intuitive definition. Let R be an $\mathbb{N}$-graded algebra which is finitely generated over $R_0 = k$. Recall that every such ring admits a *Noether Normalization:* that is, R can be viewed as finite integral extension of some (graded) *polynomial* subalgebra A. Then R is Cohen–Macaulay if and only if R is free as an A module. For example, in the case of Example 1.21, the ring $R = S^G = \mathbb{C}[x^2, xy, y^2]$ can be viewed as an extension of the regular subring $A = k[x^2, y^2]$. As an A-module, R is free with basis $\{1, xy\}$. That is, every element of R can be written *uniquely* as a sum $a + bxy$, where a and b are polynomials in x^2, y^2.

If (R, m) is not graded but is a *complete* algebra over a field, then an analog of Noether Normalization called the "Cohen-Structure theorem" holds, which allows us to write R as a finite extension of a power-series subring A. Again, R is Cohen–Macaulay if and only if R is *free* as an A-module.

We remark that in both the graded and complete cases, it is easy to find the regular subring A. In the graded case, the k-algebra generated by any homogenous system of parameters will be a Noether normalization. Likewise, in the complete case, the power series subalgebra generated by any system of parameters will work.

Even if the local ring (R, m) is not complete, this criterion of Cohen–Macaulayness can be adapted by *completing* R at its maximal ideal: it is not hard to prove that a local ring (R, m) is Cohen–Macaulay if and only if its completion $\hat{R}$ at the

[29] homogenous in the graded case

maximal ideal is Cohen–Macaulay. This follows immediately from the definition of regular sequence: since $\hat{R}$ is a faithfully flat R-algebra, the sequence

$$0 \to R/(x_1, \dots, x_{i-1}) \xrightarrow{\cdot x_i} R/(x_1, \dots, x_{i-1}) \to R/(x_1, \dots, x_i) \to 0$$

is exact if and only if the sequence

$$0 \to \hat{R}/(x_1, \dots, x_{i-1}) \xrightarrow{\cdot x_i} \hat{R}/(x_1, \dots, x_{i-1}) \to \hat{R}/(x_1, \dots, x_i) \to 0$$

is exact.

A cohomological definition well-loved by commutative algebraists. The local or graded ring (R, m) is Cohen–Macaulay if and only if the local cohomology modules $H^i_m(R)$ are all zero for $i < \dim R$. We will not launch into a long discussion of local cohomology here, which is well-known to all commutative algebraists [Bruns and Herzog 1993, §3.5]. It suffices to know that local cohomology has all the usual functorial properties of any cohomology theory, so even if you don't know the precise definition, a passing familiarity with any kind of cohomology should suffice to follow the ideas in arguments in many situations.

Acknowledgements

We are grateful to Angelica Benito, Daniel Hernández, Greg Muller, Luis Núñez, Karl Schwede, Felipe Pèrez, and Emily Witt for carefully reading a draft of this paper and making suggestions to improve it.

References

[Aberbach 2002] I. M. Aberbach, "Some conditions for the equivalence of weak and strong F-regularity", *Comm. Algebra* **30**:4 (2002), 1635–1651.

[Aberbach and Enescu 2005] I. M. Aberbach and F. Enescu, "The structure of F-pure rings", *Math. Z.* **250**:4 (2005), 791–806.

[Aberbach and Hosry 2011] I. M. Aberbach and A. Hosry, "A less restrictive Briançon–Skoda theorem with coefficients", *J. Algebra* **345** (2011), 72–80.

[Aberbach and Huneke 1996] I. M. Aberbach and C. Huneke, "A theorem of Briançon–Skoda type for regular local rings containing a field", *Proc. Amer. Math. Soc.* **124**:3 (1996), 707–713.

[Aberbach and Leuschke 2003] I. M. Aberbach and G. J. Leuschke, "The F-signature and strong F-regularity", *Math. Res. Lett.* **10**:1 (2003), 51–56.

[Atiyah and Macdonald 1969] M. F. Atiyah and I. G. Macdonald, *Introduction to commutative algebra*, Addison-Wesley, Reading, MA, 1969.

[Benito et al. 2013] A. Benito, E. Faber, and K. E. Smith, "Measuring singularities with Frobenius: the basics", pp. 57–97 in *Commutative algebra*, edited by I. Peeva, Springer, New York, 2013.

[Bhatt and Singh 2015] B. Bhatt and A. Singh, "The F-pure threshold of a Calabi–Yau hypersurface", *Math. Ann.* **362**:1-2 (2015), 551–567.

[Birkar et al. 2010] C. Birkar, P. Cascini, C. D. Hacon, and J. McKernan, "Existence of minimal models for varieties of log general type", *J. Amer. Math. Soc.* **23**:2 (2010), 405–468.

[Blickle 2004] M. Blickle, "The intersection homology D-module in finite characteristic", *Math. Ann.* **328**:3 (2004), 425–450.

[Blickle 2013] M. Blickle, "Test ideals via algebras of p^{-e}-linear maps", *J. Algebraic Geom.* **22**:1 (2013), 49–83.

[Blickle and Böckle 2011] M. Blickle and G. Böckle, "Cartier modules: Finiteness results", *J. Reine Angew. Math.* **661** (2011), 85–123.

[Blickle and Lazarsfeld 2004] M. Blickle and R. Lazarsfeld, "An informal introduction to multiplier ideals", pp. 87–114 in *Trends in commutative algebra*, edited by L. L. Avramov et al., Math. Sci. Res. Inst. Publ. **51**, Cambridge Univ. Press, 2004.

[Blickle and Schwede 2013] M. Blickle and K. Schwede, "p^{-1}-linear maps in algebra and geometry", pp. 123–205 in *Commutative algebra*, edited by I. Peeva, Springer, New York, 2013.

[Blickle and Stäler 2015] M. Blickle and A. Stäler, "Bernstein–Sato polynomials and test modules in positive characteristic", preprint, 2015. arXiv 1402.1333

[Blickle et al. 2008] M. Blickle, M. Mustaţă, and K. E. Smith, "Discreteness and rationality of F-thresholds", *Michigan Math. J.* **57** (2008), 43–61.

[Blickle et al. 2009] M. Blickle, M. Mustaţă, and K. E. Smith, "F-thresholds of hypersurfaces", *Trans. Amer. Math. Soc.* **361**:12 (2009), 6549–6565.

[Blickle et al. 2010] M. Blickle, K. Schwede, S. Takagi, and W. Zhang, "Discreteness and rationality of F-jumping numbers on singular varieties", *Math. Ann.* **347**:4 (2010), 917–949.

[Blickle et al. 2012] M. Blickle, K. Schwede, and K. Tucker, "F-signature of pairs and the asymptotic behavior of Frobenius splittings", *Adv. Math.* **231**:6 (2012), 3232–3258.

[Blickle et al. 2013] M. Blickle, K. Schwede, and K. Tucker, "F-signature of pairs: Continuity, p-fractals and minimal log discrepancies", *J. Lond. Math. Soc.* (2) **87**:3 (2013), 802–818.

[Blickle et al. 2015] M. Blickle, K. Schwede, and K. Tucker, "F-singularities via alterations", *Amer. J. Math.* **137**:1 (2015), 61–109.

[Boutot 1987] J.-F. Boutot, "Singularités rationnelles et quotients par les groupes réductifs", *Invent. Math.* **88**:1 (1987), 65–68.

[Brenner 2008] H. Brenner, "Tight closure and vector bundles", pp. 1–71 in *Three lectures on commutative algebra*, edited by G. Colomé-Nin et al., Univ. Lecture Ser. **42**, Amer. Math. Soc., Providence, RI, 2008.

[Brenner 2013] H. Brenner, "Irrational Hilbert–Kunz multiplicities", 2013. arXiv 1305.5873

[Brion and Kumar 2005] M. Brion and S. Kumar, *Frobenius splitting methods in geometry and representation theory*, Progress in Mathematics **231**, Birkhäuser, Boston, 2005.

[Bruns and Herzog 1993] W. Bruns and J. Herzog, *Cohen–Macaulay rings*, Cambridge Studies in Advanced Mathematics **39**, Cambridge Univ. Press, 1993.

[Ein et al. 2001] L. Ein, R. Lazarsfeld, and K. E. Smith, "Uniform bounds and symbolic powers on smooth varieties", *Invent. Math.* **144**:2 (2001), 241–252.

[Ein et al. 2003] L. Ein, R. Lazarsfeld, and K. E. Smith, "Uniform approximation of Abhyankar valuation ideals in smooth function fields", *Amer. J. Math.* **125**:2 (2003), 409–440.

[Ein et al. 2004] L. Ein, R. Lazarsfeld, K. E. Smith, and D. Varolin, "Jumping coefficients of multiplier ideals", *Duke Math. J.* **123**:3 (2004), 469–506.

[Elkik 1981] R. Elkik, "Rationalité des singularités canoniques", *Invent. Math.* **64**:1 (1981), 1–6.

[Enescu and Hochster 2008] F. Enescu and M. Hochster, "The Frobenius structure of local cohomology", *Algebra Number Theory* **2**:7 (2008), 721–754.

[de Fernex and Hacon 2009] T. de Fernex and C. D. Hacon, "Singularities on normal varieties", *Compos. Math.* **145**:2 (2009), 393–414.

[Fulton 1993] W. Fulton, *Introduction to toric varieties*, Annals of Mathematics Studies **131**, Princeton University Press, 1993.

[Gabber 2004] O. Gabber, "Notes on some t-structures", pp. 711–734 in *Geometric aspects of Dwork theory, II*, edited by A. Adolphson et al., Walter de Gruyter, Berlin, 2004.

[Gongyo et al. 2015] Y. Gongyo, S. Okawa, A. Sannai, and S. Takagi, "Characterization of varieties of Fano type via singularities of Cox rings", *J. Algebraic Geom.* **24**:1 (2015), 159–182.

[Haboush 1980] W. J. Haboush, "A short proof of the Kempf vanishing theorem", *Invent. Math.* **56**:2 (1980), 109–112.

[Hara 1998] N. Hara, "A characterization of rational singularities in terms of injectivity of Frobenius maps", *Amer. J. Math.* **120**:5 (1998), 981–996.

[Hara 2001] N. Hara, "Geometric interpretation of tight closure and test ideals", *Trans. Amer. Math. Soc.* **353**:5 (2001), 1885–1906.

[Hara 2005] N. Hara, "A characteristic p analog of multiplier ideals and applications", *Comm. Algebra* **33**:10 (2005), 3375–3388.

[Hara and Takagi 2004a] N. Hara and S. Takagi, "On a generalization of test ideals", *Nagoya Math. J.* **175** (2004), 59–74.

[Hara and Takagi 2004b] N. Hara and S. Takagi, "On a generalization of test ideals", *Nagoya Math. J.* **175** (2004), 59–74.

[Hara and Watanabe 2002] N. Hara and K.-I. Watanabe, "F-regular and F-pure rings vs. log terminal and log canonical singularities", *J. Algebraic Geom.* **11**:2 (2002), 363–392.

[Hara and Yoshida 2003] N. Hara and K.-I. Yoshida, "A generalization of tight closure and multiplier ideals", *Trans. Amer. Math. Soc.* **355**:8 (2003), 3143–3174.

[Hartshorne 1977] R. Hartshorne, *Algebraic geometry*, Graduate Texts in Mathematics **52**, Springer, New York-Heidelberg, 1977.

[Hernández 2014] D. J. Hernández, "F-pure thresholds of binomial hypersurfaces", *Proc. Amer. Math. Soc.* **142**:7 (2014), 2227–2242.

[Hernández 2015] D. J. Hernández, "F-invariants of diagonal hypersurfaces", *Proc. Amer. Math. Soc.* **143**:1 (2015), 87–104.

[Hernández and Texiera 2015] D. J. Hernández and P. Texiera, "F-threshold functions: Syzygy gap fractals and the 2-variable homogeneous case", preprint, 2015. arXiv 1404.5871

[Hernández et al. 2014] D. J. Hernández, L. Nuñez-Betancourt, E. Witt, and W. Zhang, "F-pure thresholds of homogeneous polynomials", preprint, 2014. arXiv 1404.3772

[Hochster and Huneke 1989] M. Hochster and C. Huneke, "Tight closure and strong F-regularity", pp. 119–133 in *Colloque en l'honneur de Pierre Samuel* (Orsay, 1987), edited by M. Raynaud and L. Szpiro, Mém. Soc. Math. France (N.S.) **38**, Société Mathématique de France, Paris, 1989.

[Hochster and Huneke 1990] M. Hochster and C. Huneke, "Tight closure, invariant theory, and the Briançon–Skoda theorem", *J. Amer. Math. Soc.* **3**:1 (1990), 31–116.

[Hochster and Huneke 1994] M. Hochster and C. Huneke, "Tight closure of parameter ideals and splitting in module-finite extensions", *J. Algebraic Geom.* **3**:4 (1994), 599–670.

[Hochster and Huneke 2002] M. Hochster and C. Huneke, "Comparison of symbolic and ordinary powers of ideals", *Invent. Math.* **147**:2 (2002), 349–369.

[Hochster and Huneke 2006] M. Hochster and C. Huneke, "Tight closure in equal characteristic zero", preprint, 2006, available at http://www.math.lsa.umich.edu/~hochster/tcz.ps.

[Hochster and Roberts 1974] M. Hochster and J. L. Roberts, "Rings of invariants of reductive groups acting on regular rings are Cohen–Macaulay", *Advances in Math.* **13** (1974), 115–175.

[Hochster and Roberts 1976] M. Hochster and J. L. Roberts, "The purity of the Frobenius and local cohomology", *Advances in Math.* **21**:2 (1976), 117–172.

[Huneke 1996] C. Huneke, *Tight closure and its applications*, CBMS Regional Conference Series in Mathematics **88**, Amer. Math. Soc., Providence, RI, 1996.

[Huneke 2013] C. Huneke, "Hilbert–Kunz multiplicity and the F-signature", pp. 485–525 in *Commutative algebra*, edited by I. Peeva, Springer, New York, 2013.

[Huneke and Leuschke 2002] C. Huneke and G. J. Leuschke, "Two theorems about maximal Cohen–Macaulay modules", *Math. Ann.* **324**:2 (2002), 391–404.

[Huneke and Swanson 2006] C. Huneke and I. Swanson, *Integral closure of ideals, rings, and modules*, London Mathematical Society Lecture Note Series **336**, Cambridge Univ. Press, 2006.

[Katzman et al. 2009] M. Katzman, G. Lyubeznik, and W. Zhang, "On the discreteness and rationality of F-jumping coefficients", *J. Algebra* **322**:9 (2009), 3238–3247.

[Kawamata 1998] Y. Kawamata, "Subadjunction of log canonical divisors, II", *Amer. J. Math.* **120**:5 (1998), 893–899.

[Kollár 1997] J. Kollár, "Singularities of pairs", pp. 221–287 in *Algebraic geometry—Santa Cruz 1995*, edited by J. Kollár et al., Proc. Sympos. Pure Math. **62**, Amer. Math. Soc., Providence, RI, 1997.

[Kovács 2000] S. J. Kovács, "A characterization of rational singularities", *Duke Math. J.* **102**:2 (2000), 187–191.

[Kumar and Mehta 2009] S. Kumar and V. B. Mehta, "Finiteness of the number of compatibly split subvarieties", *Int. Math. Res. Not.* **2009**:19 (2009), 3595–3597.

[Kunz 1969] E. Kunz, "Characterizations of regular local rings for characteristic p", *Amer. J. Math.* **91** (1969), 772–784.

[Kunz 1976] E. Kunz, "On Noetherian rings of characteristic p", *Amer. J. Math.* **98**:4 (1976), 999–1013.

[Lauritzen et al. 2006] N. Lauritzen, U. Raben-Pedersen, and J. F. Thomsen, "Global F-regularity of Schubert varieties with applications to $\mathcal{D}$-modules", *J. Amer. Math. Soc.* **19**:2 (2006), 345–355.

[Lazarsfeld 2004] R. Lazarsfeld, *Positivity in algebraic geometry, I: Classical setting: Line bundles and linear series*, Ergebnisse der Mathematik und ihrer Grenzgebiete. 3. Folge. A Series of Modern Surveys in Mathematics **48**, Springer, Berlin, 2004.

[Lyubeznik and Smith 1999] G. Lyubeznik and K. E. Smith, "Strong and weak F-regularity are equivalent for graded rings", *Amer. J. Math.* **121**:6 (1999), 1279–1290.

[Lyubeznik and Smith 2001] G. Lyubeznik and K. E. Smith, "On the commutation of the test ideal with localization and completion", *Trans. Amer. Math. Soc.* **353**:8 (2001), 3149–3180.

[Maccrimmon 1996] B. C. Maccrimmon, *Strong F-regularity and boundedness questions in tight closure*, Ph.D. thesis, University of Michigan, 1996, available at http://search.proquest.com/docview/304246031.

[McDermott 2003] M. A. McDermott, "Test ideals in diagonal hypersurface rings, II", *J. Algebra* **264**:1 (2003), 296–304.

[Mehta and Ramanathan 1985] V. B. Mehta and A. Ramanathan, "Frobenius splitting and cohomology vanishing for Schubert varieties", *Ann. of Math.* (2) **122**:1 (1985), 27–40.

[Mehta and Ramanathan 1988] V. B. Mehta and A. Ramanathan, "Schubert varieties in $G/B \times G/B$", *Compositio Math.* **67**:3 (1988), 355–358.

[Mehta and Srinivas 1997] V. B. Mehta and V. Srinivas, "A characterization of rational singularities", *Asian J. Math.* **1**:2 (1997), 249–271.

[Monsky 1983] P. Monsky, "The Hilbert–Kunz function", *Math. Ann.* **263**:1 (1983), 43–49.

[Mumford et al. 1994] D. Mumford, J. Fogarty, and F. Kirwan, *Geometric invariant theory*, 3rd ed., Ergebnisse der Mathematik und ihrer Grenzgebiete (2) **34**, Springer, Berlin, 1994.

[Mustaţă 2009] M. Mustaţă, "Bernstein–Sato polynomials in positive characteristic", *J. Algebra* **321**:1 (2009), 128–151.

[Mustaţă 2012] M. Mustaţă, "IMPANGA lecture notes on log canonical thresholds", pp. 407–442 in *Contributions to algebraic geometry*, edited by P. Pragacz, Eur. Math. Soc., Zürich, 2012.

[Mustaţă and Srinivas 2011] M. Mustaţă and V. Srinivas, "Ordinary varieties and the comparison between multiplier ideals and test ideals", *Nagoya Math. J.* **204** (2011), 125–157.

[Mustaţă and Yoshida 2009] M. Mustaţă and K.-I. Yoshida, "Test ideals vs. multiplier ideals", *Nagoya Math. J.* **193** (2009), 111–128.

[Mustaţă et al. 2005] M. Mustaţă, S. Takagi, and K.-i. Watanabe, "F-thresholds and Bernstein–Sato polynomials", pp. 341–364 in *European Congress of Mathematics*, edited by A. Laptev, Eur. Math. Soc., Zürich, 2005.

[Nagata 1961] M. Nagata, "Complete reducibility of rational representations of a matric group", *J. Math. Kyoto Univ.* **1** (1961), 87–99.

[Peskine and Szpiro 1973] C. Peskine and L. Szpiro, "Dimension projective finie et cohomologie locale: Applications à la démonstration de conjectures de M. Auslander, H. Bass et A. Grothendieck", *Inst. Hautes Études Sci. Publ. Math.* **42** (1973), 47–119.

[Ramanan and Ramanathan 1985] S. Ramanan and A. Ramanathan, "Projective normality of flag varieties and Schubert varieties", *Invent. Math.* **79**:2 (1985), 217–224.

[Ramanathan 1991] A. Ramanathan, "Frobenius splitting and Schubert varieties", pp. 497–508 in *Proceedings of the Hyderabad Conference on Algebraic Groups* (Hyderabad, 1989), edited by S. Ramanan et al., Manoj Prakashan, Madras, 1991.

[Schwede 2009a] K. Schwede, "F-adjunction", *Algebra Number Theory* **3**:8 (2009), 907–950.

[Schwede 2009b] K. Schwede, "F-injective singularities are Du Bois", *Amer. J. Math.* **131**:2 (2009), 445–473.

[Schwede 2010a] K. Schwede, "Centers of F-purity", *Math. Z.* **265**:3 (2010), 687–714.

[Schwede 2010b] K. Schwede, "A refinement of sharply F-pure and strongly F-regular pairs", *J. Commut. Algebra* **2**:1 (2010), 91–109.

[Schwede and Smith 2010] K. Schwede and K. E. Smith, "Globally F-regular and log Fano varieties", *Adv. Math.* **224**:3 (2010), 863–894.

[Schwede and Tucker 2010] K. Schwede and K. Tucker, "On the number of compatibly Frobenius split subvarieties, prime F-ideals, and log canonical centers", *Ann. Inst. Fourier (Grenoble)* **60**:5 (2010), 1515–1531.

[Schwede and Tucker 2012] K. Schwede and K. Tucker, "A survey of test ideals", pp. 39–99 in *Progress in commutative algebra 2: Closures, finiteness and factorization*, edited by C. Francisco et al., Walter de Gruyter, Berlin, 2012.

[Schwede and Tucker 2014a] K. Schwede and K. Tucker, "On the behavior of test ideals under finite morphisms", *J. Algebraic Geom.* **23**:3 (2014), 399–443.

[Schwede and Tucker 2014b] K. Schwede and K. Tucker, "Test ideals of non-principal ideals: Computations, jumping numbers, alterations and division theorems", *J. Math. Pures Appl.* (9) **102**:5 (2014), 891–929.

[Schwede et al. 2012] K. Schwede, K. Tucker, and W. Zhang, "Test ideals via a single alteration and discreteness and rationality of F-jumping numbers", *Math. Res. Lett.* **19**:1 (2012), 191–197.

[Sharp 2007] R. Y. Sharp, "Graded annihilators of modules over the Frobenius skew polynomial ring, and tight closure", *Trans. Amer. Math. Soc.* **359**:9 (2007), 4237–4258.

[Singh 2005] A. K. Singh, "The F-signature of an affine semigroup ring", *J. Pure Appl. Algebra* **196**:2-3 (2005), 313–321.

[Smith 1994] K. E. Smith, "Tight closure of parameter ideals", *Invent. Math.* **115**:1 (1994), 41–60.

[Smith 1995a] K. E. Smith, "The D-module structure of F-split rings", *Math. Res. Lett.* **2**:4 (1995), 377–386.

[Smith 1995b] K. E. Smith, "Test ideals in local rings", *Trans. Amer. Math. Soc.* **347**:9 (1995), 3453–3472.

[Smith 1997a] K. E. Smith, "F-rational rings have rational singularities", *Amer. J. Math.* **119**:1 (1997), 159–180.

[Smith 1997b] K. E. Smith, "Vanishing, singularities and effective bounds via prime characteristic local algebra", pp. 289–325 in *Algebraic geometry—Santa Cruz 1995*, edited by J. Kollár et al., Proc. Sympos. Pure Math. **62**, Amer. Math. Soc., Providence, RI, 1997.

[Smith 2000a] K. E. Smith, "Globally F-regular varieties: Applications to vanishing theorems for quotients of Fano varieties", *Michigan Math. J.* **48** (2000), 553–572.

[Smith 2000b] K. E. Smith, "The multiplier ideal is a universal test ideal", *Comm. Algebra* **28**:12 (2000), 5915–5929.

[Smith and Van den Bergh 1997] K. E. Smith and M. Van den Bergh, "Simplicity of rings of differential operators in prime characteristic", *Proc. London Math. Soc.* (3) **75**:1 (1997), 32–62.

[Swanson 2002] I. Swanson, "Ten lectures on tight closure", lecture notes, 2002, available at http://people.reed.edu/~iswanson/tciran.pdf.

[Takagi 2006] S. Takagi, "Formulas for multiplier ideals on singular varieties", *Amer. J. Math.* **128**:6 (2006), 1345–1362.

[Takagi 2008] S. Takagi, "A characteristic p analogue of plt singularities and adjoint ideals", *Math. Z.* **259**:2 (2008), 321–341.

[Takagi and Yoshida 2008] S. Takagi and K.-i. Yoshida, "Generalized test ideals and symbolic powers", *Michigan Math. J.* **57** (2008), 711–724.

[Tucker 2012] K. Tucker, "F-signature exists", *Invent. Math.* **190**:3 (2012), 743–765.

[Vassilev 1998] J. C. Vassilev, "Test ideals in quotients of F-finite regular local rings", *Trans. Amer. Math. Soc.* **350**:10 (1998), 4041–4051.

[Von Korff 2012] M. R. Von Korff, *The F-signature of toric varieties*, Ph.D. thesis, University of Michigan, 2012, available at http://search.proquest.com/docview/1149994294.

[Watanabe and Yoshida 2004] K.-i. Watanabe and K.-i. Yoshida, "Minimal relative Hilbert–Kunz multiplicity", *Illinois J. Math.* **48**:1 (2004), 273–294.

[Yao 2006] Y. Yao, "Observations on the F-signature of local rings of characteristic p", *J. Algebra* **299**:1 (2006), 198–218.

kesmith@umich.edu *Department of Mathematics, University of Michigan, 2832 East Hall, Ann Arbor, MI 48109-1109, United States*

wlzhang@umich.edu *Department of Mathematics, University of Michigan, 1859 East Hall, Ann Arbor, MI 48109-1109, United States*

Commutative Algebra and Noncommutative Algebraic Geometry, I
MSRI Publications
Volume **67**, 2015

From Briançon–Skoda to Scherk–Varchenko

DUCO VAN STRATEN

To the memory of Egbert Brieskorn

In this survey paper we try to explain how the monodromy theorem for isolated hypersurface singularities led to unexpected conjectures by J. Scherk relating the smallest power r for which f^r belongs to the jacobian ideal J_f to the size of the Jordan blocks in the vanishing cohomology. These were proven by A. Varchenko using his asymptotic mixed Hodge structure on the vanishing cohomology.

1. The monodromy transformation

The study of the ramification of integrals depending on parameters has a history that can be traced back at least to the work of Euler, Legendre and Gauss, but it seems that the systematic study of the topology of algebraic varieties and their period integrals has its roots in the nineteenth century in the work of Poincaré and Picard. To see what is involved, let us start with a well-known and basic example.

The equation

$$y^2 = (t - x^2)(1 - x)$$

describes an affine part of an elliptic curve E_t depending on a parameter t.

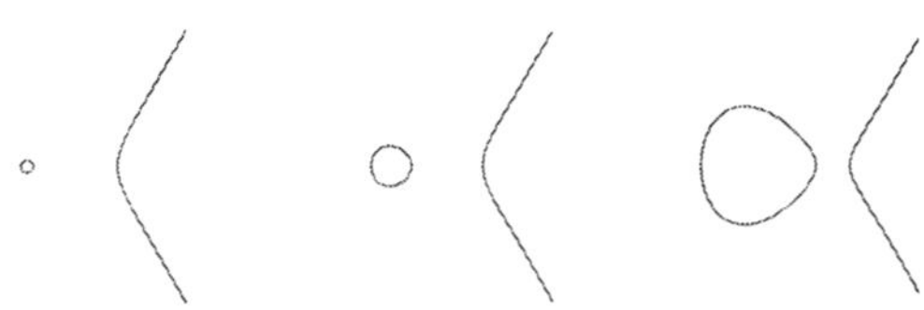

The small loop in the picture that runs between $x = -\sqrt{t}$ and $x = \sqrt{t}$ shrinks to a point for $t = 0$: it is a *vanishing cycle*. The projection to the x-line represents E_t as a double cover of the Riemann sphere, ramified over the four points

$-\sqrt{t}, \sqrt{t}, 1, \infty$:

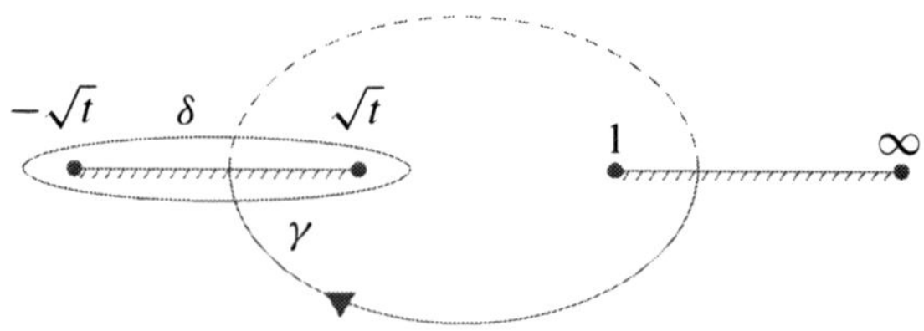

From this one can see that for general t the topology of E_t is that of a 2-torus, but for $t \to 0$, this torus degenerates to a pinched torus:

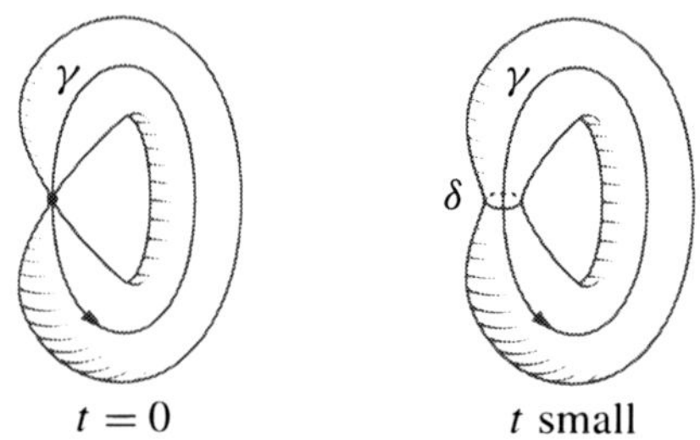

One can pick a basis for $H_1(E_t)$ consisting of the vanishing cycle $\delta = \delta(t) \in H^1(E_t)$ that runs around the points $\pm\sqrt{t}$ and a cycle $\gamma = \gamma(t)$ that survives the contraction of the vanishing cycle, but gets pinched. When we make a small detour $t = \epsilon \exp(i\theta)$, $\theta \in [0, 2\pi]$ in the complex plane around the point $t = 0$, the two branch-points $\pm\sqrt{t}$ get interchanged. When we follow the cycles by parallel transport, we obtain a *monodromy-transformation*

$$T : H_1(E_t) \to H_1(E_t).$$

For the cycles δ and γ we find $T\delta = \delta$, and $T\gamma = \gamma + \delta$, as indicated by the following pictures.

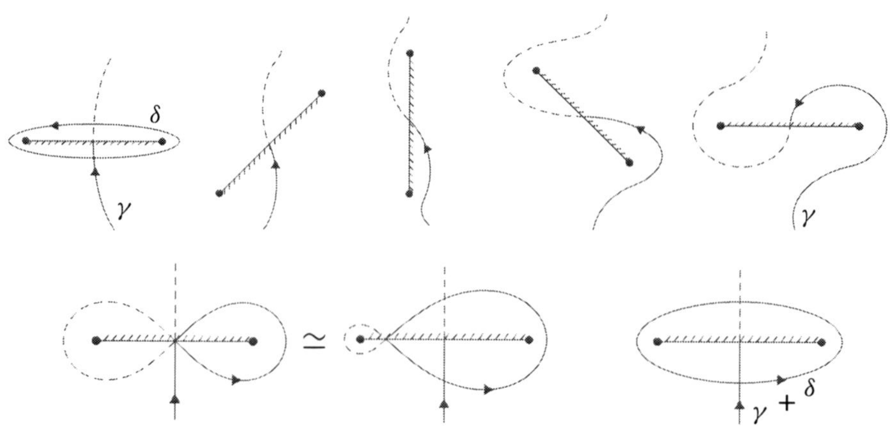

Hence the monodromy is represented by the matrix

$$T = \left(\begin{smallmatrix} 1 & 1 \\ 0 & 1 \end{smallmatrix}\right), \quad (T-1)^2 = 0.$$

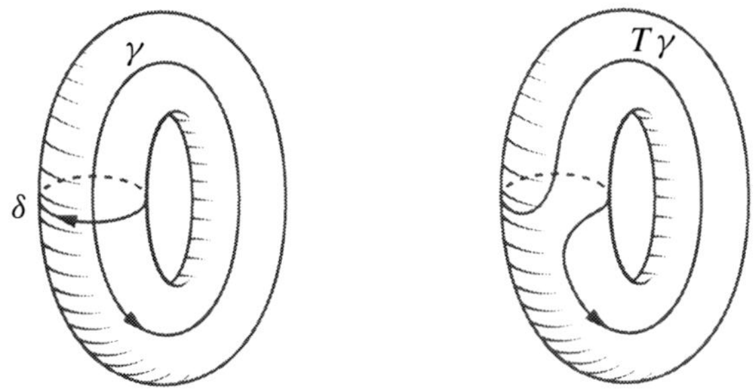

The behaviour of the cycles is reflected in the behaviour of the *period integrals*

$$\Phi_\Gamma(t) = \int_\Gamma \eta_t, \quad \eta_t = \frac{dx}{\sqrt{(t-x^2)(1-x)}}.$$

These satisfy the linear differential equation

$$(16\Theta^2 - t(4\Theta+1)(4\Theta+3))\Phi_\Gamma(t) = 0,$$

where $\Theta := t\partial/\partial t$. For the above cycles δ, γ one finds the following series expansions:

$$\Phi_\delta(t) = 2\pi\left(1 + \tfrac{3}{16}t + \tfrac{105}{1024}t^2 + \cdots\right),$$
$$2\pi i\Phi_\gamma(t) = \log(t)\Phi_\delta(t) + 2\pi\left(\tfrac{5}{8}t + \tfrac{389}{1024}t^2 + \cdots\right).$$

The analytic continuation of these period integrals exactly reflect the monodromy behaviour of the cycles δ and γ: continuation around $t = 0$ gives

$$\Phi_\delta \to \Phi_\delta, \quad \Phi_\gamma \to \Phi_\gamma + \Phi_\delta.$$

This example turns out to be part of a much more general story: for families of curves of higher genus acquiring nodes as singularities the situation is very similar and was first described in [Picard and Simart 1897, Tome I, Chapter IV, Section 19]. For an excellent account see [Brieskorn and Knörrer 1981, Section 9.3], where also an example similar to the above one is worked out in detail. The generalisation to the case of n-dimensional varieties Y_t acquiring an ordinary double point was first described by Lefschetz [1924].

The effect of the monodromy can be described by the *Picard–Lefschetz formula*

$$T : H^n(Y_t) \to H^n(Y_t), \quad \gamma \mapsto \gamma \pm \langle\gamma, \delta\rangle\delta,$$

where $\langle -, -\rangle$ denotes the intersection of cycles on Y_t, and the sign is found to be $(-1)^{(n+1)(n+2)/2}$ [Lamotke 1981; Vassiliev 2002].

In general, a holomorphic one-parameter family of compact complex n-dimensional manifolds degenerating over 0 is described by a smooth $n+1$-dimensional complex manifold $\mathcal{Y}$ with a proper holomorphic map $f := \mathcal{Y} \to D$ to the disc D, submersive on $\mathcal{Y}^* = \mathcal{Y} \setminus f^{-1}(0)$. By the Ehresmann fibration theorem, the family $f^* : \mathcal{Y}^* \to D^*$ is a differentiable fibre bundle over the punctured disc D^*. As D^* contracts to a circle, this fibre bundle is described by a geometric monodromy transformation $Y_t \to Y_t$, which induces a cohomological monodromy transformation T.

The monodromy theorem. *The cohomological monodromy transformation*

$$T : H^q(Y_t) \to H^q(Y_t)$$

is quasiunipotent. More precisely, there exists an integer e such that

$$(T^e - 1)^{q+1} = 0.$$

So the eigenvalues of T are roots of unity and the size of the Jordan blocks is bounded by $q+1$. One can write $T = S \cdot U = U \cdot S$ where S is semisimple and U is unipotent. The nilpotent operator $1 - U$ has the same Jordan type as $1 - T^e$ or as the *monodromy logarithm*

$$N := \log(U) = (U-1) - \tfrac{1}{2}(U-1)^2 + \tfrac{1}{3}(U-1)^3 + \cdots .$$

The first proof of this fundamental theorem appeared in the (unpublished) Berkeley thesis of Landman [1966] (see also [Landman 1973]). A further topological proof was given by Clemens [1969]. Many alternative proofs, avoiding resolutions of singularities and using arithmetical or Hodge theoretical arguments were given by Deligne, Grothendieck, Katz, and Borel; see [Deligne and Katz 1973; Katz 1970; 1971].

2. Isolated hypersurface singularities

Locally around any point of $\mathcal{Y}$, the map can be described by a germ

$$f : (\mathbb{C}^{n+1}, 0) \to (\mathbb{C}, 0)$$

determined by a convergent power series

$$f \in S := \mathbb{C}\{x_0, x_1, \ldots, x_n\}.$$

One speaks of an *isolated singularity* if the equations

$$\frac{\partial f}{\partial x_0} = \frac{\partial f}{\partial x_1} = \cdots = \frac{\partial f}{\partial x_n} = 0$$

have only 0 as a common solution in a neighbourhood of 0. This is equivalent to the condition that the *Jacobi ring*

$$Q_f := S/J_f, \quad J_f = (\partial_0 f, \partial_1 f, \dots, \partial_n f)$$

is of finite $\mathbb{C}$-dimension. One says that two singularities f and g are *right-equivalent*, notation $f \sim g$, if one can find a coordinate transformation

$$(\mathbb{C}^{n+1}, 0) \to (\mathbb{C}^{n+1}, 0)$$

that maps f to g. The classification up to right equivalence then starts with the famous ADE list, [Arnold 1975]. Here some pictures of some well-known singularities, together with a deformation that explains their name.

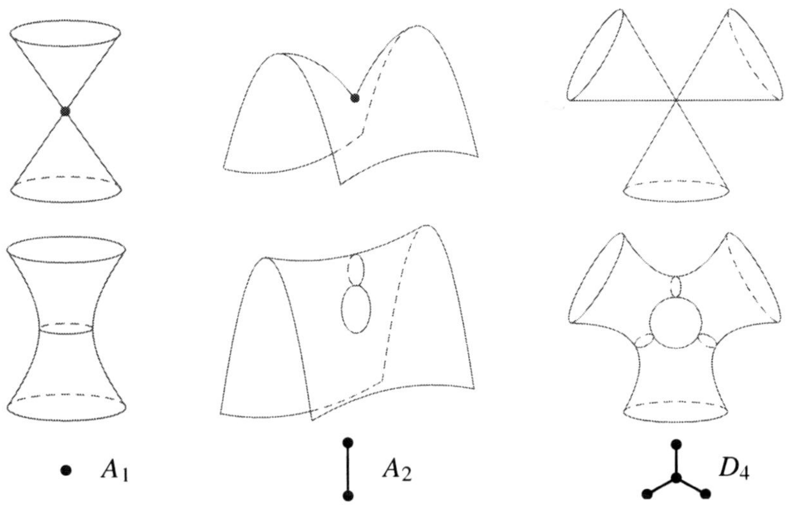

2.1. *Milnor fibration.* An isolated singularity always possesses a so-called *good representative*; see [Looijenga 1984, p. 21]. By this we mean the following. First one picks $\epsilon > 0$ so small that for all $0 < \epsilon' \le \epsilon$ the boundary $\partial B_{\epsilon'}$ is transverse to the special fibre $f^{-1}(0)$. One obtains a smooth orientable differentiable manifold

$$L = \partial B_\epsilon \cap f^{-1}(0)$$

of dimension $2n - 1$, called the *link* of the singularity. Then one picks $\eta > 0$ such that for all t with $0 < |t| \le \eta$ the fibre $f^{-1}(t)$ is transverse to ∂B_ϵ. We put $B = B_\epsilon = \{|x| \le \epsilon\}$ and $D = D_\eta = \{|t| \le \eta\}$ and let $\mathcal{X} := B \cap f^{-1}(D)$, so that f determines a map $\mathcal{X} \to D$, called good representative of the germ f. Furthermore, in such a situation we set $D^* := D \setminus \{0\}$, $\mathcal{X}^* := \mathcal{X} \setminus f^{-1}(0)$, and we obtain a map

$$f^* : \mathcal{X}^* \to D^*.$$

Again, the Ehresmann fibration theorem shows that

$$f^* : \mathcal{X}^* \to D^*$$

is a C^∞-fibre bundle. This fibration is now commonly called the *Milnor fibration*, its fibre $X_t := f^{-1}(t)$ the *Milnor fibre*.

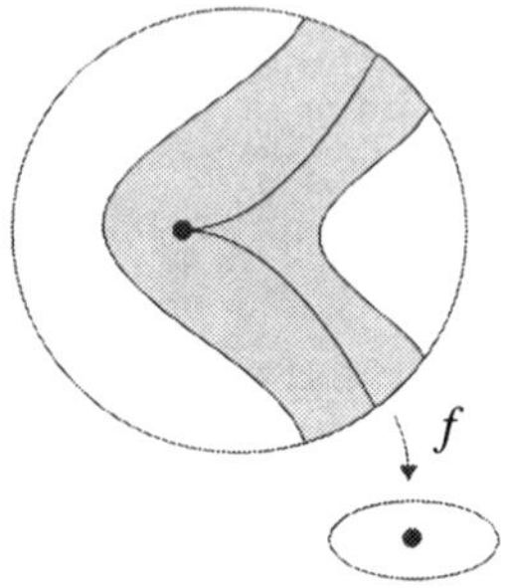

Theorem. *The Milnor fibre has the homotopy type of a bouquet of n-spheres.*

$$X_t \approx \bigvee_{i=1}^{\mu} S^n.$$

The number μ of spheres, called the Milnor number, can be computed as

$$\mu = \dim(S/J_f).$$

The spheres appearing in the first part of the statement are contracted upon approaching the fibre over 0, and are called, extending the terminology used by Lefschetz, the *vanishing cycles* of the singularity. A consequence of the bouquet-theorem is that the Milnor fibre only has one interesting cohomology group $H^n(X_t, \mathbb{Z})$, which is free of rank μ.

Although all Milnor fibres X_t are diffeomorphic, one can not speak about "the" Milnor fibre, as the manifold X_t depends on t. For some constructions it is convenient to use the *canonical Milnor fibre* X_∞, defined as the pull-back of $\mathcal{X}^*$ over the universal covering $\widetilde{D} \to D^*$ of the punctured disc

$$X_\infty = \mathcal{X}^* \times_{D^*} \widetilde{D}.$$

Then X_∞ contracts to each of the Milnor fibres X_t and we have a single group $H^n(X_\infty, \mathbb{Z})$ isomorphic to each of the $H^n(X_t, \mathbb{Z})$.

2.2. *Exotic spheres.* One of the strong motivations to study the differential topological properties of isolated hypersurface singularities came from the discoveries of Hirzebruch [1964] and Brieskorn [1966a; 1966b] that the link L

of such singularity can be a sphere with an exotic differentiable structure. The so-called *Brieskorn–Pham* polynomials of the form

$$f = x_0^{a_0} + x_1^{a_1} + \cdots + x_n^{a_n}$$

played an important role in that story. The Milnor number of f is easily seen to be

$$\mu = (a_0 - 1)(a_1 - 1) \cdots (a_n - 1).$$

Furthermore, Pham [1965] determined the cohomological monodromy T of this singularity. It is of finite order

$$e := \operatorname{lcm}(a_0, a_1, \ldots, a_n),$$

and the eigenvalues of T on $H^n(X_t)$ are the numbers

$$\omega_0 \omega_1 \ldots \omega_n,$$

where ω_i runs over all a_i-th roots of unity. A closer analysis of the topology of the Milnor fibration (see [Milnor 1968, p. 65]) shows that the link L of an isolated singularity has the integral homology of a sphere if and only if $\det(I - T) = \pm 1$, from which one can conclude for $n \neq 2$ that L in fact is *homeomorphic* to a sphere. Brieskorn [1966b] used this to show, for example, that the link of

$$x_0^2 + x_1^2 + x_2^2 + x_3^3 + x_4^{6k-1}$$

for $k = 1, 2, \ldots, 28$ represents the 28 distinct differentiable structures on the 7-sphere S^7. In fact, all exotic spheres that bound a parallelizable manifold appear as links of such *Brieskorn–Pham* singularities.

2.3. *The Brieskorn lattice.* Brieskorn [1970] described a method to determine the cohomological monodromy of an isolated hypersurface singularity and used it to give a proof of the monodromy theorem for isolated hypersurface singularities, thus answering a question of Milnor.

Monodromy Theorem for Isolated Hypersurface Singularities. *The cohomological monodromy transformation*

$$T : H^n(X_t) \to H^n(X_t)$$

is quasiunipotent: there exists e such that

$$(T^e - 1)^{n+1} = 0.$$

The idea is to look at the *cohomology bundle* over D^* with fibres $H^n(X_t, \mathbb{C})$, the cohomology of the Milnor fibre. This bundle comes with a natural flat connection defined by parallel-transport of (co)cycles: the *Gauss–Manin connection*. Brieskorn then develops a de Rham description to represent sections of this cohomology bundle and gives an explicit description of Gauss–Manin connection in local terms. The resulting system of linear differential equations describe the variation of the period integrals over the vanishing cycles and the monodromy of this differential system is identified with the cohomological monodromy T. In more detail it works as follows.

A (germ of a) differential form

$$\omega \in \Omega^{n+1} := S\, dx_0\, dx_1 \ldots dx_n = \mathbb{C}\{x_0, x_1, \ldots, x_n\}\, dx_0\, dx_1 \ldots dx_n$$

determines a section of the cohomology bundle: we obtain a family of closed differential forms on the Milnor fibres X_t by

$$\eta_t = \mathrm{Res}_{X_t}\left(\frac{\omega}{f-t}\right).$$

The forms ω that belong to the subspace $df \wedge d\Omega^{n-1}$ give rise to forms that are exact on the fibres and hence the *Brieskorn lattice* defined by

$$\mathcal{H} := \Omega^{n+1}/df \wedge d\Omega^{n-1}$$

can be thought to give families of cohomology classes on the Milnor fibration. It is called H'' in [Brieskorn 1970].

On $\mathcal{H}$ there are various important structures. First it has a natural structure as a $\mathbb{C}\{t\}$-module: the action of t on $\mathcal{H}$ is realised by multiplication of differential forms by f. In fact one has:

Theorem. *$\mathcal{H}$ is a free $\mathbb{C}\{t\}$-module of rank μ.*

The statement about the rank is due to Brieskorn, the freeness is due to Sebastiani [1970]. His and other proofs use integration and no completely algebraic proof is known to me.

Example. Consider $f = y^2 + x^3$. In the diagram at the top of the next page the dots indicate nonzero monomials

$$x^a y^b\, dx\, dy$$

in the Brieskorn lattice $\mathcal{H}$. The dotted lines indicate relations between these monomials in $\mathcal{H}$, coming from

$$df \wedge d(x^p y^q) = (3qx^{2+p} y^{q-1} - 2py^{q+1} x^{p-1})\, dx\, dy.$$

We can use the monomials $dx\, dy$ and $x\, dx\, dy$ as a $\mathbb{C}\{t\}$-basis of $\mathcal{H}$.

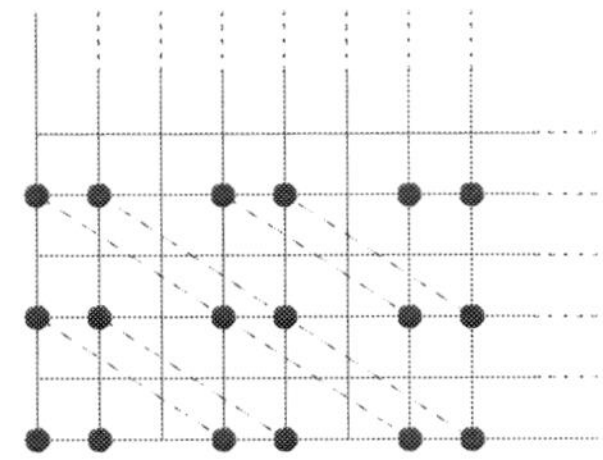

(See example on previous page).

The Brieskorn lattice $\mathcal{H}$ carries another operation called ∂^{-1}, which Brieskorn identifies as the *inverse* of the Gauss–Manin connection. It is defined as follows: if the $(n+1)$-form $\omega \in \Omega^{n+1}$ on $(\mathbb{C}^{n+1}, 0)$ represents an element of $\mathcal{H}$, we can write it as $d\eta$ for some $\eta \in \Omega^n$. One now sets

$$\partial^{-1}\omega := df \wedge \eta.$$

It is easy to check that this gives a well-defined operation on $\mathcal{H}$, which satisfies

$$t\partial^{-1} - \partial^{-1}t = \partial^{-2}.$$

The map $\partial^{-1} : \mathcal{H} \to \mathcal{H}$ is *injective* and the cokernel can be identified with

$$\mathcal{H}/\partial^{-1}\mathcal{H} = \Omega^{n+1}/df \wedge \Omega^n =: Q^f,$$

which after a choice of a volume form is isomorphic to $Q_f = S/J_f$, the Jacobi ring of $\mathbb{C}$-dimension μ. When we choose a basis $\omega_1, \omega_2, \ldots, \omega_\mu$ of $\mathcal{H}$ as $\mathbb{C}\{t\}$-module, we can write out the action of ∂^{-1} in this basis

$$\partial^{-1}\begin{pmatrix}\omega_1\\ \omega_2\\ \cdots\\ \omega_\mu\end{pmatrix} = B(t)\begin{pmatrix}\omega_1\\ \omega_2\\ \cdots\\ \omega_\mu\end{pmatrix},$$

from which one obtains a meromorphic connection matrix

$$A(t) = B(t)^{-1}(1 - B'(t))$$

for $\mathcal{H}$:

$$\partial\begin{pmatrix}\omega_1\\ \omega_2\\ \cdots\\ \omega_\mu\end{pmatrix} = A(t)\begin{pmatrix}\omega_1\\ \omega_2\\ \cdots\\ \omega_\mu\end{pmatrix}.$$

If $\delta(t)$ denotes a (multivalued) horizontal family of cycles in $H_n(X_t)$ the (in general multivalued) *period integral* is

$$\Phi(t) = \int_{\delta(t)} \eta_t.$$

For such period integrals on can prove an estimate of the form

$$|\Phi(t)| \leq O(t^{-N}),$$

which implies the *regularity theorem*: the resulting differential system is *regular singular*, hence can be transformed into a system with first order pole:

$$A(t) = \frac{A_{-1}}{t} + A_0 + A_1 t + \cdots .$$

And the monodromy $\exp(2\pi A_{-1})$ is identified with the (complexification) of the cohomological monodromy T. In this way we have a theoretical method to determine the cohomological monodromy transformation T (up to conjugacy). As T is an automorphism of the lattice $H^n(X_t, \mathbb{Z})$, the characteristic polynomial has integer coefficients, and it follows that the eigenvalues of T are algebraic numbers. From the fact that the construction is "algebraically defined", the eigenvalues α of A_{-1} are algebraic too. As by the theorem of Gelfond–Schneider for an irrational algebraic number α, the number

$$\exp(2\pi i \alpha)$$

is transcendental, Brieskorn concluded that the eigenvalues of the monodromy are roots of unity!

The period integrals expand in series of the following sort

$$\Phi(t) = \sum_{\alpha,k} A_{\alpha,k} t^{\alpha} (\log t)^k.$$

It was shown by Malgrange [1974] that in fact $A_{\alpha,k} = 0$ for $\alpha \leq -1$, which provides an alternative proof of the fact that $\mathcal{H}$ is $\mathbb{C}\{t\}$-free. (The reason is that elements $\omega \in \mathcal{H}$ in the kernel of multiplication by t belong to the space C^{-1}, defined in Section 5.)

Gauss–Manin system. It has become customary to embed the Brieskorn lattice $\mathcal{H}$ into the *Gauss–Manin system* $\mathcal{G}$ of f. This is explained by Pham [1979, pp. 153–167]: one considers the de Rham complex $\Omega^\bullet$ of (germs) of differential forms on $(\mathbb{C}^{n+1}, 0)$ and let D be a variable. By $\Omega^\bullet[D]$ we denote the set of polynomials with coefficients in $\Omega^\bullet$. On it we have a the twisted differential $\underline{d} := d + D df \wedge$:

$$\underline{d}(\omega D^k) := d\omega D^k + df \wedge \omega D^{k+1}.$$

The *Gauss–Manin system* $\mathcal{G}$ is defined as the $(n+1)$-cohomology group of the *twisted de Rham complex*:

$$\mathcal{G} := H^{n+1}(\Omega^\bullet[D], d + Ddf\wedge).$$

The element ωD^k can be thought of as standing for the family of differential forms

$$\mathrm{Res}_{X_t}\left(\frac{k!\omega}{(f-t)^{k+1}}\right)$$

on the Milnor fibres X_t. On $\mathcal{G}$ one has actions of t and ∂

$$t(\omega D^k) = f\omega D^k - k\omega D^{k-1}, \quad \partial(\omega D^k) = \omega D^{k+1},$$

which are easily checked to satisfy

$$\partial t - t\partial = 1,$$

so $\mathcal{G}$ becomes a module over $\mathcal{D} := \mathbb{C}\{t\}[\partial]$. The map $\omega \in \Omega^{n+1} \mapsto \omega D^0$ induces a well-defined embedding

$$\mathcal{H} \hookrightarrow \mathcal{G}.$$

In fact, ∂ is invertible on $\mathcal{G}$, and the restriction of the inverse ∂^{-1} coincides with the operation on $\mathcal{H}$ defined earlier.

M. Schulze has implemented Brieskorn's algorithm in [Schulze 2003]. From the computational point of view it is useful to change, as advocated by Pham [1979], to the *microlocal* point of view, that is using $s = \partial^{-1}$ as expansion parameter. This boils down to looking at the incomplete Laplace transform of the period integrals, that is to the associated *oscillatory integral*. The relevant formula is

$$\int_{\Gamma(t)} e^{-f/s}\omega = \int_0^t e^{-u/s}\int_{\delta(t)} \mathrm{Res}\left(\frac{\omega}{f-u}\right)du,$$

where $\Gamma(t)$ is trace of the vanishing cycle, also known as *Lefschetz thimble*.

3. Questions and answers

Griffiths [1970, pp. 249–250] reports on a question raised by Brieskorn and related to him by Deligne.

Problem. Is the P.-L. transformation $T : H^n(X_t) \to H^n(X_t)$ of finite order?

Here "P.-L." of course stands for "Picard–Lefschetz". Although the monodromy transformation in the global case usually has Jordan blocks, the transformation on the vanishing cohomology of the simplest singularities like the ordinary node or the Brieskorn–Pham singularities have finite order. Lê proved in 1971 that the monodromy is of finite order for *irreducible* curve singularities [Lê 1974]. There were serious attempts to prove the result in general.

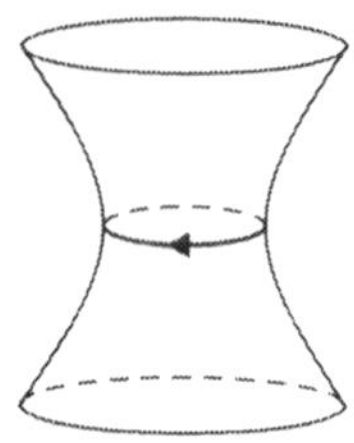

So it came somewhat as a surprise when A'Campo [1973] published the first examples of plane curve singularities where the monodromy transformations on the cohomology of the Milnor fibre had a Jordan block.

Example (A'Campo). Consider the curve singularity that consists of two cusps, with distinct tangent cones.

$$f = (x^2 + y^3)(y^2 + x^3) = x^2y^2 + x^5 + y^5 + x^3y^3 \sim x^2y^2 + x^5 + y^5.$$

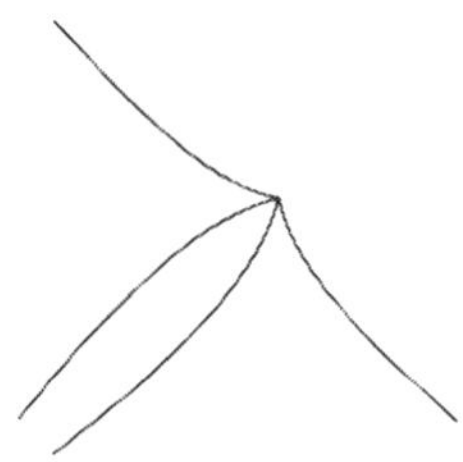

It has $\mu = 11$ and the monodromy satisfies

$$T^{10} - 1 \neq 0, \quad (T^{10} - 1)^2 = 0.$$

A good embedded resolution of $f^{-1}(0)$ is obtained by blowing up the origin and then twice in the strict transform of the two cusps. We obtain a chain of 5 exceptional divisors, with multiplicities 5, 10, 4, 10, 5; the strict transforms of the cusps pass through the components with multiplicity 10.

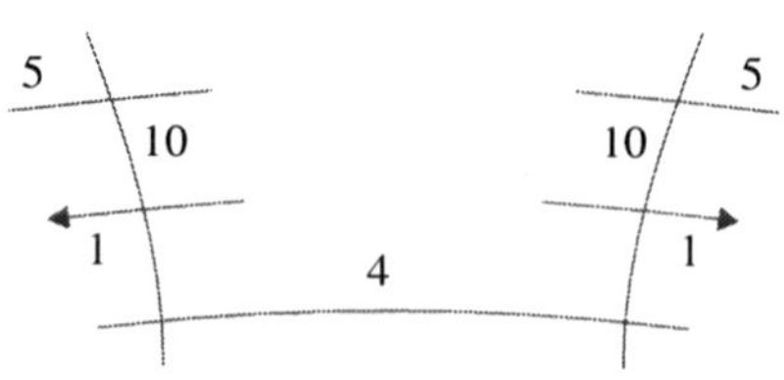

The Milnor fibre $f = t$ as a subset of the embedded resolution is a curve very close to the union of the exceptional curves and the strict transform of the two cusps. The multiplicity of each component indicates how often the Milnor fibre runs along the divisor. From this information one can build a topological model of the Milnor fibre. Usually one first performs a *semistable reduction*, which in this case amounts to replacing t by t^{10} and which comes down to taking a 10-fold cyclic cover of the embedded resolution. As Milnor fibre one obtains a Riemann surface consisting of two Riemann surfaces of genus 2, with a boundary, and glued together via two cylinders. The cycle γ indicated on the right has $(T^{10}-1)\gamma \neq 0$.

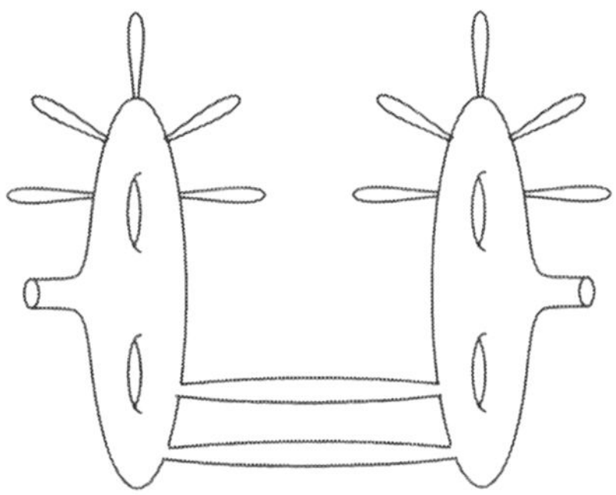

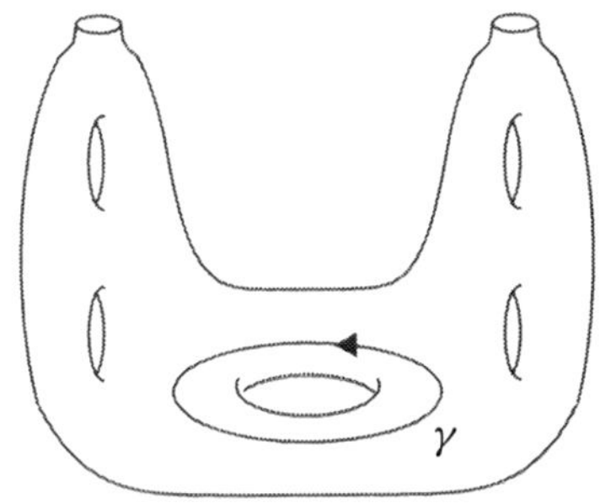

For more details we refer to [A'Campo 1973] and [Brieskorn and Knörrer 1981, p. 751].

A'Campo raised the problem of finding examples of singularities in $n+1$ variables whose cohomological monodromy had a Jordan block of maximal size $n+1$. Such examples were described by Malgrange [1973] in a letter to the editors, published front-to-back to the paper of A'Campo. Malgrange credits Hörmander for the idea.

Example [Malgrange 1973]. The singularity

$$f = (x_0x_1\dots x_n)^2 + x_0^{2n+4} + x_1^{2n+4} + \dots + x_n^{2n+4}$$

has a Jordan block of maximal size $n+1$. Let

$$E(t) := \{(x_0, x_1, \dots, x_n) \in \mathbb{R}^{n+1} \mid f \leq t\}.$$

For t small enough, this is a topological ball; its boundary

$$\delta(t) := \partial E(t)$$

is a vanishing cycle that for $n=2$ looks like the picture at the top of the next page.

Clearly:

$$\int_{\delta(t)} x_0\, dx_1 \dots dx_n = \int_{E(t)} dx_0\, dx_1 \dots dx_n = \mathrm{Vol}(E(t)).$$

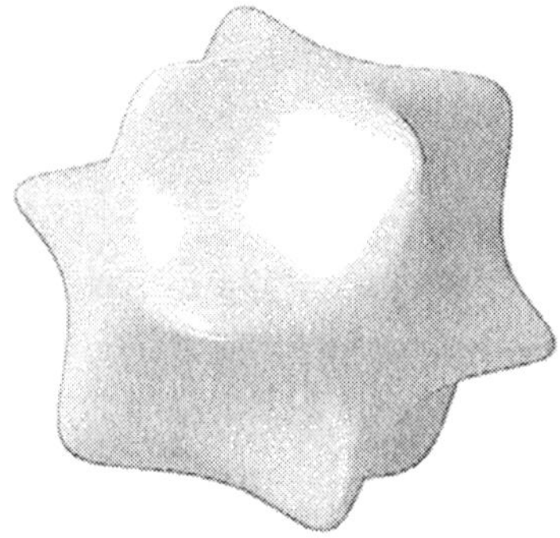

$$100\,(xyz)^2 + x^8 + y^8 + z^8 = 1.$$

Now Malgrange computes

$$\mathrm{Vol}(E(t)) \sim C t^{1/2} \log^n(t),$$

where $C \neq 0$. This shows that the vanishing cycle $\delta(t) \in H_n(X_t, \mathbb{Z})$ sits in a Jordan block of size $n+1$.

4. The Briançon–Skoda theorem and Scherk's conjecture

According to C. T. C Wall [1971] it was Mather who asked about the smallest r for which

$$f^r \in J_f.$$

Around the same time as A'Campo and Malgrange found the examples of singularities with maximal Jordan blocks in their vanishing cohomology, a strange algebraic theorem was discovered, whose proof required deep results from complex analysis.

Recall that the *integral closure* $\bar{I}$ of an ideal $I \subset S = \mathbb{C}\{x_0, x_1, \ldots, x_n\}$ consists of all functions h that satisfies an *integrality equation over* I: $h \in \bar{I}$ if and only if for some n there exist $a_k \in I^k$, $k = 1, 2, \ldots, n$ such that

$$h^n + a_1 h^{n-1} + \cdots + a_n = 0.$$

This ideal can be characterised in various other ways. For example, one has $f \in \bar{I}$ if and only if

$$\gamma^*(f) \in \gamma^* I$$

for each curve germ $\gamma : (\mathbb{C}, 0) \to (\mathbb{C}^{n+1}, 0)$ [Lipman and Teissier 1981].

Theorem [Skoda and Briançon 1974]. *If I is generated by k elements then*

$$\bar{I}^{\min(k,n+1)} \subset I.$$

This is a completely algebraic statement, but its proof was not. Lipman and Teissier [1981] wrote: "The absence of an algebraic proof has been for algebraists something like a scandal — perhaps even an insult — and certainly a challenge."

In any case, as $f \in \overline{J_f}$, it follows from this theorem that for any $f \in S$ one has

$$f^{n+1} \in J_f,$$

or equivalently, the operator

$$[f] : Q_f \to Q_f$$

induced by multiplication with f on the Jacobi ring has index of nilpotency bounded by $n+1$:

$$[f]^{n+1} = 0.$$

In [Skoda and Briançon 1974] it is also remarked that this estimate on the exponent is optimal. As an example, they give

$$f = (x_0 x_1 \dots x_n)^3 + z_0^{3n+2} + z_1^{3n+2} + \dots + z_n^{3n+2},$$

for which $f^n \notin J_f$.

So we see that to an isolated hypersurface singularity $f \in S$, one can associate two natural vector-spaces of dimension μ, each with a nilpotent endomorphism. On one hand, we have the topological space $H_f := H^n(X_\infty, \mathbb{C})$ with the endomorphism N, the monodromy logarithm. On the other hand, we have the purely algebraic Q_f with the endomorphism $[f]$. The monodromy theorem tells us that $N^{n+1} = 0$, while the theorem of Briançon–Skoda tells that $[f]^{n+1} = 0$. According to Scherk, it was Brieskorn who asked about a possible relation between the two appearances of $n+1$ in these theorems.

Conjecture 1 [Scherk 1978]. *For any isolated hypersurface singularity the following holds: If $f^{r+1} \in J_f$, then the Jordan normal form of the monodromy has blocks of size at most $(r+1)$.*

In case $r = 0$ the conjecture follows from the following two theorems.

Theorem [Saito 1971]. *If $f \in J_f$, one can find a coordinate system in which f is represented as a* quasihomogeneous polynomial.

Recall that a polynomial f is called *quasihomogeneous* if one can find positive rational weights $w_0, w_1, \dots, w_n$ such that

$$f(\lambda^{w_0} x_0, \lambda^{w_1} x_1, \dots, \lambda^{w_n} x_n) = f(x_0, x_1, \dots, x_n).$$

This is the case if and only if all monomials $x^a = x_0^{a_0} x_1^{a_1} \dots x_n^{a_n}$ appearing in f with nonzero coefficient lie in the hyperplane

$$w_0 a_0 + w_1 a_1 + \dots + w_n a_n = 1.$$

Theorem. *For a quasihomogeneous singularity with weights $w_0, w_1, \dots, w_n$, the cohomological monodromy is finite of order d, which is the least common multiple of the denominators of the w_i.*

This generalises the result of Pham on the Brieskorn–Pham singularities and can be found in [Milnor 1968, p. 71].

In the example of the $T_{p,q,r}$-singularities ($\frac{1}{p}+\frac{1}{q}+\frac{1}{r}<1$), given by

$$f(x,y,z)=x^p+y^q+z^r+xyz,$$

one has $f \notin J_f$, $f^2 \in J_f$ and indeed the monodromy has a single 2×2-block for the eigenvalue 1. In this way the conjecture may also be seen as a *refinement* of the usual monodromy theorem for isolated hypersurfaces. On the other hand, the converse of the statement is certainly not true. Scherk gives the example

$$f_a = y^6+x^4y+ax^5.$$

For $a=0$ the singularity is quasihomogeneous, so $f \in J_f$ and the monodromy is of finite order. For $a \neq 0$, the singularity is no longer quasihomogeneous and so we have $f \notin J_f$, but the monodromy is still of finite order, as the topology of the singularity does not depend on a. Similarly, one could take any quasihomogeneous singularity f and add a nontrivial term of quasihomogeneous degree > 1.

Scherk gave a proof of his conjecture [1980], using a globalisation of the Milnor fibre to a smooth projective hypersurface and using the resulting variation of Hodge structures. In that paper he also formulated a strengthening of his conjecture:

Conjecture 2 [Scherk 1980]. *For an isolated hypersurface singularity f and any integer k the following inequality takes place*:

$$\dim \operatorname{Ker}([f]^k : S/J_f \to S/J_f) \le \dim \ker(N^k : H^n(X_t) \to H^n(X_t)).$$

5. Period integrals and mixed Hodge structures

The second conjecture of Scherk was proven by Varchenko in [1981] as a consequence of a stronger theorem.

Theorem [Varchenko 1981]. *Consider an isolated hypersurface singularity $f \in S$. There exists a filtration $V^\bullet$ on S/J_f with the property that*

$$[f] : V^\alpha \mapsto V^{\alpha+1},$$

and such that

$$\{f\} := Gr_V^\bullet[f] : Gr_V^\bullet S/J_f \to Gr_V^{\bullet+1} S/J_f$$

and

$$N : H^n(X_\infty, \mathbb{C}) \to H^n(X_\infty, \mathbb{C})$$

have the same Jordan normal form.

As by going to an associated graded of a filtration kernels only can get bigger, one obtains:

$$\dim \ker[f]^k \leq \dim \ker N^k.$$

The construction of the filtration $V^\bullet$ is a bit involved. It lives naturally on the Gauss–Manin system $\mathcal{G}$ and the Brieskorn lattice

$$\mathcal{H} = \Omega^{n+1}/df \wedge d\Omega^{n-1},$$

and induces a filtration on the quotient

$$Q^f = \Omega^{n+1}/df \wedge \Omega^n = \mathcal{H}/\partial^{-1}\mathcal{H}.$$

For a differential form $\omega \in \Omega^{n+1}$ the $V^\bullet$-filtration reflects the asymptotic behaviour of the period integrals

$$\Phi(t) = \int_{\delta(t)} \operatorname{Res}\left(\frac{\omega}{f-t}\right) = \sum_{\alpha,k} A_{\alpha,k} t^\alpha log(t)^k.$$

The element ω belongs to $V^\beta\mathcal{H}$, if for all $\delta(t)$ the coefficients in the above expansion vanish for $\alpha < \beta$.

Varchenko [1980] derives the theorem from his construction of an *asymptotic mixed Hodge structure* on the vanishing cohomology $H^n(X_\infty, \mathbb{Z})$.

Recall that a mixed Hodge structure on a finite rank abelian group H is a linear algebra object that consist of two filtrations, to know an increasing *weight filtration* $W_\bullet$, defined on $H_\mathbb{Q} := H \otimes \mathbb{Q}$, and a decreasing *Hodge filtration* $F^\bullet$ defined on $H_\mathbb{C} := H \otimes \mathbb{C}$, such that $F^\bullet$ induces on the graded pieces $Gr_k^W H = W_k/W_{k-1}$ a pure Hodge structure of weight k. We refer to [Peters and Steenbrink 2008] for a more systematic account of mixed Hodge theory.

Steenbrink [1975/76; 1977] had first constructed such a mixed Hodge structure, using an embedded resolution of f. The weight-filtration is constructed using the nilpotent operator N: it is the unique increasing filtration $W_\bullet$

$$0 \subset W_0 \subset W_1 \subset W_2 \ldots \subset W_{2n-1} \subset W_{2n} = H^n(X_\infty, \mathbb{Q}),$$

such that

$$N : W_k \to W_{k-2},$$

with the property that the operator N^k induces an isomorphism from $\mathrm{Gr}_{n+k}^W H$ to $\mathrm{Gr}_{n-k}^W H$:

$$N^k : \mathrm{Gr}_{n+k}^W H \xrightarrow{\approx} \mathrm{Gr}_{n-k}^W H.$$

As this filtration is uniquely defined by the cohomological monodromy operator, it is called the *monodromy weight filtration.*

In the asymptotic mixed Hodge structure of Varchenko, the Hodge filtration $F^\bullet$ is related to the $V^\bullet$-filtration and encodes the asymptotic behaviour of the period integrals when t approaches the origin radially. So we have the nice picture that the two filtrations, in a way, arise from the decomposition in *angular* and *radial* components as the parameter $t \to 0$.

We now describe, following [Scherk and Steenbrink 1985], the construction of the asymptotic mixed Hodge structure in more detail.

The generalised eigenspaces C^α. The Gauss–Manin system $\mathcal{G}$ has the structure of a (finitely generated) regular singular $\mathbb{C}\{t\}[\partial]$-module. One defines the generalised α-eigenspace by

$$C^\alpha := \bigcup_{k>0} \ker(t\partial - \alpha)^k \subset \mathcal{G}.$$

These are finite-dimensional $\mathbb{C}$-vector spaces. Note that

$$(t\partial_t - \alpha)^{k-1}(t^\alpha \log^k t) = 0,$$

so that C^α picks out those elements of $\mathcal{G}$ that "behave like" the function $t^\alpha \log^k t$ for some k.

The structure of $\mathcal{G}$ can be schematically visualised as follows:

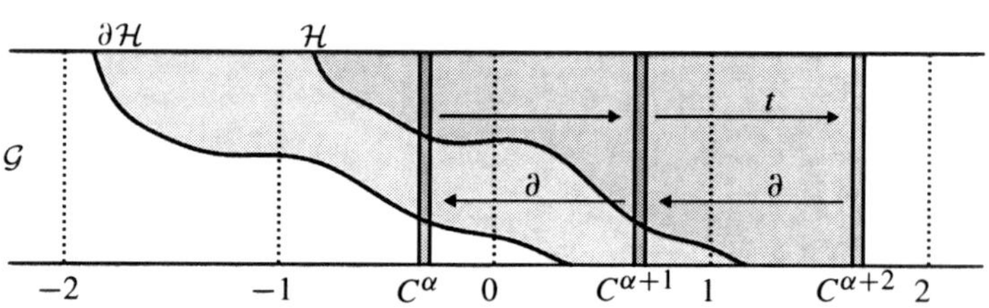

Horizontally runs the eigenvalue parameter α. The vertical bars represent the generalised eigenspaces C^α. Multiplication by t maps C^α to $C^{\alpha+1}$, whereas ∂ maps $C^{\alpha+1}$ back to C^α. The operators $t\partial - \alpha$ act "vertically" and are nilpotent on C^α. One has an isomorphism

$$H^n(X_\infty, \mathbb{C}) = \bigoplus_{-1<\alpha\leq 0} C^\alpha.$$

It follows from the regularity of the Gauss–Manin connection that the generalised $\exp(2\pi i\alpha)$-eigenspace $H^n(X_\infty, \mathbb{C})_\alpha$ of the monodromy T is isomorphic to the space C^α and the monodromy logarithm N identifies, up to a factor $2\pi i$, with the operator $t\partial - \alpha$.

$$\begin{array}{ccc} H^n(X_\infty, \mathbb{C})_\alpha & \xrightarrow{\approx} & C^\alpha \\ {\scriptstyle N}\big\downarrow & & \big\downarrow{\scriptstyle 2\pi i\,(t\partial-\alpha)} \\ H^n(X_\infty, \mathbb{C})_\alpha & \xrightarrow{\approx} & C^\alpha \end{array}$$

The position of the Brieskorn lattice $\mathcal{H}$ inside $\mathcal{G}$ contains important information and is indicated in the picture as the region to right of the wiggly curve. Note that $\mathcal{H} \subset V^{>-1}$, by the result of Malgrange. The action of ∂ moves $\mathcal{H}$ to the left.

The $V^\bullet$-filtration. The $V^\bullet$-filtration of $\mathcal{G}$ is defined as the $\mathbb{C}\{t\}$-span of the C^β with $\beta \geq \alpha$

$$V^\alpha \mathcal{G} := \langle C^\beta \mid \beta \geq \alpha \rangle,$$

and we have

$$C^\alpha \approx V^\alpha / V^{>\alpha}.$$

As $\mathcal{H} \subset \mathcal{G}$ we obtain by intersection a $V^\bullet$-filtration on the Brieskorn lattice $\mathcal{H}$.

On the quotient

$$Q^f = \Omega^{n+1}/df \wedge \Omega^n = \mathcal{H}/\partial^{-1}\mathcal{H},$$

one has a natural induced filtration by setting

$$V^\alpha Q^f := (V^\alpha \mathcal{H} + \partial^{-1}\mathcal{H})/\partial^{-1}\mathcal{H}.$$

For an important class of singularities the $V^\bullet$-filtration can be computed easily:

Theorem [Saito 1988]. *For a Newton nondegenerate f the $V^\bullet$-filtration on Q^f coincides with the Newton filtration $\mathcal{N}^\bullet$, shifted by one*:

$$V^\alpha Q^f = \mathcal{N}^{\alpha-1} Q^f.$$

Hodge filtration on $H^n(X_\infty)$. By applying the operator ∂ to $\mathcal{H} \subset \mathcal{G}$, we obtain a "Hodge filtration" on $\mathcal{G}$:

$$\mathcal{H} \subset \partial\mathcal{H} \subset \partial^2\mathcal{H} \subset \cdots \subset \mathcal{G}.$$

Using this, we define a filtration $F^\bullet$ on C^α by setting

$$F^p C^\alpha := (\partial^{n-p}\mathcal{H} \cap V^\alpha + V^{>\alpha})/V^{>\alpha} \subset C^\alpha$$

and

$$F^p H^n(X_\infty, \mathbb{C}) := \bigoplus_{-1<\alpha\leq 0} F^p C^\alpha.$$

Unwinding the definitions, one finds that the spaces $\mathrm{Gr}_F^p C^\alpha$ can be identified with certain $V^\bullet$-graded piece of Q^f:

$$\partial^{n-p} : \mathrm{Gr}_V^{\alpha+n-p} Q^f \overset{\approx}{\to} \mathrm{Gr}_F^p C^\alpha.$$

The main theorem on asymptotic mixed Hodge theory is the following.

Theorem. *The space $H^n(X_\infty)$, together with the monodromy weight-filtration $W_\bullet$ and the above defined Hodge filtration $F^\bullet$ define a* mixed Hodge structure, *isomorphic to the limiting mixed Hodge structure defined in* [Steenbrink 1977].

This theorem, in a slightly different form, was first proven by Varchenko [1980; 1982]. We basically followed here the presentation of [Scherk and Steenbrink 1985].

Although all the ingredients of the mixed Hodge structure can be defined locally, the proofs of the required Hodge properties use globalisation to a projective hypersurface in an essential way; apparently no purely local proof is known.

5.1. ***Varchenko's theorem.*** A feature of mixed Hodge theory is that all morphisms of mixed Hodge structures are *strictly compatible* with weight and Hodge filtration: going from a morphism $H \to H'$ of mixed Hodge structures to maps between the associated graded pieces, such as

$$\mathrm{Gr}_F^p \mathrm{Gr}_k^W H \to \mathrm{Gr}_F^p \mathrm{Gr}_k^W H',$$

preserves exactness properties. In our situation there is one particular interesting morphism of mixed Hodge structures, namely the morphism

$$N : H^n(X_\infty, \mathbb{Q}) \to H^n(X_\infty, \mathbb{Q}).$$

As by construction $N : W_k \to W_{k-2}$ and $N : F^p \to F^{p-1}$, N is a morphism of type $(-1, -1)$.

One now can argue as follows:

(1) From the strictness, the Jordan structure of N on $H := H^n(X_\infty, \mathbb{C})$ is the same as that of

$$\mathrm{Gr}_F N : \mathrm{Gr}_F^\bullet H \to \mathrm{Gr}_F^{\bullet-1} H.$$

(2) On the component C^α the map

$$\mathrm{Gr}_F N : \mathrm{Gr}_F^p C^\alpha \to \mathrm{Gr}_F^{p-1} C^\alpha.$$

is represented by

$$2\pi i(t\partial - \alpha) = 2\pi i t\partial \mod F^p.$$

(3) Identifying the Hodge spaces $\mathrm{Gr}_F^p C^\alpha$ with pieces of the $V^\bullet$-filtration on Q^f

we obtain a diagram

$$\begin{array}{ccc} \mathrm{Gr}_V^{\alpha+n-p}\, Q^f & \xrightarrow{\{f\}} & \mathrm{Gr}_V^{\alpha+n-p+1}\, Q^f \\ \Big\downarrow{\scriptstyle \partial^{n-p}} & & \Big\downarrow{\scriptstyle \partial^{n-p+1}} \\ \mathrm{Gr}_F^{p}\, C^\alpha & \xrightarrow{\mathrm{Gr}_F N} & \mathrm{Gr}_F^{p-1}\, C^\alpha \end{array}$$

that is commutative up to a factor $2\pi i$.

Corollary 1. *The operator $\{f\}$ on $\mathrm{Gr}_V^\bullet\, Q^f$ and N on $H^n(X_\infty, \mathbb{C})$ have the same Jordan type.*

Example. We analyse the example of A'Campo in terms of the $V^\bullet$-filtration. As the function is Newton nondegenerate, we can use the theorem of M. Saito to identify the V-filtration with the Newton filtration (shifted by one). A basis for Q^f is given by the 11 differential forms

$$dx\,dy,\ xy\,dx\,dy,\ x^2y^2dx\,dy,$$
$$x\,dx\,dy,\ x^2dx\,dy,\ x^3dx\,dy,\ x^4dx\,dy, \quad y\,dx\,dy,\ y^2dx\,dy,\ y^3dx\,dy,\ y^4dx\,dy$$

The Newton weights of these monomials can be read off from the Newton diagram as

$$\tfrac{1}{2},\ 1,\ \tfrac{3}{2},$$
$$\tfrac{7}{10},\ \tfrac{9}{10},\ \tfrac{11}{10},\ \tfrac{13}{10}, \quad \tfrac{7}{10},\ \tfrac{9}{10},\ \tfrac{11}{10},\ \tfrac{13}{10}.$$

The fractions appearing here (or diminished by 1) are called the *spectral numbers* of the singularity.

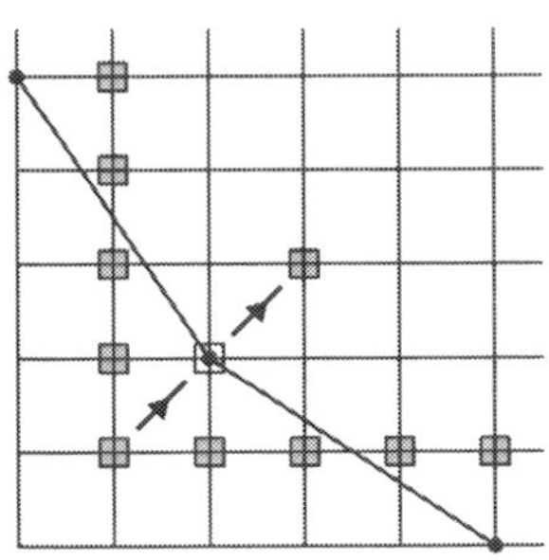

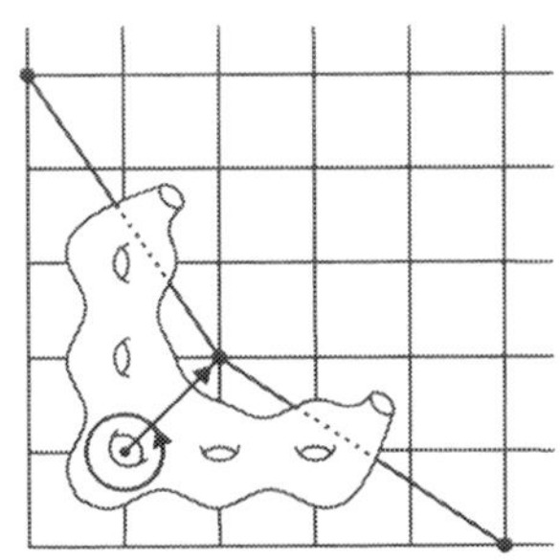

Multiplication of the monomial $dx\,dy$ of weight $\frac{1}{2}$ by f maps to the monomial $x^2y^2dx\,dy$ of weight $\frac{3}{2}$, which thus represents a nontrivial Jordan block N of the monodromy.

The picture on the right shows the Milnor fibre of f, which was described earlier and seen to be a genus 5 Riemann surface with two holes. We drew the surface around the monomials of the Newton diagram, with holes piercing through the edges of the Newton diagram. In a way that is a bit hard to explain in a precise way, one can see that the nontrivial Jordan block "hits" the cycle γ on the Riemann surface that appeared in A'Campo's example!

This concludes our account of a unique key period in the theory of isolated hypersurface singularities. Many important developments arose out of them, e.g., M. Saito's theory of *mixed Hodge modules* and applications to log-canonical thresholds, multiplier ideals, jumping coefficients, etc. For these more recent developments we refer to [Peters and Steenbrink 2008; Blickle and Lazarsfeld 2004; Ein et al. 2004; Mustaţă 2012].

Acknowledgement

I thank the organisers of the special program on commutative algebra at MSRI for giving me the opportunity (October 2012) to give a talk on the subject of this paper. Further thanks to Norbert A'Campo, John Scherk, Joseph Steenbrink and Bernard Teissier for useful exchanges on the topics of this paper.

References

[A'Campo 1973] N. A'Campo, "Sur la monodromie des singularités isolées d'hypersurfaces complexes", *Invent. Math.* **20** (1973), 147–169.

[Arnold 1975] V. I. Arnold, "Critical points of smooth functions", pp. 19–39 in *Proceedings of the International Congress of Mathematicians* (Vancouver, 1974), vol. 1, Canad. Math. Congress, Montreal, Quebec, 1975.

[Blickle and Lazarsfeld 2004] M. Blickle and R. Lazarsfeld, "An informal introduction to multiplier ideals", pp. 87–114 in *Trends in commutative algebra*, edited by L. L. Avramov et al., Math. Sci. Res. Inst. Publ. **51**, Cambridge Univ. Press, Cambridge, 2004.

[Brieskorn 1966a] E. V. Brieskorn, "Examples of singular normal complex spaces which are topological manifolds", *Proc. Nat. Acad. Sci. U.S.A.* **55** (1966), 1395–1397.

[Brieskorn 1966b] E. Brieskorn, "Beispiele zur Differentialtopologie von Singularitäten", *Invent. Math.* **2** (1966), 1–14.

[Brieskorn 1970] E. Brieskorn, "Die Monodromie der isolierten Singularitäten von Hyperflächen", *Manuscripta Math.* **2** (1970), 103–161.

[Brieskorn and Knörrer 1981] E. Brieskorn and H. Knörrer, *Ebene algebraische Kurven*, Birkhäuser, Basel, 1981.

[Clemens 1969] C. H. Clemens, Jr., "Picard–Lefschetz theorem for families of nonsingular algebraic varieties acquiring ordinary singularities", *Trans. Amer. Math. Soc.* **136** (1969), 93–108.

[Deligne and Katz 1973] P. Deligne and N. Katz, *Groupes de monodromie en géométrie algébrique*, Lecture Notes in Mathematics **340**, Springer, Berlin, 1973.

[Ein et al. 2004] L. Ein, R. Lazarsfeld, K. E. Smith, and D. Varolin, "Jumping coefficients of multiplier ideals", *Duke Math. J.* **123**:3 (2004), 469–506.

[Griffiths 1970] P. A. Griffiths, "Periods of integrals on algebraic manifolds: Summary of main results and discussion of open problems", *Bull. Amer. Math. Soc.* **76** (1970), 228–296.

[Hirzebruch 1964] F. Hirzebruch, "The topology of normal singularities of an algebraic surface", exposé no. 250, 129–137 in *Séminare Bourbaki,* 1962/1963, vol. 8, Société Mathématique de France, 1964.

[Katz 1970] N. M. Katz, "Nilpotent connections and the monodromy theorem: Applications of a result of Turrittin", *Inst. Hautes Études Sci. Publ. Math.* 39 (1970), 175–232.

[Katz 1971] N. M. Katz, "The regularity theorem in algebraic geometry", pp. 437–443 in *Actes du Congrès International des Mathématiciens* (Nice, 1970), vol. I, Gauthier-Villars, Paris, 1971.

[Lamotke 1981] K. Lamotke, "The topology of complex projective varieties after S. Lefschetz", *Topology* **20**:1 (1981), 15–51.

[Landman 1966] A. Landman, *On the Picard–Lefschetz transformation*, Ph.D. thesis, University of California, Berkeley, 1966.

[Landman 1973] A. Landman, "On the Picard–Lefschetz transformation for algebraic manifolds acquiring general singularities", *Trans. Amer. Math. Soc.* **181** (1973), 89–126.

[Lê 1974] Lê Dũng Tráng, "Finitude de la monodromie locale des courbes planes", pp. 341–346 in *Fonctions de plusieurs variables complexes*, edited by F. Norguet, Lecture Notes in Math. **409**, Springer, Berlin, 1974.

[Lefschetz 1924] S. Lefschetz, *L'analysis situs et la géométrie algébrique*, Gauthier-Villars, Paris, 1924.

[Lipman and Teissier 1981] J. Lipman and B. Teissier, "Pseudorational local rings and a theorem of Briançon–Skoda about integral closures of ideals", *Michigan Math. J.* **28**:1 (1981), 97–116.

[Looijenga 1984] E. J. N. Looijenga, *Isolated singular points on complete intersections*, London Mathematical Society Lecture Note Series **77**, Cambridge University Press, Cambridge, 1984.

[Malgrange 1973] B. Malgrange, "Letter to the editors", *Invent. Math.* **20** (1973), 171–172.

[Malgrange 1974] B. Malgrange, "Intégrales asymptotiques et monodromie", *Ann. Sci. École Norm. Sup.* (4) **7**:3 (1974), 405–430.

[Milnor 1968] J. Milnor, *Singular points of complex hypersurfaces*, Annals of Mathematics Studies **61**, Princeton University Press, 1968.

[Mustaţă 2012] M. Mustaţă, "IMPANGA lecture notes on log canonical thresholds", pp. 407–442 in *Contributions to algebraic geometry*, edited by P. Pragacz, Eur. Math. Soc., Zürich, 2012.

[Peters and Steenbrink 2008] C. A. M. Peters and J. H. M. Steenbrink, *Mixed Hodge structures*, Ergebnisse der Math. und ihrer Grenzgebiete (3) **52**, Springer, Berlin, 2008.

[Pham 1965] F. Pham, "Formules de Picard–Lefschetz généralisées et ramification des intégrales", *Bull. Soc. Math. France* **93** (1965), 333–367.

[Pham 1979] F. Pham, *Singularités des systèmes différentiels de Gauss–Manin*, Progress in Mathematics **2**, Birkhäuser, Boston, 1979.

[Picard and Simart 1897] É. Picard and G. Simart, *Théorie des fonctions algébriques de deux variables indépendantes*, vol. I, Gauthier-Villars, Paris, 1897.

[Saito 1971] K. Saito, "Quasihomogene isolierte Singularitäten von Hyperflächen", *Invent. Math.* **14** (1971), 123–142.

[Saito 1988] M. Saito, "Exponents and Newton polyhedra of isolated hypersurface singularities", *Math. Ann.* **281**:3 (1988), 411–417.

[Scherk 1978] J. Scherk, "On the Gauss–Manin connection of an isolated hypersurface singularity", *Math. Ann.* **238**:1 (1978), 23–32.

[Scherk 1980] J. Scherk, "On the monodromy theorem for isolated hypersurface singularities", *Invent. Math.* **58**:3 (1980), 289–301.

[Scherk and Steenbrink 1985] J. Scherk and J. H. M. Steenbrink, "On the mixed Hodge structure on the cohomology of the Milnor fibre", *Math. Ann.* **271**:4 (1985), 641–665.

[Schulze 2003] M. Schulze, "Monodromy of hypersurface singularities", *Acta Appl. Math.* **75**:1-3 (2003), 3–13.

[Sebastiani 1970] M. Sebastiani, "Preuve d'une conjecture de Brieskorn", *Manuscripta Math.* **2** (1970), 301–308.

[Skoda and Briançon 1974] H. Skoda and J. Briançon, "Sur la clôture intégrale d'un idéal de germes de fonctions holomorphes en un point de $\mathbf{C}^n$", *C. R. Acad. Sci. Paris Sér. A* **278** (1974), 949–951.

[Steenbrink 1975/76] J. Steenbrink, "Limits of Hodge structures", *Invent. Math.* **31**:3 (1975/76), 229–257.

[Steenbrink 1977] J. H. M. Steenbrink, "Mixed Hodge structure on the vanishing cohomology", pp. 525–563 in *Real and complex singularities* (Oslo, 1976), Sijthoff and Noordhoff, Alphen aan den Rijn, 1977.

[Varchenko 1980] A. Varchenko, "The asymptotics of holomorphic forms determine a mixed Hodge structure", *Sov. Math. Dokl.* **22**:5 (1980), 772–775.

[Varchenko 1981] A. N. Varčenko, "On the monodromy operator in vanishing cohomology and the multiplication operator by f in the local ring", *Dokl. Akad. Nauk SSSR* **260**:2 (1981), 272–276. In Russian: translated in *Soviet Math. Dokl.*, **24** (1981), no. 2, 248–252 (1982).

[Varchenko 1982] A. Varchenko, "Asymptotic Hodge structure in the vanishing cohomology", *Math. USSR. Izv.* **18** (1982), 469–512.

[Vassiliev 2002] V. A. Vassiliev, *Applied Picard–Lefschetz theory*, Mathematical Surveys and Monographs **97**, American Mathematical Society, Providence, RI, 2002.

[Wall 1971] C. T. C. Wall, "Lectures on C^∞-stability and classification", pp. 178–206 in *Proceedings of Singularities Symposium* (Liverpool 1969/70), vol. I, Lecture Notes in Mathematics **192**, Springer, Berlin, 1971.

straten@mathematik.uni-mainz.de *Institut für Mathematik, Johannes Gutenberg University, D-55099 Mainz, Germany*

Commutative Algebra and Noncommutative Algebraic Geometry, I
MSRI Publications
Volume **67**, 2015

The interplay of algebra and geometry in the setting of regular algebras

MICHAELA VANCLIFF

This article aims to motivate and describe the geometric techniques introduced by M. Artin, J. Tate and M. Van den Bergh in the 1980s at a level accessible to graduate students. Additionally, some advances in the subject since the early 1990s are discussed, including a recent generalization of complete intersection to the noncommutative setting, and the notion of graded skew Clifford algebra and its application to classifying quadratic regular algebras of global dimension at most three. The article concludes by listing some open problems.

Introduction

Many noncommutative algebraists in the 1980s were aware of the successful marriage of algebra and algebraic geometry in the commutative setting and wished to duplicate that relationship in the noncommutative setting. One such line of study was the search for a subclass of noncommutative algebras that "behave" enough like polynomial rings that a geometric theory could be developed for them. One proposal for such a class of algebras are the *regular* algebras, introduced in [Artin and Schelter 1987], that were investigated using new geometric techniques in the pivotal papers of M. Artin, J. Tate and M. Van den Bergh [Artin et al. 1990; 1991].

About the same time, advances in quantum mechanics in the 20th century had produced many new noncommutative algebras on which traditional techniques had only yielded limited success, so a need had arisen to find new techniques to study such algebras (see [Reshetikhin et al. 1989; Kapustin et al. 2001; Sklyanin 1982; 1985; Sudbery 1993]). One such algebra was the Sklyanin algebra, which had emerged from the study of quantum statistical mechanics [Sklyanin 1982; 1985]. By the early 1990s, T. Levasseur, S. P. Smith, J. T. Stafford and others

Vancliff was supported in part by NSF grants DMS-0900239 and DMS-1302050 and by general membership at MSRI.
MSC2010: 16S38, 14A22, 16E65, 16S37, 14M10.
Keywords: regular algebra, Clifford algebra, skew polynomial ring, base point, point module, quadric, complete intersection.

had solved the ten-year old open problem of completely classifying all the finite-dimensional irreducible representations (simple modules) over the Sklyanin algebra, and their methods were the geometric techniques developed by Artin, Tate and Van den Bergh [Levasseur and Smith 1993; Smith and Stafford 1992; Smith and Staniszkis 1993].

Concurrent with the above developments, another approach was considered via differential geometry and deformation theory to study the algebras produced by quantum physics. That approach is the study of certain noncommutative algebras via Poisson geometry (see [Drinfeld 1987]). At the heart of both approaches are homological and categorical techniques, so it is perhaps no surprise that the two approaches have much overlap; often, certain geometric objects from one approach are in one-to-one correspondence with various geometric objects from the other approach (depending on the algebra being studied — cf. [Vancliff 1995; 1999; 2000]). A survey of recent advances in Poisson geometry may be found in [Goodearl 2010].

Given the above developments, the early 1990s welcomed a new era in the field of noncommutative algebra in which geometric techniques took center stage. Since that time, the subject has spawned many new ideas and directions, as demonstrated by the MSRI programs in 2000 and 2013.

This article is based on a talk given by the author in the *Connections for Women* workshop held at MSRI in January 2013 and it has two objectives. The first is to motivate and describe the geometric techniques of Artin, Tate and Van den Bergh at a level accessible to graduate students, and the second is to discuss some developments towards the attempted classification of quadratic regular algebras of global dimension four, while listing open problems. An outline of the article is as follows.

Section 1 concerns the motivation and development of the subject, with emphasis on quadratic regular algebras of global dimension four. Section 2 discusses constructions of certain types of quadratic regular algebras of arbitrary finite global dimension, with focus on graded Clifford algebras and graded skew Clifford algebras. This section also discusses a new type of symmetry for square matrices called μ-symmetry. We conclude this section by revisiting the classification of quadratic regular algebras of global dimension at most three, since almost all such algebras may be formed from regular graded skew Clifford algebras. In Section 3, we discuss geometric techniques that apply to graded Clifford algebras and graded skew Clifford algebras in order to determine when those algebras are regular. This section also considers the issue of complete intersection in the noncommutative setting. We conclude with Section 4 which lists some open problems and related topics.

Although the main objects of study from [Artin et al. 1990; 1991] are discussed in this article, several topics from the same are omitted; for surveys of those topics, the reader is referred to [Stafford 2002; Stafford and van den Bergh 2001] and to D. Rogalski's lecture notes [2014] from the graduate workshop "Noncommutative Algebraic Geometry" at MSRI in June 2012.

1. The geometric objects

In this section, we discuss the motivation and development of the subject, with emphasis on quadratic regular algebras of global dimension at most four.

Throughout this section, $\Bbbk$ denotes an algebraically closed field and, for any graded algebra B, the span of the homogeneous elements of degree i will be denoted by B_i.

1A. *Motivation.* Consider the $\Bbbk$-algebra, S, on generators $z_1, \ldots, z_n$ with defining relations

$$z_j z_i = \mu_{ij} z_i z_j, \qquad \text{for all distinct } i,\ j,$$

where $0 \neq \mu_{ij} \in \Bbbk$ for all i, j, and $\mu_{ij}\mu_{ji} = 1$ for all distinct i, j. If $\mu_{ij} = 1$, for all i, j, then S is the commutative polynomial ring and has a rich subject of algebraic geometry associated with it; in particular, by the (projective) Nullstellensatz, the points of $\mathbb{P}(S_1^*)$ are in one-to-one correspondence with certain ideals of S via $(\alpha_1, \ldots, \alpha_n) \leftrightarrow \langle \alpha_i z_1 - \alpha_1 z_i, \ldots, \alpha_i z_n - \alpha_n z_i \rangle$, where $\alpha_i \neq 0$. Before continuing, we first observe that for such an ideal I, the graded module S/I has the property that its Hilbert series is $H(t) = 1/(1-t)$ and that S/I is a 1-critical (with respect to GK-dimension) graded cyclic module over S.

However, if $\mu_{ij} \neq 1$ for any i, j, then S still "feels" close to commutative, and one would expect there to be a way to relate algebraic geometry to it. The geometric objects in [Artin et al. 1990] are modeled on the module S/I above; instead of using actual points or lines etc, certain graded modules are used as follows.

1B. *Points, lines, etc.*

Definition 1.1. [Artin et al. 1990] Let $A = \bigoplus_{i=0}^{\infty} A_i$ denote an $\mathbb{N}$-graded, connected (meaning $A_0 = \Bbbk$) $\Bbbk$-algebra generated by A_1 where $\dim(A_1) = n < \infty$. A graded right A-module $M = \bigoplus_{i=0}^{\infty} M_i$ is called a *right point module* (respectively, *line module*) if

(a) M is cyclic with $M = M_0 A$, and

(b) $\dim_{\Bbbk}(M_i) = 1$ for all i (respectively, $\dim_{\Bbbk}(M_i) = i + 1$) for all i.

If A is the polynomial ring S, then the module S/I from Section 1A is a point module. In general, one may associate some geometry to point and line modules as follows. Condition (a) implies that A maps onto M via $a \mapsto ma$, for all $a \in A$, where $\{m\}$ is a $\Bbbk$-basis for M_0, and this map restricts via the grading to a linear map $\theta : A_1 \to M_1$. Let $K \subset A_1$ denote the kernel of θ. Condition (b) implies that $\dim_{\Bbbk}(K) = n-1$ (respectively, $n-2$), so that $K^{\perp} \subset A_1{}^*$ has dimension one (respectively, two). Thus, $\mathbb{P}(K^{\perp})$ is a point (respectively, a line) in the geometric space $\mathbb{P}(A_1{}^*)$.

The Hilbert series of a point module is $H(t) = 1/(1-t)$, whereas the Hilbert series of a line module is $1/(1-t)^2$. Hence, a plane module is defined as in Definition 1.1 but condition (b) is replaced by the requirement that the module have Hilbert series $1/(1-t)^3$ (see [Artin et al. 1990]). Similarly, one may define d-linear modules, where the definition is modeled on Definition 1.1, but the module has Hilbert series $1/(1-t)^{d+1}$ (see [Shelton and Vancliff 2002]).

For many algebras, d-linear modules are $(d+1)$-critical with respect to GK-dimension. This leads to the following generalization of a point module.

Definition 1.2 [Cassidy and Vancliff 2014]. With A as in Definition 1.1, we define a *right base-point module* over A to be a graded 1-critical (with respect to GK-dimension) right A-module M such that $M = \bigoplus_{i=0}^{\infty} M_i = M_0 A$ and M has Hilbert series $H_M(t) = c/(1-t)$ for some $c \in \mathbb{N}$.

If $c = 1$ in Definition 1.2, then the module is a point module; whereas if $c \geq 2$, then the module is called a fat point module [Artin 1992]. The only base-point modules over the polynomial ring are point modules. On the other hand, in general, the algebra S from Section 1A can have fat point modules, so fat point modules are viewed as generalizations of points, and this is made more precise in [Artin 1992].

Artin, Tate and Van den Bergh [1990] proved that, under certain conditions, the point modules are parametrized by a scheme; that is, there is a scheme that represents the functor of point modules. Later, in [Vancliff and Van Rompay 1997], this scheme was called the *point scheme*. A decade later, in [Shelton and Vancliff 2002], it was proved by B. Shelton and the author that (under certain conditions) d-linear modules are parametrized by a scheme; that is, there is a scheme that represents the functor of d-linear modules. If $d = 0$, then this scheme is isomorphic to the point scheme; if $d = 1$, the scheme is called the *line scheme*.

By factoring out a nonzero graded submodule from a point module, one obtains a truncated point module as follows.

Definition 1.3 [Artin et al. 1990]. With A as in Definition 1.1, we define a *truncated right point module of length m* to be a graded right A-module $M = \bigoplus_{i=0}^{m-1} M_i$ such that M is cyclic, $M = M_0 A$ and $\dim_{\Bbbk}(M_i) = 1$ for all $i = 0, \ldots, m-1$.

For many quadratic algebras A, there exists a one-to-one correspondence between the truncated point modules over A of length three and the point modules over A. Moreover, if the algebra A in Definition 1.3 is quadratic, then the truncated point modules of length three are in one-to-one correspondence with the zero locus in $\mathbb{P}(A_1^*) \times \mathbb{P}(A_1^*)$ of the defining relations of A. To see this, we fix a $\Bbbk$-basis $\{x_1, \ldots, x_n\}$ for A_1, and use T to denote the free $\Bbbk$-algebra on $x_1, \ldots, x_n$, and let $Z \subset \mathbb{P}(A_1^*) \times \mathbb{P}(A_1^*)$ denote the zero locus of the defining relations of A. Viewing each x_i as the i-th coordinate function on A_1^*, let $p = (\alpha_i) \in \mathbb{P}(A_1^*)$ and $r = (\beta_i) \in \mathbb{P}(A_1^*)$, where $\alpha_i, \beta_i \in \Bbbk$ for all $i = 1, \ldots, n$. Let $M = \Bbbk v_0 \oplus \Bbbk v_1 \oplus \Bbbk v_2$ denote a three-dimensional vector space that is a T-module via the action determined by

$$v_0 x_i = \alpha_i v_1, \quad v_1 x_i = \beta_i v_2, \quad v_2 x_i = 0,$$

for all i. It follows that M is a truncated point module over T of length three. If $g \in T_2$, then $v_1 g = 0 = v_2 g$ and $v_0 g = g(p, r) v_2$. In particular, if $f \in T_2$ is a defining relation of A, then $Mf = 0$ if and only if $f(p, r) = 0$. Hence, M is an A-module if and only if $(p, r) \in Z$. This one-to-one correspondence between Z and truncated point modules of length three also exists at the level of schemes; the reason being that the scheme Z represents the functor of truncated point modules of length three. The method of proof of this is to repeat the preceding argument for a truncated point module of length three over $R \otimes_{\Bbbk} T$ and $R \otimes_{\Bbbk} A$, where R is a commutative $\Bbbk$-algebra, together with localization techniques; for details the reader is referred to [Artin et al. 1990, Proposition 3.9], its proof, and the paragraph preceding that result. This correspondence will be revisited in Section 1D.

For completeness, we finish this subsection with some technical definitions that play minor roles throughout the text. The reader is referred to [Levasseur 1992; Levasseur and Smith 1993] for details and for results concerning algebras satisfying these definitions.

Definition 1.4 [Levasseur 1992, Definition 2.1]. A noetherian ring B is called *Auslander-regular* (respectively, *Auslander–Gorenstein*) if

(a) the global homological dimension (respectively, (left and right) injective dimension) of B is finite, and

(b) every finitely generated B-module M satisfies the *Auslander condition*, namely, for every $i \geq 0$ and for every B-submodule N of $\operatorname{Ext}_B^i(M, B)$, we have $j(N) \geq i$, where $j(N) = \inf\{\ell : \operatorname{Ext}_B^\ell(N, B) \neq 0\}$.

Definition 1.5 [Levasseur 1992, Definition 5.8]. A noetherian $\Bbbk$-algebra B of integral GK-dimension n satisfies the *Cohen–Macaulay property* if $\operatorname{GKdim}(M) + j(M) = n$ for all nonzero finitely generated B-modules M.

1C. *Regular algebras.* The goal of [Artin et al. 1990] was to classify, in a user-friendly way, the generic regular algebras of global dimension three that were first analyzed in [Artin and Schelter 1987]. Using geometric techniques developed for the purpose, those algebras were shown in [Artin et al. 1990] to be noetherian. Regular algebras are often viewed as noncommutative analogues of polynomial rings and are defined as follows.

Definition 1.6 [Artin and Schelter 1987]. A finitely generated, $\mathbb{N}$-graded, connected $\Bbbk$-algebra $A = \bigoplus_{i=0}^{\infty} A_i$, generated by A_1, is *regular* (or *AS-regular*) of global dimension r if

(a) it has global homological dimension $r < \infty$, and

(b) it has polynomial growth (i.e., there exist positive real numbers c and δ such that $\dim_{\Bbbk}(A_i) \le c i^{\delta}$ for all i), and

(c) it satisfies the *Gorenstein condition*, namely, a minimal projective resolution of the left trivial module ${}_A\Bbbk$ consists of finitely generated modules and dualizing this resolution yields a minimal projective resolution of the right trivial module $\Bbbk_A[e]$, shifted by some degree e.

Although all three conditions in Definition 1.6 are satisfied by the polynomial ring, the main reason a regular algebra is viewed as a noncommutative analogue of a polynomial ring is due to condition (c), since it imposes a symmetry condition on the algebra that replaces the symmetry condition of commutativity. The reader should note that, in the literature, (c) is sometimes replaced by an equivalent condition that makes the symmetry property less obvious; namely, $\mathrm{Ext}_A^i({}_A\Bbbk, A) \cong \delta_r^i \, \Bbbk_A[e]$, where δ_r^i is the Kronecker delta. An $\mathbb{N}$-graded connected $\Bbbk$-algebra that is generated by degree-1 elements and which is Auslander-regular with polynomial growth is AS-regular [Levasseur 1992]. For a notion of regular algebra where the algebra is not generated by degree-1 elements; see [Cassidy 1999; 2003; Stephenson 1996; 1997; 2000a; 2000b].

Examples 1.7. (a) The algebra S from Section 1A is regular.

(b) If $\Bbbk = \mathbb{C}$, then many algebras from physics are regular. In particular, homogenizations of universal enveloping algebras of finite-dimensional Lie algebras, the coordinate ring of quantum affine n-space, the coordinate ring of quantum $m \times n$ matrices, and the coordinate ring of quantum symplectic n-space are all regular [Le Bruyn and Smith 1993; Le Bruyn and Van den Bergh 1993; Levasseur and Stafford 1993].

(c) If the global dimension of a regular algebra is one, then the algebra is the polynomial ring on one variable. However, by [Artin and Schelter 1987], if the global dimension is two, then there are two types of such algebra as follows. For both types, the algebra has two generators, x, y, of degree one and one defining relation f, where either $f = xy - yx - x^2$ (*Jordan plane*) or $f = xy - qyx$ (*quantum affine plane*), where $q \in \Bbbk$ can be any nonzero scalar.

However, if the global dimension is three, then the situation is much richer; some of the algebras are quadratic with three generators and three defining relations, whereas the rest have two generators and two cubic relations [Artin and Schelter 1987]. Such algebras that are generic are classified in [Artin et al. 1990] according to their point schemes, and in all cases, the point scheme is the graph of an automorphism σ. Moreover, the algebra is a finite module over its center if and only if σ has finite order.

1D. *Global dimension four.* Although many regular algebras of global dimension four have been extensively studied, there is no classification yet. Recently, the progress towards classifying nonquadratic regular algebras of global dimension four made good headway via the work in [Lu et al. 2007; Rogalski and Zhang 2012]. However, quadratic regular algebras of global dimension four constitute most of the regular algebras of global dimension four, so their attempted classification is one of the motivating problems that drives the subject forward. We end this section by summarizing some key results for this latter case; in this setting, the algebra has four generators and six relations.

In unpublished work, Van den Bergh proved in the mid-1990s that any quadratic (not necessarily regular) algebra A on four generators with six generic defining relations has twenty (counted with multiplicity) nonisomorphic truncated point modules of length three. Hence, A has at most twenty nonisomorphic point modules. He also proved that if, additionally, A is Auslander-regular of global dimension four, then A has a 1-parameter family of line modules. For lack of a suitable reference, we outline the proof of these results. Let $M(4, \Bbbk)$ denote the space of 4×4 matrices with entries in $\Bbbk$. For the first result, we write points of $\mathbb{P}(A_1^*)$ as columns and, by mapping $(a, b) \in \mathbb{P}(A_1^*) \times \mathbb{P}(A_1^*)$ to the matrix $ab^T \in M(4, \Bbbk)$, we have that $\mathbb{P}(A_1^*) \times \mathbb{P}(A_1^*)$ is isomorphic to the scheme Ω_1 of rank-1 elements in $\mathbb{P}(M(4, \Bbbk))$. Correspondingly, the defining relations of A map to homogeneous degree-1 polynomial functions on $M(4, \Bbbk)$, and their zero locus $Z' \subset \mathbb{P}(M(4, \Bbbk))$ can be identified with a $\mathbb{P}^9$. With these identifications, the zero locus $Z \subset \mathbb{P}(A_1^*) \times \mathbb{P}(A_1^*)$ of the defining relations of A is isomorphic to $\Omega_1 \cap Z' \subset \mathbb{P}(M(4, \Bbbk))$. Since Ω_1 has dimension six and degree twenty, $\dim(Z) \geq 6 + 9 - 15 = 0$, and, by Bézout's Theorem, $\deg(Z) = 20$. Hence, generically, Z is finite with twenty points, so the first result follows by using the discussion after Definition 1.3. For the second result, we identify $A_1 \otimes_{\Bbbk} A_1$ with $M(4, \Bbbk)$, and the assumption on regularity allows the application of [Levasseur and Smith 1993, Proposition 2.8], so that the line modules are in one-to-one correspondence with the elements in the span of the defining relations of A that have rank at most two. In particular, we compute $\dim(\Omega_2 \cap \Delta)$ in $\mathbb{P}(M(4, \Bbbk))$, where Ω_2 denotes the elements in $\mathbb{P}(M(4, \Bbbk))$ of rank at most two and Δ denotes the projectivization of the image in $\mathbb{P}(M(4, \Bbbk))$ of the span of the defining relations of A. Since $\Delta \cong \mathbb{P}^5$ and $\dim(\Omega_2) = 11$, the dimension is thus at least equal to $11+5-15=1$, so, generically, A has a 1-parameter family of line modules.

In spite of Van den Bergh's work, it was still not clear that a regular algebra satisfying the hypotheses from the preceding paragraph could have both a finite point scheme (especially one of cardinality twenty) and a 1-dimensional line scheme simultaneously. However, Vancliff, Van Rompay and Willaert [Vancliff et al. 1998] proved that there exists a quadratic regular algebra of global dimension four on four generators with six defining relations that has exactly one point module (up to isomorphism) and a 1-parameter family of line mods.

Some years later, Shelton and Vancliff [2002] proved that if a quadratic algebra on four generators with six defining relations has a *finite* scheme of truncated point modules of length three, then that scheme determines the defining relations of the algebra. One should note that this result assumes no hypothesis of regularity nor of any other homological data. Moreover, by [Van Rompay 1996] this result is false in general if the scheme is infinite, even if the algebra is assumed to be regular and noetherian.

Shelton and Vancliff [2002] also proved that if a quadratic regular algebra of global dimension four (that satisfies a few other homological conditions) has four generators and six defining relations and a 1-dimensional line scheme, then that scheme determines the defining relations of the algebra.

These last two results are counterintuitive, since they seem to be saying that if the point scheme (respectively, line scheme) is *as small as possible*, then the defining relations can be recovered from it.

However, by the start of 2001, it was still unclear whether or not any quadratic regular algebra exists that has global dimension four, four generators, six defining relations, exactly twenty nonisomorphic point modules and a 1-dimensional line scheme. Fortunately, this was resolved by Shelton and Tingey [2001] in the affirmative. Sadly, their method to produce their example used much trial and error on a computer, which they and others were unable to duplicate to produce more examples. This hurdle likely had a negative impact on the development of the subject, since it is difficult to make conjectures if there is only one known example. Hence, a quest began to find an algorithm to construct such algebras, but it was another several years before this situation was remedied, and that is discussed in the next section.

2. Graded Clifford algebras, graded skew Clifford algebras and quantum planes

This section describes a construction of a certain type of regular algebra of arbitrary finite global dimension; such an algebra is called a graded skew Clifford algebra as it is modeled on the construction of a graded Clifford algebra. If the global dimension is four, then this construction is able to produce regular algebras that have the desired properties described at the end of the previous section. We conclude this section by revisiting the classification of quadratic regular algebras of global dimension three, and show that almost all such algebras may be obtained from regular graded skew Clifford algebras.

We continue to assume that $\Bbbk$ is algebraically closed; we additionally assume $\mathrm{char}(\Bbbk) \neq 2$. We write $M(n, \Bbbk)$ for the space of $n \times n$ matrices with entries in $\Bbbk$, and M_{ij} for the entry in the $n \times n$ matrix M that is in row i and column j.

2A. *Graded Clifford algebras.*

Definition 2.1 [Aubry and Lemaire 1985; Le Bruyn 1995]. Let

$$M_1, \ldots, M_n \in M(n, \Bbbk)$$

denote symmetric matrices. A *graded Clifford algebra* (GCA) is the $\Bbbk$-algebra C on degree-one generators $x_1, \ldots, x_n$ and on degree-two generators $y_1, \ldots, y_n$ with defining relations given by

(i) (degree-2 relations) $x_i x_j + x_j x_i = \sum_{k=1}^{n} (M_k)_{ij}\, y_k$ for all $i, j = 1, \ldots, n$, and

(ii) degree-3 and degree-4 relations that guarantee y_k is central in C for all $k = 1, \ldots, n$.

In general, GCAs need not be quadratic nor regular, as demonstrated by the next example.

Example 2.2. Let $M_1 = \left[\begin{smallmatrix} 2 & -1 \\ -1 & 0 \end{smallmatrix}\right]$ and $M_2 = \left[\begin{smallmatrix} 0 & -1 \\ -1 & 2 \end{smallmatrix}\right]$. The corresponding GCA is the $\Bbbk$-algebra on degree-one generators $x_1,\ x_2$ with defining relations

$$x_1 x_2 + x_2 x_1 = -x_1^2 - x_2^2, \quad x_1^2 x_2 = x_2 x_1^2,$$

so this algebra is not quadratic nor regular (as $(x_1 + x_2)^2 = 0$). For more details on this algebra, the reader may consult [Vancliff 2015, Example 2.4].

GCAs C are noetherian by [Artin et al. 1990, Lemma 8.2], since

$$\dim_{\Bbbk}(C/\langle y_1, \ldots, y_n \rangle) < \infty.$$

Moreover, since each matrix M_k in the definition is symmetric, we may associate a quadratic form to M_k, and thereby associate a quadric in $\mathbb{P}^{n-1}$ to M_k for each

k. This means that for each GCA, as in Definition 2.1, there is an associated quadric system $\mathfrak{Q}$ in $\mathbb{P}^{n-1}$. Quadric systems are said to be base-point free if they yield a complete intersection; that is, the intersection of all the quadrics in the quadric system is empty. Although Example 2.2 demonstrates that a GCA need not be quadratic nor regular, if $\mathfrak{Q}$ is base-point free, it determines these properties of the associated GCA as follows.

Theorem 2.3 [Aubry and Lemaire 1985; Le Bruyn 1995]. *The GCA C is quadratic, Auslander-regular of global dimension n and satisfies the Cohen–Macaulay property with Hilbert series $1/(1-t)^n$ if and only if the associated quadric system is base-point free; in this case, C is regular and a domain.*

In spite of this result, regular GCAs of global dimension four are not candidates for generic quadratic regular algebras of global dimension four, since, although their point schemes can be finite [Stephenson and Vancliff 2007; Vancliff et al. 1998], the symmetry of their relations prevents their line schemes from having dimension one [Shelton and Vancliff 2002]. The standard argument to prove this for a quadratic regular GCA C of global dimension four exploits the symmetry of the defining relations of C to move the computation of Section 1D inside $\mathbb{P}(W)$, where W is the 10-dimensional subspace of $M(4, \Bbbk)$ consisting of all symmetric matrices. Hence, using the notation from Section 1D, $\Delta \subset \mathbb{P}(W)$ and the line modules are parametrized by $(\Omega_2 \cap \mathbb{P}(W)) \cap \Delta \subset \mathbb{P}(W)$; thus the dimension is at least $6+5-9$, so it is at least two.

Hence, a modification of the definition of GCA is desired in such a way that enough symmetry is retained so as to allow an analogue of Theorem 2.3 to hold, while, at the same time, losing some symmetry so that the line scheme might have dimension one.

2B. ***Graded skew Clifford algebras.*** In order to generalize the notion of GCA and to have a result analogous to Theorem 2.3, we need to generalize the notions of symmetric matrix and quadric system and make use of normalizing sequences. For any $\mathbb{N}$-graded $\Bbbk$-algebra B, a sequence $\{g_1, \dots, g_m\}$ of homogeneous elements of positive degree is called *normalizing* if g_1 is a normal element in B and, for each $k = 1, \dots, m-1$, the image of g_{k+1} in $B/\langle g_1, \dots, g_k\rangle$ is a normal element.

We write $\Bbbk^\times$ for $\Bbbk \setminus \{0\}$.

Definition 2.4 [Cassidy and Vancliff 2010].

(a) Let $\mu \in M(n, \Bbbk^\times)$ satisfy $\mu_{ij}\mu_{ji} = 1$ for all distinct i, j. We say a matrix $M \in M(n, \Bbbk)$ is *μ-symmetric* if $M_{ij} = \mu_{ij} M_{ji}$ for all $i, j = 1, \dots, n$. We write $M^\mu(n, \Bbbk)$ for the subspace of $M(n, \Bbbk)$ consisting of all μ-symmetric matrices.

(b) Fix μ as in (a) and additionally assume $\mu_{ii} = 1$ for all i. Let $M_1, \dots, M_n \in M^{\mu}(n, \Bbbk)$. A *graded skew Clifford algebra* (GSCA) associated to μ and $M_1, \dots, M_n$ is a graded $\Bbbk$-algebra $A = A(\mu, M_1, \dots, M_n)$ on degree-one generators $x_1, \dots, x_n$ and on degree-two generators $y_1, \dots, y_n$ with defining relations given by

(i) (degree-2 relations) $x_i x_j + \mu_{ij} x_j x_i = \sum_{k=1}^{n} (M_k)_{ij} y_k$ for all $i, j = 1, \dots, n$, and

(ii) degree-3 and degree-4 relations that guarantee the existence of a normalizing sequence $\{y_1', \dots, y_n'\}$ that spans $\sum_{k=1}^{n} \Bbbk y_k$.

Clearly, symmetric matrices and skew-symmetric matrices are μ-symmetric matrices for appropriate μ, and GCAs are GSCAs. Moreover, by [Artin et al. 1990, Lemma 8.2], GSCAs A are noetherian since $\dim_{\Bbbk}(A/\langle y_1, \dots, y_n \rangle) < \infty$. Furthermore, in Definition 2.4(b)(i), for all i, j, the ji-relation can be deduced from the ij-relation by the μ-symmetry of the M_k.

Examples 2.5. (a) With μ as in Definition 2.4(b), skew polynomial rings on generators $x_1, \dots, x_n$ with relations $x_i x_j = -\mu_{ij} x_j x_i$, for all $i \neq j$, are GSCAs.

(b) (quantum affine plane) Let $n = 2$, and $M_1 = \left[\begin{smallmatrix} 2 & 0 \\ 0 & 0 \end{smallmatrix}\right]$ and $M_2 = \left[\begin{smallmatrix} 0 & 0 \\ 0 & 2 \end{smallmatrix}\right]$. The degree-2 relations of $A(\mu, M_1, M_2)$ have the form

$$2x_1^2 = 2y_1, \quad 2x_2^2 = 2y_2, \quad x_1 x_2 + \mu_{12} x_2 x_1 = 0,$$

so that $\Bbbk\langle x_1, x_2 \rangle / \langle x_1 x_2 + \mu_{12} x_2 x_1 \rangle \twoheadrightarrow A(\mu, M_1, M_2)$. By Theorem 2.6 below, this map is an isomorphism (see Examples 3.2(a)).

(c) ("Jordan" plane) Let $n = 2$, and $M_1 = \left[\begin{smallmatrix} 2 & 1 \\ \mu_{21} & 0 \end{smallmatrix}\right]$ and $M_2 = \left[\begin{smallmatrix} 0 & 0 \\ 0 & 2 \end{smallmatrix}\right]$. The degree-2 relations of $A(\mu, M_1, M_2)$ have the form

$$2x_1^2 = 2y_1, \quad 2x_2^2 = 2y_2, \quad x_1 x_2 + \mu_{12} x_2 x_1 = y_1 = x_1^2,$$

so that $\Bbbk\langle x_1, x_2 \rangle / \langle x_1 x_2 + \mu_{12} x_2 x_1 - x_1^2 \rangle \twoheadrightarrow A(\mu, M_1, M_2)$. By Theorem 2.6 below, this map is an isomorphism (see Examples 3.2(b)). Depending on the choice of μ_{12}, this family of examples contains the Jordan plane and some quantum affine planes.

(d) The quadratic regular algebra of global dimension four found by Shelton and Tingey [2001], and discussed above in Section 1D, that has exactly twenty nonisomorphic point modules and a 1-dimensional line scheme is a GSCA [Cassidy and Vancliff 2010].

One can associate a noncommutative "quadric" to each μ-symmetric matrix M_k and, in so doing, there is also a notion of "base-point free". These ideas are discussed in Section 3B below, and yield a generalization of Theorem 2.3:

Theorem 2.6 [Cassidy and Vancliff 2010]. *The GSCA A is quadratic, Auslander-regular of global dimension n and satisfies the Cohen–Macaulay property with Hilbert series* $1/(1-t)^n$ *if and only if the associated quadric system is normalizing and base-point free; in this case, A is regular and a domain and uniquely determined, up to isomorphism, by the data* $\mu, M_1, \ldots, M_n$.

Theorem 2.6 allowed the production in [Cassidy and Vancliff 2010] of many algebras that are candidates for generic quadratic regular algebras of global dimension four. In particular, there exist quadratic regular GSCAs of global dimension four on four generators with six defining relations that have exactly twenty nonisomorphic point modules and a 1-dimensional line scheme.

It is an open problem to describe the 1-dimensional line schemes of the regular GSCAs of global dimension four in [Cassidy and Vancliff 2010] that have exactly twenty nonisomorphic point modules.

By Examples 1.7(c) and 2.5(b)(c), the regular algebras of global dimension at most two are GSCAs, and, by Section 2C, almost all quadratic regular algebras of global dimension three are determined by GSCAs, so GSCAs promise to be very helpful in the classification of all quadratic regular algebras of global dimension four.

2C. *Quadratic quantum planes.* In the language of [Artin 1992], a regular algebra of global dimension three that is generated by degree-1 elements is sometimes called a *quantum plane* or *quantum projective plane* or a *quantum* $\mathbb{P}^2$. The classification of the *generic* quantum planes is in [Artin and Schelter 1987; Artin et al. 1990; 1991]. In this subsection, we summarize the results of [Nafari et al. 2011], in which all *quadratic* quantum planes are classified by using GSCAs.

We continue to assume that $\Bbbk$ is algebraically closed, but its characteristic is arbitrary unless specifically stated otherwise.

Let D denote a quadratic quantum plane and let $X \subset \mathbb{P}^2$ denote its point scheme. By [Artin et al. 1990, Proposition 4.3; Nafari et al. 2011, Lemma 2.1], there are, in total, four cases to consider:

- X contains a line, or
- X is a nodal cubic curve in $\mathbb{P}^2$, or
- X is a cuspidal cubic curve in $\mathbb{P}^2$, or
- X is a (nonsingular) elliptic curve in $\mathbb{P}^2$.

Theorem 2.7 [Nafari et al. 2011]. *Suppose* $\operatorname{char}(\Bbbk) \neq 2$. *If X contains a line, then either D is a twist, by an automorphism, of a GSCA, or D is a twist, by a twisting system, of an Ore extension of a regular GSCA of global dimension two.*

Theorem 2.8 [Nafari et al. 2011]. *If X is a nodal cubic curve, then D is isomorphic to a $\Bbbk$-algebra on generators x_1, x_2, x_3 with defining relations*

$$\lambda x_1x_2 = x_2x_1, \quad \lambda x_2x_3 = x_3x_2 - x_1^2, \quad \lambda x_3x_1 = x_1x_3 - x_2^2, \tag{$*$}$$

where $\lambda \in \Bbbk$ and $\lambda^3 \notin \{0,\ 1\}$. Conversely, for any such λ, any quadratic algebra with defining relations $(*)$ is a quantum plane and its point scheme is a nodal cubic curve in $\mathbb{P}^2$. Moreover, if $\operatorname{char}(\Bbbk) \neq 2$, *then D is an Ore extension of a regular GSCA of global dimension two; in particular, if $\lambda^3 = -1$, then D is a GSCA.*

Theorem 2.9 [Nafari et al. 2011]. *If* $\operatorname{char}(\Bbbk) = 3$, *then X is not a cuspidal cubic curve in $\mathbb{P}^2$. If* $\operatorname{char}(\Bbbk) \neq 3$ *and if X is a cuspidal cubic curve in $\mathbb{P}^2$, then D is isomorphic to a $\Bbbk$-algebra on generators x_1, x_2, x_3 with defining relations*

$$\begin{aligned} x_1x_2 = x_2x_1 + x_1^2, \quad x_3x_1 = x_1x_3 + x_1^2 + 3x_2^2,& \\ x_3x_2 = x_2x_3 - 3x_2^2 - 2x_1x_3 - 2x_1x_2.& \end{aligned} \tag{$\dagger$}$$

Moreover, any quadratic algebra with defining relations $(\dagger)$ is a quantum plane; it has point scheme given by a cuspidal cubic curve in $\mathbb{P}^2$ if and only if $\operatorname{char}(\Bbbk) \neq 3$. *If* $\operatorname{char}(\Bbbk) \neq 2$, *then any quadratic algebra with defining relations given by $(\dagger)$ is an Ore extension of a regular GSCA of global dimension two.*

It remains to discuss the case that X is an elliptic curve. In [Artin and Schelter 1987; Artin et al. 1990], such algebras are classified into types A, B, E, H, where some members of each type might not have an elliptic curve as their point scheme, but a generic member does.

Theorem 2.10 [Nafari et al. 2011]. *Suppose that* $\operatorname{char}(\Bbbk) \neq 2$ *and that X is an elliptic curve.*

(a) *Quadratic quantum planes of type H are GSCAs.*

(b) *Quadratic quantum planes of type B are GSCAs.*

(c) *As in* [Artin and Schelter 1987; Artin et al. 1990], *a quadratic quantum plane D of type A is given by a $\Bbbk$-algebra on generators x, y, z with defining relations*

$$axy + byx + cz^2 = 0, \quad ayz + bzy + cx^2 = 0, \quad azx + bxz + cy^2 = 0,$$

where $a, b, c \in \Bbbk^\times$, $(3abc)^3 \neq (a^3 + b^3 + c^3)^3$, $\operatorname{char}(\Bbbk) \neq 3$, *and either $a^3 \neq b^3$, or $a^3 \neq c^3$, or $b^3 \neq c^3$. In the case that $a^3 = b^3 \neq c^3$, D is a GSCA; whereas in the case $a^3 \neq b^3 = c^3$ (respectively, $a^3 = c^3 \neq b^3$), D is a twist, by an automorphism, of a GSCA.*

In (c) of the last result, the case that $a^3 \neq b^3 \neq c^3 \neq a^3$ is still open. Moreover, the case when D is of type E is still open, but this case only consists of one algebra, up to isomorphism and antiisomorphism. However, both type A and type E have the property that the Koszul dual of D is a quotient of a regular GSCA; so, in this sense, such algebras are weakly related to GSCAs.

3. Complete intersections

In this section, we define the geometric terms used in Theorem 2.6. That discussion leads naturally into a consideration of a notion of noncommutative complete intersection that mimics the commutative definition.

We continue to assume that the field $\Bbbk$ is algebraically closed.

3A. ***Commutative complete intersection and quadric systems.*** Let R denote the commutative polynomial ring on n generators of degree one. If $f_1, \ldots, f_m$ are homogeneous elements of R of positive degree, then $\{f_1, \ldots, f_m\}$ is a regular sequence in R if and only if $\operatorname{GKdim}(R/\langle f_1, \ldots, f_k\rangle) = n - k \geq 0$, for all $k = 1, \ldots, m$. Geometrically, this corresponds to the zero locus in $\mathbb{P}(R_1^{\,*})$ of the ideal $J_k = \langle f_1, \ldots, f_k\rangle$ having dimension $n-1-k \geq -1$ for all k. If $\{f_1, \ldots, f_m\}$ is a regular sequence, then the zero locus of J_m (respectively, R/J_m) is called a *complete intersection* (see [Eisenbud 1995]).

In the setting of Definition 2.1, a quadric system $\mathfrak{Q}$ is associated to symmetric matrices $M_1, \ldots, M_n$. In that setting, $\mathfrak{Q}$ corresponds to a regular sequence in R if and only if $\mathfrak{Q}$ is a complete intersection, that is, if and only if $\mathfrak{Q}$ has no base points (a base point is a point that lies on all the quadrics in $\mathfrak{Q}$). A noncommutative analogue of this is needed for Theorem 2.6.

3B. ***Noncommutative complete intersection and quadric systems.*** The following result uses the notion of base-point module defined in Definition 1.2.

Proposition 3.1 [Cassidy and Vancliff 2010; 2014]. *Let S denote the skew polynomial ring from Section 1A, and let $f_1, \ldots, f_n$ denote homogeneous elements of S of positive degree. If $\{f_1, \ldots, f_n\}$ is a normalizing sequence in S, then the following are equivalent*:

(a) $\{f_1, \ldots, f_n\}$ *is a regular sequence in S.*

(b) $\dim_{\Bbbk}(S/\langle f_1, \ldots, f_n\rangle) < \infty$.

(c) *For each $k = 1, \ldots, n$, we have* $\operatorname{GKdim}(S/\langle f_1, \ldots, f_k\rangle) = n - k$.

(d) *The factor ring $S/\langle f_1, \ldots, f_n\rangle$ has no right base-point modules.*

(e) *The factor ring $S/\langle f_1, \ldots, f_n\rangle$ has no left base-point modules.*

Such a sequence $\{f_1, \dots, f_n\}$ (respectively, $S/\langle f_1, \dots, f_n\rangle$) satisfying the equivalent conditions (a)-(e) from Proposition 3.1 is called a *complete intersection* in [Cassidy and Vancliff 2014].

In the setting of Section 2B, one associates S to the GSCA by using μ. The isomorphism $M^{\mu}(n, \Bbbk) \to S_2$ defined by $M \mapsto (z_1, \dots, z_n)M(z_1, \dots, z_n)^T$ associates a quadric system $\mathfrak{Q}$ to the μ-symmetric matrices $M_1, \dots, M_n$; that is, $\mathfrak{Q}$ is the span in S_2 of the images of the M_k under this map. If $\mathfrak{Q}$ is given by a normalizing sequence in S, then it is called a *normalizing quadric system*. By Proposition 3.1, if $\mathfrak{Q}$ is normalizing, then it corresponds to a regular sequence in S if and only if it is a complete intersection, that is, if and only if $S/\langle\mathfrak{Q}\rangle$ has no right (respectively, left) base-point modules; this is the meaning of *base-point free* in Theorem 2.6.

Examples 3.2. (a) [Cassidy and Vancliff 2010] We revisit the quantum affine plane from Example 2.5(b), where $n = 2$. In that case, $M_i \mapsto q_i = 2z_i^2 \in S_2$, for $i = 1, 2$. The sequence $\{q_1, q_2\}$ is normalizing in S and $\dim(S/\langle q_1, q_2\rangle) < \infty$. Thus, by Proposition 3.1, the corresponding quadric system is base-point free.

(b) [Cassidy and Vancliff 2010] For Example 2.5(c), $n = 2$ and $M_1 \mapsto q_1 = 2(z_1^2 + z_1z_2)$ and $M_2 \mapsto q_2 = 2z_2^2$. Here, the sequence $\{q_2, q_1\}$ is normalizing in S and $\dim(S/\langle q_2, q_1\rangle) < \infty$, so by Proposition 3.1, the corresponding quadric system is base-point free.

Proposition 3.1 has recently been extended in [Vancliff 2015] to a family of algebras that contains the skew polynomial ring S from Section 1A. In particular, an analogue of Proposition 3.1 holds for regular GSCAs, many quantum groups, and homogenizations of finite-dimensional Lie algebras.

Theorem 3.3 [Vancliff 2015]. *Let $A = \bigoplus_{i=0}^{\infty} A_i$ denote a connected, $\mathbb{N}$-graded $\Bbbk$-algebra that is generated by A_1. Suppose A is Auslander–Gorenstein of finite injective dimension and satisfies the Cohen–Macaulay property, and that there exists a normalizing sequence $\{y_1, \dots, y_\nu\} \subset A \setminus \Bbbk$ consisting of homogeneous elements such that* $\operatorname{GKdim}(A/\langle y_1, \dots, y_\nu\rangle) = 1$. *If* $\operatorname{GKdim}(A) = n \in \mathbb{N}$, *and if $F = \{f_1, \dots, f_n\} \subset A \setminus \Bbbk^\times$ is a normalizing sequence of homogeneous elements, then the following are equivalent*:

(a) *F is a regular sequence in A.*

(b) $\dim_{\Bbbk}(A/\langle F\rangle) < \infty$.

(c) *For each $k = 1, \dots, n$, we have* $\operatorname{GKdim}(A/\langle f_1, \dots, f_k\rangle) = n - k$.

(d) *The factor ring $A/\langle F\rangle$ has no right base-point modules.*

(e) *The factor ring $A/\langle F\rangle$ has no left base-point modules.*

The reader should note that other notions of complete intersection abound in the literature, with most emphasizing a homological approach, such as the recent work in [Kirkman et al. 2013].

4. Conclusion

In this section, we list some open problems and related topics. The open problems are not listed in any particular order in regards to difficulty, and many challenge levels are included, with some quite computational in nature, and so accessible to junior researchers.

4A. ***Some open problems.*** (1) As stated at the end of Section 2, it is still open whether or not quadratic quantum planes of type A with $a^3 \neq b^3 \neq c^3 \neq a^3$ are directly related to GSCAs; the analogous problem is also open for type E.

(2) Is it possible to classify cubic quantum planes by using GSCAs, or by using an appropriate analogue of a GSCA?

(3) Is it possible to classify quadratic regular algebras of global dimension four by using GSCAs? Presumably, such a classification will use both the point scheme and the line scheme.

(4) Can standard results on commutative quadratic forms and quadrics be extended to noncommutative quadratic forms and quadrics? For example, P. Veerapen and Vancliff [2013] have extended the notion of rank of a (commutative) quadratic form to noncommutative quadratic forms on n generators, where $n = 2,\ 3$; can this be done for $n \geq 4$?

(5) Can results concerning GCAs be carried over to GSCAs? In particular, Veerapen and the author applied their aforementioned generalization of rank to GSCAs in a way that is analogous to that used for the traditional notion of rank with GCAs in [Vancliff et al. 1998]. They proved [2014] that various results in [Vancliff et al. 1998] concerning point modules over GCAs apply to point modules over GSCAs.

(6) Can standard results concerning symmetric matrices be extended or generalized to μ-symmetric matrices?

(7) Can the results in [Vancliff 2015], mentioned above at the end of Section 3, on complete intersections be extended to an even larger family of algebras?

(8) By combining results in [Cassidy and Vancliff 2010] and [Stephenson and Vancliff 2007], it is known that regular GSCAs of global dimension four can have exactly N nonisomorphic point modules, where $N \notin \{2,\ 19\}$; it is not yet known if $N \in \{2,\ 19\}$ is possible. In fact, by [Stephenson and Vancliff 2006], $N = 2$ is possible if the algebra is quadratic and regular of global dimension four

but is not a GSCA, but it is not known if $N = 19$ is possible, even if the algebra is not a GSCA.

(9) What is the line scheme of some known quadratic regular algebras of global dimension four? Such as those in [Cassidy and Vancliff 2010, Section 5], double Ore extensions in [Zhang and Zhang 2008; 2009], generalized Laurent polynomial rings in [Cassidy et al. 2006], etc.

(10) Does the line scheme of a generic quadratic regular algebra of global dimension four have a particular form? Perhaps a union of elliptic curves? Or, perhaps it contains at least one elliptic curve?

(11) Suppose A is as in Definition 1.1 and F is as in Theorem 3.3. Let $I_k = \langle f_1, \dots, f_k \rangle$ for all $k \leq n$, and let $\widehat{\mathcal{V}(I_k)}$ denote the set of isomorphism classes of right base-point modules over A/I_k. If A is commutative, then, for each k, $\widehat{\mathcal{V}(I_k)}$ is a scheme, and so has a dimension. In particular, if A is the polynomial ring, then F is regular if and only if $\dim(\widehat{\mathcal{V}(I_k)}) = n - k - 1$, for all $k \leq n$. However, if A is not commutative, is there an analogous statement and under what hypotheses on A could it hold?

4B. ***Related topics.*** Since the publication of [Artin et al. 1990], the subject has branched out in many directions, the key topics being: classification of regular algebras; classification of projective surfaces; seeing which commutative techniques (e.g., blowing-up, blowing-down) carry over to the noncommutative setting; and connections with differential geometry (e.g., via Poisson geometry). Module categories and homological algebra provide a unifying umbrella. These directions are highlighted in the references cited in the Introduction and throughout the text, and in some of the articles in this volume.

New directions continue to emerge, with one of the most recent trends being the study of regular algebras and Hopf algebras together via the consideration of Hopf actions on regular algebras, such as the work in [Chan et al. 2014]. However, perhaps the most recent exciting triumph of the subject is when the universal enveloping algebra of the Witt algebra was viewed through the geometric lens of [Artin et al. 1990] by Sierra and Walton [2014], enabling them to solve the long-standing problem of whether or not that algebra is noetherian.

In view of all these advances, it is now clear that the marriage of noncommutative algebra and algebraic geometry, à la [Artin et al. 1990], is a dynamic and evolving field of research.

Acknowledgements

The author is grateful to MSRI for general membership in 2013 and for providing an office during her visits. Additionally, special thanks are extended to the organizers of the workshop "Connections for Women" in January 2013 for their invitation to give a talk in the workshop. The environment at MSRI was particularly stimulating during the program "Noncommutative Algebraic Geometry and Representation Theory" in Spring 2013, and for that the author warmly thanks the organizers and participants of the program, especially T. Cassidy, M. Van den Bergh and J. T. Stafford.

References

[Artin 1992] M. Artin, "Geometry of quantum planes", pp. 1–15 in *Azumaya algebras, actions, and modules* (Bloomington, IN, 1990), edited by D. Haile and J. Osterburg, Contemp. Math. **124**, Amer. Math. Soc., Providence, RI, 1992.

[Artin and Schelter 1987] M. Artin and W. F. Schelter, "Graded algebras of global dimension 3", *Adv. in Math.* **66**:2 (1987), 171–216.

[Artin et al. 1990] M. Artin, J. Tate, and M. Van den Bergh, "Some algebras associated to automorphisms of elliptic curves", pp. 33–85 in *The Grothendieck Festschrift, I*, edited by P. Cartier et al., Progr. Math. **86**, Birkhäuser, Boston, 1990.

[Artin et al. 1991] M. Artin, J. Tate, and M. Van den Bergh, "Modules over regular algebras of dimension 3", *Invent. Math.* **106**:2 (1991), 335–388.

[Aubry and Lemaire 1985] M. Aubry and J.-M. Lemaire, "Zero divisors in enveloping algebras of graded Lie algebras", *J. Pure Appl. Algebra* **38**:2-3 (1985), 159–166.

[Cassidy 1999] T. Cassidy, "Global dimension 4 extensions of Artin–Schelter regular algebras", *J. Algebra* **220**:1 (1999), 225–254.

[Cassidy 2003] T. Cassidy, "Central extensions of Stephenson's algebras", *Comm. Algebra* **31**:4 (2003), 1615–1632.

[Cassidy and Vancliff 2010] T. Cassidy and M. Vancliff, "Generalizations of graded Clifford algebras and of complete intersections", *J. Lond. Math. Soc.* (2) **81**:1 (2010), 91–112.

[Cassidy and Vancliff 2014] T. Cassidy and M. Vancliff, "Corrigendum to 'Generalizations of graded Clifford algebras and of complete intersections'", *J. Lond. Math. Soc. (2)* **90**:2 (2014), 631–636.

[Cassidy et al. 2006] T. Cassidy, P. Goetz, and B. Shelton, "Generalized Laurent polynomial rings as quantum projective 3-spaces", *J. Algebra* **303**:1 (2006), 358–372.

[Chan et al. 2014] K. Chan, C. Walton, Y. H. Wang, and J. J. Zhang, "Hopf actions on filtered regular algebras", *J. Algebra* **397** (2014), 68–90.

[Drinfeld 1987] V. G. Drinfeld, "Quantum groups", pp. 798–820 in *Proceedings of the International Congress of Mathematicians* (Berkeley, 1986), vol. 1, edited by A. M. Gleason, Amer. Math. Soc., Providence, RI, 1987.

[Eisenbud 1995] D. Eisenbud, *Commutative algebra with a view toward algebraic geometry*, Graduate Texts in Mathematics **150**, Springer, New York, 1995.

[Goodearl 2010] K. R. Goodearl, "Semiclassical limits of quantized coordinate rings", pp. 165–204 in *Advances in ring theory*, edited by D. V. Huynh and S. R. López-Permouth, Birkhäuser/Springer, Basel, 2010.

[Kapustin et al. 2001] A. Kapustin, A. Kuznetsov, and D. Orlov, "Noncommutative instantons and twistor transform", *Comm. Math. Phys.* **221**:2 (2001), 385–432.

[Kirkman et al. 2013] E. Kirkman, J. Kuzmanovich, and J. J. Zhang, "Noncommutative complete intersections", preprint, 2013. arXiv 1302.6209

[Le Bruyn 1995] L. Le Bruyn, "Central singularities of quantum spaces", *J. Algebra* **177**:1 (1995), 142–153.

[Le Bruyn and Smith 1993] L. Le Bruyn and S. P. Smith, "Homogenized $\mathfrak{sl}(2)$", *Proc. Amer. Math. Soc.* **118**:3 (1993), 725–730.

[Le Bruyn and Van den Bergh 1993] L. Le Bruyn and M. Van den Bergh, "On quantum spaces of Lie algebras", *Proc. Amer. Math. Soc.* **119**:2 (1993), 407–414.

[Levasseur 1992] T. Levasseur, "Some properties of noncommutative regular graded rings", *Glasgow Math. J.* **34**:3 (1992), 277–300.

[Levasseur and Smith 1993] T. Levasseur and S. P. Smith, "Modules over the 4-dimensional Sklyanin algebra", *Bull. Soc. Math. France* **121**:1 (1993), 35–90.

[Levasseur and Stafford 1993] T. Levasseur and J. T. Stafford, "The quantum coordinate ring of the special linear group", *J. Pure Appl. Algebra* **86**:2 (1993), 181–186.

[Lu et al. 2007] D.-M. Lu, J. H. Palmieri, Q.-S. Wu, and J. J. Zhang, "Regular algebras of dimension 4 and their A_∞-Ext-algebras", *Duke Math. J.* **137**:3 (2007), 537–584.

[Nafari et al. 2011] M. Nafari, M. Vancliff, and J. Zhang, "Classifying quadratic quantum $\mathbb{P}^2$s by using graded skew Clifford algebras", *J. Algebra* **346** (2011), 152–164.

[Reshetikhin et al. 1989] N. Y. Reshetikhin, L. A. Takhtadzhyan, and L. D. Faddeev, "Quantization of Lie groups and Lie algebras", *Algebra i Analiz* **1**:1 (1989), 178–206. In Russian; translated in *Leningrad Math. J.* **1**:1 (1990), 193–225.

[Rogalski 2014] D. Rogalski, "An introduction to noncommutative projective algebraic geometry", preprint, 2014. arXiv 1403.3065

[Rogalski and Zhang 2012] D. Rogalski and J. J. Zhang, "Regular algebras of dimension 4 with 3 generators", pp. 221–241 in *New trends in noncommutative algebra*, edited by P. Ara et al., Contemp. Math. **562**, Amer. Math. Soc., Providence, RI, 2012.

[Shelton and Tingey 2001] B. Shelton and C. Tingey, "On Koszul algebras and a new construction of Artin–Schelter regular algebras", *J. Algebra* **241**:2 (2001), 789–798.

[Shelton and Vancliff 2002] B. Shelton and M. Vancliff, "Schemes of line modules, I", *J. London Math. Soc.* (2) **65**:3 (2002), 575–590.

[Sierra and Walton 2014] S. J. Sierra and C. Walton, "The universal enveloping algebra of the Witt algebra is not noetherian", *Adv. Math.* **262** (2014), 239–260.

[Sklyanin 1982] E. K. Sklyanin, "Some algebraic structures connected with the Yang–Baxter equation", *Funktsional. Anal. i Prilozhen.* **16**:4 (1982), 27–34. In Russian; translated in *Functional Anal. Appl.* **16**:4 (1982), 263–270.

[Sklyanin 1985] E. K. Sklyanin, "On an algebra generated by quadratic relations", *Uspekhi Mat. Nauk* **40**:2 (1985), 214. Abstract of lecture given at the Moscow Mathematical Society. In Russian.

[Smith and Stafford 1992] S. P. Smith and J. T. Stafford, "Regularity of the four-dimensional Sklyanin algebra", *Compositio Math.* **83**:3 (1992), 259–289.

[Smith and Staniszkis 1993] S. P. Smith and J. M. Staniszkis, "Irreducible representations of the 4-dimensional Sklyanin algebra at points of infinite order", *J. Algebra* **160**:1 (1993), 57–86.

[Stafford 2002] J. T. Stafford, "Noncommutative projective geometry", pp. 93–103 in *Proceedings of the International Congress of Mathematicians, II* (Beijing, 2002), edited by T. Li, Higher Ed. Press, Beijing, 2002.

[Stafford and van den Bergh 2001] J. T. Stafford and M. van den Bergh, "Noncommutative curves and noncommutative surfaces", *Bull. Amer. Math. Soc.* (*N.S.*) **38**:2 (2001), 171–216.

[Stephenson 1996] D. R. Stephenson, "Artin–Schelter regular algebras of global dimension three", *J. Algebra* **183**:1 (1996), 55–73.

[Stephenson 1997] D. R. Stephenson, "Algebras associated to elliptic curves", *Trans. Amer. Math. Soc.* **349**:6 (1997), 2317–2340.

[Stephenson 2000a] D. R. Stephenson, "Quantum planes of weight $(1, 1, n)$", *J. Algebra* **225**:1 (2000), 70–92.

[Stephenson 2000b] D. R. Stephenson, "Corrigendum to 'Quantum planes of weight $(1, 1, n)$'", *J. Algebra* **234**:1 (2000), 277–278.

[Stephenson and Vancliff 2006] D. R. Stephenson and M. Vancliff, "Some finite quantum $\mathbb{P}^3$s that are infinite modules over their centers", *J. Algebra* **297**:1 (2006), 208–215.

[Stephenson and Vancliff 2007] D. R. Stephenson and M. Vancliff, "Constructing Clifford quantum $\mathbb{P}^3$s with finitely many points", *J. Algebra* **312**:1 (2007), 86–110.

[Sudbery 1993] A. Sudbery, "Matrix-element bialgebras determined by quadratic coordinate algebras", *J. Algebra* **158**:2 (1993), 375–399.

[Van Rompay 1996] K. Van Rompay, "Segre product of Artin–Schelter regular algebras of dimension 2 and embeddings in quantum $\mathbf{P}^3$'s", *J. Algebra* **180**:2 (1996), 483–512.

[Vancliff 1995] M. Vancliff, "The defining relations of quantum $n \times n$ matrices", *J. London Math. Soc.* (2) **52**:2 (1995), 255–262.

[Vancliff 1999] M. Vancliff, "Primitive and Poisson spectra of twists of polynomial rings", *Algebr. Represent. Theory* **2**:3 (1999), 269–285.

[Vancliff 2000] M. Vancliff, "Non-commutative spaces for graded quantum groups and graded Clifford algebras", pp. 303–320 in *Clifford algebras and their applications in mathematical physics, I* (Ixtapa-Zihuatanejo, 1999), edited by R. Abłamowicz and B. Fauser, Progr. Phys. **18**, Birkhäuser, Boston, 2000.

[Vancliff 2015] M. Vancliff, "On the notion of complete intersection outside the setting of skew polynomial rings", *Comm. Algebra* **43**:2 (2015), 460–470.

[Vancliff and Van Rompay 1997] M. Vancliff and K. Van Rompay, "Embedding a quantum nonsingular quadric in a quantum $\mathbf{P}^3$", *J. Algebra* **195**:1 (1997), 93–129.

[Vancliff and Veerapen 2013] M. Vancliff and P. P. Veerapen, "Generalizing the notion of rank to noncommutative quadratic forms", pp. 241–250 in *Noncommutative birational geometry, representations and combinatorics*, edited by A. Berenstein and V. Retakh, Contemp. Math. **592**, Amer. Math. Soc., Providence, RI, 2013.

[Vancliff and Veerapen 2014] M. Vancliff and P. P. Veerapen, "Point modules over regular graded skew Clifford algebras", *J. Algebra* **420** (2014), 54–64.

[Vancliff et al. 1998] M. Vancliff, K. Van Rompay, and L. Willaert, "Some quantum $\mathbf{P}^3$s with finitely many points", *Comm. Algebra* **26**:4 (1998), 1193–1208.

[Zhang and Zhang 2008] J. J. Zhang and J. Zhang, "Double Ore extensions", *J. Pure Appl. Algebra* **212**:12 (2008), 2668–2690.

[Zhang and Zhang 2009] J. J. Zhang and J. Zhang, "Double extension regular algebras of type (14641)", *J. Algebra* **322**:2 (2009), 373–409.

vancliff@uta.edu

*Department of Mathematics,
University of Texas at Arlington, P.O. Box 19408,
Arlington, TX 76019-0408, United States*

Commutative Algebra and Noncommutative Algebraic Geometry, I
MSRI Publications
Volume **67**, 2015

Survey on the D-module f^s

ULI WALTHER
APPENDIX BY ANTON LEYKIN

We discuss various aspects of the singularity invariants with differential origin derived from the D-module generated by f^s.

In this survey we discuss various aspects of the singularity invariants with differential origin derived from the D-module generated by f^s. We should like to point the reader to some other works: [Saito 2007] for V-filtration, Bernstein–Sato polynomials, multiplier ideals; [Budur 2012b] for all these and Milnor fibers; [Torrelli 2007] and [Narváez-Macarro 2008] for homogeneity and free divisors; [Suciu 2014] on details of arrangements, specifically their Milnor fibers, although less focused on D-modules.

This material is based in part upon work supported by the National Science Foundation under grant no. 0932078 000, while Walther was in residence at the Mathematical Science Research Institute (MSRI) in Berkeley, California, during the spring of 2013. Walther was supported in part by the NSF under grants DMS-0901123 and DMS-1401392. Leykin was supported in part by the NSF under grant DMS 1151297.

MSC2010: 14F10, 14N20, 13D45, 32S22, 58A10, 14F40, 14J17, 32C38.

Keywords: Bernstein–Sato polynomial, b-function, hyperplane, arrangement, zeta function, logarithmic comparison theorem, multiplier ideal, Milnor fiber, algorithmic, free.

1. Introduction

Notation 1.1. In this article, X will denote a complex manifold. Unless indicated otherwise, X will be $\mathbb{C}^n$.

Throughout, let $R = \mathbb{C}[x_1, \dots, x_n]$ be the ring of polynomials in n variables over the complex numbers. We denote by $D = R\langle \partial_1, \dots, \partial_n \rangle$ the Weyl algebra. In particular, ∂_i denotes the partial differentiation operator with respect to x_i. If X is a general manifold, $\mathscr{O}_X$ (the sheaf of regular functions) and $\mathscr{D}_X$ (the sheaf of $\mathbb{C}$-linear differential operators on $\mathscr{O}_X$) take the places of R and D.

If $X = \mathbb{C}^n$ we use Roman letters to denote rings and modules; in the general case we use calligraphic letters to denote corresponding sheaves.

By the ideal J_f we mean the $\mathscr{O}_X$-ideal generated by the partial derivatives $\partial f/\partial x_1, \dots, \partial f/\partial x_n$; this ideal varies with the choice of coordinate system in which we calculate. In contrast, the Jacobian ideal $\operatorname{Jac}(f) = J_f + (f)$ is independent.

The ring D (resp. the sheaf $\mathscr{D}_X$) is coherent, and both left- and right-Noetherian; it has only trivial two-sided ideals [Björk 1993, Theorem 1.2.5]. Introductions to the theory of D-modules as we use them here can be found in [Kashiwara 2003; Bernstein ca. 1997; Björk 1993; 1979].

The ring D admits the order filtration induced by the weight $x_i \to 0$, $\partial_i \to 1$. The order filtration (and other good filtrations) leads to graded objects $\operatorname{gr}_{(0,1)}(-)$; see [Schapira 1985]. The graded objects obtained from ideals are ideals in the polynomial ring $\mathbb{C}[x, \xi]$, homogeneous in the symbols of the differentiation operators; their radicals are closed under the Poisson bracket, and thus the corresponding varieties are involutive [Kashiwara 1975; Kashiwara and Kawai 1981a]. For a D-module M and a component C of the support of $\operatorname{gr}_{(0,1)}(M)$, attach to the pair (M, C) the multiplicity $\mu(M, C)$ of $\operatorname{gr}_{(0,1)}(M)$ along C. The characteristic cycle of M is $\operatorname{charC}(M) = \sum_C \mu(M, C) \cdot C$, an element of the Chow ring on $T^*\mathbb{C}^n$. The module is *holonomic* if it is zero or if its characteristic variety is of dimension n, the minimal possible value.

Throughout, f will be a regular function on X, with divisor $\operatorname{Var}(f)$. We distinguish several homogeneity conditions on f:

- f is *locally (strongly) Euler-homogeneous* if for all $p \in \operatorname{Var}(f)$ there is a vector field θ_p defined near p with $\theta_p \bullet (f) = f$ (and θ_p vanishes at p).
- f is *locally (weakly) quasihomogeneous* if near all $p \in \operatorname{Var}(f)$ there is a local coordinate system $\{x_i\}$ and a positive (resp. nonnegative) weight vector $a = \{a_1, \dots, a_n\}$ with respect to which $f = \sum_{i=1}^n a_i x_i \partial_i(f)$.
- We reserve *homogeneous* and *quasihomogeneous* for the case when $X = \mathbb{C}^n$ and f is globally homogeneous or quasihomogeneous.

To any nonconstant $f \in R$, one can attach several invariants that measure the singularity structure of the hypersurface $f = 0$. In this article, we are primarily interested in those derived from the (parametric) annihilator $\operatorname{ann}_{D[s]}(f^s)$ of f^s:

Definition 1.2. Let s be a new variable, and denote by $R_f[s] \cdot f^s$ the free module generated by f^s over the localized ring $R_f[s] = R[f^{-1}, s]$. Via the chain rule

$$\partial_i \bullet \left(\frac{g}{f^k} f^s \right) = \partial_i \bullet \left(\frac{g}{f^k} \right) f^s + \frac{sg}{f^{k+1}} \cdot \frac{\partial f}{\partial x_i} f^s \tag{1-1}$$

for each $g(x, s) \in R[s]$, $R_f[s] \cdot f^s$ acquires the structure of a left $D[s]$-module. Denote by

$$\operatorname{ann}_{D[s]}(f^s) = \{P \in D[s] \mid P \bullet f^s = 0\}$$

the *parametric annihilator*, and by

$$\mathscr{M}_f(s) = D[s]/\operatorname{ann}_{D[s]}(f^s)$$

the cyclic $D[s]$-module generated by $1 \cdot f^s \in R_f[s] \cdot f^s$.

Bernstein's functional equation [1972] asserts the existence of a differential operator $P(x, \partial, s)$ and a nonzero polynomial $b_{f,P}(s) \in \mathbb{C}[s]$ such that

$$P(x, \partial, s) \bullet f^{s+1} = b_{f,P}(s) \cdot f^s, \tag{1-2}$$

i.e., the existence of the element $P \cdot f - b_{f,P}(s) \in \operatorname{ann}_{D[s]}(f^s)$. Bernstein's result implies that $D[s] \bullet f^s$ is D-coherent (while $R_f[s] f^s$ is not).

Definition 1.3. The monic generator of the ideal in $\mathbb{C}[s]$ generated by all $b_{f,P}(s)$ appearing in an equation (1-2) is the *Bernstein–Sato polynomial* $b_f(s)$. Denote $\rho_f \subseteq \mathbb{C}$ the set of roots of $b_f(s)$.

Note that the operator P in the functional equation is only determined up to $\operatorname{ann}_{D[s]}(f^s)$. See [Björk 1979] for an elementary proof of the existence of $b_f(s)$. Alternative (and more general) proofs are given in [Kashiwara 2003]; see also [Bernstein ca. 1997; Mebkhout and Narváez-Macarro 1991; Núñez-Betancourt 2013].

The $\mathbb{C}[s]$-module $\mathscr{M}_f(s)/\mathscr{M}_f(s+1)$ is precisely annihilated by $b_f(s)$. It is an interesting problem to determine for any $q(s) \in \mathbb{C}[s]$ the ideals

$$\mathfrak{a}_{f,q(s)} = \left\{ g \in R \mid q(s) g f^s \in D[s] \bullet f^{s+1} \right\}$$

from [Walther 2005]. By [Malgrange 1975],

$$\mathfrak{a}_{f,s+1} = R \cap (\operatorname{ann}_{D[s]}(f^s) + D[s] \cdot (f, J_f)).$$

Question 1.4. Is $\mathfrak{a}_{f,s+1} = J_f + (f)$?

A positive answer would throw light on connections between $b_f(s)$ and cohomology of Milnor fibers.

Remark 1.5. At the 1954 International Congress of Mathematics in Amsterdam, I. M. Gelfand asked the following question. Given a real analytic function $f\colon \mathbb{R}^n \to \mathbb{R}$, the assignment ($s \in \mathbb{C}$)

$$f(x)^s_+ = \begin{cases} f(x)^s & \text{if } f(x) > 0, \\ 0 & \text{if } f(x) \le 0. \end{cases}$$

is continuous in x and analytic in s where the real part of s is positive. Can one analytically continue $f(x)^s_+$? Sato introduced $b_f(s)$ in order to answer Gelfand's question; Bernstein [1972] established their existence in general.

Remark 1.6. Let $m \in M$ be a nonzero section of a holonomic D-module. Generalizing the case $1 \in R$ there is a functional equation

$$P(x, \partial, s) \bullet (m f^{s+1}) = b_{f,P;m}(s) \cdot m f^s$$

with $b_{f,P;m}(s) \in \mathbb{C}[s]$ nonzero. The monic generator of the ideal $\{b_{f,P;m}(s)\}$ is the b-function $b_{f;m}(s)$ [Kashiwara 1976].

2. Parameters and numbers

For any complex number γ, the expression f^γ represents, locally outside $\operatorname{Var}(f)$, a multivalued analytic function. Via the chain rule as in (1-1), the cyclic R_f-module $R_f \cdot f^\gamma$ becomes a left D-module, and we set

$$\mathscr{M}_f(\gamma) = D \bullet f^\lambda \cong D/\operatorname{ann}_D(f^\gamma).$$

There are natural $D[s]$-linear maps

$$\operatorname{ev}_f(\gamma)\colon \mathscr{M}_f(s) \to \mathscr{M}_f(\gamma), \qquad P(x, \partial, s) \bullet f^s \mapsto P(x, \partial, \gamma) \bullet f^\gamma,$$

and D-linear inclusions

$$\operatorname{inc}_f(s)\colon \mathscr{M}_f(s+1) \to \mathscr{M}_f(s), \qquad P(x, \partial, s) \bullet f^{s+1} \mapsto P(x, \partial, s) \cdot f \bullet f^s$$

with cokernel $\mathscr{N}_f(s) = \mathscr{M}_f(s)/\mathscr{M}_f(s+1) \cong D[s]/(\operatorname{ann}_{D[s]}(f^s) + D[s]f)$, and

$$\operatorname{inc}_f(\gamma)\colon \mathscr{M}_f(\gamma+1) \to \mathscr{M}_f(\gamma), \qquad P(x, \partial) \bullet f^{\lambda+1} \mapsto P(x, \partial) \cdot f \bullet f^\lambda$$

with cokernel $\mathscr{N}_f(\gamma) = \mathscr{M}_f(\gamma)/\mathscr{M}_f(\gamma+1) \cong D/(\operatorname{ann}_D(f^\gamma) + D \cdot f)$.

The kernel of the morphism $\operatorname{ev}_f(\gamma)$ contains the (two-sided) ideal $D[s](s-\gamma)$; the containment can be proper, for example if $\gamma = 0$. If $\{\gamma - 1, \gamma - 2, \ldots\}$ is disjoint from the root set ρ_f then $\ker \operatorname{ev}_f(\gamma) = D[s] \cdot (s - \gamma)$ [Kashiwara 1976]. If $\gamma \notin \rho_f$ then $\operatorname{inc}_f(\gamma)$ is an isomorphism because of the functional equation; if

$\gamma = -1$, or if $b_f(\gamma) = 0$ while ρ_f does not meet $\{\gamma-1, \gamma-2, \dots\}$ then $\mathrm{inc}_f(\gamma)$ is not surjective [Walther 2005].

Question 2.1. Does $\mathrm{inc}_f(\gamma)$ fail to be an isomorphism for all $\gamma \in \rho_f$?

In contrast, the induced maps $\mathscr{M}_f(s)/(s-\gamma-1) \to \mathscr{M}_f(s)/(s-\gamma)$ are isomorphisms exactly when $\gamma \notin \rho_f$ [Björk 1993, 6.3.15]. The morphism $\mathrm{inc}_f(s)$ is never surjective as $s+1$ divides $b_f(s)$. One sets

$$\tilde{b}_f(s) = \frac{b_f(s)}{s+1}.$$

By [Torrelli 2009, 4.2], the following are equivalent for a section $m \neq 0$ of a holonomic module:

- the smallest integral root of $b_{f;m}(s)$ is at least $-\ell$;
- $(D \bullet m) \otimes_R R[f^{-1}]$ is generated by $m/f^\ell = m \otimes 1/f^\ell$;
- $(D \bullet m) \otimes_R R[f^{-1}]/D \bullet (m \otimes 1)$ is generated by m/f^ℓ;
- $D[s] \bullet mf^s \to (D \bullet m) \otimes_R R[f^{-1}]$, $P(s) \bullet (mf^s) \mapsto P(-\ell) \bullet (m/f^\ell)$ is an epimorphism with kernel $D[s] \cdot (s+\ell)mf^s$.

Definition 2.2. We say that f satisfies condition

(A_1) (resp. (A_s)) if $\mathrm{ann}_D(1/f)$ (resp. $\mathrm{ann}_D(f^s)$) is generated by operators of order one;

(B_1) if R_f is generated by $1/f$ over D.

Condition (A_1) implies (B_1) in any case [Torrelli 2004]. Local Euler-homogeneity, (A_s) and (B_1) combined imply (A_1) [Torrelli 2007], and for Koszul free divisors (see Definition 4.7 below) this implication can be reversed [Torrelli 2004].

Condition (A_1) does not imply (A_s): $f = xy(x+y)(x+yz)$ is free (see Definition 4.1), and locally Euler-homogeneous and satisfies (A_1) and (B_1) [Calderón-Moreno 1999; Calderón-Moreno et al. 2002; Calderón-Moreno and Narváez-Macarro 2002b; Castro-Jiménez and Ucha 2001; Torrelli 2004], but $\mathrm{ann}_{D[s]}(f^s)$ and $\mathrm{ann}_D(f^s)$ require a second order generator.

Condition (A_1) implies local Euler-homogeneity if f has isolated singularities [Torrelli 2002], or if it is Koszul-free or of the form $z^n - g(x,y)$ for reduced g [Torrelli 2004]. In [Castro-Jiménez et al. 2007] it is shown that for certain locally weakly quasihomogeneous free divisors $\mathrm{Var}(f)$, (A_1) holds for high powers of f, and even for f itself by [Narváez-Macarro 2008, Remark 1.7.4].

For an isolated singularity, f has (A_1) if and only if it has (B_1) and is quasihomogeneous [Torrelli 2002]. For example, a reduced plane curve (has automatically (B_1) and) has (A_1) if and only if it is quasihomogeneous. See [Schulze 2007] for further results.

Condition (B_1) is equivalent to $\mathrm{inc}_f(-2), \mathrm{inc}_f(-3), \ldots$ all being isomorphisms, and also to -1 being the only integral root of $b_f(s)$ [Kashiwara 1976]. Locally quasihomogeneous free divisors satisfy condition (B_1) at any point [Castro-Jiménez and Ucha 2002].

3. V-filtration and Bernstein–Sato polynomials

3A. *V-filtration.* The articles [Saito 1994; Maisonobe and Mebkhout 2004; Budur 2005; Budur 2012b] are recommended for material on V-filtrations.

3A1. *Definition and basic properties.* Let Y be a smooth complex manifold (or variety), and let X be a closed submanifold (or -variety) of Y defined by the ideal sheaf $\mathscr{I}$. The V-filtration on $\mathscr{D}_Y$ along X is, for $k \in \mathbb{Z}$, given by

$$V^k(\mathscr{D}_Y) = \{P \in \mathscr{D}_Y \mid P \bullet \mathscr{I}^{k'} \subseteq \mathscr{I}^{k+k'} \text{ for all } k' \in \mathbb{Z}\},$$

with the understanding that $\mathscr{I}^{k'} = \mathscr{O}_Y$ for $k' \leq 0$. The associated graded sheaf of rings $\mathrm{gr}_V(\mathscr{D}_Y)$ is isomorphic to the sheaf of rings of differential operators on the normal bundle $T_X(Y)$, algebraic in the fiber of the bundle.

Suppose that $Y = \mathbb{C}^n \times \mathbb{C}$ with coordinate function t on $\mathbb{C}$, and let X be the hyperplane $t = 0$. Then $V^k(D_Y)$ is spanned by $\{x^u \partial^v t^a \partial_t^b \mid a - b \geq k\}$. Given a coherent holonomic D_Y-module M with regular singularities in the sense of [Kashiwara and Kawai 1981b], Kashiwara [1983] and Malgrange [1983] define an exhaustive decreasing rationally indexed filtration on M that is compatible with the V-filtration on D_Y and has the following properties:

(1) Each $V^\alpha(M)$ is coherent over $V^0(D_Y)$ and the set of α with nonzero $\mathrm{gr}_V^\alpha(M) = V^\alpha(M)/V^{>\alpha}(M)$ has no accumulation point.

(2) For $\alpha \gg 0$, $V^1(D_Y)V^\alpha(M) = V^{\alpha+1}(M)$.

(3) $t\partial_t - \alpha$ acts nilpotently on $\mathrm{gr}_V^\alpha(M)$.

The V-filtration is unique and can be defined in somewhat greater generality [Budur 2005]. Of special interest is the following case considered in [Malgrange 1983; Kashiwara 1983].

Notation 3.1. Denote $R_{x,t}$ the polynomial ring $R[t]$, t a new indeterminate, and let $D_{x,t}$ be the corresponding Weyl algebra. Fix $f \in R$ and consider the regular $D_{x,t}$-module

$$\mathscr{B}_f = H^1_{f-t}(R[t]),$$

the unique local cohomology module of $R[t]$ supported in $f - t$. Then $\mathscr{B}_f$ is naturally isomorphic as $D_{x,t}$-module to the direct image (in the D-category) $i_+(R)$ of R under the graph embedding

$$i : X \to X \times \mathbb{C}, \quad x \mapsto (x, f(x)).$$

Moreover, extending (1-1) via

$$t \bullet (g(x,s) f^{s-k}) = g(x,s+1) f^{s+1-k},$$
$$\partial_t \bullet (g(x,s) f^{s-k}) = -s g(x,s-1) f^{s-1-k},$$

the module $R_f[s] \otimes f^s$ becomes a $D_{x,t}$-module extending the $D[s]$-action where $-\partial_t t$ acts as s.

The existence of the V-filtration on $\mathscr{B}_f = i_+(R)$ is equivalent to the existence of generalized b-functions $b_{f;\eta}(s)$ in the sense of [Kashiwara 1976]; see [Kashiwara 1978; Malgrange 1983]. In fact, one can recover one from the other:

$$V^\alpha(\mathscr{B}_f) = \{\eta \in \mathscr{B}_f \mid [b_{f;\eta}(-c) = 0] \Rightarrow [\alpha \le c]\}$$

and the multiplicity of $b_{f;\eta}(s)$ at α is the degree of the minimal polynomial of $s-\alpha$ on $\mathrm{gr}_V^\alpha(D[s]\eta f^s / D[s]\eta f^{s+1})$ [Sabbah 1987a]. For more on this "microlocal approach", see [Saito 1994].

3B. ***The log-canonical threshold.*** By [Kollár 1997] (see also [Lichtin 1989; Yano 1978]), the absolute value of the largest root of $b_f(s)$ is the *log-canonical threshold* $\mathrm{lct}(f)$ given by the supremum of all numbers s such that the local integrals

$$\int_{U \ni p} \frac{|dx|}{|f|^{2s}}$$

converge for all $p \in X$ and all small open U around p. Smaller lct corresponds to worse singularities; the best one can hope for is $\mathrm{lct}(f) = 1$ as one sees by looking at a smooth point. The notion goes back to Arnol'd, who called it (essentially) the complex singular index [Arnold et al. 1985].

The point of *multiplier ideals* is to force the finiteness of the integral by allowing moderating functions in the integral:

$$\mathscr{I}(f,\lambda)_p = \{g \in \mathscr{O}_X \mid g/f^\lambda \text{ is } L^2\text{-integrable near } p \in \mathrm{Var}(f)\},$$

for $\lambda \in \mathbb{R}$. By [Ein et al. 2004], there is a finite collection of *jumping numbers* for f of rational numbers $0 = \alpha_0 < \alpha_1 < \cdots < \alpha_\ell = 1$ such that $\mathscr{I}(f,\alpha)$ is constant on $[\alpha_i, \alpha_{i+1})$ but $\mathscr{I}(f,\alpha_i) \neq \mathscr{I}(f,\alpha_{i+1})$. The log-canonical threshold appears as α_1. These ideas had appeared before in [Lipman 1982; Loeser and Vaquié 1990].

Generalizing Kollar's approach, each α_i is a root of $b_f(s)$ [Ein et al. 2004]. In [Saito 2007, Theorem 4.4] a partial converse is shown for locally Euler-homogeneous divisors. Extending the idea of jumping numbers to the range $\alpha > 1$ one sees that α is a jumping number if and only if $\alpha + 1$ is a jumping number, but the connection to the Bernstein–Sato polynomial is lost in general. For example, if $f(x,y) = x^2 + y^3$ then jumping numbers are $\{5/6, 1\} + \mathbb{N}$ while $b_f(s) = (s+5/6)(s+1)(s+7/6)$.

3C. *Bernstein–Sato polynomial.* The roots of $b_f(s)$ relate to an astounding number of other invariants, see for example [Kollár 1997] for a survey. However, besides the functional equation there is no known way to describe ρ_f.

3C1. *Fundamental results.* Let $p \in \mathbb{C}^n$ be a closed point, cut out by the maximal ideal $\mathfrak{m} \subseteq R$. Extending R to the localization $R_{\mathfrak{m}}$ (or even the ring of holomorphic functions at p) one arrives at potentially larger sets of polynomials $b_{f,P}(s)$ that satisfy a functional equation (1-2) with $P(x, \partial, s)$ now in the correspondingly larger ring of differential operators. The *local* (resp. *local analytic*) Bernstein–Sato polynomial $b_{f,p}(s)$ (resp. $b_{f,p^{an}}(s)$) is the generator of the resulting ideal generated by the $b_{f,P}(s)$ in $\mathbb{C}[s]$. We denote by $\rho_{f,p}$ (resp. $\tilde{\rho}_{f,p}$) the root set of $b_{f,p^{an}}(s)$ (resp. $b_{f,p^{an}}(s)/(s+1)$). From the definitions and [Lyubeznik 1997b; Briançon et al. 2000; Briançon and Maynadier 1999] we have

$$b_{f,p^{an}}(s)|b_{f,p}(s)|b_f(s) = \operatorname{lcm}_{p\in\operatorname{Var}(f)} b_{f,p}(s) = \operatorname{lcm}_{p\in\operatorname{Var}(f)} b_{f,p^{an}}(s), \quad (3\text{-}1)$$

and the function $\mathbb{C}^n \ni p \mapsto \operatorname{Var}(b_f(s))$, counting with multiplicity, is upper semi-continuous in the sense that for p' sufficiently near p one has $b_{f,p'}(s)\,|\,b_{f,p}(s)$. The underlying reason is the coherence of D.

The Bernstein–Sato polynomial $b_f(s)$ factors over $\mathbb{Q}$ into linear factors, $\rho_f \subseteq \mathbb{Q}$, and all roots are negative [Malgrange 1975; Kashiwara 1976]. The proof uses resolution of singularities over $\mathbb{C}$ in order to reduce to simple normal crossing divisors, where rationality and negativity of the roots is evident. For this Kashiwara proves a comparison theorem [Kashiwara 1976, Theorem 5.1] that establishes $b_f(s)$ as a divisor of a shifted product of the least common multiple of the local Bernstein–Sato polynomials of the pullback of f under the resolution map. There is a refinement by Lichtin [1989] for plane curves. The roots of $b_f(s)$, besides being negative, are always greater than $-n$, n being the minimum number of variables required to express f locally analytically [Varchenko 1981; Saito 1994].

3C2. *Constructible sheaves from f^s.* Let $V = V(n, d)$ be the vector space of all complex polynomials in $x_1 \dots, x_n$ of degree at most d. Consider the function $\beta\colon V \ni f \mapsto b_f(s)$. By [Lyubeznik 1997b; Briançon and Maynadier 1999], there is an algebraic stratification of V such that on each stratum the function β is constant. For varying n, d these stratifications can be made to be compatible.

3C3. *Special cases.* If p is a smooth point of $\operatorname{Var}(f)$ then f can be used as an analytic coordinate near p, hence $b_{f,p^{an}}(s) = s+1$, and so $b_f(s) = s+1$ for all smooth hypersurfaces. By Proposition 2.6 in [Briançon and Maisonobe 1996], an extension of [Briançon et al. 1991], the equation $b_f(s) = s+1$ implies smoothness of $\operatorname{Var}(f)$. Explicit formulas for the Bernstein–Sato polynomial are rare; here are some classes of examples.

- $f = \prod x_i^{a_i}$: $P = \prod \partial_i^{a_i}$ up to a scalar, $b_f(s) = \prod_i \prod_{j=1}^{a_i}(s + j/a_j)$.
- f (quasi)homogeneous with isolated singularity at zero:

$$\tilde{b}_f(s) = \operatorname{lcm}\left(s + \frac{\deg(g\,\mathrm{d}x)}{\deg(f)}\right),$$

  where g runs through a (quasi)homogeneous standard basis for J_f by work of Kashiwara, Sato, Miwa, Malgrange, Kochman [Malgrange 1975; Yano 1978; Torrelli 2005; Kochman 1976]. Note that the Jacobian ring of such a singularity is an Artinian Gorenstein ring, whose duality operator implies symmetry of ρ_f.
- $f = \det(x_{i,j})_1^n$: $P = \det(\partial_{i,j})_1^n$, $b_f(s) = (s+1)\cdots(s+n)$. This is attributed to Cayley, but see the comments in [Caracciolo et al. 2013].
- For some hyperplane arrangements, $b_f(s)$ is known; see [Walther 2005; Budur et al. 2011c].
- A long list of examples is worked out in [Yano 1978].

If V is a complex vector space, G a reductive group acting linearly on V with open orbit U such that $V \smallsetminus U$ is a divisor $\operatorname{Var}(f)$, Sato's theory of prehomogeneous vectors spaces [Sato and Shintani 1974; Muro 1988; Sato 1990; Yano 1977] yields a factorization for $b_f(s)$. For reductive linear free divisors, Granger and Schulze [2010] and Sevenheck [2011] discuss symmetry properties of Bernstein–Sato polynomials. In [Narvaez-Macarro 2013] this theme is taken up again, investigating specifically symmetry properties of ρ_f when $D[s] \bullet f^s$ has a Spencer logarithmic resolution (see [Castro-Jiménez and Ucha 2002] for definitions). This covers locally quasihomogeneous free divisors, and more generally free divisors whose Jacobian is of linear type. The motivation is the fact that roots of $b_f(s)$ seem to come in strands, and whenever roots can be understood the strands appear to be linked to Hodge-theory.

There are several results on ρ_f for other divisors of special shape. Trivially, if $f(x) = g(x_1, \dots, x_k) \cdot h(x_{k+1}, \dots, x_n)$ then $b_f(s) \mid b_g(s) \cdot b_h(s)$; the question of equality appears to be open. In contrast, $b_f(s)$ cannot be assembled from the Bernstein–Sato polynomials of the factors of f in general, even if the factors are hyperplanes and one has some control on the intersection behavior; see Section 8 below. If $f(x) = g(x_1, \dots, x_k) + h(x_{k+1}, \dots, x_n)$ and at least one is locally Euler-homogeneous, then there are Thom–Sebastiani type formulas [Saito 1994]. In particular, diagonal hypersurfaces are completely understood.

3C4. *Relation to intersection homology module.* Suppose

$$Y = \operatorname{Var}(f_1, \dots, f_k) \subseteq X$$

is a complete intersection and denote by $\mathscr{H}^k_Y(\mathscr{O}_X)$ the unique (algebraic) local cohomology module of $\mathscr{O}_X$ along Y. Brylinski [1983; 1985], continuing work of Kashiwara, defined $\mathscr{L}(Y, X) \subseteq \mathscr{H}^k_Y(\mathscr{O}_X)$, the *intersection homology $\mathscr{D}_X$-module of Y*, the smallest $\mathscr{D}_X$-module equal to $\mathscr{H}^k_Y(\mathscr{O}_X)$ in the generic point(s). See also [Barlet and Kashiwara 1986]. The module $\mathscr{L}(X, Y)$ contains the fundamental class of Y in X [Barlet 1980].

Question 3.2. When is $\mathscr{L}(X, Y) = \mathscr{H}^k_Y(\mathscr{O}_X)$?

Equality is equivalent to $\mathscr{H}^k_Y(\mathscr{O}_X)$ being generated by the cosets of $\Delta/\prod_{i=1}^k f_i$ over $\mathscr{D}_X$ where Δ is the ideal generated by the k-minors of the Jacobian matrix of $f_1, \dots, f_k$. A necessary condition is that $1/\prod_{i=1}^k f_i$ generates $\mathscr{H}^k_Y(\mathscr{O}_X)$, but this is not sufficient: consider $xy(x+y)(x+yz)$, where $\rho_f = -\{\frac{1}{2}, \frac{3}{4}, 1, 1, 1, \frac{5}{4}\}$. Indeed by [Torrelli 2009], equality can be characterized in terms of functional equations, as the following are equivalent at $p \in X$:

(1) $\mathscr{L}(X, Y) = \mathscr{H}^k_Y(\mathscr{O}_X)$ in the stalk;

(2) $\tilde{\rho}_{f,p} \cap \mathbb{Z} = \varnothing$;

(3) 1 is not an eigenvalue of the monodromy operator on the reduced cohomology of the Milnor fibers near p.

If $1/\prod_{i=1}^k f_i$ generates $R[1/\prod f_i]$ and $1/\prod_{i=1}^k f_i \in \mathscr{L}(X, Y)$, then $\tilde{b}_f(-1) \neq 0$ [Torrelli 2009]. It seems unknown whether (irrespective of $1/\prod_{i=1}^k f_i$ generating $R[1/\prod f_i]$) the condition $\tilde{b}_f(-1) \neq 0$ is equivalent to $1/\prod_{i=1}^k f_i$ being in $\mathscr{L}(X, Y)$. See also [Massey 2009] for a topological viewpoint. (By the Riemann–Hilbert correspondence of [Kashiwara 1984] and [Mebkhout 1984], $\mathscr{L}(X, Y)$ corresponds to the intersection cohomology complex of Y on X [Brylinski 1983] and $\mathscr{H}^k_Y(\mathscr{O}_X)$ to $\mathbb{C}_Y[n-k]$ [Grothendieck 1966; Kashiwara 1976; Mebkhout 1977]. Equality then says: the link is a rational homology sphere). Barlet [1999] characterizes property (3) above in terms of currents for complexified real f. Equivalence of (1) and (3) for isolated singularities can be derived from [Milnor 1968; Brieskorn 1970]; the general case can be shown using [Saito 1990, 4.5.8] and the formalism of weights. For the case $k = 1$, (1) requires irreducibility; in general, there is a criterion in terms of b-functions [Torrelli 2009, 1.6, 1.10].

4. LCT and logarithmic ideal

4A. ***Logarithmic forms.*** Let $X = \mathbb{C}^n$ be the analytic manifold, f a holomorphic function on X, and $Y = \operatorname{Var}(f)$ a divisor in X with $j : U = X \smallsetminus Y \hookrightarrow X$ the embedding. Let $\Omega^\bullet_X(*Y)$ denote the complex of differential forms on X that are (at worst) meromorphic along Y. By [Grothendieck 1966], $\Omega^\bullet_X(*Y) \to \mathbb{R}j_*\mathbb{C}_U$ is a quasiisomorphism.

A form ω is *logarithmic* along Y if $f\omega$ and $f d\omega$ are holomorphic; these ω form the logarithmic de Rham complex $\Omega_X^\bullet(\log Y)$ on X along Y. The complex $\Omega_X^\bullet(\log Y)$ was first used with great effect on normal crossing divisors by Deligne [1971; 1974] in order to establish mixed Hodge structures, and later by Esnault and Viehweg [1992] in order to prove vanishing theorems. A major reason for the success of normal crossings is that in that case $\Omega_X^i(\log Y)$ is a locally free module over $\mathscr{O}_X$. The logarithmic de Rham complex was introduced in [Saito 1980] for general divisors.

4B. *Free divisors.*

Definition 4.1. A divisor $\mathrm{Var}(f)$ is *free* if (locally) $\Omega_X^1(\log f)$ is a free $\mathscr{O}_X$-module.

For a nonsmooth locally Euler-homogeneous divisor, freeness is equivalent to the Jacobian ring $\mathscr{O}_X/J_f$ being a codimension-2 Cohen–Macaulay $\mathscr{O}_X$-module; in general, freeness is equivalent to the Tjurina algebra $R/(f, \partial f/\partial x_1, \dots, \partial f/\partial x_n)$ being of projective dimension 2 or less over R. See [Saito 1980; Aleksandrov 1986] for relations to determinantal equations. Free divisors have rather big singular locus, and are in some ways at the opposite end from isolated singularities in the singularity zoo. If $\Omega_X^1(\log f)$ is (locally) free, then $\Omega_X^i(\log f) \cong \bigwedge^i \Omega_X^i(\log f)$ and also (locally) free [Saito 1980]. A weakening is

Definition 4.2. A divisor $\mathrm{Var}(f)$ is *tame* if, for all $i \in \mathbb{N}$, (locally) $\Omega_X^i(\log f)$ has projective dimension at most i as a $\mathscr{O}_X$-module.

Plane curves are trivially free; surfaces in 3-space are trivially tame. Normal crossing divisors are easily shown to be free. Discriminants of (semi)versal deformations of an isolated complete intersection singularity (and some others) are free [Aleksandrov 1986; 1990; Looijenga 1984; Saito 1981; Damon 1998; Buchweitz et al. 2009]. Unitary reflection arrangements are free [Terao 1981].

Definition 4.3. The *logarithmic derivations* $\mathrm{Der}_X(-\log f)$ along $Y = \mathrm{Var}(f)$ are the $\mathbb{C}$-linear derivations $\theta \in \mathrm{Der}(\mathscr{O}_X; \mathbb{C})$ that satisfy $\theta \bullet f \in (f)$.

A derivation θ is logarithmic along Y if and only it is so along each component of the reduced divisor to Y [Saito 1980]. The modules $\mathrm{Der}_X(-\log f)$ and $\Omega_X^1(\log f)$ are reflexive and mutually dual over R. Moreover, $\Omega_X^i(\log f)$ and $\Omega_X^{n-i}(\log f)$ are dual.

4C. *LCT.*

Definition 4.4. If

$$\Omega_X^\bullet(\log Y) \to \Omega_X^\bullet(*Y) \tag{4-1}$$

is a quasiisomorphism, we say that *LCT holds for* Y.

We recommend [Narváez-Macarro 2008].

Remark 4.5. (1) This "logarithmic comparison theorem", a property of a divisor, is very hard to check explicitly. No general algorithms are known, even in $\mathbb{C}^3$ (but see [Castro-Jiménez and Takayama 2009] for $n = 2$).

(2) LCT fails for rather simple divisors such as $f = x_1x_2 + x_3x_4$.

(3) If Y is a reduced normal crossing divisor, Deligne [1970] proved (4-1) to be a filtered (by pole filtration) quasiisomorphism; this provided a crucial step in the development of the theory of mixed Hodge structures [Deligne 1971; 1974].

(4) Limiting the order of poles in forms needed to capture all cohomology of U started with the seminal [Griffiths 1969a; 1969b] and continues; see for example [Deligne and Dimca 1990; Dimca 1991; Karpishpan 1991].

(5) The free case was studied for example in [Castro-Jiménez et al. 1996]. But even in this case, LCT is not understood.

(6) If f is quasihomogeneous with an isolated singularity at the origin, then LCT for f is equivalent to a topological condition (the link of f at the origin being a rational homology sphere), as well as an arithmetic one on the Milnor algebra of f [Holland and Mond 1998]. In [Schulze 2010], using the Gauss–Manin connection, this is extended to a list of conditions on an isolated hypersurface singularity, each one of which forces the implication [D has LCT] $\Rightarrow$ [D is quasihomogeneous].

(7) For a version regarding more general connections, see [Calderón-Moreno and Narváez-Macarro 2009].

A plane curve satisfies LCT if and only it is locally quasihomogeneous [Calderón-Moreno et al. 2002]. By [Castro-Jiménez et al. 1996], free locally quasihomogeneous divisors satisfy LCT in any dimension. By [Granger and Schulze 2006a], in dimension three, free divisors with LCT must be locally Euler-homogeneous. Conjecturally, LCT implies local Euler-homogeneity [Calderón-Moreno et al. 2002]. The converse is false, see for example [Castro-Jiménez and Ucha 2005]. The classical example of rotating lines with varying cross-ratio $f = xy(x+y)(x+yz)$ is free, satisfies LCT and is locally Euler-homogeneous, but only weakly quasihomogeneous [Calderón-Moreno et al. 2002]. In [Castro-Jiménez et al. 2007], the effect of the Spencer property on LCT is discussed in the presence of homogeneity conditions. For locally quasihomogeneous divisors (or if the nonfree locus is zero-dimensional), LCT implies (B_1) [Castro-Jiménez and Ucha 2002; Torrelli 2007]. In particular, LCT implies (B_1) for divisors with isolated singularities. In [Granger and Schulze 2006b] quasihomogeneity of isolated singularities is characterized in terms of a map of local cohomology modules of logarithmic differentials.

A free divisor is *linear free* if the (free) module $\mathrm{Der}_X(-\log f)$ has a basis of linear vector fields. In [Granger et al. 2009], linear free divisors in dimension at most 4 are classified, and for these divisors LCT holds at least on global sections. In the process, it is shown that LCT is implied if the Lie algebra of linear logarithmic vector fields is reductive. The example of $n \times n$ invertible upper triangular matrices acting on symmetric matrices [Granger et al. 2009, Example 5.1] shows that LCT may hold without the reductivity assumption. Linear free divisors appear naturally, for example in quiver representations and in the theory of prehomogeneous vector spaces and castling transformations [Buchweitz and Mond 2006; Sato and Kimura 1977; Granger et al. 2011]. Linear freeness is related to unfoldings and Frobenius structures [de Gregorio et al. 2009].

Denote by $\mathrm{Der}_{X,0}(-\log f)$ the derivations θ with $\theta \bullet f = 0$. In the presence of a global Euler-homogeneity E on Y there is a splitting

$$\mathrm{Der}_X(-\log f) \cong R \cdot E \oplus \mathrm{Der}_{X,0}(-\log f).$$

Reading derivations as operators of order one,

$$\mathrm{Der}_{X,0}(-\log f) \subseteq \mathrm{ann}_D(f^s).$$

We write S for $\mathrm{gr}_{(0,1)}(D)$; if y_i is the symbol of ∂_i then we have $S = R[y]$.

Definition 4.6. The inclusion $\mathrm{Der}_{X,0}(-\log f) \hookrightarrow \mathrm{ann}_D(f^s)$, via the order filtration, defines a subideal of $\mathrm{gr}_{(0,1)}(\mathrm{ann}_D(f^s)) \subseteq \mathrm{gr}_{(0,1)}(D) = S$ called the *logarithmic ideal* L_f of $\mathrm{Var}(f)$.

Note that the symbols of $\mathrm{Der}_X(-\log f)$ are in the ideal $R \cdot y$ of height n.

Definition 4.7. If $\mathrm{Der}_X(-\log f)$ has a generating set (as an R-module) whose symbols form a regular sequence on S, then Y is called *Koszul free*.

As $\mathrm{Der}_X(-\log f)$ has rank n, a Koszul free divisor is indeed free. Divisors in the plane [Saito 1980] and locally quasihomogeneous free divisors [Calderón-Moreno and Narváez-Macarro 2002b; 2002a] are Koszul free. In the case of normal crossings, this has been used to make resolutions for $D[s] \bullet f^s$ and $D[s]/D[s](\mathrm{ann}_{D[s]} f^s, f)$ [Gros and Narváez-Macarro 2000]. A way to distill invariants from resolutions of $D[s] \bullet f^s$ is given in [Arcadias 2010]. The logarithmic module $\widetilde{M}^{\log f} = D/D \cdot \mathrm{Der}_X(-\log f)$ has in the Spencer case (see [Castro-Jiménez and Ucha 2002; Calderón-Moreno and Narváez-Macarro 2005]) a natural free resolution of Koszul type.

For Koszul-free divisors, the ideal $D \cdot \mathrm{Der}_X(-\log f)$ is holonomic [Calderón-Moreno 1999]. By [Granger et al. 2009, Theorem 7.4], in the presence of freeness, the Koszul property is equivalent to the local finiteness of Saito's logarithmic stratification. This yields an algorithmic way to certify (some) free divisors as not locally quasihomogeneous, since free locally quasihomogeneous divisors are Koszul free. Based on similar ideas, one may devise a test for strong local Euler-homogeneity [Granger et al. 2009, Lemma 7.5]. See [Calderón-Moreno 1999] and [Torrelli 2007, Section 2] for relations of Koszul freeness to perversity of the logarithmic de Rham complex.

Castro-Jiménez and Ucha established conditions for $Y = \mathrm{Var}(f)$ to have LCT in terms of D-modules [Castro-Jiménez and Ucha 2001; 2002; 2004b] for certain free f. For example, LCT is equivalent to (A_1) for Spencer free divisors. Calderón-Moreno and Narváez-Macarro [2005] proved that free divisors have LCT if and only if the natural morphism $\mathscr{D}_X \otimes^L_{V^0(\mathscr{D}_X)} \mathscr{O}_X(Y) \to \mathscr{O}_X(*Y)$ is a quasiisomorphism, $\mathscr{O}_X(Y)$ being the meromorphic functions with simple pole along f. For Koszul free Y, one has at least

$$\mathscr{D}_X \otimes^L_{V^0(\mathscr{D}_X)} \mathscr{O}_X(Y) \cong \mathscr{D}_X \otimes_{V^0(\mathscr{D}_X)} \mathscr{O}_X(Y).$$

A similar condition ensures that the logarithmic de Rham complex is perverse [Calderón-Moreno 1999; Calderón-Moreno and Narváez-Macarro 2005]. The two results are related by duality between logarithmic connections on $\mathscr{D}_X$ and the V-filtration [Castro-Jiménez and Ucha 2002; 2004a; Calderón-Moreno and Narváez-Macarro 2005].

It is unknown how LCT is related to (A_1) in general, but for quasihomogeneous polynomials with isolated singularities the two conditions are equivalent [Torrelli 2007].

4D. *Logarithmic linearity.*

Definition 4.8. We say that $f \in R$ satisfies (L_s) if the characteristic ideal of $\mathrm{ann}_D(f^s)$ is generated by symbols of derivations.

Condition (L_s) holds for isolated singularities [Yano 1978], locally quasihomogeneous free divisors [Calderón-Moreno and Narváez-Macarro 2002b], and locally strongly Euler-homogeneous holonomic tame divisors [Walther 2015]. Also, (L_s) plus (B_1) yields (A_1) for locally Euler-homogeneous f by [Kashiwara 1976]; see [Torrelli 2007].

The logarithmic ideal supplies an interesting link between $\Omega^\bullet_X(\log f)$ and $\mathrm{ann}_D(f^s)$ via approximation complexes: if f is holonomic, strongly locally Euler-homogeneous and also tame then the complex $(\Omega^\bullet_X(\log f)[y], y\,\mathrm{d}x)$ is a resolution of the logarithmic ideal L_f, and S/L_f is a Cohen–Macaulay domain of dimension $n+1$; if f is in fact free, S/L_f is a complete intersection [Narváez-Macarro 2008; Walther 2015].

Question 4.9. For locally Euler-homogeneous divisors, is $\mathrm{ann}_D(f^s)$ related to the cohomology of $(\Omega^\bullet_X(\log f)[y], y\,\mathrm{d}x)$?

5. Characteristic variety

We continue to assume that $X = \mathbb{C}^n$. For $f \in R$ let U_f be the open set defined by $\mathrm{d}f \neq 0 \neq f$. Because of the functional equation, $\mathscr{M}_f(s)$ is coherent over D [Bernstein 1972; Kashiwara 1976], and the restriction of $\mathrm{charV}(D[s] \bullet f^s)$ to U_f is the Zariski closure of

$$\left\{ \left(\xi, s \frac{\mathrm{d}f(\xi)}{f(\xi)} \right) \middle| \xi \in U_f, s \in \mathbb{C} \right\}; \tag{5-1}$$

it is an $(n+1)$-dimensional involutive subvariety of T^*U_f [Kashiwara 2003]. Ginsburg [1986] gives a formula for the characteristic cycle of $D[s] \bullet m f^s$ in terms of an intersection process for holonomic sections m.

In favorable cases, more can be said. By [Calderón-Moreno and Narváez-Macarro 2002b], if the divisor is reduced, free and locally quasihomogeneous then $\mathrm{ann}_{D[s]}(f^s)$ is generated by derivations, both $\mathscr{M}_f(s)$ and $\mathscr{N}_f(s)$ have Koszul–Spencer type resolutions, and so the characteristic varieties are complete intersections. In the more general case where f is locally strongly Euler-homogeneous, holonomic and tame, $\mathrm{ann}_D(f^s)$ is still generated by order one operators and the ideal of symbols of $\mathrm{ann}_D(f^s)$ (and hence the characteristic ideal of $\mathscr{M}_f(s)$ as well) is a Cohen–Macaulay prime ideal [Walther 2015]. Under these hypotheses, the characteristic ideal of $\mathscr{N}_f(s)$ is Cohen–Macaulay but usually not prime.

5A. *Stratifications.* By [Kashiwara and Schapira 1979], the resolution theorem of Hironaka can be used to show that there is a stratification of $\mathbb{C}^n$ such that for each holonomic D-module M, $\mathrm{charC}(M) = \bigsqcup_{\sigma \in \Sigma} \mu(M, \sigma) T^*_\sigma$ where T^*_σ is the closure of the conormal bundle of the smooth stratum σ in $\mathbb{C}^n$ and $\mu(M, \sigma) \in \mathbb{N}$.

For $D[s] \bullet f^s / D[s] \bullet f^{s+1}$ Kashiwara proved that if one considers a Whitney stratification S for f (for example the "canonical" stratification in [Damon and Mond 1991]) then the characteristic variety of the D-module $\mathscr{N}_f(s)$ is in the union of the conormal varieties of the strata $\sigma \in S$ [Yano 1978].

If one slices a pair (X, D) of a smooth space and a divisor with a hyperplane, various invariants of the divisor will behave well provided that the hyperplane is not "special". A prime example are Bertini and Lefschetz theorems. For D-modules, Kashiwara defined the notion of *noncharacteristic restriction*: the smooth hypersurface H is noncharacteristic for the D-module M if it meets each component of the characteristic variety of M transversally (see [Pham 1979] for an exposition). The condition assures that the inverse image functor attached to the embedding $H \hookrightarrow X$ has no higher derived functors for M. In [Dimca et al. 2006] these ideas are used to show that the V-filtration, and hence the multiplier ideals as well as nearby and vanishing cycle sheaves, behave nicely under noncharacteristic restriction.

5B. ***Deformations.*** Varchenko proved, via establishing constancy of Hodge numbers, that in a μ-constant family of isolated singularities, the spectrum is constant [Varchenko 1982]. In [Dimca et al. 2006] it is shown that the formation of the spectrum along the divisor $Y \subseteq X$ commutes with the intersection with a hyperplane transversal to any stratum of a Whitney regular stratification of D, and a weak generalization of Varchenko's constancy results for certain deformations of nonisolated singularities is derived.

In contrast, the Bernstein–Sato polynomial may not be constant along μ-constant deformations. Suppose $f(x) + \lambda g(x)$ is a 1-parameter family of plane curves with isolated singularities at the origin. If the Milnor number $\dim_{\mathbb{C}}(R/J_{(f+\lambda g)})$ is constant in the family, the singularity germs in the family are topologically equivalent [Tráng and Ramanujam 1976]; for discussion, see [Dimca 1992, Section 2]. However, in such a family $b_f(s)$ can vary, as it is a differential (but not a topological) invariant. Indeed, $f + \lambda g = x^4 + y^5 + \lambda x y^4$ has constant Milnor number 20, and the general curve (not quasihomogeneous in any coordinate system, as $\rho_{f+\lambda g}$ is not symmetric about -1; see Section 3C) has $-\rho_{f+\lambda g} = \{1\} \cup \frac{1}{20}\{9, 11, 13, 14, 17, 18, 19, 21, 22, 23, 26, 27\}$ while the special curve has $-\rho_f = -\rho_{f+\lambda g} \cup \{-31/20\} \smallsetminus \{-11/20\}$. See [Cassou-Noguès 1986] for details and similar examples based on Newton polytope considerations, and [Stahlke 1997] for deformations of plane diagonal curves.

6. Milnor fiber and monodromy

6A. ***Milnor fibers.*** Let $B(p, \varepsilon)$ denote the ε-ball around $p \in \operatorname{Var}(f) \subseteq \mathbb{C}^n$. Milnor [1968] proved that the diffeomorphism type of the open real manifold

$$M_{p,t_0,\varepsilon} = B(p, \varepsilon) \cap \operatorname{Var}(f - t_0)$$

is independent of ε, t_0 as long as $0 < |t_0| \ll \varepsilon \ll 1$. For $0 < \tau \ll \varepsilon \ll 1$ denote by M_p the fiber of the bundle $B(p, \varepsilon) \cap \{q \in \mathbb{C}^n \mid 0 < |f(q)| < \tau\} \to f(q)$.

The direct image functor for D-modules to the projection $\mathbb{C}^n \times \mathbb{C} \to \mathbb{C}$, $(x, t) \mapsto t$ turns the $D_{x,t}$-module $\mathscr{B}_f$ into the *Gauss–Manin system* $\mathscr{H}_f$. The D-module restriction of $H^k(\mathscr{H}_f)$ to $t = t_0$ is the k-th cohomology of the Milnor fibers along $\operatorname{Var}(f)$ for $0 < |t_0| < \tau$.

Fix a k-cycle $\sigma \in H_p(\operatorname{Var}(f - t_0))$ and choose $\eta \in H^k(\mathscr{H}_f)$. Deforming σ to a k-cycle over t using the Milnor fibration, one can evaluate $\int_{\sigma_t} \eta$. The Gauss–Manin system has Fuchsian singularities and these periods are in the Nilsson class [Malgrange 1974]. For example, the classical Gauss hypergeometric function saw the light of day the first time as solution to a system of differential equations attached to the variation of the Hodge structure on an elliptic curve (expressed as integrals of the first and second kind) [Brieskorn and Knörrer

1981]. In [Pham 1979] this point of view is taken to be the starting point. The techniques explained there form the foundation for many connections between f^s and singularity invariants attached to $\mathrm{Var}(f)$.

In [Budur 2003], a bijection (for $0 < \alpha \leq 1$) is established between a subset of the jumping numbers of f at $p \in \mathrm{Var}(f)$ and the support of the *Hodge spectrum* [Steenbrink 1989]

$$\mathrm{Sp}(f) = \sum_{\alpha \in \mathbb{Q}} n_\alpha(f) t^\alpha,$$

with $n_\alpha(f)$ determined by the size of the α-piece of Hodge component of the cohomology of the Milnor fiber of f at p. See also [Saito 1993; Varchenko 1981], and [Steenbrink 1987] for a survey on Hodge invariants. We refer to [Budur 2012b; Saito 2009] for many more aspects of this part of the story.

6B. *Monodromy.* The vector spaces $H^k(M_{p,t_0,\varepsilon}, \mathbb{C})$ form a smooth vector bundle over a punctured disk $\mathbb{C}^*$. The linear transformation $\mu_{f,p,k}$ on $H^k(M_{p,t_0,\varepsilon}, \mathbb{C})$ induced by $p \mapsto p \cdot \exp(2\pi i \lambda)$ is the k-th monodromy of f at p. Let $\chi_{f,p,k}(t)$ denote the characteristic polynomial of $\mu_{f,p,k}$, set

$$e_{f,p,k} = \{\gamma \in \mathbb{C} \mid \gamma \text{ is an eigenvalue of } \mu_{f,p,k}\}$$

and put $e_{f,p} = \bigcup e_{f,p,k}$.

For most (in a quantifiable sense) divisors f with given Newton diagram, a combinatorial recipe can be given that determines the alternating product $\prod(\chi_{f,p,k}(t))^{(-1)^k}$ [Varchenko 1976], similarly to A'Campo's formula in terms of an embedded resolution [A'Campo 1975].

6C. *Degrees, eigenvalues, and Bernstein–Sato polynomial.* By [Malgrange 1983; Kashiwara 1983], the exponential function maps the root set of the local analytic Bernstein–Sato polynomial of f at p onto $e_{f,p}$. The set $\exp(-2\pi i \tilde{\rho}_{f,p})$ is the set of eigenvalues of the monodromy on the Grothendieck–Deligne vanishing cycle sheaf $\phi_f(\mathbb{C}_{X,p})$. This was shown in [Saito 1994] by algebraic microlocalization.

If f is an isolated singularity, the Milnor fiber M_f is a bouquet of spheres, and $H^{n-1}(M_f, \mathbb{C})$ can be identified with the Jacobian ring R/J_f. Moreover, if f is quasihomogeneous, then under this identification R/J_f is a $\mathbb{Q}[s]$-module, s acting via the Euler operator, and $\tilde{\rho}_f$ is in bijection with the degree set of the nonzero quasihomogeneous elements in R/J_f. For nonisolated singularities, most of this breaks down, since R/J_f is not Artinian in that case. However, for homogeneous f, consider the *Jacobian module*

$$H^0_{\mathfrak{m}}(R/J_f) = \{g + J_f \mid \exists k \in \mathbb{N}, \forall i, x_i^k g \in J_f\}$$

and the canonical $(n-1)$-form

$$\eta=\sum_i x_i \mathrm{d}x_1\wedge\cdots\wedge\widehat{\mathrm{d}x_i}\wedge\cdots\wedge \mathrm{d}x_n.$$

Every class in $H^{n-1}(M_f;\mathbb{C})$ is of the form $g\eta$ for suitable $g\in R$, and there is a filtration on $H^{n-1}(M_f,\mathbb{C})$ induced by integration of $\mathscr{B}_f$ along $\partial_1,\dots,\partial_n$, with the following property: if $g\in R$ is the smallest degree homogeneous polynomial such that $g\eta$ represents a chosen element of $H^{n-1}(M_f,\mathbb{C})$ then $b_f(-(\deg(g\eta))/\deg(f))=0$ [Walther 2005]. Moreover, let $\bar g\neq 0$ by a homogeneous element in the Jacobian module and suppose that its degree $\deg(g\eta)=\deg(g)+\sum_i\deg(x_i)$ is between d and $2d$. Then, by [Walther 2015], $g\eta$ represents a nonzero cohomology class in $H^{n-1}(M_f,\mathbb{C})$ as in the isolated case.

6D. *Zeta functions.* The zeta function $Z_f(s)$ attached to a divisor $f\in R$ is the rational function

$$Z_f(s)=\sum_{I\subseteq S}\chi(E_I^*)\prod_{i\in I}\frac{1}{N_i s+\nu_i},$$

where $\pi:\left(Y,\bigcup_I E_i\right)\to(\mathbb{C}^n,\mathrm{Var}(f))$ is an embedded resolution of singularities, and N_i (resp. ν_i-1) are the multiplicities of E_i in $\pi^*(f)$ (resp. in the Jacobian of π). By results of Denef and Loeser [1992], $Z_f(s)$ is independent of the resolution.

Conjecture 6.1 (Topological Monodromy Conjecture).

(SMC) *Any pole of $Z_f(s)$ is a root of the Bernstein–Sato polynomial $b_f(s)$.*

(MC) *Any pole of $Z_f(s)$ yields under exponentiation an eigenvalue of the monodromy operator at some $p\in\mathrm{Var}(f)$.*

The strong version (SMC) implies (MC) by [Malgrange 1975; Kashiwara 1983]. Each version allows a generalization to ideals.

(SMC), formulated by Igusa [2000] and Denef–Loeser [1992] holds for

- reduced curves by [Loeser 1988] with a discussion on the nature of the poles by Veys [1993; 1990; 1995];
- certain Newton-nondegenerate divisors by [Loeser 1990];
- some hyperplane arrangements (see Section 8);
- monomial ideals in any dimension by [Howald et al. 2007].

Additionally, (MC) holds for

- bivariate ideals by Van Proeyen and Veys [2010];
- all hyperplane arrangements by [Budur et al. 2011b; 2011c];

- some partial cases: [Artal Bartolo et al. 2002; Lemahieu and Veys 2009] some surfaces; [Artal Bartolo et al. 2005] quasiordinary power series; [Lichtin and Meuser 1985; Loeser 1990] in certain Newton nondegenerate cases; [Igusa 1992; Kimura et al. 1990] for invariants of prehomogeneous vector spaces; [Lemahieu and Van Proeyen 2011] for nondegenerate surfaces.

Strong evidence for (MC) for $n = 3$ is procured in [Veys 2006]. The articles [Rodrigues 2004; Némethi and Veys 2012] explore what (MC) could mean on a normal surface as ambient space and gives some results and counterexamples to naive generalizations. See also [Denef 1991] and the introductions of [Bories 2013b; 2013] for more details in survey format.

A root of $b_f(s)$, a monodromy eigenvalue, and a pole of $Z_f(s)$ may have multiplicity; can the monodromy conjecture be strengthened to include multiplicities? This version of (SMC) was proved for reduced bivariate f in [Loeser 1988]; in [Melle-Hernández et al. 2009; 2010] it is proved for certain nonreduced bivariate f, and for some trivariate ones.

A different variation, due to Veys, of the conjecture is the following. Vary the definition of $Z_f(s)$ to $Z_{f;g}(s) = \sum_{I\subseteq S} \chi(E_I^*) \prod_{i\in I} 1/(N_i s + \nu_i')$, where ν_i' is the multiplicity of E_i in the pullback along π of some differential form g. (The standard case is when g is the volume form). Two questions arise: (1) varying over a suitable set G of forms g, can one generate all roots of $b_f(s)$ as poles of the resulting zeta functions? And if so, can one (2) do this such that the pole sets of all zeta functions so constructed are always inside ρ_f, so that

$$\rho_f = \{\alpha \mid \text{ there exists } g \in G, \lim_{s\to\alpha} Z_{f;g}(s) = \infty\}?$$

Némethi and Veys [2010; 2012] prove a weak version: if $n = 2$ then monodromy eigenvalues are exponentials of poles of zeta functions from differential forms.

The following is discussed in [Bories 2013a]. For some ideals with $n = 2$, (1) is false for the topological zeta function (even for divisors: consider $xy^5 + x^3y^2 + x^4y$). For monomial ideals with two generators in $n = 2$, (1) is correct; with more than two generators it can fail. Even in the former case, (2) can be false.

7. Multivariate versions

If $f = (f_1, \dots, f_r)$ defines a map $f\colon \mathbb{C}^n \to \mathbb{C}^r$, several b-functions can be defined:

(1) the univariate Bernstein–Sato polynomial $b_f(s)$ attached to the ideal $(f) \subseteq R$ from [Budur et al. 2006a];

(2) the multivariate Bernstein–Sato polynomials $b_{f,i}(s)$ of all elements $b(s)$ of $\mathbb{C}[s_1, \dots, s_r]$ such that there is an equation $P(x, \partial, s) \bullet f_i f^s = b(s) f^s$ in multiindex notation;

(3) the multivariate Bernstein–Sato ideal $B_{f,\mu}(s)$ for $\mu \in \mathbb{N}^r$ of all $b(s) \in \mathbb{C}[s_1, \dots, s_r]$ such that there is an equation $P(x, \partial, s) \bullet f^{s+\mu} = b(s) f^s$ in multiindex notation. The most interesting case is $\mu = \mathbf{1} = (1, \dots, 1)$;

(4) the multivariate Bernstein–Sato ideal $B_{f,\Sigma}(s)$ of all $b(s) \in \mathbb{C}[s_1, \dots, s_r]$ that multiply f^s into $\sum D[s] f_i f^s$ in multiindex notation.

The Bernstein–Sato polynomial in (1) above has been studied in the case of a monomial ideal in [Budur et al. 2006b] and more generally from the point of view of the Newton polygon in [Budur et al. 2006c]. While the roots for monomial ideals do not depend just on the Newton polygon, their residue classes modulo $\mathbb{Z}$ do.

Nontriviality of the quantities in (2)–(4) have been established in [Sabbah 1987c; 1987d; 1987b], but see also [Bahloul 2005]. The ideals $B_{f,\mu}(s)$ and $B_{f,\Sigma}(s)$ do not have to be principal [Ucha and Castro-Jiménez 2004; Bahloul and Oaku 2010]. In [Sabbah 1987c; Gyoja 1993] it is shown that $B_{f,\mu}(s)$ contains a polynomial that factors into linear forms with nonnegative rational coefficients and positive constant term. Bahloul and Oaku [2010] show that these ideals are local in the sense of (3-1).

The following would generalize Kashiwara's result in the univariate case as well as the results of Sabbah and Gyoja above.

Conjecture 7.1 [Budur 2012a]. *The Bernstein–Sato ideal $B_{f,\mu}(s)$ is generated by products of linear forms $\sum \alpha_i s_i + a$ with α_i, a nonnegative rational and $a > 0$.*

For $n = 2$, partial results by Cassou-Noguès and Libgober exist [2011]. In [Budur 2012a] it is further conjectured that the Malgrange–Kashiwara result, exponentiating $\rho_{f,p}$ gives $e_{f,p}$, generalizes: monodromy in this case is defined in [Verdier 1983], and Sabbah's specialization functor ψ_f from [1990] takes on the rôle of the nearby cycle functor, and conjecturally exponentiating the variety of $B_{f,p}(s)$ yields the uniform support (near p) of Sabbah's functor. The latter conjecture would imply Conjecture 7.1.

Similarly to the one-variable case, if $V(n, d, m)$ is the vector space of (ordered) m-tuples of polynomials in $x_1, \dots, x_n$ of degree at most d, there is an algebraic stratification of $V(n, d, m)$ such that on each stratum the function $V \ni f = (f_1, \dots, f_m) \mapsto b_f(s)$ is constant. Corresponding results for the Bernstein–Sato ideal $B_{f,1}(s)$ hold by [Briançon et al. 2000].

8. Hyperplane arrangements

A *hyperplane arrangement* is a divisor of the form

$$\mathscr{A} = \prod_{i \in I} \alpha_i$$

where each α_i is a polynomial of degree one. We denote $H_i = \mathrm{Var}(\alpha_i)$. Essentially all information we are interested in is of local nature, so we assume that each α_i is a form so that $\mathscr{A}$ is *central*. If there is a coordinate change in $\mathbb{C}^n$ such that $\mathscr{A}$ becomes the product of polynomials in disjoint sets of variables, the arrangement is *decomposable*, otherwise it is *indecomposable*.

A *flat* is any (set-theoretic) intersection $\bigcap_{i \in J} H_i$ where $J \subseteq I$. The *intersection lattice* $L(\mathscr{A})$ is the partially ordered set consisting of the collection of all flats, with order given by inclusion.

8A. *Numbers and parameters.* Hyperplane arrangements satisfy (B_1) everywhere [Walther 2005]. Arrangements satisfy (A_1) everywhere if they decompose into a union of a generic and a hyperbolic arrangement [Torrelli 2004], and if they are tame [Walther 2015]. Terao conjectured that all hyperplane arrangements satisfy (A_1); some of them fail (A_s) [Walther 2015].

Apart from recasting various of the previously encountered problems in the world of arrangements, a popular study is the following: choose a discrete invariant I of a divisor. Does the function $\mathscr{A} \mapsto I(\mathscr{A})$ factor through the map $\mathscr{A} \mapsto L(\mathscr{A})$? Randell showed that if two arrangements are connected by a one-parameter family of arrangements which have the same intersection lattice, the complements are diffeomorphic [Randell 1989] and the isomorphism type of the Milnor fibration is constant [Randell 1997]. Rybnikov [2011] (see also [Artal Bartolo et al. 2006]) showed on the other hand that there are arrangements (even in the projective plane) with equal lattice but different complement. In particular, not all isotopic arrangements can be linked by a smooth deformation.

8B. *LCT and logarithmic ideal.* The most prominent positive result is one by Brieskorn [1973]: the *Orlik–Solomon algebra* $\mathrm{OS}(\mathscr{A}) \subseteq \Omega^\bullet(\log \mathscr{A})$ generated by the forms $\mathrm{d}\alpha_i/\alpha_i$ is quasiisomorphic to $\Omega^\bullet(*\mathscr{A})$, hence to the singular cohomology algebra of $U_{\mathscr{A}}$. The relation with combinatorics was given in [Orlik and Solomon 1980; Orlik and Terao 1992]. For a survey on the Orlik–Solomon algebra, see [Yuzvinsky 2001]. The best known open problem in this area is this:

Conjecture 8.1 [Terao 1978]. $\mathrm{OS}(\mathscr{A}) \to \Omega^\bullet(\log \mathscr{A})$ *is a quasiisomorphism.*

While the general case remains open, Wiens and Yuzvinsky [1997] proved it for tame arrangements, and also if $n \leq 4$. The techniques are based on [Castro-Jiménez et al. 1996].

8C. ***Milnor fibers.*** There is a survey article by Suciu on complements, Milnor fibers, and cohomology jump loci [Suciu 2014], and [Budur 2012b] contains further information on the topic. It is not known whether $L(\mathscr{A})$ determines the Betti numbers (even less the Hodge numbers) of the Milnor fiber of an arrangement. The first Betti number of the Milnor fiber $M_{\mathscr{A}}$ at the origin is stable under intersection with a generic hyperplane (if $n > 2$). But it is unknown whether the first Betti number of an arrangement in 3-space is a function of the lattice alone. By [Dimca et al. 2013], this is so for collections of up to 14 lines with up to 5-fold intersections in the projective plane. See also [Libgober 2012] for the origins of the approach. By [Budur et al. 2011a], a lower combinatorial bound for the λ-eigenspace of $H^1(M_{\mathscr{A}})$ is given under favorable conditions on L. If L satisfies stronger conditions, the bound is shown to be exact. In any case, [Budur et al. 2011a] gives an algebraic, although perhaps noncombinatorial, formula for the Hodge pieces in terms of multiplier ideals.

By [Orlik and Randell 1993], the Betti numbers of $M_{\mathscr{A}}$ are combinatorial if $\mathscr{A}$ is generic. See also [Cohen and Suciu 1995].

8D. ***Multiplier ideals.*** Mustaţă gave a formula for the multiplier ideals of arrangements, and used it to show that the log-canonical threshold is a function of $L(\mathscr{A})$. The formula is somewhat hard to use for showing that each jumping number is a lattice invariant; this problem was solved in [Budur and Saito 2010]. Explicit formulas in low dimensional cases follow from the spectrum formulas given there and in [Yoon 2013]. Teitler [2008] improved Mustaţă's formula [2006] for multiplier ideals to not necessarily reduced hyperplane arrangements.

8E. ***Bernstein–Sato polynomials.*** By [Walther 2005], $\rho_{\mathscr{A}} \cap \mathbb{Z} = \{-1\}$; by [Saito 2006], $\rho_{\mathscr{A}} \subseteq (-2, 0)$. There are few classes of arrangements with explicit formulas for their Bernstein–Sato polynomial:

- Boolean (a normal crossing arrangement, locally given by $x_1 \cdots x_k$);
- hyperbolic (essentially an arrangement in two variables);
- generic (central, and all intersections of n hyperplanes equal the origin).

The first case is trivial, the second is easy, the last is [Walther 2005] with assistance from [Saito 2007]. Some interesting computations are in [Budur et al. 2011c], and [Budur 2012a] has a partial confirmation of the multivariable Kashiwara–Malgrange theorem. The Bernstein–Sato polynomial is not determined by the intersection lattice [Walther 2015].

8F. ***Zeta functions.*** Budur, Mustaţă and Teitler [Budur et al. 2011b] show: (MC) holds for arrangements, and in order to prove (SMC), it suffices to show the following conjecture.

Conjecture 8.2. *For all indecomposable central arrangements with d planes in n-space,* $b_{\mathcal{A}}(-n/d) = 0$.

The idea is to use the resolution of singularities obtained by blowing up the dense edges from [Schechtman et al. 1995]. The corresponding computation of the zeta function is inspired from [Igusa 1974; 1975]. The number $-n/d$ does not have to be the log-canonical threshold. By [Budur et al. 2011b], Conjecture 8.2 holds in a number of cases, including reduced arrangements in dimension 3. By [Walther 2015] it holds for tame arrangements.

Examples of Veys (in 4 variables) show that (SMC) may hold even if Conjecture 8.2 were false in general, since $-n/d$ is not always a pole of the zeta function [Budur et al. 2011c]. However, in these examples, $-n/d$ is in fact a root of $b_f(s)$.

For arrangements, each monodromy eigenvalue can be captured by zeta functions in the sense of Némethi and Veys (see Section 6D), but not necessarily all of $\rho_{\mathcal{A}}$ (Veys and Walther, unpublished).

9. Positive characteristic

Let here $\mathbb{F}$ denote a field of characteristic $p > 0$. The theory of D-modules is rather different in positive characteristic compared to their behavior over the complex numbers. There are several reasons for this:

(1) On the downside, the ring D_p of $\mathbb{F}$-linear differential operators on $R_p = \mathbb{F}[x_1, \ldots, x_n]$ is no longer finitely generated: as an $\mathbb{F}$-algebra it is generated by the elements $\partial^{(\alpha)}$, $\alpha \in \mathbb{N}^n$, which act via $\partial^{(\alpha)} \bullet (x^\beta) = \binom{\beta}{\alpha} x^{\beta-\alpha}$.

(2) As a trade-off, one has access to the Frobenius morphism $x_i \mapsto x_i^p$, as well as the Frobenius functor $F(M) = R' \otimes_R M$ where R' is the $R-R$-bimodule on which R acts via the identity on the left, and via the Frobenius on the right. Lyubeznik [1997a] created the category of F-finite F-modules and proved striking finiteness results. The category includes many interesting D_p-modules, and all F-modules are D_p-modules.

(3) Holonomicity is more complicated; see [Bögvad 2002].

A most surprising consequence of Lyubeznik's ideas is that in positive characteristic the property (B_1) is meaningless: it holds for every $f \in R_p$ [Alvarez-Montaner et al. 2005]. The proof uses in significant ways the difference between D_p and the Weyl algebra. In particular, the theory of Bernstein–Sato polynomials is rather different in positive characteristic. In [Mustaţă 2009] a sequence of Bernstein–Sato polynomials is attached to a polynomial f assuming that the Frobenius morphism is finite on R (e.g., if $\mathbb{F}$ is finite or algebraically closed); these polynomials are then linked to test ideals, the finite characteristic counterparts to multiplier ideals. In [Blickle et al. 2009] variants of our modules $\mathcal{M}_f(\gamma)$ are introduced and [Núñez-Betancourt and Pérez 2013] shows that simplicity of these modules detects the F-thresholds from [Mustaţă et al. 2005]. These are cousins of the jumping numbers of multiplier ideals and related to the Bernstein–Sato polynomial via base-p-expansions. The Kashiwara–Brylinski intersection homology module was shown to exist in positive characteristic by Blickle in his thesis [Blickle 2004].

Appendix: Computability

by Anton Leykin

Computations around f^s can be carried out by hand in special cases. Generally, the computations are enormous and computers are required (although not often sufficient). One of the earliest such approaches are in [Briançon et al. 1989; Aleksandrov and Kistlerov 1992], but at least implicitly Buchberger's algorithm in a Weyl algebra was discussed as early as [Castro-Jimenez 1984]. An algorithmic approach to the isolated singularities case [Maisonobe 1994] preceded the general algorithms based on Gröbner bases in a noncommutative setting outlined below.

10A. ***Gröbner bases.*** The *monomials* $x^\alpha \partial^\beta$ with $\alpha, \beta \in \mathbb{N}^n$ form a $\mathbb{C}$-basis of D; expressing $p \in D$ as linear combination of monomials leads to its *normal form*. The monomial orders on the commutative monoid $[x, \partial]$ for which for all $i \in [n]$ the leading monomial of $\partial_i x_i = x_i \partial_i + 1$ is $x_i \partial_i$, can be used to run Buchberger's algorithm in D. Modifications are needed in improvements that exploit commutativity, but the naïve Buchberger's algorithm works without any changes. See [Kandri-Rody and Weispfenning 1990] for more general settings in polynomial rings of solvable type. Surprisingly, the worst case complexity of Gröbner bases computations in Weyl algebras is *not* worse than in the commutative polynomial case: it is doubly exponential in the number of indeterminates [Aschenbrenner and Leykin 2009; Grigoriev and Chistov 2008].

10B. ***Characteristic variety.*** A weight vector $(u, v) \in \mathbb{Z}^n \times \mathbb{Z}^n$ with $u + v \geq 0$ induces a filtration of D,

$$F_i = \mathbb{C} \cdot \{x^\alpha \partial^\beta \mid u \cdot \alpha + v \cdot \beta \leq i\}, \quad i \in \mathbb{Z}.$$

The (u, v)-*Gröbner deformation* of a left ideal $I \subseteq D$ is

$$\mathrm{in}_{(u,v)}(I) = \mathbb{C} \cdot \{\mathrm{in}_{(u,v)}(P) \mid P \in I\} \subseteq \mathrm{gr}_{(u,v)} D,$$

the ideal of *initial forms* of elements of I with respect to the given weight in the associated graded algebra. It is possible to compute Gröbner deformations in the homogenized Weyl algebra

$$D^h = D\langle h\rangle / \langle \partial_i x_i - x_i \partial_i - h^2, x_i h - h x_i, \partial_i h - h \partial_i \mid 1 \le i \le n\rangle;$$

see [Castro-Jimenez and Narváez-Macarro 1997; Oaku and Takayama 2001b]. Gröbner deformations are the main topic of [Saito et al. 2000].

10C. ***Annihilator.*** Recall the construction appearing in the beginning of Section 6A: $D_{x,t}$ acts on $D[s]f^s$; in particular, the operator $-\partial_t t$ acts as multiplication by s. It is this approach that lead Oaku to an algorithm for $\operatorname{ann}_{D[s]}(f^s)$, $\operatorname{ann}_D(f^s)$ and $b_f(s)$ [Oaku 1997]. We outline the ideas.

Malgrange observed that

$$\operatorname{ann}_{D[s]}(f^s) = \operatorname{ann}_{D_{x,t}}(f^s) \cap D[s], \tag{10-1}$$

$$\text{with} \quad \operatorname{ann}_{D_{x,t}}(f^s) = \left\langle t - f, \partial_1 + \frac{\partial f}{\partial x_1}\partial_t, \ldots, \partial_n + \frac{\partial f}{\partial x_n}\partial_t \right\rangle \subseteq D_{x,t}. \tag{10-2}$$

The former can be found from the latter by eliminating t and ∂_t from the ideal

$$\langle s + t\partial_t\rangle + \operatorname{ann}_{D_{x,t}}(f^s) \subseteq D_{x,t}\langle s\rangle; \tag{10-3}$$

of course $s = -\partial_t t$ does not commute with t, ∂_t here.

Oaku's method for $\operatorname{ann}_{D[s]}(f^s)$ accomplished the elimination by augmenting two commuting indeterminates:

$$\begin{aligned} &\operatorname{ann}_{D[s]}(f^s) = I'_f \cap D[s], \\ &I'_f = \left\langle t - uf, \partial_1 + u\frac{\partial f}{\partial x_1}\partial_t, \ldots, \partial_n + u\frac{\partial f}{\partial x_n}\partial_t, uv - 1 \right\rangle \subseteq D_{x,t}[u, v]. \end{aligned} \tag{10-4}$$

Now outright eliminate u, v. Note that I'_f is quasihomogeneous if the weights are $t, u \rightsquigarrow -1$ and $\partial_t, v \rightsquigarrow 1$, all other variables having weight zero. The homogeneity of the input and the relation $[\partial_t, t] = 1$ assures the termination of the computation. The operators of weight 0 in the output (with $-\partial_t t$ replaced by s) generate $I'_f \cap D[s]$.

A modification given in [Briançon and Maisonobe 2002] and used, e.g., in [Ucha and Castro-Jiménez 2004], reduces the number of algebra generators by one. Consider the subalgebra $D\langle s, \partial_t\rangle \subset D_{x,t}$; the relation $[s, \partial_t] = \partial_t$ shows that it is of solvable type. According to [Briançon and Maisonobe 2002],

$$\begin{aligned} &\operatorname{ann}_{D[s]}(f^s) = I''_f \cap D[s], \\ &I''_f = \left\langle s + f\partial_t, \partial_1 + \frac{\partial f}{\partial x_1}\partial_t, \ldots, \partial_n + \frac{\partial f}{\partial x_n}\partial_t \right\rangle \subset D\langle s, \partial_t\rangle. \end{aligned} \tag{10-5}$$

Note that $I''_f = \operatorname{ann}_{D_{x,t}}(f^s) \cap D\langle s, \partial_t\rangle$. The elimination step is done as in [Oaku 1997]; the decrease of variables usually improves performance. An algorithm to decide (A_1) for arrangements is given in [Àlvarez Montaner et al. 2007].

10D. ***Algorithms for the Bernstein–Sato polynomial.*** As the minimal polynomial of s on $\mathcal{N}_f(s)$, $b_f(s)$ can be obtained by means of linear algebra as a syzygy for the normal forms of powers of s modulo $\operatorname{ann}_{D[s]}(f^s) + D[s] \cdot f$ with respect to any fixed monomial order on $D[s]$. Most methods follow this path, starting with [Oaku 1997]. Variations appear in [Walther 1999; Oaku and Takayama 2001a; Oaku et al. 2000]; see also [Saito et al. 2000].

A different approach is to compute $b_f(s)$ without recourse to $\operatorname{ann}_{D[s]}(f^s)$, via a Gröbner deformation of the ideal $I_f = \operatorname{ann}_{D_{x,t}}(f^s)$ in (10-2) with respect to the weight $(-w, w)$ with $w = (0^n, 1) \in \mathbb{N}^{n+1}$: $\langle b_f(s)\rangle = \operatorname{in}_{(-w,w)}(I_f) \cap \mathbb{Q}[-\partial_t t]$. Here again, computing the minimal polynomial using linear algebra tends to provide some savings in practice.

In [Levandovskyy and Martín-Morales 2012] the authors give a method to check specific numbers for being in ρ_f. A method for $b_f(s)$ in the prehomogeneous vector space setup is in [Muro 2000].

10E. ***Stratification from $b_f(s)$.*** The Gröbner deformation $\operatorname{in}_{(-w,w)}(I_f)$ in the previous subsection can be refined as follows; see [Berkesch and Leykin 2010, Theorem 2.2]. Let $b(x, s)$ be nonzero in the polynomial ring $\mathbb{C}[x, s]$. Then $b(x, s) \in (\operatorname{in}_{(-w,w)} I_f) \cap \mathbb{C}[x, s]$ if and only if there exists $P \in D[s]$ satisfying the functional equation $b(x, s) f^s = P f f^s$. From this one can design an algorithm not only for computing the *local* Bernstein–Sato polynomial $b_{f,p}(s)$ for $p \in \operatorname{Var}(f)$, but also the stratification of $\mathbb{C}^n$ according to local Bernstein–Sato polynomials; see [Nishiyama and Noro 2010; Berkesch and Leykin 2010] for various approaches. Moreover, one can compute the stratifications from Section 3C2; see [Leykin 2001].

For the ideal case, [Andres et al. 2009] gives a method to compute an intersection of a left ideal of an associative algebra over a field with a subalgebra, generated by a single element. An application is a method for the computation of the Bernstein-Sato polynomial of an ideal. Another such was given by Bahloul [2001], and a version on general varieties was given by the same in [2003].

10F. ***Multiplier ideals.*** Consider polynomials $f_1, \dots, f_r \in \mathbb{C}[x]$, let f stand for $(f_1, \dots, f_r)$, s for $s_1, \dots, s_r$, and f^s for $\prod_{i=1}^r f_i^{s_i}$. In this subsection, let $D_{x,t} = \mathbb{C}\langle x, t, \partial_x, \partial_t\rangle$ be the $(n+r)$-th Weyl algebra.

Consider $D_{x,t}(s) \bullet f^s \subseteq R_{x,t}[f^{-1}, s] f^s$ and put

$$t_j \bullet h(x, s_1, \dots, s_j, \dots, s_r) f^s = h(x, s_1, \dots, s_j + 1, \dots, s_r) f_j f^s,$$

$$\partial_{t_j} \bullet h(x, s_1, \dots, s_j, \dots, s_r) f^s = -s_j h(x, s_1, \dots, s_j - 1, \dots, s_r) f_j^{-1} f^s,$$

for $h \in \mathbb{C}[x][f^{-1}, s]$, generalizing the univariate constructions.

The *generalized Bernstein–Sato polynomial* $b_{f,g}(\sigma)$ of f at $g \in \mathbb{C}[x]$ is the monic univariate polynomial b of the lowest degree for which there exist $P_k \in D_{x,t}$ such that

$$b(\sigma)gf^s = \sum_{k=1}^{r} P_k g f_k f^s, \quad \sigma = -\Big(\sum_{i=1}^{r} \partial_{t_i} t_i\Big). \tag{10-6}$$

An algorithm for $b_{f,g}(\sigma)$ is an essential ingredient for the algorithms in [Shibuta 2011; Berkesch and Leykin 2010] that compute the jumping numbers and corresponding multiplier ideals for $I = \langle f_1, \ldots, f_r \rangle$. That $b_{f,g}(\sigma)$ is related to multiplier ideals was worked out in [Budur et al. 2006a].

There are algorithms for special cases: monomial ideals [Howald 2001], hyperplane arrangements [Mustaţă 2006], and determinantal ideals [Johnson 2003]. A Macaulay2 package *MultiplierIdeals* by Teitler collects all implementations available in Macaulay2. See also [Budur 2005].

10G. *Software.* Algorithms for computing Bernstein–Sato polynomials have been implemented in *kan/sm1* [Takayama], *Risa/Asir* [Noro et al.], the *dmod_lib* library [Levandovskyy and Morales] for Singular [Decker et al. 2012], and the *D-modules* package [Leykin and Tsai] for Macaulay2 [Grayson and Stillman]. The best source of information of these is documentation in the current versions of the corresponding software. A relatively recent comparison of the performance for several families of examples is given in [Levandovskyy and Martín Morales 2008].

The following are articles by developers discussing their implementations: [Noro 2002; Nishiyama and Noro 2010; Oaku and Takayama 2001a; Andres et al. 2010; Levandovskyy and Morales; Leykin 2002; Berkesch and Leykin 2010].

Acknowledgements

We are greatly indebted to Nero Budur, Francisco J. Castro-Jiménez, Luis Narváez-Macarro, Wim Veys and an unknown referee for their careful reading of early versions of this article and their relentless hunt for errors. We claim ownership, and apologize for, all surviving mistakes, oversights and omissions.

References

[A'Campo 1975] N. A'Campo, "La fonction zêta d'une monodromie", *Comment. Math. Helv.* **50** (1975), 233–248. 6B

[Aleksandrov 1986] A. G. Aleksandrov, "Euler-homogeneous singularities and logarithmic differential forms", *Ann. Global Anal. Geom.* **4**:2 (1986), 225–242. 4B, 4B

[Aleksandrov 1990] A. G. Aleksandrov, "Nonisolated hypersurface singularities", pp. 211–246 in *Theory of singularities and its applications*, edited by V. I. Arnold, Adv. Soviet Math. **1**, Amer. Math. Soc., Providence, RI, 1990. 4B

[Aleksandrov and Kistlerov 1992] A. G. Aleksandrov and V. L. Kistlerov, "Computer method in calculating b-functions of nonisolated singularities", pp. 319–335 in *Proceedings of the International Conference on Algebra, part 3* (Novosibirsk, 1989), edited by L. A. Bokut et al., Contemp. Math. **131**, Amer. Math. Soc.,, Providence, RI, 1992. 9

[Alvarez-Montaner et al. 2005] J. Alvarez-Montaner, M. Blickle, and G. Lyubeznik, "Generators of D-modules in positive characteristic", *Math. Res. Lett.* **12**:4 (2005), 459–473. 9

[Àlvarez Montaner et al. 2007] J. Àlvarez Montaner, F. J. Castro-Jiménez, and J. M. Ucha, "Localization at hyperplane arrangements: combinatorics and $\mathscr{D}$-modules", *J. Algebra* **316**:2 (2007), 662–679. 10C

[Andres et al. 2009] D. Andres, V. Levandovskyy, and J. M. Morales, "Principal intersection and Bernstein–Sato polynomial of an affine variety", pp. 231–238 in *ISSAC 2009 — Proceedings of the 2009 International Symposium on Symbolic and Algebraic Computation*, edited by J. P. May, ACM, New York, 2009. 10E

[Andres et al. 2010] D. Andres, M. Brickenstein, V. Levandovskyy, J. Martín-Morales, and H. Schönemann, "Constructive D-module theory with SINGULAR", *Math. Comput. Sci.* **4**:2-3 (2010), 359–383. 10G

[Arcadias 2010] R. Arcadias, "Minimal resolutions of geometric $\mathscr{D}$-modules", *J. Pure Appl. Algebra* **214**:8 (2010), 1477–1496. 4C

[Arnold et al. 1985] V. I. Arnold, S. M. Gusein-Zade, and A. N. Varchenko, *Singularities of differentiable maps, I: the classification of critical points, caustics and wave fronts*, Monographs in Mathematics **82**, Birkhäuser, Boston, 1985. 3B

[Artal Bartolo et al. 2002] E. Artal Bartolo, P. Cassou-Noguès, I. Luengo, and A. Melle Hernández, "Monodromy conjecture for some surface singularities", *Ann. Sci. École Norm. Sup.* (4) **35** (2002), 605–640. 6D

[Artal Bartolo et al. 2005] E. Artal Bartolo, P. Cassou-Noguès, I. Luengo, and A. Melle Hernández, *Quasi-ordinary power series and their zeta functions*, Mem. Amer. Math. Soc. **841**, 2005. 6D

[Artal Bartolo et al. 2006] E. Artal Bartolo, J. Carmona Ruber, J. I. Cogolludo Agustín, and M. Á. Marco Buzunáriz, "Invariants of combinatorial line arrangements and Rybnikov's example", pp. 1–34 in *Singularity theory and its applications*, edited by S. Izumiya et al., Adv. Stud. Pure Math. **43**, Math. Soc. Japan, Tokyo, 2006. 8A

[Aschenbrenner and Leykin 2009] M. Aschenbrenner and A. Leykin, "Degree bounds for Gröbner bases in algebras of solvable type", *J. Pure Appl. Algebra* **213**:8 (2009), 1578–1605. 10A

[Bahloul 2001] R. Bahloul, "Algorithm for computing Bernstein–Sato ideals associated with a polynomial mapping: effective methods in rings of differential operators", *J. Symbolic Comput.* **32**:6 (2001), 643–662. 10E

[Bahloul 2003] R. Bahloul, "Global generic Bernstein–Sato polynomial on an irreducible affine scheme", *Proc. Japan Acad. Ser. A Math. Sci.* **79**:9 (2003), 146–149. 10E

[Bahloul 2005] R. Bahloul, "Démonstration constructive de l'existence de polynômes de Bernstein–Sato pour plusieurs fonctions analytiques", *Compos. Math.* **141**:1 (2005), 175–191. 7

[Bahloul and Oaku 2010] R. Bahloul and T. Oaku, "Local Bernstein–Sato ideals: algorithm and examples", *J. Symbolic Comput.* **45**:1 (2010), 46–59. 7

[Barlet 1980] D. Barlet, "Familles analytiques de cycles et classes fondamentales relatives", pp. 1–24 in *Functions of several complex variables, IV*, edited by F. Norguet, Lecture Notes in Math. **807**, Springer, Berlin, 1980. In French (Sem. François Norguet, 1977–1979). 3C4

[Barlet 1999] D. Barlet, "Multiple poles at negative integers for $\int_A f^\lambda \Box$ in the case of an almost isolated singularity", *Publ. Res. Inst. Math. Sci.* **35**:4 (1999), 571–584. 3C4

[Barlet and Kashiwara 1986] D. Barlet and M. Kashiwara, "Le réseau L^2 d'un système holonome régulier", *Invent. Math.* **86**:1 (1986), 35–62. 3C4

[Berkesch and Leykin 2010] C. Berkesch and A. Leykin, "Algorithms for Bernstein–Sato polynomials and multiplier ideals", pp. 99–106 in *ISSAC 2010 — Proceedings of the 2010 International Symposium on Symbolic and Algebraic Computation*, edited by S. M. Watt, ACM, New York, 2010. 10E, 10F, 10G

[Bernstein 1972] I. N. Bernstein, "Analytic continuation of generalized functions with respect to a parameter", *Funkcional. Anal. i Priložen.* **6**:4 (1972), 26–40. In Russian; translated in *Functional Anal. Appl.*, **6** (1972), 273–285 (1973). 1, 1.5, 5

[Bernstein ca. 1997] J. Bernstein, "Algebraic theory of D-modules", ca. 1997, available at http://math.columbia.edu/~khovanov/resources/Bernstein-dmod.pdf. lecture notes. 1, 1

[Björk 1979] J.-E. Björk, *Rings of differential operators*, North-Holland Mathematical Library **21**, North-Holland, Amsterdam, 1979. 1, 1

[Björk 1993] J.-E. Björk, *Analytic $\mathscr{D}$-modules and applications*, Mathematics and its Applications **247**, Kluwer Academic Publishers Group, Dordrecht, 1993. 1, 2

[Blickle 2004] M. Blickle, "The intersection homology D-module in finite characteristic", *Math. Ann.* **328**:3 (2004), 425–450. 9

[Blickle et al. 2009] M. Blickle, M. Mustaţă, and K. E. Smith, "F-thresholds of hypersurfaces", *Trans. Amer. Math. Soc.* **361**:12 (2009), 6549–6565. 9

[Bögvad 2002] R. Bögvad, "An analogue of holonomic $\mathscr{D}$-modules on smooth varieties in positive characteristics", *Homology Homotopy Appl.* **4**:2, part 1 (2002), 83–116. 9

[Bories 2013a] B. Bories, "Zeta functions and Bernstein–Sato polynomials for ideals in dimension two", *Rev. Mat. Complut.* **26**:2 (2013), 753–772. 6D

[Bories 2013b] B. Bories, *Zeta functions, Bernstein–Sato polynomials, and the monodromy conjecture*, Ph.D. thesis, Katholieke Universiteit Leuven, 2013. 6D

[Bories and Veys 2013] B. Bories and W. Veys, "Igusa's p-adic local zeta function and the monodromy conjecture for non-degenerated surface singularities", (2013). arXiv 1306.6012 6D

[Briançon and Maisonobe 2002] J. Briançon and P. Maisonobe, "Remarques sur l'idéal de Bernstein associé à des polynomes", preprint 650, Université de Nice Sophia-Antipolis, 2002. 10C

[Briançon and Maisonobe 1996] J. Briançon and P. Maisonobe, "Caractérisation géométrique de l'existence du polynôme de Bernstein relatif", pp. 215–236 in *Algebraic geometry and singularities* (La Rábida, 1991), edited by A. Campillo López and L. Narváez Macarro, Progr. Math. **134**, Birkhäuser, Basel, 1996. 3C3

[Briançon and Maynadier 1999] J. Briançon and H. Maynadier, "Équations fonctionnelles généralisées: transversalité et principalité de l'idéal de Bernstein–Sato", *J. Math. Kyoto Univ.* **39**:2 (1999), 215–232. 3C1, 3C2

[Briançon et al. 1989] J. Briançon, M. Granger, P. Maisonobe, and M. Miniconi, "Algorithme de calcul du polynôme de Bernstein: cas non dégénéré", *Ann. Inst. Fourier* (*Grenoble*) **39**:3 (1989), 553–610. 9

[Briançon et al. 1991] J. Briançon, Y. Laurent, and P. Maisonobe, "Sur les modules différentiels holonomes réguliers, cohérents relativement à une projection", *C. R. Acad. Sci. Paris Sér. I Math.* **313**:5 (1991), 285–288. 3C3

[Briançon et al. 2000] J. Briançon, P. Maisonobe, and M. Merle, "Constructibilité de l'idéal de Bernstein", pp. 79–95 in *Singularities* (Sapporo, 1998), edited by J.-P. Brasselet and T. Suwa, Adv. Stud. Pure Math. **29**, Kinokuniya, Tokyo, 2000. 3C1, 7

[Brieskorn 1970] E. Brieskorn, "Die Monodromie der isolierten Singularitäten von Hyperflächen", *Manuscripta Math.* **2** (1970), 103–161. 3C4

[Brieskorn 1973] E. Brieskorn, "Sur les groupes de tresses [d'après V. I. Arnol'd]", pp. exposé no. 401, 21–44 in *Séminaire Bourbaki, 24ème année (1971/1972)*, Lecture Notes in Math. **317**, Springer, Berlin, 1973. 8B

[Brieskorn and Knörrer 1981] E. Brieskorn and H. Knörrer, *Ebene algebraische Kurven*, Birkhäuser, Basel, 1981. 6A

[Brylinski 1983] J.-L. Brylinski, "Modules holonomes à singularités régulières et filtration de Hodge, II", pp. 75–117 in *Analysis and topology on singular spaces, II, III* (Luminy, 1981), Astérisque **101**, Soc. Math. France, Paris, 1983. 3C4, 3C4

[Brylinski 1985] J.-L. Brylinski, "La classe fondamentale d'une variété algébrique engendre le $\mathcal{D}$-module qui calcule sa cohomologie d'intersection", pp. 260–271 in *Differential systems and singularities* (Luminy, 1983), Astérisque **130**, 1985. 3C4

[Buchweitz and Mond 2006] R.-O. Buchweitz and D. Mond, "Linear free divisors and quiver representations", pp. 41–77 in *Singularities and computer algebra*, edited by C. Lossen and G. Pfister, London Math. Soc. Lecture Note Ser. **324**, Cambridge Univ. Press, Cambridge, 2006. 4C

[Buchweitz et al. 2009] R.-O. Buchweitz, W. Ebeling, and H.-C. Graf von Bothmer, "Low-dimensional singularities with free divisors as discriminants", *J. Algebraic Geom.* **18**:2 (2009), 371–406. 4B

[Budur 2003] N. Budur, "On Hodge spectrum and multiplier ideals", *Math. Ann.* **327**:2 (2003), 257–270. 6A

[Budur 2005] N. Budur, "On the V-filtration of $\mathcal{D}$-modules", pp. 59–70 in *Geometric methods in algebra and number theory*, edited by F. Bogomolov and Y. Tschinkel, Progr. Math. **235**, Birkhäuser, Boston, 2005. 3A, 3A1, 10F

[Budur 2012a] N. Budur, "Bernstein–Sato ideals and local systems", 2012. arXiv 1209.3725 7.1, 7, 8E

[Budur 2012b] N. Budur, "Singularity invariants related to Milnor fibers: survey", pp. 161–187 in *Zeta functions in algebra and geometry*, edited by A. Campillo et al., Contemp. Math. **566**, Amer. Math. Soc., Providence, RI, 2012. 0, 3A, 6A, 8C

[Budur and Saito 2010] N. Budur and M. Saito, "Jumping coefficients and spectrum of a hyperplane arrangement", *Math. Ann.* **347**:3 (2010), 545–579. 8D

[Budur et al. 2006a] N. Budur, M. Mustaţă, and M. Saito, "Bernstein–Sato polynomials of arbitrary varieties", *Compos. Math.* **142**:3 (2006), 779–797. 7, 10F

[Budur et al. 2006b] N. Budur, M. Mustaţă, and M. Saito, "Combinatorial description of the roots of the Bernstein–Sato polynomials for monomial ideals", *Comm. Algebra* **34**:11 (2006), 4103–4117. 7

[Budur et al. 2006c] N. Budur, M. Mustaţă, and M. Saito, "Roots of Bernstein–Sato polynomials for monomial ideals: a positive characteristic approach", *Math. Res. Lett.* **13**:1 (2006), 125–142. 7

[Budur et al. 2011a] N. Budur, A. Dimca, and M. Saito, "First Milnor cohomology of hyperplane arrangements", pp. 279–292 in *Topology of algebraic varieties and singularities*, edited by J. I. Cogolludo-Agustín and E. Hironaka, Contemp. Math. **538**, Amer. Math. Soc., Providence, RI, 2011. 8C

[Budur et al. 2011b] N. Budur, M. Mustaţă, and Z. Teitler, "The monodromy conjecture for hyperplane arrangements", *Geom. Dedicata* **153** (2011), 131–137. 6D, 8F, 8F

[Budur et al. 2011c] N. Budur, M. Saito, and S. Yuzvinsky, "On the local zeta functions and the b-functions of certain hyperplane arrangements", *J. Lond. Math. Soc.* (2) **84**:3 (2011), 631–648. 3C3, 6D, 8E, 8F

[Calderón-Moreno 1999] F. J. Calderón-Moreno, "Logarithmic differential operators and logarithmic de Rham complexes relative to a free divisor", *Ann. Sci. École Norm. Sup.* (4) **32**:5 (1999), 701–714. 2, 4C

[Calderón-Moreno and Narváez-Macarro 2002a] F. Calderón-Moreno and L. Narváez-Macarro, "Locally quasi-homogeneous free divisors are Koszul free", *Tr. Mat. Inst. Steklova* **238** (2002), 81–85. In Russian; translated in *Proc. Steklov Inst. Math.* **238** 3 (2002), 72–76. 4C

[Calderón-Moreno and Narváez-Macarro 2002b] F. Calderón-Moreno and L. Narváez-Macarro, "The module $\mathscr{D}f^s$ for locally quasi-homogeneous free divisors", *Compositio Math.* **134**:1 (2002), 59–74. 2, 4C, 4D, 5

[Calderón-Moreno and Narváez-Macarro 2005] F. J. Calderón-Moreno and L. Narváez-Macarro, "Dualité et comparaison sur les complexes de de Rham logarithmiques par rapport aux diviseurs libres", *Ann. Inst. Fourier (Grenoble)* **55**:1 (2005), 47–75. 4C

[Calderón-Moreno and Narváez-Macarro 2009] F. J. Calderón-Moreno and L. Narváez-Macarro, "On the logarithmic comparison theorem for integrable logarithmic connections", *Proc. Lond. Math. Soc.* (3) **98**:3 (2009), 585–606. 4.5

[Calderón-Moreno et al. 2002] F. J. Calderón-Moreno, D. Mond, L. Narváez-Macarro, and F. J. Castro-Jiménez, "Logarithmic cohomology of the complement of a plane curve", *Comment. Math. Helv.* **77**:1 (2002), 24–38. 2, 4C

[Caracciolo et al. 2013] S. Caracciolo, A. D. Sokal, and A. Sportiello, "Algebraic/combinatorial proofs of Cayley-type identities for derivatives of determinants and Pfaffians", *Adv. in Appl. Math.* **50**:4 (2013), 474–594. 3C3

[Cassou-Noguès 1986] P. Cassou-Noguès, "Racines de polyômes de Bernstein", *Ann. Inst. Fourier (Grenoble)* **36**:4 (1986), 1–30. 5B

[Cassou-Noguès and Libgober 2011] P. Cassou-Noguès and A. Libgober, "Multivariable Hodge theoretical invariants of germs of plane curves", *J. Knot Theory Ramifications* **20**:6 (2011), 787–805. 7

[Castro-Jimenez 1984] F. Castro-Jimenez, *Théorème de division pour les opérateurs différentiels et calcul des multiplicités*, Ph.D. thesis, Université Denis Diderot Paris VII, 1984. Thèse de troisième cycle. 9

[Castro-Jiménez and Ucha 2001] F. J. Castro-Jiménez and J. M. Ucha-Enríquez, "Explicit comparison theorems for $\mathscr{D}$-modules: effective methods in rings of differential operators", *J. Symbolic Comput.* **32**:6 (2001), 677–685. 2, 4C

[Castro-Jiménez and Ucha 2004a] F. J. Castro-Jiménez and J. M. Ucha-Enríquez, "Quasi-free divisors and duality", *C. R. Math. Acad. Sci. Paris* **338**:6 (2004), 461–466. 4C

[Castro-Jiménez and Ucha 2004b] F. J. Castro-Jiménez and J. M. Ucha-Enríquez, "Testing the logarithmic comparison theorem for free divisors", *Experiment. Math.* **13**:4 (2004), 441–449. 4C

[Castro-Jiménez and Ucha 2005] F. J. Castro-Jiménez and J. M. Ucha-Enríquez, "Logarithmic comparison theorem and some Euler homogeneous free divisors", *Proc. Amer. Math. Soc.* **133**:5 (2005), 1417–1422. 4C

[Castro-Jimenez and Narváez-Macarro 1997] F. Castro-Jimenez and L. Narváez-Macarro, "Homogenizing differential operators", preprint, no. 36, Facultad de Matematicas de la Universidad de Sevilla, 1997. 10B

[Castro-Jiménez and Takayama 2009] F.-J. Castro-Jiménez and N. Takayama, "The computation of the logarithmic cohomology for plane curves", *J. Algebra* **322**:11 (2009), 3839–3851. 4.5

[Castro-Jiménez and Ucha 2002] F. J. Castro-Jiménez and J. M. Ucha, "Free divisors and duality for $\mathscr{D}$-modules", *Tr. Mat. Inst. Steklova* **238**:Monodromiya v Zadachakh Algebr. Geom. i Differ. Uravn. (2002), 97–105. 2, 3C3, 4C, 4C

[Castro-Jiménez et al. 1996] F. J. Castro-Jiménez, L. Narváez-Macarro, and D. Mond, "Cohomology of the complement of a free divisor", *Trans. Amer. Math. Soc.* **348**:8 (1996), 3037–3049. 4.5, 4C, 8B

[Castro-Jiménez et al. 2007] F.-J. Castro-Jiménez, J. Gago-Vargas, M.-I. Hartillo-Hermoso, and J.-M. Ucha, "A vanishing theorem for a class of logarithmic $\mathscr{D}$-modules", pp. 171–184 in *Actas del XVI Coloquio Latinoamericano de Álgebra* (Colonia del Sacramento, Uruguay, 2005), edited by W. Ferrer Santos et al., Rev. Mat. Iberoamericana, Madrid, 2007. In Spanish. 2, 4C

[Cohen and Suciu 1995] D. C. Cohen and A. I. Suciu, "On Milnor fibrations of arrangements", *J. London Math. Soc.* (2) **51**:1 (1995), 105–119. 8C

[Damon 1998] J. Damon, "On the legacy of free divisors: discriminants and Morse-type singularities", *Amer. J. Math.* **120**:3 (1998), 453–492. 4B

[Damon and Mond 1991] J. Damon and D. Mond, "$\mathscr{A}$-codimension and the vanishing topology of discriminants", *Invent. Math.* **106**:2 (1991), 217–242. 5A

[Decker et al. 2012] W. Decker, G.-M. Greuel, G. Pfister, and H. Schönemann, "SINGULAR: a computer algebra system for polynomial computations (version 3-1-6)", 2012, available at http://www.singular.uni-kl.de. 10G

[Deligne 1970] P. Deligne, *Équations différentielles à points singuliers réguliers*, Lecture Notes in Mathematics **163**, Springer, Berlin-New York, 1970. 4.5

[Deligne 1971] P. Deligne, "Théorie de Hodge, II", *Inst. Hautes Études Sci. Publ. Math.* **40** (1971), 5–57. 4A, 4.5

[Deligne 1974] P. Deligne, "Théorie de Hodge, III", *Inst. Hautes Études Sci. Publ. Math.* 44 (1974), 5–77. 4A, 4.5

[Deligne and Dimca 1990] P. Deligne and A. Dimca, "Filtrations de Hodge et par l'ordre du pôle pour les hypersurfaces singulières", *Ann. Sci. École Norm. Sup.* (4) **23**:4 (1990), 645–656. 4.5

[Denef 1991] J. Denef, "Report on Igusa's local zeta function", exposé no. 741, 359–386 in *Séminaire Bourbaki,* 1990/91, Astérisque **201–203**, 1991. 6D

[Denef and Loeser 1992] J. Denef and F. Loeser, "Caractéristiques d'Euler–Poincaré, fonctions zêta locales et modifications analytiques", *J. Amer. Math. Soc.* **5**:4 (1992), 705–720. 6D, 6D

[Dimca 1991] A. Dimca, "On the de Rham cohomology of a hypersurface complement", *Amer. J. Math.* **113**:4 (1991), 763–771. 4.5

[Dimca 1992] A. Dimca, *Singularities and topology of hypersurfaces*, Springer, New York, 1992. 5B

[Dimca et al. 2006] A. Dimca, P. Maisonobe, M. Saito, and T. Torrelli, "Multiplier ideals, V-filtrations and transversal sections", *Math. Ann.* **336**:4 (2006), 901–924. 5A, 5B

[Dimca et al. 2013] A. Dimca, D. Ibadula, and D. A. Macinic, "Pencil type line arrangements of low degree: classification and monodromy", preprint, 2013. arXiv 1305.5092 8C

[Ein et al. 2004] L. Ein, R. Lazarsfeld, K. E. Smith, and D. Varolin, "Jumping coefficients of multiplier ideals", *Duke Math. J.* **123**:3 (2004), 469–506. 3B

[Esnault and Viehweg 1992] H. Esnault and E. Viehweg, *Lectures on vanishing theorems*, DMV Seminar **20**, Birkhäuser, Basel, 1992. 4A

[Ginsburg 1986] V. Ginsburg, "Characteristic varieties and vanishing cycles", *Invent. Math.* **84**:2 (1986), 327–402. 5

[Granger and Schulze 2006a] M. Granger and M. Schulze, "On the formal structure of logarithmic vector fields", *Compos. Math.* **142**:3 (2006), 765–778. 4C

[Granger and Schulze 2006b] M. Granger and M. Schulze, "Quasihomogeneity of isolated hypersurface singularities and logarithmic cohomology", *Manuscripta Math.* **121**:4 (2006), 411–416. 4C

[Granger and Schulze 2010] M. Granger and M. Schulze, "On the symmetry of b-functions of linear free divisors", *Publ. Res. Inst. Math. Sci.* **46**:3 (2010), 479–506. 3C3

[Granger et al. 2009] M. Granger, D. Mond, A. Nieto-Reyes, and M. Schulze, "Linear free divisors and the global logarithmic comparison theorem", *Ann. Inst. Fourier (Grenoble)* **59**:2 (2009), 811–850. 4C, 4C

[Granger et al. 2011] M. Granger, D. Mond, and M. Schulze, "Free divisors in prehomogeneous vector spaces", *Proc. Lond. Math. Soc.* (3) **102**:5 (2011), 923–950. 4C

[Grayson and Stillman] D. R. Grayson and M. E. Stillman, "Macaulay 2, a software system for research in algebraic geometry", available at http://www.math.uiuc.edu/Macaulay2/. 10G

[de Gregorio et al. 2009] I. de Gregorio, D. Mond, and C. Sevenheck, "Linear free divisors and Frobenius manifolds", *Compos. Math.* **145**:5 (2009), 1305–1350. 4C

[Griffiths 1969a] P. A. Griffiths, "On the periods of certain rational integrals, I", *Ann. of Math.* (2) **90** (1969), 496–541. 4.5

[Griffiths 1969b] P. A. Griffiths, "On the periods of certain rational integrals, II", *Ann. of Math.* (2) **90** (1969), 460–495. 4.5

[Grigoriev and Chistov 2008] D. Y. Grigoriev and A. L. Chistov, "Complexity of the standard basis of a D-module", *Algebra i Analiz* **20**:5 (2008), 41–82. 10A

[Gros and Narváez-Macarro 2000] M. Gros and L. Narváez-Macarro, "Cohomologie évanescente p-adique: calculs locaux", *Rend. Sem. Mat. Univ. Padova* **104** (2000), 71–90. 4C

[Grothendieck 1966] A. Grothendieck, "On the de Rham cohomology of algebraic varieties", *Inst. Hautes Études Sci. Publ. Math.* **29** (1966), 95–103. 3C4, 4A

[Gyoja 1993] A. Gyoja, "Bernstein–Sato's polynomial for several analytic functions", *J. Math. Kyoto Univ.* **33**:2 (1993), 399–411. 7

[Holland and Mond 1998] M. P. Holland and D. Mond, "Logarithmic differential forms and the cohomology of the complement of a divisor", *Math. Scand.* **83**:2 (1998), 235–254. 4.5

[Howald 2001] J. A. Howald, "Multiplier ideals of monomial ideals", *Trans. Amer. Math. Soc.* **353**:7 (2001), 2665–2671. 10F

[Howald et al. 2007] J. Howald, M. Mustaţă, and C. Yuen, "On Igusa zeta functions of monomial ideals", *Proc. Amer. Math. Soc.* **135**:11 (2007), 3425–3433. 6D

[Igusa 1974] J.-i. Igusa, "Complex powers and asymptotic expansions, I: functions of certain types", *J. Reine Angew. Math.* **268/269** (1974), 110–130. 8F

[Igusa 1975] J.-i. Igusa, "Complex powers and asymptotic expansions, II: asymptotic expansions", *J. Reine Angew. Math.* **278/279** (1975), 307–321. 8F

[Igusa 1992] J.-i. Igusa, "Local zeta functions of certain prehomogeneous vector spaces", *Amer. J. Math.* **114**:2 (1992), 251–296. 6D

[Igusa 2000] J.-i. Igusa, *An introduction to the theory of local zeta functions*, AMS/IP Studies in Advanced Mathematics **14**, American Mathematical Society; International Press, Providence, RI; Cambridge, MA, 2000. 6D

[Johnson 2003] A. A. Johnson, *Multiplier ideals of determinantal ideals*, Ph.D. thesis, University of Michigan, Ann Arbor, MI, 2003. 10F

[Kandri-Rody and Weispfenning 1990] A. Kandri-Rody and V. Weispfenning, "Noncommutative Gröbner bases in algebras of solvable type", *J. Symbolic Comput.* **9**:1 (1990), 1–26. 10A

[Karpishpan 1991] Y. Karpishpan, "Pole order filtration on the cohomology of algebraic links", *Compositio Math.* **78**:2 (1991), 213–226. 4.5

[Kashiwara 1975] M. Kashiwara, "On the maximally overdetermined system of linear differential equations, I", *Publ. Res. Inst. Math. Sci.* **10** (1975), 563–579. 1

[Kashiwara 1976] M. Kashiwara, "B-functions and holonomic systems: rationality of roots of B-functions", *Invent. Math.* **38**:1 (1976), 33–53. 1.6, 2, 2, 3A1, 3C1, 3C4, 4D, 5

[Kashiwara 1978] M. Kashiwara, "On the holonomic systems of linear differential equations, II", *Invent. Math.* **49**:2 (1978), 121–135. 3A1

[Kashiwara 1983] M. Kashiwara, "Vanishing cycle sheaves and holonomic systems of differential equations", pp. 134–142 in *Algebraic geometry* (Tokyo/Kyoto, 1982), edited by M. Raynaud and T. Shioda, Lecture Notes in Math. **1016**, Springer, Berlin, 1983. 3A1, 3A1, 6C, 6D

[Kashiwara 1984] M. Kashiwara, "The Riemann–Hilbert problem for holonomic systems", *Publ. Res. Inst. Math. Sci.* **20**:2 (1984), 319–365. 3C4

[Kashiwara 2003] M. Kashiwara, *D-modules and microlocal calculus*, Translations of Mathematical Monographs **217**, American Mathematical Society, Providence, RI, 2003. 1, 1, 5

[Kashiwara and Kawai 1981a] M. Kashiwara and T. Kawai, "On holonomic systems of microdifferential equations, III: Systems with regular singularities", *Publ. Res. Inst. Math. Sci.* **17**:3 (1981), 813–979. 1

[Kashiwara and Kawai 1981b] M. Kashiwara and T. Kawai, "On holonomic systems of microdifferential equations, III: Systems with regular singularities", *Publ. Res. Inst. Math. Sci.* **17**:3 (1981), 813–979. 3A1

[Kashiwara and Schapira 1979] M. Kashiwara and P. Schapira, "Micro-hyperbolic systems", *Acta Math.* **142**:1-2 (1979), 1–55. 5A

[Kimura et al. 1990] T. Kimura, F. Satō, and X.-W. Zhu, "On the poles of p-adic complex powers and the b-functions of prehomogeneous vector spaces", *Amer. J. Math.* **112**:3 (1990), 423–437. 6D

[Kochman 1976] F. Kochman, "Bernstein polynomials and Milnor algebras", *Proc. Nat. Acad. Sci. U.S.A.* **73**:8 (1976), 2546. 3C3

[Kollár 1997] J. Kollár, "Singularities of pairs", pp. 221–287 in *Algebraic geometry* (Santa Cruz 1995), edited by J. Kollár et al., Proc. Sympos. Pure Math. **62**, Amer. Math. Soc., Providence, RI, 1997. 3B, 3C

[Lemahieu and Van Proeyen 2011] A. Lemahieu and L. Van Proeyen, "Monodromy conjecture for nondegenerate surface singularities", *Trans. Amer. Math. Soc.* **363**:9 (2011), 4801–4829. 6D

[Lemahieu and Veys 2009] A. Lemahieu and W. Veys, "Zeta functions and monodromy for surfaces that are general for a toric idealistic cluster", *Int. Math. Res. Not.* **2009**:1 (2009), Art. ID rnn122, 11–62. 6D

[Levandovskyy and Martín Morales 2008] V. Levandovskyy and J. Martín Morales, "Computational D-module theory with SINGULAR, comparison with other systems and two new algorithms", pp. 173–180 in *ISSAC 2008*, edited by D. J. Jeffrey, ACM, New York, 2008. 10G

[Levandovskyy and Martín-Morales 2012] V. Levandovskyy and J. Martín-Morales, "Algorithms for checking rational roots of b-functions and their applications", *J. Algebra* **352** (2012), 408–429. 10D

[Levandovskyy and Morales] V. Levandovskyy and J. M. Morales, "Algorithms for algebraic D-modules", available at http://www.singular.uni-kl.de. 10G

[Leykin 2001] A. Leykin, "Constructibility of the set of polynomials with a fixed Bernstein–Sato polynomial: an algorithmic approach", *J. Symbolic Comput.* **32**:6 (2001), 663–675. 10E

[Leykin 2002] A. Leykin, "D-modules for Macaulay 2", pp. 169–179 in *Mathematical software* (Beijing, 2002), edited by A. M. Cohen et al., World Sci. Publ., River Edge, NJ, 2002. 10G

[Leykin and Tsai] A. Leykin and H. Tsai, "Software package "D-modules for Macaulay 2"", available at http://people.math.gatech.edu/~aleykin3/Dmodules. 10G

[Libgober 2012] A. Libgober, "On combinatorial invariance of the cohomology of the Milnor fiber of arrangements and the Catalan equation over function fields", pp. 175–187 in *Arrangements of hyperplanes* (Sapporo, 2009), edited by H. Terao and S. Yuzvinsky, Adv. Stud. Pure Math. **62**, Math. Soc. Japan, Tokyo, 2012. 8C

[Lichtin 1989] B. Lichtin, "Poles of $|f(z,w)|^{2s}$ and roots of the b-function", *Ark. Mat.* **27**:2 (1989), 283–304. 3B, 3C1

[Lichtin and Meuser 1985] B. Lichtin and D. Meuser, "Poles of a local zeta function and Newton polygons", *Compositio Math.* **55**:3 (1985), 313–332. 6D

[Lipman 1982] J. Lipman, "Equimultiplicity, reduction, and blowing up", pp. 111–147 in *Commutative algebra* (Fairfax, Va., 1979), edited by R. N. Draper, Lecture Notes in Pure and Appl. Math. **68**, Dekker, New York, 1982. 3B

[Loeser 1988] F. Loeser, "Fonctions d'Igusa p-adiques et polynômes de Bernstein", *Amer. J. Math.* **110**:1 (1988), 1–21. 6D

[Loeser 1990] F. Loeser, "Fonctions d'Igusa p-adiques, polynômes de Bernstein, et polyèdres de Newton", *J. Reine Angew. Math.* **412** (1990), 75–96. 6D

[Loeser and Vaquié 1990] F. Loeser and M. Vaquié, "Le polynôme d'Alexander d'une courbe plane projective", *Topology* **29**:2 (1990), 163–173. 3B

[Looijenga 1984] E. J. N. Looijenga, *Isolated singular points on complete intersections*, London Mathematical Society Lecture Note Series **77**, Cambridge University Press, 1984. 4B

[Lyubeznik 1997a] G. Lyubeznik, "F-modules: applications to local cohomology and D-modules in characteristic $p > 0$", *J. Reine Angew. Math.* **491** (1997), 65–130. 9

[Lyubeznik 1997b] G. Lyubeznik, "On Bernstein–Sato polynomials", *Proc. Amer. Math. Soc.* **125**:7 (1997), 1941–1944. 3C1, 3C2

[Maisonobe 1994] P. Maisonobe, "$\mathcal{D}$-modules: an overview towards effectivity", pp. 21–55 in *Computer algebra and differential equations (1992)*, edited by E. Tournier, London Math. Soc. Lecture Note Ser. **193**, Cambridge Univ. Press, Cambridge, 1994. 9

[Maisonobe and Mebkhout 2004] P. Maisonobe and Z. Mebkhout, "Le théorème de comparaison pour les cycles évanescents", pp. 311–389 in *Éléments de la théorie des systèmes différentiels géométriques*, edited by P. Maisonobe and L. Narváez Macarro, Sémin. Congr. **8**, Soc. Math. France, Paris, 2004. 3A

[Malgrange 1974] B. Malgrange, "Intégrales asymptotiques et monodromie", *Ann. Sci. École Norm. Sup.* (4) **7** (1974), 405–430. 6A

[Malgrange 1975] B. Malgrange, "Le polynôme de Bernstein d'une singularité isolée", pp. 98–119 in *Fourier integral operators and partial differential equations* (Nice, 1974), Lecture Notes in Math. **459**, Springer, Berlin, 1975. 1, 3C1, 3C3, 6D

[Malgrange 1983] B. Malgrange, "Polynômes de Bernstein–Sato et cohomologie évanescente", pp. 243–267 in *Analysis and topology on singular spaces, II, III* (Luminy, 1981), Astérisque **101–102**, Soc. Math. France, Paris, 1983. 3A1, 3A1, 3A1, 6C

[Massey 2009] D. B. Massey, "Intersection cohomology, monodromy and the Milnor fiber", *Internat. J. Math.* **20**:4 (2009), 491–507. 3C4

[Mebkhout 1977] Z. Mebkhout, "Local cohomology of analytic spaces", pp. 247–256 in *Proceedings of the Oji Seminar on Algebraic Analysis and the RIMS Symposium on Algebraic Analysis* (Kyoto, 1976), vol. 12, 1977. 3C4

[Mebkhout 1984] Z. Mebkhout, "Une autre équivalence de catégories", *Compositio Math.* **51**:1 (1984), 63–88. 3C4

[Mebkhout and Narváez-Macarro 1991] Z. Mebkhout and L. Narváez-Macarro, "La théorie du polynôme de Bernstein–Sato pour les algèbres de Tate et de Dwork–Monsky–Washnitzer", *Ann. Sci. École Norm. Sup.* (4) **24**:2 (1991), 227–256. 1

[Melle-Hernández et al. 2009] A. Melle-Hernández, T. Torrelli, and W. Veys, "On 'maximal' poles of zeta functions, roots of b-functions, and monodromy Jordan blocks", *J. Topol.* **2**:3 (2009), 517–526. 6D

[Melle-Hernández et al. 2010] A. Melle-Hernández, T. Torrelli, and W. Veys, "Monodromy Jordan blocks, b-functions and poles of zeta functions for germs of plane curves", *J. Algebra* **324**:6 (2010), 1364–1382. 6D

[Milnor 1968] J. Milnor, *Singular points of complex hypersurfaces*, Annals of Mathematics Studies **61**, Princeton University Press, Princeton, 1968. 3C4, 6A

[Muro 1988] M. Muro, "A note on the holonomic system of invariant hyperfunctions on a certain prehomogeneous vector space", pp. 493–503 in *Algebraic analysis, II*, edited by M. Kashiwara and T. Kawai, Academic Press, Boston, MA, 1988. 3C3

[Muro 2000] M. Muro, "An algorithm to compute the b_P-functions via Gröbner bases of invariant differential operators on prehomogeneous vector spaces", pp. 68–81 in *Algorithms for D-modules: Proceedings of a symposium held at the Research Institute for Mathematical Sciences* (Kyoto, 2000), Sūrikaisekikenkyūsho Kōkyūroku **1171**, 2000. In Japanese. 10D

[Mustaţă 2006] M. Mustaţă, "Multiplier ideals of hyperplane arrangements", *Trans. Amer. Math. Soc.* **358**:11 (2006), 5015–5023. 8D, 10F

[Mustaţă 2009] M. Mustaţă, "Bernstein–Sato polynomials in positive characteristic", *J. Algebra* **321**:1 (2009), 128–151. 9

[Mustaţă et al. 2005] M. Mustaţă, S. Takagi, and K.-i. Watanabe, "F-thresholds and Bernstein–Sato polynomials", pp. 341–364 in *European Congress of Mathematics*, edited by A. Laptev, Eur. Math. Soc., Zürich, 2005. 9

[Narváez-Macarro 2008] L. Narváez-Macarro, "Linearity conditions on the Jacobian ideal and logarithmic-meromorphic comparison for free divisors", pp. 245–269 in *Singularities, I*, edited by J.-P. Brasselet et al., Contemp. Math. **474**, Amer. Math. Soc., Providence, RI, 2008. 0, 2, 4C, 4D

[Narvaez-Macarro 2013] L. Narvaez-Macarro, "A duality approach to the symmetry of Bernstein–Sato polynomials of free divisors", preprint, 2013. arXiv 1201.3594 3C3

[Némethi and Veys 2010] A. Némethi and W. Veys, "Monodromy eigenvalues are induced by poles of zeta functions: the irreducible curve case", *Bull. Lond. Math. Soc.* **42**:2 (2010), 312–322. 6D

[Némethi and Veys 2012] A. Némethi and W. Veys, "Generalized monodromy conjecture in dimension two", *Geom. Topol.* **16**:1 (2012), 155–217. 6D

[Nishiyama and Noro 2010] K. Nishiyama and M. Noro, "Stratification associated with local b-functions", *J. Symbolic Comput.* **45**:4 (2010), 462–480. 10E, 10G

[Noro 2002] M. Noro, "An efficient modular algorithm for computing the global b-function", pp. 147–157 in *Mathematical software* (Beijing, 2002), edited by A. M. Cohen et al., World Sci. Publ., River Edge, NJ, 2002. 10G

[Noro et al.] M. Noro, T. Shimoyama, and T. Takeshima, "Computer algebra system Risa/Asir", available at http://www.math.kobe-u.ac.jp/Asir/index.html. 10G

[Núñez-Betancourt 2013] L. Núñez-Betancourt, "On certain rings of differentiable type and finiteness properties of local cohomology", *J. Algebra* **379** (2013), 1–10. 1

[Núñez-Betancourt and Pérez 2013] L. Nuñez Betancourt and F. Pérez, "F-jumping and F-Jacobian ideals for hypersurfaces", preprint, 2013. arXiv 1302.3327 9

[Oaku 1997] T. Oaku, "Algorithms for the b-function and D-modules associated with a polynomial", *J. Pure Appl. Algebra* **117/118** (1997), 495–518. 10C, 10C, 10D

[Oaku and Takayama 2001a] T. Oaku and N. Takayama, "Algorithms for D-modules — restriction, tensor product, localization, and local cohomology groups", *J. Pure Appl. Algebra* **156**:2-3 (2001), 267–308. 10D, 10G

[Oaku and Takayama 2001b] T. Oaku and N. Takayama, "Minimal free resolutions of homogenized D-modules", *J. Symbolic Comput.* **32**:6 (2001), 575–595. 10B

[Oaku et al. 2000] T. Oaku, N. Takayama, and U. Walther, "A localization algorithm for D-modules", *J. Symbolic Comput.* **29**:4-5 (2000), 721–728. 10D

[Orlik and Randell 1993] P. Orlik and R. Randell, "The Milnor fiber of a generic arrangement", *Ark. Mat.* **31**:1 (1993), 71–81. 8C

[Orlik and Solomon 1980] P. Orlik and L. Solomon, "Combinatorics and topology of complements of hyperplanes", *Invent. Math.* **56**:2 (1980), 167–189. 8B

[Orlik and Terao 1992] P. Orlik and H. Terao, *Arrangements of hyperplanes*, Grundlehren der Math. Wissenschaften **300**, Springer, Berlin, 1992. 8B

[Pham 1979] F. Pham, *Singularités des systèmes différentiels de Gauss–Manin*, Progress in Mathematics **2**, Birkhäuser, Boston, 1979. 5A, 6A

[Randell 1989] R. Randell, "Lattice-isotopic arrangements are topologically isomorphic", *Proc. Amer. Math. Soc.* **107**:2 (1989), 555–559. 8A

[Randell 1997] R. Randell, "Milnor fibrations of lattice-isotopic arrangements", *Proc. Amer. Math. Soc.* **125**:10 (1997), 3003–3009. 8A

[Rodrigues 2004] B. Rodrigues, "On the monodromy conjecture for curves on normal surfaces", *Math. Proc. Cambridge Philos. Soc.* **136**:2 (2004), 313–324. 6D

[Rybnikov 2011] G. L. Rybnikov, "On the fundamental group of the complement of a complex hyperplane arrangement", *Funktsional. Anal. i Prilozhen.* **45**:2 (2011), 71–85. 8A

[Sabbah 1987a] C. Sabbah, "D-modules et cycles évanescents (d'après B. Malgrange et M. Kashiwara)", pp. 53–98 in *Géométrie algébrique et applications, III* (La Rábida, 1984), edited by J.-M. Aroca et al., Travaux en Cours **24**, Hermann, Paris, 1987. 3A1

[Sabbah 1987b] C. Sabbah, "Polynômes de Bernstein–Sato à plusieurs variables", pp. exposé no. XIX, 6 in *Séminaire sur les équations aux dérivées partielles 1986–1987*, École Polytech., Palaiseau, 1987. 7

[Sabbah 1987c] C. Sabbah, "Proximité évanescente, I: La structure polaire d'un $\mathcal{D}$-module", *Compositio Math.* **62**:3 (1987), 283–328. 7

[Sabbah 1987d] C. Sabbah, "Proximité évanescente, II: équations fonctionnelles pour plusieurs fonctions analytiques", *Compositio Math.* **64**:2 (1987), 213–241. 7

[Sabbah 1990] C. Sabbah, "Modules d'Alexander et $\mathcal{D}$-modules", *Duke Math. J.* **60**:3 (1990), 729–814. 7

[Saito 1980] K. Saito, "Theory of logarithmic differential forms and logarithmic vector fields", *J. Fac. Sci. Univ. Tokyo Sect. IA Math.* **27**:2 (1980), 265–291. 4A, 4B, 4B, 4C

[Saito 1981] K. Saito, "Primitive forms for a universal unfolding of a function with an isolated critical point", *J. Fac. Sci. Univ. Tokyo Sect. IA Math.* **28**:3 (1981), 775–792. 4B

[Saito 1990] M. Saito, "Mixed Hodge modules", *Publ. Res. Inst. Math. Sci.* **26**:2 (1990), 221–333. 3C4

[Saito 1993] M. Saito, "On b-function, spectrum and rational singularity", *Math. Ann.* **295**:1 (1993), 51–74. 6A

[Saito 1994] M. Saito, "On microlocal b-function", *Bull. Soc. Math. France* **122**:2 (1994), 163–184. 3A, 3A1, 3C1, 3C3, 6C

[Saito 2006] M. Saito, "Bernstein–Sato polynomials of hyperplane arrangements", preprint, 2006. arXiv 0602527 8E

[Saito 2007] M. Saito, "Multiplier ideals, b-function, and spectrum of a hypersurface singularity", *Compos. Math.* **143**:4 (2007), 1050–1068. 0, 3B, 8E

[Saito 2009] M. Saito, "On b-function, spectrum and multiplier ideals", pp. 355–379 in *Algebraic analysis and around*, edited by T. Miwa et al., Adv. Stud. Pure Math. **54**, Math. Soc. Japan, Tokyo, 2009. 6A

[Saito et al. 2000] M. Saito, B. Sturmfels, and N. Takayama, *Gröbner deformations of hypergeometric differential equations*, Algorithms and Computation in Mathematics **6**, Springer, Berlin, 2000. 10B, 10D

[Sato 1990] M. Sato, "Theory of prehomogeneous vector spaces (algebraic part) — the English translation of Sato's lecture from Shintani's note", *Nagoya Math. J.* **120** (1990), 1–34. 3C3

[Sato and Kimura 1977] M. Sato and T. Kimura, "A classification of irreducible prehomogeneous vector spaces and their relative invariants", *Nagoya Math. J.* **65** (1977), 1–155. 4C

[Sato and Shintani 1974] M. Sato and T. Shintani, "On zeta functions associated with prehomogeneous vector spaces", *Ann. of Math.* (2) **100** (1974), 131–170. 3C3

[Schapira 1985] P. Schapira, *Microdifferential systems in the complex domain*, Grundlehren der Math. Wissenschaften **269**, Springer, Berlin, 1985. 1

[Schechtman et al. 1995] V. Schechtman, H. Terao, and A. Varchenko, "Local systems over complements of hyperplanes and the Kac–Kazhdan conditions for singular vectors", *J. Pure Appl. Algebra* **100**:1-3 (1995), 93–102. 8F

[Schulze 2007] M. Schulze, "A criterion for the logarithmic differential operators to be generated by vector fields", *Proc. Amer. Math. Soc.* **135**:11 (2007), 3631–3640. 2

[Schulze 2010] M. Schulze, "Logarithmic comparison theorem versus Gauss–Manin system for isolated singularities", *Adv. Geom.* **10**:4 (2010), 699–708. 4.5

[Sevenheck 2011] C. Sevenheck, "Bernstein polynomials and spectral numbers for linear free divisors", *Ann. Inst. Fourier (Grenoble)* **61**:1 (2011), 379–400. 3C3

[Shibuta 2011] T. Shibuta, "Algorithms for computing multiplier ideals", *J. Pure Appl. Algebra* **215**:12 (2011), 2829–2842. 10F

[Stahlke 1997] C. Stahlke, *Bernstein-Polynom und Tjurinazahl von μ-konstant-Deformationen der Singularitäten $x^a + y^b$*, Ph.D. thesis, Rheinische Friedrich-Wilhelms-Universität Bonn, 1997. Bonner Mathematische Schriften **305**. 5B

[Steenbrink 1987] J. Steenbrink, "Mixed Hodge structures and singularities: a survey", pp. 99–123 in *Géométrie algébrique et applications, III* (La Rábida, 1984), edited by J.-M. Aroca et al., Travaux en Cours **24**, Hermann, Paris, 1987. 6A

[Steenbrink 1989] J. H. M. Steenbrink, "The spectrum of hypersurface singularities", pp. 163–184 in *Actes du Colloque de Théorie de Hodge* (Luminy, 1987), Astérisque **179–180**, 1989. 6A

[Suciu 2014] A. I. Suciu, "Hyperplane arrangements and Milnor fibrations", *Ann. Fac. Sci. Toulouse Math.* (6) **23**:2 (2014), 417–481. 0, 8C

[Takayama] N. Takayama, "`kan/sm1`: a computer algebra system for algebraic analysis", available at http://www.math.sci.kobe-u.ac.jp/KAN/. 10G

[Teitler 2008] Z. Teitler, "A note on Mustaţă's computation of multiplier ideals of hyperplane arrangements", *Proc. Amer. Math. Soc.* **136**:5 (2008), 1575–1579. 8D

[Terao 1978] H. Terao, "Forms with logarithmic pole and the filtration by the order of the pole", pp. 673–685 in *Proceedings of the international symposium on algebraic geometry* (Kyoto, 1977), edited by M. Nagata, Kinokuniya Book Store, Tokyo, 1978. 8.1

[Terao 1981] H. Terao, "Generalized exponents of a free arrangement of hyperplanes and Shepherd–Todd–Brieskorn formula", *Invent. Math.* **63**:1 (1981), 159–179. 4B

[Torrelli 2002] T. Torrelli, "Polynômes de Bernstein associés à une fonction sur une intersection complète à singularité isolée", *Ann. Inst. Fourier (Grenoble)* **52**:1 (2002), 221–244. 2

[Torrelli 2004] T. Torrelli, "On meromorphic functions defined by a differential system of order 1", *Bull. Soc. Math. France* **132**:4 (2004), 591–612. 2, 8A

[Torrelli 2005] T. Torrelli, "Closed formulae for a b-function associated with a weighted homogeneous isolated complete intersection singularity", *J. Algebra* **286**:1 (2005), 35–40. 3C3

[Torrelli 2007] T. Torrelli, "Logarithmic comparison theorem and $\mathscr{D}$-modules: an overview", pp. 995–1009 in *Singularity theory*, edited by D. Chéniot et al., World Sci. Publ., Hackensack, NJ, 2007. 0, 2, 4C, 4C, 4D

[Torrelli 2009] T. Torrelli, "Intersection homology $\mathscr{D}$-module and Bernstein polynomials associated with a complete intersection", *Publ. Res. Inst. Math. Sci.* **45**:2 (2009), 645–660. 2, 3C4, 3C4

[Tráng and Ramanujam 1976] L. D. Tráng and C. P. Ramanujam, "The invariance of Milnor's number implies the invariance of the topological type", *Amer. J. Math.* **98**:1 (1976), 67–78. 5B

[Ucha and Castro-Jiménez 2004] J. M. Ucha and F. J. Castro-Jiménez, "On the computation of Bernstein–Sato ideals", *J. Symbolic Comput.* **37**:5 (2004), 629–639. 7, 10C

[Van Proeyen and Veys 2010] L. Van Proeyen and W. Veys, "The monodromy conjecture for zeta functions associated to ideals in dimension two", *Ann. Inst. Fourier (Grenoble)* **60**:4 (2010), 1347–1362. 6D

[Varchenko 1976] A. N. Varchenko, "Zeta-function of monodromy and Newton's diagram", *Invent. Math.* **37**:3 (1976), 253–262. 6B

[Varchenko 1981] A. N. Varchenko, "Asymptotic Hodge structure on vanishing cohomology", *Izv. Akad. Nauk SSSR Ser. Mat.* **45**:3 (1981), 540–591, 688. In Russian. 3C1, 6A

[Varchenko 1982] A. N. Varchenko, "The complex singularity index does not change along the stratum $\mu = \text{const}$", *Funktsional. Anal. i Prilozhen.* **16**:1 (1982), 1–12, 96. 5B

[Verdier 1983] J.-L. Verdier, "Spécialisation de faisceaux et monodromie modérée", pp. 332–364 in *Analysis and topology on singular spaces, II, III* (Luminy, 1981), Astérisque **101–102**, Soc. Math. France, Paris, 1983. 7

[Veys 1990] W. Veys, "On the poles of Igusa's local zeta function for curves", *J. London Math. Soc.* (2) **41**:1 (1990), 27–32. 6D

[Veys 1993] W. Veys, "Poles of Igusa's local zeta function and monodromy", *Bull. Soc. Math. France* **121**:4 (1993), 545–598. 6D

[Veys 1995] W. Veys, "Determination of the poles of the topological zeta function for curves", *Manuscripta Math.* **87**:4 (1995), 435–448. 6D

[Veys 2006] W. Veys, "Vanishing of principal value integrals on surfaces", *J. Reine Angew. Math.* **598** (2006), 139–158. 6D

[Walther 1999] U. Walther, "Algorithmic computation of local cohomology modules and the local cohomological dimension of algebraic varieties", *J. Pure Appl. Algebra* **139**:1-3 (1999), 303–321. 10D

[Walther 2005] U. Walther, "Bernstein–Sato polynomial versus cohomology of the Milnor fiber for generic hyperplane arrangements", *Compos. Math.* **141**:1 (2005), 121–145. 1, 2, 3C3, 6C, 8A, 8E

[Walther 2015] U. Walther, "The Jacobian module, the Milnor fiber, and the D-module generated by f^s", preprint, 2015. arXiv 1504.07164 4D, 5, 6C, 8A, 8E, 8F

[Wiens and Yuzvinsky 1997] J. Wiens and S. Yuzvinsky, "De Rham cohomology of logarithmic forms on arrangements of hyperplanes", *Trans. Amer. Math. Soc.* **349**:4 (1997), 1653–1662. 8B

[Yano 1977] T. Yano, "On the holonomic system of f^s and b-function", pp. 469–480 in *Proceedings of the Oji Seminar on Algebraic Analysis and the RIMS Symposium on Algebraic Analysis* (Kyoto, 1976), vol. 12, 1977. 3C3

[Yano 1978] T. Yano, "On the theory of b-functions", *Publ. Res. Inst. Math. Sci.* **14**:1 (1978), 111–202. 3B, 3C3, 4D, 5A

[Yoon 2013] Y. Yoon, *Spectrum of hyperplane arrangements in four variables*, Ph.D. thesis, University of Notre Dame, Ann Arbor, MI, 2013. arXiv 1211.1689 8D

[Yuzvinsky 2001] S. Yuzvinsky, "Orlik–Solomon algebras in algebra and topology", *Uspekhi Mat. Nauk* **56**:2(338) (2001), 87–166. In Russian; translated in *Russian Math. Surveys* **56** (2001), no. 2, 293–364. 8B

walther@math.purdue.edu *Department of Mathematics, Purdue University, 150 North University Street, West Lafayette, IN 47907-2067, United States*

leykin@math.gatech.edu *School of Mathematics, Georgia Institute of Technology, 686 Cherry Street, Atlanta, GA 30332-0160, United States*

Commutative Algebra and Noncommutative Algebraic Geometry, I
MSRI Publications
Volume **67**, 2015

Introduction to derived categories

AMNON YEKUTIELI

Derived categories were invented by Grothendieck and Verdier around 1960, not very long after the "old" homological algebra (of derived functors between abelian categories) was established. This "new" homological algebra, of derived categories and derived functors between them, provides a significantly richer and more flexible machinery than the "old" homological algebra. For instance, the important concepts of *dualizing complex* and *tilting complex* do not exist in the "old" homological algebra.

1. The homotopy category

Suppose M is an *abelian category*. The main examples for us are these:

- A is a ring, and $\mathsf{M} = \operatorname{Mod} A$, the category of left A-modules.
- $(X, \mathcal{A})$ is a ringed space, and $\mathsf{M} = \operatorname{Mod} \mathcal{A}$, the category of sheaves of left $\mathcal{A}$-modules.

A *complex* in M is a diagram

$$M = \left(\cdots \to M^{-1} \xrightarrow{\mathrm{d}_M^{-1}} M^0 \xrightarrow{\mathrm{d}_M^0} M^1 \to \cdots \right)$$

This paper is an edited version of the notes for a two-lecture minicourse given at MSRI in January 2013. Sections 1–5 are about the general theory of derived categories, and the material is taken from my manuscript "A course on derived categories" (available online). Sections 6–9 are on more specialized topics, leaning towards noncommutative algebraic geometry.

in M such that $\mathrm{d}_M^{i+1} \circ \mathrm{d}_M^i = 0$. A morphism of complexes $\phi : M \to N$ is a commutative diagram

$$\begin{array}{ccccccccc}
\cdots & \longrightarrow & M^{-1} & \xrightarrow{\mathrm{d}_M^{-1}} & M^0 & \xrightarrow{\mathrm{d}_M^0} & M^1 & \longrightarrow & \cdots \\
 & & \downarrow{\scriptstyle \phi^{-1}} & & \downarrow{\scriptstyle \phi^0} & & \downarrow{\scriptstyle \phi^1} & & \\
\cdots & \longrightarrow & N^{-1} & \xrightarrow[\mathrm{d}_N^{-1}]{} & N^0 & \xrightarrow[\mathrm{d}_N^0]{} & N^1 & \longrightarrow & \cdots
\end{array} \tag{1.1}$$

in M. Let us denote by $\mathbf{C}(\mathsf{M})$ *the category of complexes* in M. It is again an abelian category; but it is also a *differential graded category*, as we now explain.

Given $M, N \in \mathbf{C}(\mathsf{M})$ we let

$$\mathrm{Hom}_{\mathsf{M}}(M, N)^i := \prod_{j \in \mathbb{Z}} \mathrm{Hom}_{\mathsf{M}}(M^j, N^{j+i})$$

and

$$\mathrm{Hom}_{\mathsf{M}}(M, N) := \bigoplus_{i \in \mathbb{Z}} \mathrm{Hom}_{\mathsf{M}}(M, N)^i.$$

For $\phi \in \mathrm{Hom}_{\mathsf{M}}(M, N)^i$ we let

$$\mathrm{d}(\phi) := \mathrm{d}_N \circ \phi - (-1)^i \cdot \phi \circ \mathrm{d}_M.$$

In this way $\mathrm{Hom}_{\mathsf{M}}(M, N)$ becomes a complex of abelian groups, i.e. a DG (differential graded) $\mathbb{Z}$-module. Given a third complex $L \in \mathbf{C}(\mathsf{M})$, composition of morphisms in M induces a homomorphism of DG $\mathbb{Z}$-modules

$$\mathrm{Hom}_{\mathsf{M}}(L, M) \otimes_{\mathbb{Z}} \mathrm{Hom}_{\mathsf{M}}(M, N) \to \mathrm{Hom}_{\mathsf{M}}(L, N).$$

Compare Section 5; a DG algebra is a DG category with one object.

Note that the abelian structure of $\mathbf{C}(\mathsf{M})$ can be recovered from the DG structure as follows:

$$\mathrm{Hom}_{\mathbf{C}(\mathsf{M})}(M, N) = \mathrm{Z}^0\big(\mathrm{Hom}_{\mathsf{M}}(M, N)\big),$$

the set of 0-cocycles. Indeed, for $\phi : M \to N$ of degree 0 the condition $\mathrm{d}(\phi) = 0$ is equivalent to the commutativity of the diagram (1.1).

Next we define the *homotopy category* $\mathbf{K}(\mathsf{M})$. Its objects are the complexes in M (same as $\mathbf{C}(\mathsf{M})$), and

$$\mathrm{Hom}_{\mathbf{K}(\mathsf{M})}(M, N) = \mathrm{H}^0\big(\mathrm{Hom}_{\mathsf{M}}(M, N)\big).$$

In other words, these are homotopy classes of morphisms $\phi : M \to N$ in $\mathbf{C}(\mathsf{M})$.

There is an additive functor $\mathbf{C}(\mathsf{M}) \to \mathbf{K}(\mathsf{M})$, which is the identity on objects and surjective on morphisms. The additive category $\mathbf{K}(\mathsf{M})$ is no longer abelian – it is a *triangulated category*. Let me explain what this means.

Suppose K is an additive category, with an automorphism T called the *translation* (or shift, or suspension). A *triangle* in K is a diagram of morphisms of this sort:

$$L \xrightarrow{\alpha} M \xrightarrow{\beta} N \xrightarrow{\gamma} \mathrm{T}(L).$$

The name comes from the alternative typesetting

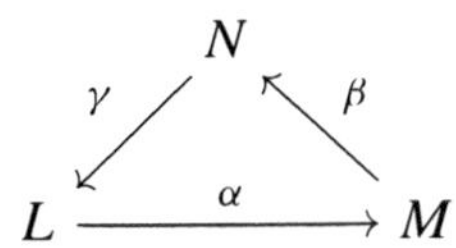

A triangulated category structure on K is a set of triangles called *distinguished triangles*, satisfying a list of axioms (that are not so important for us). Details can be found in [Yekutieli 2012; Schapira 2015; Hartshorne 1966; Weibel 1994; Kashiwara and Schapira 1990; Neeman 2001; Lipman and Hashimoto 2009].

The translation T of the category $\mathbf{K}(\mathsf{M})$ is defined as follows. On objects we take $\mathrm{T}(M)^i := M^{i+1}$ and $\mathrm{d}_{\mathrm{T}(M)} := -\mathrm{d}_M$. On morphisms it is $\mathrm{T}(\phi)^i := \phi^{i+1}$. For $k \in \mathbb{Z}$, the k-th translation of M is denoted by $M[k] := \mathrm{T}^k(M)$.

Given a morphism $\alpha : L \to M$ in $\mathbf{C}(\mathsf{M})$, its *cone* is the complex

$$\operatorname{cone}(\alpha) := \mathrm{T}(L) \oplus M = \begin{bmatrix} \mathrm{T}(L) \\ M \end{bmatrix}$$

with differential (in matrix notation)

$$\mathrm{d} := \begin{bmatrix} \mathrm{d}_{\mathrm{T}(L)} & 0 \\ \alpha & \mathrm{d}_M \end{bmatrix},$$

where α is viewed as a degree 1 morphism $\mathrm{T}(L) \to M$. There are canonical morphisms $M \to \operatorname{cone}(\alpha)$ and $\operatorname{cone}(\alpha) \to \mathrm{T}(L)$ in $\mathbf{C}(\mathsf{M})$.

A triangle in $\mathbf{K}(\mathsf{M})$ is distinguished if it is isomorphic, as a diagram in $\mathbf{K}(\mathsf{M})$, to the triangle

$$L \xrightarrow{\alpha} M \to \operatorname{cone}(\alpha) \to \mathrm{T}(L)$$

for some morphism $\alpha : L \to M$ in $\mathbf{C}(\mathsf{M})$. A calculation shows that $\mathbf{K}(\mathsf{M})$ is indeed triangulated (i.e. the axioms that I did not specify are satisfied).

The relation between distinguished triangles and exact sequences will be mentioned later.

Suppose K and K' are triangulated categories. A *triangulated functor* $F : \mathsf{K} \to \mathsf{K}'$ is an additive functor that commutes with the translations, and sends distinguished triangles to distinguished triangles.

Example 1.2. Let $F : \mathsf{M} \to \mathsf{M}'$ be an additive functor (not necessarily exact) between abelian categories. Extend F to a functor

$$\mathbf{C}(F) : \mathbf{C}(\mathsf{M}) \to \mathbf{C}(\mathsf{M}')$$

in the obvious way, namely

$$\mathbf{C}(F)(M)^i := F(M^i)$$

for a complex $M = \{M^i\}_{i \in \mathbb{Z}}$. The functor $\mathbf{C}(F)$ respects homotopies, so we get an additive functor

$$\mathbf{K}(F) : \mathbf{K}(\mathsf{M}) \to \mathbf{K}(\mathsf{M}').$$

This is a triangulated functor.

2. The derived category

As before M is an abelian category. Given a complex $M \in \mathbf{C}(\mathsf{M})$, we can consider its cohomologies

$$\mathrm{H}^i(M) := \ker(\mathrm{d}^i_M)/\operatorname{im}(\mathrm{d}^{i-1}_M) \in \mathsf{M}.$$

Since the cohomologies are homotopy-invariant, we get additive functors

$$\mathrm{H}^i : \mathbf{K}(\mathsf{M}) \to \mathsf{M}.$$

A morphism $\psi : M \to N$ in $\mathbf{K}(\mathsf{M})$ is called a *quasi-isomorphism* if $\mathrm{H}^i(\psi)$ are isomorphisms for all i. Let us denote by $\mathbf{S}(\mathsf{M})$ the set of all quasi-isomorphisms in $\mathbf{K}(\mathsf{M})$. Clearly $\mathbf{S}(\mathsf{M})$ is a multiplicatively closed set, i.e. the composition of two quasi-isomorphisms is a quasi-isomorphism. A calculation shows that $\mathbf{S}(\mathsf{M})$ is a left and right denominator set (as in ring theory). It follows that the Ore localization $\mathbf{K}(\mathsf{M})_{\mathbf{S}(\mathsf{M})}$ exists. This is an additive category, with object set

$$\mathrm{Ob}(\mathbf{K}(\mathsf{M})_{\mathbf{S}(\mathsf{M})}) = \mathrm{Ob}(\mathbf{K}(\mathsf{M})).$$

There is a functor

$$\mathrm{Q} : \mathbf{K}(\mathsf{M}) \to \mathbf{K}(\mathsf{M})_{\mathbf{S}(\mathsf{M})}$$

called the localization functor, which is the identity on objects. Every morphism $\chi : M \to N$ in $\mathbf{K}(\mathsf{M})_{\mathbf{S}(\mathsf{M})}$ can be written as

$$\chi = \mathrm{Q}(\phi_1) \circ \mathrm{Q}(\psi_1^{-1}) = \mathrm{Q}(\psi_2^{-1}) \circ \mathrm{Q}(\phi_2)$$

for some $\phi_i \in \mathbf{K}(\mathsf{M})$ and $\psi_i \in \mathbf{S}(\mathsf{M})$.

The category $\mathbf{K}(\mathsf{M})_{\mathbf{S}(\mathsf{M})}$ inherits a triangulated structure from $\mathbf{K}(\mathsf{M})$, and the localization functor Q is triangulated. There is a universal property: given a triangulated functor

$$F : \mathbf{K}(\mathsf{M}) \to \mathsf{E}$$

to a triangulated category E, such that $F(\psi)$ is an isomorphism for every ψ in $\mathbf{S}(\mathsf{M})$, there exists a unique triangulated functor

$$F_{\mathbf{S}(\mathsf{M})} : \mathbf{K}(\mathsf{M})_{\mathbf{S}(\mathsf{M})} \to \mathsf{E}$$

such that

$$F_{\mathbf{S}(\mathsf{M})} \circ \mathrm{Q} = F.$$

Definition 2.1. The *derived category* of the abelian category M is the triangulated category

$$\mathbf{D}(\mathsf{M}) := \mathbf{K}(\mathsf{M})_{\mathbf{S}(\mathsf{M})}.$$

The derived category was introduced by Grothendieck and Verdier around 1960. The first published material is the book *Residues and duality*, written by Hartshorne [1966] following notes by Grothendieck.

Let $\mathbf{D}(\mathsf{M})^0$ be the full subcategory of $\mathbf{D}(\mathsf{M})$ consisting of the complexes whose cohomology is concentrated in degree 0.

Proposition 2.2. The obvious functor $\mathsf{M} \to \mathbf{D}(\mathsf{M})^0$ is an equivalence.

This allows us to view M as an additive subcategory of $\mathbf{D}(\mathsf{M})$. It turns out that the abelian structure of M can be recovered from this embedding.

Proposition 2.3. Consider a sequence

$$0 \to L \xrightarrow{\alpha} M \xrightarrow{\beta} N \to 0$$

in M. This sequence is exact if and only if there is a morphism $\gamma : N \to L[1]$ in $\mathbf{D}(\mathsf{M})$ such that

$$L \xrightarrow{\alpha} M \xrightarrow{\beta} N \xrightarrow{\gamma} L[1]$$

is a distinguished triangle.

3. Derived functors

As before M is an abelian category. Recall the localization functor

$$\mathrm{Q} : \mathbf{K}(\mathsf{M}) \to \mathbf{D}(\mathsf{M}).$$

It is a triangulated functor, which is the identity on objects, and inverts quasi-isomorphisms.

Suppose E is some triangulated category, and $F : \mathbf{K}(\mathsf{M}) \to \mathsf{E}$ a triangulated functor. We now introduce the right and left derived functors of F. These are triangulated functors

$$\mathrm{R}F, \mathrm{L}F : \mathbf{D}(\mathsf{M}) \to \mathsf{E}$$

satisfying suitable universal properties.

Definition 3.1. A *right derived functor* of F is a triangulated functor

$$\mathrm{R}F : \mathsf{D}(\mathsf{M}) \to \mathsf{E},$$

together with a morphism

$$\eta : F \to \mathrm{R}F \circ \mathrm{Q}$$

of triangulated functors $\mathsf{K}(\mathsf{M}) \to \mathsf{E}$, satisfying this condition:

(∗) The pair $(\mathrm{R}F, \eta)$ is initial among all such pairs.

Being initial means that if (G, η') is another such pair, then there is a unique morphism of triangulated functors $\theta : \mathrm{R}F \to G$ s.t. $\eta' = \theta \circ \eta$. The universal condition implies that if a right derived functor $(\mathrm{R}F, \eta)$ exists, then it is unique, up to a unique isomorphism of triangulated functors.

Definition 3.2. A *left derived functor* of F is a triangulated functor

$$\mathrm{L}F : \mathsf{D}(\mathsf{M}) \to \mathsf{E},$$

together with a morphism

$$\eta : \mathrm{L}F \circ \mathrm{Q} \to F$$

of triangulated functors $\mathsf{K}(\mathsf{M}) \to \mathsf{E}$, satisfying this condition:

(∗) The pair $(\mathrm{L}F, \eta)$ is terminal among all such pairs.

Again, if $(\mathrm{L}F, \eta)$ exists, then it is unique up to a unique isomorphism.

There are various modifications. One of them is a contravariant triangulated functor $F : \mathsf{K}(\mathsf{M}) \to \mathsf{E}$. This can be handled using the fact that $\mathsf{K}(\mathsf{M})^{\mathrm{op}}$ is triangulated, and $F : \mathsf{K}(\mathsf{M})^{\mathrm{op}} \to \mathsf{E}$ is covariant.

We will also want to derive bifunctors. Namely to a bitriangulated bifunctor

$$F : \mathsf{K}(\mathsf{M}) \times \mathsf{K}(\mathsf{M}') \to \mathsf{E}$$

we will want to associate bitriangulated bifunctors

$$\mathrm{R}F, \mathrm{L}F : \mathsf{D}(\mathsf{M}) \times \mathsf{D}(\mathsf{M}') \to \mathsf{E}.$$

This is done similarly, and I won't give details.

4. Resolutions

Consider an additive functor $F : \mathsf{M} \to \mathsf{M}'$ between abelian categories, and the corresponding triangulated functor $\mathsf{K}(F) : \mathsf{K}(\mathsf{M}) \to \mathsf{K}(\mathsf{M}')$, as in Example 1.2. By slight abuse we write F instead of $\mathsf{K}(F)$. We want to construct (or prove existence) of the derived functors

$$\mathrm{R}F, \mathrm{L}F : \mathsf{D}(\mathsf{M}) \to \mathsf{D}(\mathsf{M}').$$

If F is exact (namely F sends exact sequences to exact sequences), then $\mathrm{R}F = \mathrm{L}F = F$. (This is an easy exercise.) Otherwise we need *resolutions*.

The DG structure of $\mathsf{C}(\mathsf{M})$ gives, for every $M, N \in \mathsf{C}(\mathsf{M})$, a complex of abelian groups $\mathrm{Hom}_{\mathsf{M}}(M, N)$. Recall that a complex N is called acyclic if $\mathrm{H}^i(N) = 0$ for all i; i.e. N is an exact sequence in M.

Definition 4.1. (1) A complex $I \in \mathsf{K}(\mathsf{M})$ is called *K-injective* if for every acyclic $N \in \mathsf{K}(\mathsf{M})$, the complex $\mathrm{Hom}_{\mathsf{M}}(N, I)$ is also acyclic.

(2) Let $M \in \mathsf{K}(\mathsf{M})$. A *K-injective resolution* of M is a quasi-isomorphism $M \to I$ in $\mathsf{K}(\mathsf{M})$, where I is K-injective.

(3) We say that $\mathsf{K}(\mathsf{M})$ *has enough K-injectives* if every $M \in \mathsf{K}(\mathsf{M})$ has some K-injective resolution.

Theorem 4.2. If $\mathsf{K}(\mathsf{M})$ has enough K-injectives, then every triangulated functor $F : \mathsf{K}(\mathsf{M}) \to \mathsf{E}$ has a right derived functor $(\mathrm{R}F, \eta)$. Moreover, for every K-injective complex $I \in \mathsf{K}(\mathsf{M})$, the morphism $\eta_I : F(I) \to \mathrm{R}F(I)$ in E is an isomorphism.

The proof/construction goes like this: for every $M \in \mathsf{K}(\mathsf{M})$ we choose a K-injective resolution $\zeta_M : M \to I_M$, and we define

$$\mathrm{R}F(M) := F(I_M)$$

and

$$\eta_M := F(\zeta_M) : F(M) \to F(I_M)$$

in E.

Regarding existence of K-injective resolutions:

Proposition 4.3. A bounded below complex of injective objects of M is a K-injective complex.

This is the type of injective resolution used in [Hartshorne 1966]. The most general statement I know is this (see [Kashiwara and Schapira 2006, Theorem 14.3.1]):

Theorem 4.4. If M is a Grothendieck abelian category, then $\mathsf{K}(\mathsf{M})$ has enough K-injectives.

This includes $\mathsf{M} = \mathsf{Mod}\, A$ for a ring A, and $\mathsf{M} = \mathsf{Mod}\, \mathcal{A}$ for a sheaf of rings $\mathcal{A}$. Actually in these cases the construction of K-injective resolutions can be done very explicitly, and it is not so difficult.

Example 4.5. Let $f : (X, \mathcal{A}_X) \to (Y, \mathcal{A}_Y)$ be a map of ringed spaces. (For instance a map of schemes $f : (X, \mathcal{O}_X) \to (Y, \mathcal{O}_Y)$.) The map f induces an additive functor

$$f_* : \mathsf{Mod}\, \mathcal{A}_X \to \mathsf{Mod}\, \mathcal{A}_Y$$

called push-forward, which is usually not exact (it is left exact though). Since $\mathsf{K}(\mathsf{Mod}\, \mathscr{A}_X)$ has enough K-injectives, the right derived functor

$$\mathrm{R}f_* : \mathsf{D}(\mathsf{Mod}\, \mathscr{A}_X) \to \mathsf{D}(\mathsf{Mod}\, \mathscr{A}_Y)$$

exists.

For $\mathcal{M} \in \mathsf{Mod}\, \mathscr{A}_X$ we can use an injective resolution $\mathcal{M} \to \mathcal{J}$ (in the "classical" sense), and therefore

$$\mathrm{H}^q(\mathrm{R}f_*(\mathcal{M})) \cong \mathrm{H}^q(f_*(\mathcal{J})) \cong \mathrm{R}^q f_*(\mathcal{M}),$$

where the latter is the "classical" right derived functor.

Analogously we have:

Definition 4.6. (1) A complex $P \in \mathsf{K}(\mathsf{M})$ is called *K-projective* if for every acyclic $N \in \mathsf{K}(\mathsf{M})$, the complex $\mathrm{Hom}_{\mathsf{M}}(P, N)$ is also acyclic.

(2) Let $M \in \mathsf{K}(\mathsf{M})$. A *K-projective resolution* of M is a quasi-isomorphism $P \to M$ in $\mathsf{K}(\mathsf{M})$, where P is K-projective.

(3) We say that $\mathsf{K}(\mathsf{M})$ *has enough K-projectives* if every $M \in \mathsf{K}(\mathsf{M})$ has some K-projective resolution.

Theorem 4.7. If $\mathsf{K}(\mathsf{M})$ has enough K-projectives, then every triangulated functor $F : \mathsf{K}(\mathsf{M}) \to \mathsf{E}$ has a left derived functor $(\mathrm{L}F, \eta)$. Moreover, for every K-projective complex $P \in \mathsf{K}(\mathsf{M})$, the morphism $\eta_P : \mathrm{L}F(P) \to F(P)$ in E is an isomorphism.

The construction of $\mathrm{L}F$ is by K-projective resolutions.

Proposition 4.8. A bounded above complex of projective objects of M is a K-projective complex.

Proposition 4.9. Let A be a ring. Then $\mathsf{K}(\mathsf{Mod}\, A)$ has enough K-projectives.

The construction of K-projective resolutions in this case can be done very explicitly, and it is not difficult.

The concepts of K-injective and K-projective complexes were introduced in [Spaltenstein 1988]. They were independently discovered by others at about the same time; see [Keller 1994; Bökstedt and Neeman 1993], for example. (Some authors used the term *homotopically injective complex*.)

Example 4.10. Suppose $\mathbb{K}$ is a commutative ring and A is a $\mathbb{K}$-algebra (i.e. A is a ring and there is a homomorphism $\mathbb{K} \to \mathrm{Z}(A)$). Consider the bi-additive bifunctor

$$\mathrm{Hom}_A(-,-) : (\mathsf{Mod}\, A)^{\mathrm{op}} \times \mathsf{Mod}\, A \to \mathsf{Mod}\, \mathbb{K}.$$

We have seen how to extend this functor to complexes (this is sometimes called "product totalization"), giving rise to a bitriangulated bifunctor

$$\mathrm{Hom}_A(-,-) : \mathbf{K}(\mathsf{Mod}\, A)^{\mathrm{op}} \times \mathbf{K}(\mathsf{Mod}\, A) \to \mathbf{K}(\mathsf{Mod}\, \mathbb{K}).$$

The right derived bifunctor

$$\mathrm{RHom}_A(-,-) : \mathbf{D}(\mathsf{Mod}\, A)^{\mathrm{op}} \times \mathbf{D}(\mathsf{Mod}\, A) \to \mathbf{D}(\mathsf{Mod}\, \mathbb{K})$$

can be constructed/calculated by a K-injective resolution in either the first or the second argument. Namely given $M, N \in \mathbf{K}(\mathsf{Mod}\, A)$ we can choose a K-injective resolution $N \to I$, and let

$$\mathrm{RHom}_A(M, N) := \mathrm{Hom}_A(M, I) \in \mathbf{D}(\mathsf{Mod}\, \mathbb{K}). \tag{4.11}$$

Or we can choose a K-injective resolution $M \to P$ in $\mathbf{K}(\mathsf{Mod}\, A)^{\mathrm{op}}$, which is really a K-projective resolution $P \to M$ in $\mathbf{K}(\mathsf{Mod}\, A)$, and let

$$\mathrm{RHom}_A(M, N) := \mathrm{Hom}_A(P, N) \in \mathbf{D}(\mathsf{Mod}\, \mathbb{K}). \tag{4.12}$$

The two complexes (4.11) and (4.12) are canonically related by the quasi-isomorphisms

$$\mathrm{Hom}_A(P, N) \to \mathrm{Hom}_A(P, I) \leftarrow \mathrm{Hom}_A(M, I).$$

If $M, N \in \mathsf{Mod}\, A$ then of course

$$\mathrm{H}^q\big(\mathrm{RHom}_A(M, N)\big) \cong \mathrm{Ext}^q_A(M, N),$$

where the latter is "classical" Ext.

K-projective and K-injective complexes are good also for understanding the structure of $\mathbf{D}(\mathsf{M})$.

Proposition 4.13. Suppose $P \in \mathbf{K}(\mathsf{M})$ is K-projective and $I \in \mathbf{K}(\mathsf{M})$ is K-injective. Then for any $M \in \mathbf{K}(\mathsf{M})$ the homomorphisms

$$\mathrm{Q} : \mathrm{Hom}_{\mathbf{K}(\mathsf{M})}(P, M) \to \mathrm{Hom}_{\mathbf{D}(\mathsf{M})}(P, M)$$

and

$$\mathrm{Q} : \mathrm{Hom}_{\mathbf{K}(\mathsf{M})}(M, I) \to \mathrm{Hom}_{\mathbf{D}(\mathsf{M})}(M, I)$$

are bijective.

Let us denote by $\mathbf{K}(\mathsf{M})_{\mathrm{prj}}$ and $\mathbf{K}(\mathsf{M})_{\mathrm{inj}}$ the full subcategories of $\mathbf{K}(\mathsf{M})$ on the K-projective and the K-injective complexes respectively.

Corollary 4.14. The triangulated functors

$$\mathrm{Q} : \mathsf{K}(\mathsf{M})_{\mathrm{prj}} \to \mathsf{D}(\mathsf{M})$$

and

$$\mathrm{Q} : \mathsf{K}(\mathsf{M})_{\mathrm{inj}} \to \mathsf{D}(\mathsf{M})$$

are fully faithful.

Corollary 4.15. (1) If $\mathsf{K}(\mathsf{M})$ has enough K-projectives, then the triangulated functor $\mathrm{Q} : \mathsf{K}(\mathsf{M})_{\mathrm{prj}} \to \mathsf{D}(\mathsf{M})$ is an equivalence.

(2) If $\mathsf{K}(\mathsf{M})$ has enough K-injectives, then the triangulated functor $\mathrm{Q} : \mathsf{K}(\mathsf{M})_{\mathrm{inj}} \to \mathsf{D}(\mathsf{M})$ is an equivalence.

Exercise 4.16. Let $\mathbb{K}$ be a nonzero commutative ring and $A := \mathbb{K}[t]$ the polynomial ring. We view $\mathbb{K}$ as an A-module via $t \mapsto 0$. Find a nonzero morphism $\chi : \mathbb{K} \to \mathbb{K}[1]$ in $\mathsf{D}(\mathsf{Mod}\, A)$. Show that $\mathrm{H}^q(\chi) = 0$ for all $q \in \mathbb{Z}$.

When $\mathsf{M} = \mathsf{Mod}\, A$ for a ring A, we can also talk about *K-flat* complexes. A complex P is K-flat if for any acyclic complex $N \in \mathsf{Mod}\, A^{\mathrm{op}}$ the complex $N \otimes_A P$ is acyclic. Any K-projective complex is K-flat. The left derived bifunctor $N \otimes^{\mathrm{L}}_A M$ can be constructed using K-flat resolutions of either argument:

$$N \otimes^{\mathrm{L}}_A M \cong N \otimes_A P \cong Q \otimes_A M$$

for any K-flat resolutions $P \to M$ in $\mathsf{K}(\mathsf{Mod}\, A)$ and $Q \to N$ in $\mathsf{K}(\mathsf{Mod}\, A^{\mathrm{op}})$.

5. DG algebras

A DG algebra (or DG ring) is a graded ring $A = \bigoplus_{i \in \mathbb{Z}} A^i$, with differential d of degree 1, satisfying the graded Leibniz rule

$$\mathrm{d}(a \cdot b) = \mathrm{d}(a) \cdot b + (-1)^i \cdot a \cdot \mathrm{d}(b)$$

for $a \in A^i$ and $b \in A^j$.

A left DG A-module is a left graded A-module $M = \bigoplus_{i \in \mathbb{Z}} M^i$, with differential d of degree 1, satisfying the graded Leibniz rule. Denote by $\mathsf{DGMod}\, A$ the category of left DG A-modules.

As in the ring case, for any $M, N \in \mathsf{DGMod}\, A$ there is a complex of $\mathbb{Z}$-modules $\mathrm{Hom}_A(M, N)$, and

$$\mathrm{Hom}_{\mathsf{DGMod}\, A}(M, N) = \mathrm{Z}^0\big(\mathrm{Hom}_A(M, N)\big).$$

The homotopy category is $\tilde{\mathsf{K}}(\mathsf{DGMod}\, A)$, with

$$\mathrm{Hom}_{\tilde{\mathsf{K}}(\mathsf{DGMod}\, A)}(M, N) = \mathrm{H}^0\big(\mathrm{Hom}_A(M, N)\big).$$

After inverting the quasi-isomorphisms in $\tilde{\mathsf{K}}(\mathsf{DGMod}\, A)$ we obtain the derived category $\tilde{\mathsf{D}}(\mathsf{DGMod}\, A)$. These are triangulated categories.

Example 5.1. Suppose A is a ring (i.e. $A^i = 0$ for $i \neq 0$). Then $\mathsf{DGMod}\, A = \mathsf{C}(\mathsf{Mod}\, A)$ and $\tilde{\mathsf{D}}(\mathsf{DGMod}\, A) = \mathsf{D}(\mathsf{Mod}\, A)$.

Derived functors are defined as in the ring case, and there are enough K-injectives, K-projectives and K-flats in $\tilde{\mathsf{K}}(\mathsf{DGMod}\, A)$.

Let $A \to B$ be a homomorphism of DG algebras. There are additive functors

$$B \otimes_A - : \ \mathsf{DGMod}\, A \rightleftarrows \mathsf{DGMod}\, B \ : \mathrm{rest}_{B/A},$$

where $\mathrm{rest}_{B/A}$ is the forgetful functor. These are adjoint. We get induced derived functors

$$B \otimes^{\mathsf{L}}_A - : \ \tilde{\mathsf{D}}(\mathsf{DGMod}\, A) \rightleftarrows \tilde{\mathsf{D}}(\mathsf{DGMod}\, B) \ : \mathrm{rest}_{B/A} \tag{5.2}$$

that are also adjoint.

Proposition 5.3. If $A \to B$ is a quasi-isomorphism, then the functors (5.2) are equivalences.

We say that A is *strongly commutative* if $b \cdot a = (-1)^{i+j} \cdot a \cdot b$ and $c^2 = 0$ for all $a \in A^i$, $b \in A^j$ and $c \in A^k$, where k is odd. We call A *nonpositive* if $A^i = 0$ for al $i > 0$.

Let $f : A \to B$ be a homomorphism between nonpositive strongly commutative DG algebras. A K-flat DG algebra resolution of B relative to A is a factorization of f into $A \xrightarrow{g} \tilde{B} \xrightarrow{h} B$, where h is a quasi-isomorphism, and $\tilde{B}$ is a K-flat DG A-module. Such resolutions exist.

Example 5.4. Take $A = \mathbb{Z}$ and $B := \mathbb{Z}/(6)$. We can take $\tilde{B}$ to be the Koszul complex

$$\tilde{B} := (\cdots 0 \to \mathbb{Z} \xrightarrow{6} \mathbb{Z} \to 0 \cdots)$$

concentrated in degrees -1 and 0.

Example 5.5. For a homomorphism of commutative rings $A \to B$, the Hochschild cohomology of B relative to A is the cohomology of the complex

$$\mathrm{RHom}_{\tilde{B} \otimes_A \tilde{B}}(B, B),$$

where $\tilde{B}$ is a K-flat resolution as above.

6. Commutative dualizing complexes

I will talk about dualizing complexes over commutative rings. There is a richer theory for schemes, but there is not enough time for it. See [Hartshorne 1966;

Yekutieli 1992b; Neeman 1996; Yekutieli 2010; Alonso Tarrío et al. 1999; Lipman and Hashimoto 2009], for example.

Let A be a noetherian commutative ring. We denote by $\mathsf{D}^{\mathrm{b}}_{\mathrm{f}}(\mathsf{Mod}\,A)$ the subcategory of $\mathsf{D}(\mathsf{Mod}\,A)$ consisting of bounded complexes whose cohomologies are finitely generated A-modules. This is a full triangulated subcategory.

A complex $M \in \mathsf{D}(\mathsf{Mod}\,A)$ is said to have *finite injective dimension* if it has a bounded injective resolution. Namely there is a quasi-isomorphism $M \to I$ for some bounded complex of injective A-modules I. Note that such I is a K-injective complex.

Take any $M \in \mathsf{D}(\mathsf{Mod}\,A)$. Because A is commutative, we have a triangulated functor

$$\mathrm{RHom}_A(-, M) : \mathsf{D}(\mathsf{Mod}\,A)^{\mathrm{op}} \to \mathsf{D}(\mathsf{Mod}\,A).$$

Compare Example 4.10.

Definition 6.1. A *dualizing complex* over A is a complex $R \in \mathsf{D}^{\mathrm{b}}_{\mathrm{f}}(\mathsf{Mod}\,A)$ with finite injective dimension, such that the canonical morphism

$$A \to \mathrm{RHom}_A(R, R)$$

in $\mathsf{D}(\mathsf{Mod}\,A)$ is an isomorphism.

If we choose a bounded injective resolution $R \to I$, then there is an isomorphism of triangulated functors

$$\mathrm{RHom}_A(-, R) \cong \mathrm{Hom}_A(-, I).$$

Example 6.2. Assume A is a *Gorenstein ring*, namely the free module $R := A$ has finite injective dimension. There are plenty of Gorenstein rings; for instance any regular ring is Gorenstein. Then $R \in \mathsf{D}^{\mathrm{b}}_{\mathrm{f}}(\mathsf{Mod}\,A)$, and the reflexivity condition holds:

$$\mathrm{RHom}_A(R, R) \cong \mathrm{Hom}_A(A, A) \cong A.$$

We see that the module $R = A$ is a dualizing complex over the ring A.

Here are several important results from [Hartshorne 1966].

Theorem 6.3 (duality). Suppose R is a dualizing complex over A. Then the triangulated functor

$$\mathrm{RHom}_A(-, R) : \mathsf{D}^{\mathrm{b}}_{\mathrm{f}}(\mathsf{Mod}\,A)^{\mathrm{op}} \to \mathsf{D}^{\mathrm{b}}_{\mathrm{f}}(\mathsf{Mod}\,A)$$

is an equivalence.

Theorem 6.4 (uniqueness). Suppose R and R' are dualizing complexes over A, and Spec A is connected. Then there is an invertible module P and an integer n such that $R' \cong R \otimes_A P[n]$ in $\mathsf{D}^{\mathrm{b}}_{\mathrm{f}}(\mathsf{Mod}\,A)$.

Theorem 6.5 (existence). If A has a dualizing complex, and B is a finite type A-algebra, then B has a dualizing complex.

7. Noncommutative dualizing complexes

In the last three sections of the paper we concentrate on noncommutative rings. Before going into the technicalities, here is a brief motivational preface.

Recall that one of the important tools of commutative ring theory is *localization at prime ideals*. For instance, a noetherian local commutative ring A, with maximal ideal $\mathfrak{m}$, is called a *regular local ring* if

$$\dim A = \operatorname{rank}_{A/\mathfrak{m}}(\mathfrak{m}/\mathfrak{m}^2).$$

(Here dim is Krull dimension.) A noetherian commutative ring A is called *regular* if all its local rings $A_{\mathfrak{p}}$ are regular local rings.

It is known that regularity can be described in homological terms. Indeed, if $\dim A < \infty$, then it is regular if and only if it has *finite global cohomological dimension*. Namely there is a natural number d, such that $\operatorname{Ext}^i_A(M, N) = 0$ for all $i > d$ and $M, N \in \mathsf{Mod}\, A$.

Now consider a noetherian noncommutative ring A. (This is short for: A is not-necessarily-commutative, and left-and-right noetherian.) Localization of A is almost never possible (for good reasons). A very useful substitute for localization (and other tools of commutative rings theory) is *noncommutative homological algebra*. By this we mean the study of the derived functors $\operatorname{RHom}_A(-,-)$, $\operatorname{RHom}_{A^{\mathrm{op}}}(-,-)$ and $- \otimes^{\mathrm{L}}_A -$ of formulas (7.2), (7.3) and (8.1) respectively. Here A^{op} is the opposite ring (the same addition, but multiplication is reversed). The homological criterion of regularity from the commutative framework is made the definition of regularity in the noncommutative framework – see Definition 7.1 below. This definition is the point of departure of *noncommutative algebraic geometry* of M. Artin et. al. (see the survey paper [Stafford and van den Bergh 2001]). A surprising amount of structure can be expressed in terms of noncommutative homological algebra. A few examples are sprinkled in the text, and many more are in the references.

Definition 7.1. A noncommutative ring A is called *regular* if there is a natural number d, such that $\operatorname{Ext}^i_A(M, N) = 0$ and $\operatorname{Ext}^i_{A^{\mathrm{op}}}(M', N') = 0$ for all $i > d$, $M, N \in \mathsf{Mod}\, A$ and $M', N' \in \mathsf{Mod}\, A^{\mathrm{op}}$.

For the rest of this section A is a noncommutative noetherian ring. For technical reasons we assume that it is an algebra over a field $\mathbb{K}$.

We denote by $A^{\mathrm{e}} := A \otimes_{\mathbb{K}} A^{\mathrm{op}}$ the enveloping algebra. Thus $\mathsf{Mod}\, A^{\mathrm{op}}$ is the category of right A-modules, and $\mathsf{Mod}\, A^{\mathrm{e}}$ is the category of $\mathbb{K}$-central A-bimodules.

Any $M \in \mathsf{Mod}\, A^{\mathrm{e}}$ gives rise to $\mathbb{K}$-linear functors

$$\mathrm{Hom}_A(-, M) : (\mathsf{Mod}\, A)^{\mathrm{op}} \to \mathsf{Mod}\, A^{\mathrm{op}}$$

and

$$\mathrm{Hom}_{A^{\mathrm{op}}}(-, M) : (\mathsf{Mod}\, A^{\mathrm{op}})^{\mathrm{op}} \to \mathsf{Mod}\, A.$$

These functors can be right derived, yielding $\mathbb{K}$-linear triangulated functors

$$\mathrm{RHom}_A(-, M) : \mathbf{D}(\mathsf{Mod}\, A)^{\mathrm{op}} \to \mathbf{D}(\mathsf{Mod}\, A^{\mathrm{op}}) \tag{7.2}$$

and

$$\mathrm{RHom}_{A^{\mathrm{op}}}(-, M) : \mathbf{D}(\mathsf{Mod}\, A^{\mathrm{op}})^{\mathrm{op}} \to \mathbf{D}(\mathsf{Mod}\, A). \tag{7.3}$$

One way to construct these derived functors is to choose a K-injective resolution $M \to I$ in $\mathbf{K}(\mathsf{Mod}\, A^{\mathrm{e}})$. Then (because A is flat over $\mathbb{K}$) the complex I is K-injective over A and over A^{op}, and we get

$$\mathrm{RHom}_A(-, M) \cong \mathrm{Hom}_A(-, I)$$

and

$$\mathrm{RHom}_{A^{\mathrm{op}}}(-, M) \cong \mathrm{Hom}_{A^{\mathrm{op}}}(-, I).$$

Note that even if A is commutative, this setup is still meaningful – not all A-bimodules are A-central!

Definition 7.4 ([Yekutieli 1992a]). A *noncommutative dualizing complex* over A is a complex $R \in \mathbf{D}^{\mathrm{b}}(\mathsf{Mod}\, A^{\mathrm{e}})$ satisfying these three conditions:

(i) The cohomology modules $\mathrm{H}^q(R)$ are finitely generated over A and over A^{op}.

(ii) The complex R has finite injective dimension over A and over A^{op}.

(iii) The canonical morphisms

$$A \to \mathrm{RHom}_A(R, R)$$

and

$$A \to \mathrm{RHom}_{A^{\mathrm{op}}}(R, R)$$

in $\mathbf{D}(\mathsf{Mod}\, A^{\mathrm{e}})$ are isomorphisms.

Condition (ii) implies that R has a "bounded bi-injective resolution", namely there is a quasi-isomorphism $R \to I$ in $\mathbf{K}(\mathsf{Mod}\, A^{\mathrm{e}})$, with I a bounded complex of bimodules that are injective on both sides.

Theorem 7.5 (duality [Yekutieli 1992a]). Suppose R is a noncommutative dualizing complex over A. Then the triangulated functor

$$\mathrm{RHom}_A(-, R) : \mathbf{D}^{\mathrm{b}}_{\mathrm{f}}(\mathsf{Mod}\, A)^{\mathrm{op}} \to \mathbf{D}^{\mathrm{b}}_{\mathrm{f}}(\mathsf{Mod}\, A^{\mathrm{op}})$$

is an equivalence, with quasi-inverse $\mathrm{RHom}_{A^{\mathrm{op}}}(-, R)$.

Existence and uniqueness are much more complicated than in the noncommutative case. I will talk about them later.

Example 7.6. The noncommutative ring A is called *Gorenstein* if the bimodule A has finite injective dimension on both sides. It is not hard to see that A is Gorenstein if and only if it has a noncommutative dualizing complex of the form $P[n]$, for some integer n and *invertible bimodule* P. Here invertible bimodule is in the sense of Morita theory, namely there is another bimodule $P^\vee$ such that

$$P \otimes_A P^\vee \cong P^\vee \otimes_A P \cong A$$

in $\mathsf{Mod}\, A^{\mathrm{e}}$. Any regular ring is Gorenstein.

For more results about noncommutative Gorenstein rings see [Jørgensen 1999] and [Jørgensen and Zhang 2000].

8. Tilting complexes and derived Morita theory

Let A and B be noncommutative algebras over a field $\mathbb{K}$. Suppose $M \in \mathbf{D}(\mathsf{Mod}\, A \otimes_{\mathbb{K}} B^{\mathrm{op}})$ and $N \in \mathbf{D}(\mathsf{Mod}\, B \otimes_{\mathbb{K}} A^{\mathrm{op}})$. The left derived tensor product

$$M \otimes^{\mathsf{L}}_B N \in \mathbf{D}(\mathsf{Mod}\, A \otimes_{\mathbb{K}} A^{\mathrm{op}}) \tag{8.1}$$

exists. It can be constructed by choosing a resolution $P \to M$ in $\mathbf{K}(\mathsf{Mod}\, A \otimes_{\mathbb{K}} B^{\mathrm{op}})$, where P is a complex that's K-projective over B^{op}; or by choosing a resolution $Q \to N$ in $\mathbf{K}(\mathsf{Mod}\, B \otimes_{\mathbb{K}} A^{\mathrm{op}})$, where Q is a complex that's K-projective over B.

Here is a definition generalizing the notion of invertible bimodule. It is due to Rickard [1989; 1991].

Definition 8.2. A complex

$$T \in \mathbf{D}(\mathsf{Mod}\, A \otimes_{\mathbb{K}} B^{\mathrm{op}})$$

is called a *two-sided tilting complex* over A-B if there exists a complex

$$T^\vee \in \mathbf{D}(\mathsf{Mod}\, B \otimes_{\mathbb{K}} A^{\mathrm{op}})$$

such that

$$T \otimes^{\mathsf{L}}_B T^\vee \cong A$$

in $\mathbf{D}(\mathsf{Mod}\, A^{\mathrm{e}})$, and

$$T^\vee \otimes^{\mathsf{L}}_A T \cong B$$

in $\mathbf{D}(\mathsf{Mod}\, B^{\mathrm{e}})$.

When $B = A$ we say that T is a two-sided tilting complex over A.

The complex $T^\vee$ is called a quasi-inverse of T. It is unique up to isomorphism in $\mathsf{D}(\mathsf{Mod}\, B \otimes_{\mathbb{K}} A^{\mathrm{op}})$. Indeed we have this result:

Proposition 8.3. Let T be a two-sided tilting complex.

(1) The quasi-inverse $T^\vee$ is isomorphic to $\mathrm{RHom}_A(T, A)$.

(2) T has a bounded bi-projective resolution $P \to T$.

Definition 8.4. The algebras A and B are said to be *derived Morita equivalent* if there is a $\mathbb{K}$-linear triangulated equivalence

$$\mathsf{D}(\mathsf{Mod}\, A) \approx \mathsf{D}(\mathsf{Mod}\, B).$$

Theorem 8.5 [Rickard 1991]. The $\mathbb{K}$-algebras A and B are derived Morita equivalent if and only if there exists a two-sided tilting complex over A-B.

Here is a result relating dualizing complexes and tilting complexes.

Theorem 8.6 (uniqueness [Yekutieli 1999]). Suppose R and R' are noncommutative dualizing complexes over A. Then the complex

$$T := \mathrm{RHom}_A(R, R')$$

is a two-sided tilting complex over A, and

$$R' \cong R \otimes^{\mathsf{L}}_A T$$

in $\mathsf{D}(\mathsf{Mod}\, A^{\mathrm{e}})$.

It is easy to see that if T_1 and T_2 are two-sided tilting complexes over A, then so is $T_1 \otimes^{\mathsf{L}}_A T_2$. This leads to the next definition.

Definition 8.7 [Yekutieli 1999]. Let A be a noncommutative $\mathbb{K}$-algebra. The *derived Picard group* of A is the group $\mathrm{DPic}_{\mathbb{K}}(A)$ whose elements are the isomorphism classes (in $\mathsf{D}(\mathsf{Mod}\, A^{\mathrm{e}})$) of two-sided tilting complexes. The multiplication is induced by the operation $T_1 \otimes^{\mathsf{L}}_A T_2$, and the identity element is the class of A.

Here is a consequence of Theorem 8.6.

Corollary 8.8. Suppose the noncommutative $\mathbb{K}$-algebra A has at least one dualizing complex. Then operation $R \otimes^{\mathsf{L}}_A T$ induces a simply transitive right action of the group $\mathrm{DPic}_{\mathbb{K}}(A)$ on the set of isomorphism classes of dualizing complexes.

It is natural to ask about the structure of the group $\mathrm{DPic}(A)$.

Theorem 8.9 [Rouquier and Zimmermann 2003; Yekutieli 1999]. If the ring A is either commutative (with nonempty connected spectrum) or local, then

$$\mathrm{DPic}_{\mathbb{K}}(A) \cong \mathrm{Pic}_{\mathbb{K}}(A) \times \mathbb{Z}.$$

Here $\operatorname{Pic}_{\mathbb{K}}(A)$ is the noncommutative Picard group of A, made up of invertible bimodules. If A is commutative, then

$$\operatorname{Pic}_{\mathbb{K}}(A) \cong \operatorname{Aut}_{\mathbb{K}}(A) \ltimes \operatorname{Pic}_A(A),$$

where $\operatorname{Pic}_A(A)$ is the usual (commutative) Picard group of A. A noncommutative ring A is said to be local if $A/\mathfrak{r}$ is a simple artinian ring, where $\mathfrak{r}$ is the Jacobson radical.

For nonlocal noncommutative rings the group $\operatorname{DPic}_{\mathbb{K}}(A)$ is bigger. See the paper [Miyachi and Yekutieli 2001] for some calculations. These calculations are related to CY-dimensions of some rings; see Example 9.7.

9. Rigid dualizing complexes

The material in this final section is largely due to Van den Bergh [van den Bergh 1997]. His results were extended by J. Zhang and myself. Again A is a noetherian noncommutative algebra over a field $\mathbb{K}$, and $A^{\mathrm{e}} = A \otimes_{\mathbb{K}} A^{\mathrm{op}}$.

Take $M \in \mathsf{Mod}\, A^{\mathrm{e}}$. Then the $\mathbb{K}$-module $M \otimes_{\mathbb{K}} M$ has four commuting actions by A, which we arrange as follows. The algebra $A^{\mathrm{e;\,in}} := A^{\mathrm{e}}$ acts on $M \otimes_{\mathbb{K}} M$ by

$$(a_1 \otimes a_2) \cdot_{\mathrm{in}} (m_1 \otimes m_2) := (m_1 \cdot a_2) \otimes (a_1 \cdot m_2),$$

and the algebra $A^{\mathrm{e;\,out}} := A^{\mathrm{e}}$ acts by

$$(a_1 \otimes a_2) \cdot_{\mathrm{out}} (m_1 \otimes m_2) := (a_1 \cdot m_1) \otimes (m_2 \cdot a_2).$$

The bimodule A is viewed as an object of $\mathsf{D}(\mathsf{Mod}\, A^{\mathrm{e}})$ in the obvious way.

Now take $M \in \mathsf{D}(\mathsf{Mod}\, A^{\mathrm{e}})$. We define the *square* of M to be the complex

$$\operatorname{Sq}_{A/\mathbb{K}}(M) := \operatorname{RHom}_{A^{\mathrm{e;\,out}}}(A, M \otimes_{\mathbb{K}} M) \in \mathsf{D}(\mathsf{Mod}\, A^{\mathrm{e;\,in}}).$$

We get a functor

$$\operatorname{Sq}_{A/\mathbb{K}} : \mathsf{D}(\mathsf{Mod}\, A^{\mathrm{e}}) \to \mathsf{D}(\mathsf{Mod}\, A^{\mathrm{e}}).$$

This is not an additive functor. Indeed, it is a quadratic functor: given an element $a \in \mathrm{Z}(A)$ and a morphism $\phi : M \to N$ in $\mathsf{D}(\mathsf{Mod}\, A^{\mathrm{e}})$, one has

$$\operatorname{Sq}_{A/\mathbb{K}}(a \cdot \phi) = \operatorname{Sq}_{A/\mathbb{K}}(\phi \cdot a) = a^2 \cdot \operatorname{Sq}_{A/\mathbb{K}}(\phi).$$

Note that the cohomologies of $\operatorname{Sq}_{A/\mathbb{K}}(M)$ are

$$\mathrm{H}^j(\operatorname{Sq}_{A/\mathbb{K}}(M)) = \operatorname{Ext}^j_{A^{\mathrm{e}}}(A, M \otimes_{\mathbb{K}} M),$$

so they are precisely the Hochschild cohomologies of $M \otimes_{\mathbb{K}} M$.

A *rigid complex* over A (relative to $\mathbb{K}$) is a pair (M, ρ) consisting of a complex $M \in \mathsf{D}(\mathsf{Mod}\, A^{\mathrm{e}})$, and an isomorphism

$$\rho : M \xrightarrow{\simeq} \mathrm{Sq}_{A/\mathbb{K}}(M)$$

in $\mathsf{D}(\mathsf{Mod}\, A^{\mathrm{e}})$.

Let (M, ρ) and (N, σ) be rigid complexes over A. A *rigid morphism*

$$\phi : (M, \rho) \to (N, \sigma)$$

is a morphism $\phi : M \to N$ in $\mathsf{D}(\mathsf{Mod}\, A^{\mathrm{e}})$, such that the diagram

$$\begin{array}{ccc} M & \xrightarrow{\rho} & \mathrm{Sq}_{A/\mathbb{K}}(M) \\ {\scriptstyle \phi}\downarrow & & \downarrow{\scriptstyle \mathrm{Sq}_{A/\mathbb{K}}(\phi)} \\ N & \xrightarrow{\sigma} & \mathrm{Sq}_{A/\mathbb{K}}(N) \end{array}$$

is commutative.

Definition 9.1 ([van den Bergh 1997]). A *rigid dualizing complex* over A (relative to $\mathbb{K}$) is a rigid complex (R, ρ) such that R is a dualizing complex.

Theorem 9.2 (uniqueness [van den Bergh 1997; Yekutieli 1999]). Suppose (R, ρ) and (R', ρ') are both rigid dualizing complexes over A. Then there is a unique rigid isomorphism

$$\phi : (R, \rho) \xrightarrow{\simeq} (R', \rho').$$

As for existence, let me first give an easy case.

Proposition 9.3. If A is finite over its center $\mathrm{Z}(A)$, and $\mathrm{Z}(A)$ is finitely generated as $\mathbb{K}$-algebra, then A has a rigid dualizing complex.

Actually, in this case it is quite easy to write down a formula for the rigid dualizing complex.

In the next existence result, by a filtration $F = \{F_i(A)\}_{i \in \mathbb{N}}$ of the algebra A we mean an ascending exhaustive filtration by $\mathbb{K}$-submodules, such that $1 \in F_0(A)$ and $F_i(A) \cdot F_j(A) \subset F_{i+j}(A)$. Such a filtration gives rise to a graded $\mathbb{K}$-algebra

$$\mathrm{gr}^F(A) = \bigoplus_{i \geq 0} \mathrm{gr}_i^F(A).$$

Theorem 9.4 (existence, [van den Bergh 1997; Yekutieli and Zhang 2005]). Suppose A admits a filtration F, such that $\mathrm{gr}^F(A)$ is finite over its center $\mathrm{Z}(\mathrm{gr}^F(A))$, and $\mathrm{Z}(\mathrm{gr}^F(A))$ is finitely generated as $\mathbb{K}$-algebra. Then A has a rigid dualizing complex.

This theorem applies to the ring of differential operators $\mathcal{D}(C)$, where C is a smooth commutative $\mathbb{K}$-algebra (and char $\mathbb{K} = 0$). It also applies to any quotient of the universal enveloping algebra $\mathrm{U}(\mathfrak{g})$ of a finite dimensional Lie algebra $\mathfrak{g}$.

I will finish with some examples.

Example 9.5. Let A be a noetherian $\mathbb{K}$-algebra satisfying these two conditions:

- A is smooth, namely the A^{e}-module A has finite projective dimension.
- There is an integer n such that

$$\mathrm{Ext}^j_{A^{\mathrm{e}}}(A, A^{\mathrm{e}}) \cong \begin{cases} A & \text{if } j = n, \\ 0 & \text{otherwise.} \end{cases}$$

Then A is a regular ring (Definition 7.1), and the complex $R := A[n]$ is a rigid dualizing complex over A. Such an algebra A is called an *n-dimensional Artin–Schelter regular algebra*, or an *n-dimensional Calabi–Yau algebra*.

Example 9.6. Let $\mathfrak{g}$ be an n-dimensional Lie algebra, and $A := \mathrm{U}(\mathfrak{g})$, the universal enveloping algebra. Then the rigid dualizing complex of A is $R := A^{\sigma}[n]$, where A^{σ} is the trivial bimodule A, twisted on the right by an automorphism σ. Using the Hopf structure of A we can express A^{σ} like this:

$$A^{\sigma} \cong \mathrm{U}(\mathfrak{g}) \otimes_{\mathbb{K}} \textstyle\bigwedge^n(\mathfrak{g}),$$

the twist by the 1-dimensional representation $\bigwedge^n(\mathfrak{g})$. See [Yekutieli 2000]. So A is a *twisted Calabi–Yau* algebra. If $\mathfrak{g}$ is semi-simple then there is no twist, and A is Calabi–Yau. This was used by Van den Bergh in his duality for Hochschild (co)homology [van den Bergh 1998].

Example 9.7. Let

$$A := \begin{bmatrix} \mathbb{K} & \mathbb{K} \\ 0 & \mathbb{K} \end{bmatrix}$$

the 2×2 upper triangular matrix algebra. The rigid dualizing complex here is

$$R := \mathrm{Hom}_{\mathbb{K}}(A, \mathbb{K}).$$

It is known that

$$R \otimes^{\mathrm{L}}_A R \otimes^{\mathrm{L}}_A R \cong A[1]$$

in $\mathsf{D}(\mathsf{Mod}\, A^{\mathrm{e}})$. So A is a *Calabi–Yau algebra of dimension* $\frac{1}{3}$. See [Yekutieli 1999; Miyachi and Yekutieli 2001].

References

[Alonso Tarrío et al. 1999] L. Alonso Tarrío, A. Jeremías López, and J. Lipman, "Duality and flat base change on formal schemes", pp. 3–90 in *Studies in duality on Noetherian formal schemes and non–Noetherian ordinary schemes*, Contemporary Mathematics **244**, Amer. Math. Soc., Providence, RI, 1999.

[van den Bergh 1997] M. van den Bergh, "Existence theorems for dualizing complexes over non-commutative graded and filtered rings", *J. Algebra* **195**:2 (1997), 662–679.

[van den Bergh 1998] M. van den Bergh, "A relation between Hochschild homology and cohomology for Gorenstein rings", *Proc. Amer. Math. Soc.* **126**:5 (1998), 1345–1348.

[Bökstedt and Neeman 1993] M. Bökstedt and A. Neeman, "Homotopy limits in triangulated categories", *Compositio Math.* **86**:2 (1993), 209–234.

[Hartshorne 1966] R. Hartshorne, *Residues and duality*, Lecture Notes in Mathematics **20**, Springer, Berlin, 1966.

[Jørgensen 1999] P. Jørgensen, "Gorenstein homomorphisms of noncommutative rings", *J. Algebra* **211**:1 (1999), 240–267.

[Jørgensen and Zhang 2000] P. Jørgensen and J. J. Zhang, "Gourmet's guide to Gorensteinness", *Adv. Math.* **151**:2 (2000), 313–345.

[Kashiwara and Schapira 1990] M. Kashiwara and P. Schapira, *Sheaves on manifolds*, Grundlehren der Mathematischen Wissenschaften **292**, Springer, Berlin, 1990.

[Kashiwara and Schapira 2006] M. Kashiwara and P. Schapira, *Categories and sheaves*, Grundlehren der Mathematischen Wissenschaften **332**, Springer, Berlin, 2006.

[Keller 1994] B. Keller, "Deriving DG categories", *Ann. Sci. École Norm. Sup.* (4) **27**:1 (1994), 63–102.

[Lipman and Hashimoto 2009] J. Lipman and M. Hashimoto, *Foundations of Grothendieck duality for diagrams of schemes*, Lecture Notes in Mathematics **1960**, Springer, Berlin, 2009.

[Miyachi and Yekutieli 2001] J.-i. Miyachi and A. Yekutieli, "Derived Picard groups of finite-dimensional hereditary algebras", *Compositio Math.* **129**:3 (2001), 341–368.

[Neeman 1996] A. Neeman, "The Grothendieck duality theorem via Bousfield's techniques and Brown representability", *J. Amer. Math. Soc.* **9**:1 (1996), 205–236.

[Neeman 2001] A. Neeman, *Triangulated categories*, Annals of Mathematics Studies **148**, Princeton University Press, 2001.

[Rickard 1989] J. Rickard, "Morita theory for derived categories", *J. London Math. Soc.* (2) **39**:3 (1989), 436–456.

[Rickard 1991] J. Rickard, "Derived equivalences as derived functors", *J. London Math. Soc.* (2) **43**:1 (1991), 37–48.

[Rouquier and Zimmermann 2003] R. Rouquier and A. Zimmermann, "Picard groups for derived module categories", *Proc. London Math. Soc.* (3) **87**:1 (2003), 197–225.

[Schapira 2015] P. Schapira, "Categories and homological algebra", Course notes, 2015, http://webusers.imj-prg.fr/~pierre.schapira/lectnotes/HomAl.pdf.

[Spaltenstein 1988] N. Spaltenstein, "Resolutions of unbounded complexes", *Compositio Math.* **65**:2 (1988), 121–154.

[Stafford and van den Bergh 2001] J. T. Stafford and M. van den Bergh, "Noncommutative curves and noncommutative surfaces", *Bull. Amer. Math. Soc.* (*N.S.*) **38**:2 (2001), 171–216.

[Weibel 1994] C. A. Weibel, *An introduction to homological algebra*, Cambridge Studies in Advanced Mathematics **38**, Cambridge University Press, 1994.

[Yekutieli 1992a] A. Yekutieli, "Dualizing complexes over noncommutative graded algebras", *J. Algebra* **153**:1 (1992), 41–84.

[Yekutieli 1992b] A. Yekutieli, *An explicit construction of the Grothendieck residue complex*, Astérisque **208**, Société Mathématique de France, Paris, 1992.

[Yekutieli 1999] A. Yekutieli, "Dualizing complexes, Morita equivalence and the derived Picard group of a ring", *J. London Math. Soc.* (2) **60**:3 (1999), 723–746.

[Yekutieli 2000] A. Yekutieli, "The rigid dualizing complex of a universal enveloping algebra", *J. Pure Appl. Algebra* **150**:1 (2000), 85–93.

[Yekutieli 2010] A. Yekutieli, "Rigid dualizing complexes via differential graded algebras (survey)", pp. 452–463 in *Triangulated categories*, edited by T. Holm et al., London Math. Soc. Lecture Note Ser. **375**, Cambridge Univ. Press, 2010.

[Yekutieli 2012] A. Yekutieli, "A course on derived categories", preprint, 2012. arXiv 1206.6632

[Yekutieli and Zhang 2005] A. Yekutieli and J. J. Zhang, "Dualizing complexes and perverse modules over differential algebras", *Compos. Math.* **141**:3 (2005), 620–654.

amyekut@math.bgu.ac.il

Department of Mathematics, Ben Gurion University, Be'er Sheva 84105, 62381 Beersheva, Israel

CPSIA information can be obtained
at www.ICGtesting.com
Printed in the USA
LVOW12*1721150916
504779LV00014B/144/P

9 781107 065628